The ethics of consumption

The citizen, the market and the law

The ethics of consumption

The citizen, the market and the law

EurSafe 2013
Uppsala, Sweden
11-14 September 2013

edited by:
Helena Röcklinsberg
Per Sandin

Wageningen Academic
Publishers

ISBN: 978-90-8686-231-3
e-ISBN: 978-90-8686-784-4
DOI: 10.3921/978-90-8686-784-4

First published, 2013

Acknowledgements

Scientific work is a collaborative undertaking, and the production of this volume would not have been possible without the contribution of expert knowledge, time, and effort by several of our eminent colleagues. The editors and the further members of the organization committee (Anne Algers, Bo Algers, Stefan Gunnarsson and Jan Hultgren) wish to thank our colleagues on the scientific committee, listed below, for helping us in reviewing, selecting and improving the contributions. We are truly impressed with and most grateful for their critical and constructive reading of abstracts and full papers. Our special thanks also go to the EurSafe board whose members have contributed not only as scientific committee but also with support and assistance in planning the conference. The Swedish Research Council Formas and the Faculty of Veterinary Medicine and Animal Science at SLU have supported the conference financially. This support is gratefully acknowledged. But most of all we thank the contributors and participants of the 11th EurSafe Congress for making this congress and volume possible, and for contributing to this important discussion. Thank you all!

Scientific committee

Table of contents

Section 2. Labelling and market forces

Section 3. Legal issues in food production and consumption

Section 4. Citizenship and consumers

Section 5. Animal husbandry and innovative meat production

Section 6. Animal philosophy

Section 7. Sustainable aquaculture

Section 8. Food philosophy

Section 9. Technology, society and ethics

Section 10. Animal welfare

Section 11. Food in the public sphere

Section 12. Community and culture

Section 13. Teaching veterinarians

Section 14. Poster session

Preface

In bringing about the 11th Congress of the European Society for Agricultural and Food Ethics and the present volume, several things have been consumed: Food and drink, paper, jet fuel, numerous services, electricity, ink, and a host of other things. This has consequences of moral importance.

The theme for the congress and this volume is 'The ethics of consumption'. We are all consumers in the sense that we need to consume food, water and other nutrients that we ingest to keep our bodies functioning. But we are also consumers in a slightly different sense: we *buy* things with money – not only food and agricultural products, but a host of other items as well.

While arguably remarkably efficient, the present system for agriculture and food production involves a number of detrimental consequences for human health, the environment, and animal welfare. Great challenges lie ahead as we are facing population growth and climate change and reduced availability of fossil fuels. In the future, how we manage agriculture, agricultural land and biobased goods and services, including biofuels, as well as water resources, can be expected to become even more important than today. Arguably, agriculture and forestry contain some of the sources of these challenges as well as some of the most promising ways of meeting them. In addition, organic products and other products marketed with reference to ethical concerns have increased their market share, and for instance, some beef, pig, poultry and egg producers strive to make animal welfare a competitive advantage.

It is frequently argued that one key to meeting those challenges is changing consumption patterns among individual as well as institutions, for instance through reducing meat consumption, switching to organic or fair trade products, boycotting or 'buycotting' certain products, or consuming less overall. There is considerable disagreement regarding how to bring this about, whose responsibility it is, and even whether it is desirable. Is it a question of political initiatives, retailers and producers or the virtues and vices of individual consumers in the developed world, or something else?

Many of these issues pose profound intellectual challenges at the intersection of ethics, political philosophy, economics, sociology and several other fields. They touch upon problems of liberty and paternalism, distributive justice and fairness, responsibility and care, knowledge and uncertainty, the gap between knowing and acting as well as calling for definitions of key concepts like harm, welfare, integrity, value and worth – to mention but a few.

This is reflected in the contributions to this volume. EurSafe congresses have a tradition of being multidisciplinary meeting points, and to acknowledge also the disciplinary breadth – this year explicit in the sub title of the congress: 'the citizen, the market, and the law'. Hence, the 11th EurSafe Congress follows this tradition, and we hope this volume can be an inspiration for continued discussions among academics, practitioners and the general public.

Helena Röcklinsberg and Per Sandin

Keynote contributions

Economization of animals: the case of marketization of halal foods

M. Miele
School of Planning and Geography, Cardiff University, Glamorgan Building, King Edward VII Avenue, Cardiff CF10 3WA, Wales, United Kingdom; mielem@cardiff.ac.uk

Abstract

Over the last 15 years the demand of halal meat has increased significantly worldwide and dedicated markets for halal certified meat have emerged in a number of European countries. While ethnic stores still constitute the major retail outlet for halal meat in most countries, 'halal' labelled meat and meat products are increasingly available in supermarkets and fast food restaurants. Market expansion has also facilitated the rise of new certification bodies, each with their own marketing strategies and interpretations of what constitutes authentic 'halal', who question the reliability of certification policies that allow the practice of stunning before slaughter. This paper, based on research carried out during the EU funded Dialrel project, I have attempted to investigate which activities, behaviours and fields are established as being economic while dealing with nonhuman animals, in other words, I have addressed the economization of animals via marketization of halal meat.

Keywords: Halal meat, religious slaughter, economization and marketization

Introduction to economization of animals via marketization

In a two part article in *Economy and Society* Caliskan and Callon (2009, 2010) have argued that the development of economic activities, their organization and their change-producing forces cannot be analysed and understood without taking into account the work of economists, in the most comprehensive use of the term. This argument is at the core of what is now called the performativity programme (Callon, 1998). In the last ten years an interesting line of research has been developed looking at the formative relationship between economic sciences and markets and it has helped to advancing the understanding of economic phenomena (e.g. Callon *et al.*, 2007, Mackenzie *et al.*, 2007, among others). Studies such as these have focused attention on a number of new and important issues, one of which has been named 'economization' by Callon (1998). This word is used to indicate the series of actions that form the behaviours, organizations, institutions and, generally, the objects in a particular society which are defined as 'economic' by experts and ordinary people, even though there might be controversies around these qualifications. As Caliskan and Callon specify 'the construction of action(-ization) into the word implies that the economy is an achievement rather than a starting point or a pre-existing reality that can simply be revealed and acted upon'. Economization is a multifaceted process and the study of economization involves 'investigating the processes through which activities, behaviours and spheres or fields are established as being economic 'whether or not there is consensus about the content of such qualifications' (Caliskan and Callon, 2009: 370).

In economics both formalism and substantivism suggest questions about the mechanisms by which either values and methods of valuations or institutional arrangements can contribute to processes of economization (Caliskan and Callon, 2009: 378). Marketization, with its monetary values and techniques of valuations, is one (even though not the only one) of the modalities for achieving economization.

The quality of life of animals, especially farm animals, is one of the most widely shared concern among European citizens (see Evans and Miele, 2007a,b). These concerns, however, are often associated to anxieties for the safety and quality of the food that is generated by these animals (Miele *et al.*, 2005). In this paper I want to propose to investigate this ambivalent relation to animals and which activities, behaviours and fields are established as being economic while dealing with nonhuman animals, in other words, I want to address the economization of animals. More specifically I will look at the processes and devices through which nonhuman animals become economic objects (e.g. foods) via *marketization*. I will argue that the economization of animals entails processes of market creation and market maintenance[1] with processes of consumers' *qualculations* (Cochy, 2008) and objects' qualification for example via food standards (e.g. halal). I will argue that animal food standards are powerful devices in the economization of animals. These standards are based on negotiations between different actors who speak of 'care for animals' (NGOs, animal scientists, members of the EU public....) or 'religious rules' (religious authorities), but are also affected by the meat industry and other actors in the meat supply chains who speak of 'efficiency' and 'efficacy', as well as new technologies in slaughterhouses and new intermediaries (Meyer, 2010), such as the accredited certifying bodies and their marketing devices. I will make this point empirically by addressing how the process of economization of animals is articulated in the case of marketization of production of halal meat.

The growing market for halal meat and other animal products provides a particularly good example of the complexities of the process of marketization of animals (Lever and Miele, 2012). It also exemplifies the multi-dimensional character of both 'halal' and 'animal welfare' definitions and the controversies that the qualification of 'halal' might generate when religious interpretations of animal welfare (e.g. 'animal welfare' already ensured by applying the rules written in sacred texts') are contrasted with the improvements of slaughter practices in terms of reducing pain, proposed by the scientific authorities (e.g. stunning).

This paper is based on research carried out in the European research project Dialrel, which started in October 2006 and ended in June 2010[2]. The project aimed at establishing a dialogue between the religious authorities and other stakeholders (NGOs, scientists, representatives of the meat industry, policy makers, consumers....) around the welfare of animals slaughtered for production of halal and kosher meat in order to promote best practices of religious slaughter and to benchmark the practices proposed by current halal and kosher standards (Miele *et al.*, 2010). Current practices of religious slaughter are increasingly a source of concerns among animals' advocates and part of the European citizens, especially for the growth of the demand of *non-stunned* halal meat, presented by some certifying bodies as more authentic 'halal', both in Europe and worldwide. The project researched the cultural and socio-economic questions characterising the growth of this particular market, as well as the ethico-political problems involved in addressing the welfare of animals at time of killing while respecting religious freedom. The initiative not only involved academic research but also the development of practical procedures, such as consumer forums and ways of gathering concerned parties – religious authorities, animal welfare scientists, consumers, commercial players.

In this paper I will present some of the project's findings from the double perspective of its contribution to the dialogic articulation between different positions as well as an intervention in the material (re) configuration of the 'object' itself (i.e. halal labelled meat). Here the 'product' is inseparable from the process and the agencies of the actors involved. I will address these issues in conversation with two processual theoretical frameworks, Callon, Meadel and Rabeharisoa (2002) approach to the

[1] As Callon has pointed out: 'The belief used to be that markets were quasi-natural realities, and theoreticians were content to identify the conditions of their viability ... We now realize that they have to be sometimes created from scratch, and that they are in reality fragile and complicated socio-technical artefacts. It is therefore necessary to reconsider the following basic questions: what are markets made of?' (Callon, 2009: 539).

[2] For a project description see www.dialrel.eu.

'qualification' of goods and Stengers' notion of Cosmopolitics (2005). From these perspectives I hope to reveal the ethico-political maze of concerns that are composing contemporary discussions around the certification of halal meat in Europe and to trace the process of economization of animals via marketization of halal meat.

References

Caliskan, K. and Callon, M. (2009). Economization, part 1: shifting attention from the economy towards processes of economization, Economy and Society 38(3): 369-398.

Caliskan, K. and Callon, M. (2010). Economization, part 2: a research programme for the study of markets. Economy and Society 39(1): 1-32.

Callon, M. (2007). What does it mean to say that economics is performative? In: MacKenzie, D., Muniesa, F. and Siu, L. (eds.) Do economists make markets? Princeton University Press, Princeton, CA, USA, pp. 311-357.

Callon, M. (2009). Civilizing markets: carbon trading between *in vitro* and *in vivo* experiments. Accounting, Organizations and Society 34(3_4): 535-549.

Callon, M. (ed.) (1998). The laws of the markets (=Sociological Review Monographs). Blackwell, Oxford/Malden, UK.

Callon, M., Lascoumes, P. and Barthe, Y. (2009). Acting in an uncertain world. an essay on technical democracy. MIT Press, Cambridge, MA, USA.

Callon, M., Meadel, C. and Rabehariosa, V. (2002). The economy of qualities. Economy and Society 31(2): 194-217.

Callon, M., Millo, Y. and Muniesa, F. (eds.) (2007). Market devices. Blackwell, Oxford, UK.

Cochoy, F. (2008). Calculation, qualculation, calqulation: shopping cart arithmetic, equipped cognition and the clustered consumer. Marketing Theory 8(1): 15-44.

Evans, A. and Miele, M (2007b). Consumers' views about farm animal welfare, part 2: comparative report based on focus group research. Welfare Quality® reports no 5. Cardiff University, Cardiff, UK.

Evans, A. and Miele, M. (2012). Between food and flesh: how animals are made to matter (and not to matter) within food consumption practices. Environment and Planning D- Society and Space 30(2): 298-314.

Evans, A. and Miele, M. (eds.) (2007a). Consumers' views about farm animal welfare, part i: national reports based on focus group research. Welfare Quality® reports no. 4. Cardiff University, Cardiff, UK.

Lever, J. and Miele, M. (2012). The growth of the halal meat markets in Europe: an exploration of the supply side theory of religion. Journal of Rural Studies 28(4): 528-537.

Mackenzie, D.D., Muniesa, F. and Siu, L. (eds.) (2007). Do economists make markets?: on the performativity of economics. Princeton University Press, Princeton, CA, USA.

Meyer, M. (2010). The rise of the knowledge broker. Science Communication 32: 118-127.

Miele, M. (2011). The taste of happiness: free range chicken. Environment and Planning A 43(9): 2070-2090.

Miele, M. and Lever, J. (2013). Civilizing the market for welfare friendly products? The techno-ethic of the Welfare Quality© assessment. Geoforum 48: 63-72.

Miele, M., Higgin, M. and Evans, A. (2010). Dubbi dei consumatori, normativa e trasparenza del mercato (Consumer's Concerns, regulation and market transparency). In: Cenci-Goga, B. and Fermani, A.G. (ed.) La Macellazione rituale nella produzione igienica della carne. Point Veterinaire Italie, Milan, Italy, pp. 69-81.

Miele, M., Murdoch, J. and Roe, E. (2005). Animals and ambivalence, governing farm animal welfare in the European food sector. In: Higgins, V. and Lawrence, G. (eds.) Agricultural governance. Routledge, London, UK, pp. 169-185.

Stengers, I. (2005). The cosmopolitical proposal in making things public. In: Latour, B and Weibel, P. (eds.) Atmospheres of democracy. MIT Press, Cambridge, MA, USA, pp. 994-1003.

The virtue of simplicity[3]

P. Cafaro and J. Gambrel
Philosophy Department, Colorado State University, Fort Collins, CO 80523, USA; philip.cafaro@colostate.edu

Abstract

This essay describes and defends material simplicity as a virtue disposing us to act appropriately within the sphere of our consumer decisions. Simplicity is a conscientious and restrained attitude toward material goods that typically includes: (1) decreased consumption; and (2) a more conscious consumption; hence (3) greater deliberation regarding our consumer decisions; (4) a more focused life in general; and (5) a greater and a more nuanced appreciation for other things besides material goods. It is to be distinguished from simple-mindedness, a return to nature, or poverty. These ideas are illustrated through a consideration of food consumption. Simplicity is a virtue because it furthers human flourishing, both individual and social, and sustains nature's ecological flourishing. Cultivating simplicity can make important contributions to basic individual and societal flourishing, to individual freedom and autonomy, to living meaningfully and to securing the flourishing of nonhuman beings. The proven failure of materialism to further human happiness strongly argues that individuals try voluntary simplicity, in food consumption and within other areas of their lives. It also supports efforts to redirect politics in developed societies away from the pursuit of increased material wealth and toward the higher goals.

Keywords: temperance, flourishing, wisdom, necessities, luxuries

In her seminal article 'Non-relative virtues: an Aristotelian approach,' Martha Nussbaum provides a useful way to define and distinguish the virtues. 'Isolate a sphere of human experience that figures in more or less any human life, and in which more or less any human being will have to make *some* choices rather than others, and act in *some* way rather than some other,' she suggests. Crucially, to require the specification of a virtue or a range of virtuous behavior in this area, these choices must be important to people's well-being or flourishing. 'The 'thin account' of each virtue is that it is whatever being stably disposed to act appropriately in that sphere consists in' (Nussbaum, 1993: 245). The 'full or 'thick' description' of the specifies the characteristic thought processes, habituation and emotional development, ways of looking at the world, and other aspects of human character and training that help us choose well in that particular sphere. Those who reject ethical anthropocentrism, such as Louke van Wensveen (2000) and Ronald Sandler (2007), amend Nussbaum's definition to include a virtue's contribution to the flourishing of both human and nonhuman life.

Following Nussbaum's schema thus amended, we define simplicity as the virtue disposing us to act appropriately within the sphere of our consumer decisions, from food and drink to stereo and housing purchases to cars and airplane travel. As we understand it, simplicity is a conscientious and restrained attitude toward material goods. It typically includes: (1) decreased consumption; and (2) a more conscious consumption; hence (3) greater deliberation regarding our consumer decisions; (4) a more focused life in general; and (5) a greater and more nuanced appreciation for other things besides material goods; and also for (6) material goods themselves.

As Aristotle noted long ago, people may be much more likely to err in one direction rather than another in particular spheres of human choice, either due to human nature or to the pathologies of their particular societies (*Nicomachean ethics*, Book II, Chapter 8). In Athens in Aristotle's day, men were apparently

[3] This is a revised and condensed version of Joshua Gambrel and Philip Cafaro, 'The Virtue of Simplicity.' Journal of Agricultural and Environmental Ethics 23 (2010): 85-108.

more likely to err on the side of irascibility than 'inirascibility'; hence Aristotle named the virtue with regard to anger 'mildness' (Book IV, Chapter 5). In wealthy western democracies today, people are more likely to err on the side of overconsumption than underconsumption. Hence the term 'simplicity' is arguably a good one for this virtue (as long as we remember that underconsumption can also be a problem).

Simplicity overlaps with such traditional virtues as temperance (moderation in food and drink), frugality (the responsible and restrained use of wealth), prudence and self-control. Within the philosophical tradition, writers once routinely claimed that temperance, frugality and simplicity were keys to living justly and wisely. They were right. We see it as a glaring weakness of contemporary discussions of justice and wisdom that they rarely make this connection.

It is often helpful to consider simplicity as a virtuous mean between vicious extremes. However, like other complex virtues, simplicity appears to be a mean along several axes. Some of its associated vices have obvious names, others do not, perhaps owing to their rarity:

Vice	Virtue	Vice
	Simplicity	
underconsumption (poverty?)		overconsumption (gluttony)
unthinking consumption (carelessness)		overthinking consumption (obsession)
none; or crude consumption (asceticism, 'monkish virtue')		luxurious consumption
inefficient or pointless consumption (wastefulness)		hyper-efficient consumption (penny-pinching)
immoral consumption (callous, disproportional)		none; or, moral finickiness ('moral foppery')

Obviously, there is more than one way to go wrong in our stance toward consumption.

Treating simplicity as a virtue presupposes that through reflection, we can discover our deeper, more significant needs and goals; recognize some goals as ignoble, foolish, or trivial, and replace them with better ones; and pursue our goals more efficiently, with less waste and harm to others. By way of illustration, consider some steps a person might take to practice voluntary simplicity in relation to food consumption, as these relate to the six aspects of simplicity noted above.

Americans consume on average 25% more calories than necessary, on a conservative estimate (Putnam *et al.*, 2002). Today, three out of five Americans are overweight and one out of five is obese. This excessive consumption of food harms our health and quality of life (US Department of Health and Human Services, 2001). Food overconsumption also causes direct and indirect environmental degradation, through habitat loss and increased pollution from agricultural fertilizers and pesticides (Cafaro *et al.*, 2006). Approximately 20% of American greenhouse gas emissions come from growing and transporting our food (Pollan, 2007). So here simplicity clearly demands decreased consumption (aspect 1).

However, whether we are talking about personal health, healthy communities, or healthy land, consuming less is not enough. We also need to consume *differently*. We may buy and prepare more healthy foods for ourselves; buy organic foods and local foods and eat less meat, all of which decrease environmental harms; purchase more food directly from farmers at farmers' markets or as part of community-supported agriculture co-ops, to support small farmers and keep local agriculture vibrant. Such changes demand attention: a more conscious consumption, involving greater deliberation about our consumer decisions (aspects 2 and 3).

Many food simplifiers combine more conscious consumption with greater participation in food *production*: gardening, raising chickens or keeping bees, joining CSA cooperatives. Research shows that food produced in these ways is more environmentally sustainable and often more nutritious than conventionally-grown food (Felice, 2007). These activities are also often enjoyable and interesting, and connect people to their neighbors and to the earth. Similarly, taking time to prepare our own food and eating meals together offer important opportunities to connect to loved ones. Consciously taking such steps leads to a more focused life (aspect 4). It can further gratitude toward the many other species that sustain us; tune us in to nature's rhythms and details; and enrich our relationships with other people. In these ways, food simplifiers explore and sustain a wide range of nonmaterial goods (aspect 5) and come to better understand and appreciate the material realities of food production and consumption (aspect 6).

The example should begin to suggest how simplicity can contribute to human and nonhuman flourishing in important ways. It also illustrates several important points about simplicity as a virtue.

First, living simply is not necessarily simple. It requires deliberation: thinking through our choices and acting on our best judgment, rather than following the herd, or the blandishments of advertisers, or doing what we have always done, or what comes easiest. Thinking about our food consumption and improving it typically involve research and planning, and some of what we learn about how our food is grown will probably be discouraging or disgusting. Still, it is better to know the ugly facts and act in full consciousness of what we are a part of, rather than in ignorance. Simplicity is better than simple-mindedness, the default setting of many food consumers (an ignorance that the food industry spends many millions of dollars a year cultivating).

Second, though, simplicity *is* often simple: it often involves working our way back to simpler, less convoluted ways of doing things. When we plant and tend a garden, ride a bicycle and fix it ourselves, or sing songs with our children on family outings, these are relatively simple ways of satisfying some of our food, transportation and entertainment needs. The simplicity of such activities makes them less likely to stray from their goals and more likely to involve thoughtful activity rather than passive consumption. Their simplicity may make them particularly appropriate vessels for finding meaning, or expressing happiness and gratitude.

Third, simplicity is not a call to 'return to nature' in any romantic or primitivist sense. Old ways can be wasteful, or harmful; new ways can be an improvement. Similarly, simplicity is not opposed to technology, or to new technologies. It just asks that we consciously develop and appropriately incorporate technologies into our lives with reference to our real purposes and to their full effects on the world around us. Hydroponics has a role to play, along with sharing heirloom tomato seeds with the neighbors.

Fourth, simplicity is not poverty. Poverty is a state defined by lack, where people find it difficult to obtain the means to satisfy even the essential human needs – food, water, shelter, basic physical safety – let alone higher needs for self-actualization or creative personal development. Poverty means living in deprivation, against one's will. Simplicity is consciously and freely chosen. It provides greater opportunities than conventional materialism to achieve human flourishing, while poverty limits those opportunities.

Fifth, simplicity is a process, not an endpoint. Although we are *arguing* here for simplicity, we should not forget Aristotle's reminders that habituation is more important than arguments in developing virtue and that virtue demands *phronesis*: practical wisdom, applied to the details of life. Anyone who has tried to cultivate simplicity in their own life knows that Aristotle was right. Creating a character, a personal infrastructure, and daily habits that regularly result in less consumption and less dumb consumption are difficult and ongoing affairs. Hence it is a mistake to look for particular markers that indicate the presence of this virtue (although it is not a mistake, but good practice, to set down markers for ourselves and strive to achieve them).

Sixth, simplicity is not uniformity. There are as many ways to cultivate simplicity in our food or other consumption decisions as there are ways to complexify them. Different people will focus on different aspects of these problems, and our solutions should play to our individual interests and strengths (maybe you'd rather brew beer than raise tomatoes; perhaps you're the cook, not the gardener, in the family). Hence lives and lifestyles will legitimately differ. Simplicity need not limit diversity.

Seventh, simplicity, like all the virtues, needs to be cultivated by individuals and families, but also encouraged and sometimes mandated by society, if we hope to secure human and nonhuman flourishing. The very term *voluntary* simplicity emphasizes voluntarism, while most of the literature on material simplicity focuses on individual and small-group action. But this is arguably a failure of this literature (Claxton, 1994). Jerome Segal (1999) argues convincingly that creating less materialistic societies will demand fundamental political changes. Discussing the United States, Segal emphasizes changes in economic policy that would help safeguard basic physical and economic security, and thus make it easier for individuals to freely choose less materialistic paths. Because we often 'consume because others consume' (Lichtenberg, 1998) and because 'what counts as necessary [consumption] in a given society' depends in part on 'what the poorest members of society require for credible social standing' (Schudson, 1998), enacting simplicity has an important political component.

Eighth (and at the risk of sounding grandiose), this short discussion of food simplicity suggests that material simplicity does indeed further justice and wisdom, as philosophers have long maintained. Modern industrial agriculture is callous toward farmers and farm communities, and grossly unjust toward its animal 'production units.' These injustices are sustained in large part by the ignorance of consumers. Voluntary food simplicity can help reverse this process, as we learn about food and act on what we have learned, try to appreciate the processes involved in feeding us, and honor the various participants in those processes (Berry, 1990). To the extent we use resources, take life, or cause pain when we raise or eat food, simplicity enables us to do so consciously and honestly. This opens up a space within which we *may* act justly and wisely. Note that simplicity does not *guarantee* justice and wisdom in this important area of our lives: it makes them possible. Casual participants in the industrial food status quo, however, cannot act justly or wisely in their food consumption decisions. Those options are not on the menu.

Some might say that what are really needed are better rules for how farmers should be compensated, food animals treated, and so forth. Then we could follow the rules and eat whatever we wanted, with a clean conscience. We certainly agree on the need for better rules: they are essential to furthering material simplicity politically, and thus helping create more virtuous and just societies. But rules will only get us so far. The world is an unjust place and seems likely remain so for the foreseeable future; hence we cannot completely rely on 'the rules' to tell us how to behave. Further, the idea of purely economic spheres of life, where we can choose freely – that is, without the need to consult anything but our own desires and whims, perhaps restrained by a few basic moral rules – is deeply flawed; part of the economistic view of life that has given us modern industrial agriculture in the first place. Setting up such 'duty free zones' blinds us to both responsibilities and opportunities. We think we are increasing our options and our freedom of action; instead, we find we have lost the ability to distinguish right from wrong, or quantity from quality. But acting on such distinctions, in all areas of our lives, is the very definition of wisdom.

References

Berry, W. (1990). What are people for? North Point Press, New York, USA.

Cafaro, P.J., Primack, R.B. and Zimdahl, R.L. (2006). The fat of the land: linking American food overconsumption, obesity, and biodiversity loss. Journal of Agricultural and Environmental Ethics 19: 541-561.

Claxton, G. (1994). Involuntary simplicity: changing dysfunctional habits of consumption. Environmental Values 3: 71-78.

Felice, J. (2007). Food for thought. Acta Neuropsychiatrica 19 (5).

Lichtenberg, J. (1998). Consuming because others consume. In: Crocker, D. and Linden, T. (eds.) Ethics of consumption: the good life, justice, and global stewardship. Rowman & Littlefield, Lanham, MD, USA.

Nussbaum, M. (1993). Non-relative virtue: an Aristotelian approach.' In: Nussbaum, M. and Sen, A. (eds.) The quality of life. Oxford University Press, United Kingdom.

Pollan, M. (2007). The omnivore's dilemma: a natural history of four meals. Penguin Books, New York, USA.

Putnam, J., Allshouse, J. and Kantor, L.S. (2002). U.S. per capita food supply trends: more calories, refined carbohydrates, and fats.'FoodReview 25 (3): 2-15.

Sandler, R. L. (2007). Character and environment: a virtue-oriented approach to environmental ethics. Columbia University Press, New York, USA.

Schudson, M. (1998). Delectable materialism: second thoughts on consumer culture. In: Crocker, D. and Linden, T. (eds.) Ethics of consumption: the good life, justice, and global stewardship. Rowman & Littlefield, Lanham, MD, USA.

Segal, J. (1999). Graceful simplicity: toward a philosophy and politics of simple living. Henry Holt, New York, USA.

U.S. Department of Health and Human Services (2001). The surgeon general's call to action to prevent and decrease overweight and obesity. Office of the Surgeon General, Rockville, MD, USA.

Wensveen, L.V. (2000). Dirty virtues: the emergence of ecological virtue ethics. Prometheus Press, Amherst, NY, USA.

Placing and scaling ethical choices: ethical consumption and ethical public procurement

D. Kleine
Department of Geography & ICT4D Centre, Royal Holloway, University of London Egham, TW20 0EX, United Kingdom; dorothea.kleine@rhul.ac.uk

Ethical consumption, or a consumption pattern that goes beyond considerations of price to include for example social and environmental criteria, has been a significant trend, with the market for so-called ethical products in the UK alone expanding from £13.5bn (1999) to £47.2bn (2012)[4].

Within the broader ethical consumption trend, Fair Trade has been a particularly prominent and successful movement. However Fair Trade has been criticised for originally dividing the world into producer and consumer countries. Further, like many other certification schemes, Fairtrade International (FLO)'s certification scheme has been criticised for its highly centralised system of setting standards. Meanwhile, certification schemes generally are proliferating, for example those related to Fair Trade, organic and energy efficiency. These certificates represent different degrees of rigour and ethicality, and have been criticised as confusing to consumers. In some cases they have been accused of being not much more than an industry-sponsored 'whitewash' (only seeming to pursue and promote social responsibility) or 'greenwash' (only seeming to pursue and promote environmental responsibility).

Ethical consumption research has a strong bias of concentrating on (and generalising from) the experience of Europe and North America. However, phenomena which could be labelled ethical consumption are neither entirely new nor are they fully expressed in the global North experiences. In addition, there is the particular irony in the fact that these are countries which in aggregate represent highly environmentally and socially unsustainable lifestyle models.

What is considered ethical consumption is negotiated in culturally specific ways and varies in different country contexts. More research is needed into this contextualisation, or 'placing' of ethical consumption discourses and practices in different countries and cultures. Middle-income countries with their growth in educated middle-class consumers also represent emerging markets for ethical consumption.

Most recently I have been involved in the Choices Project (www.sustainablechoices.info). This was acollaboration between academics and NGOs from the UK, Chile and Brazil, exploring interpretations and practices of ethical consumption in the two Latin American countries. The project was funded by the UK Economics and Social Science Research Council and the Department for International Development and methods included focus groups and large scale surveys in each country. We found that some of the 'new ethical consumers' embrace Northern ideas of ethical consumption as a way of symbolising status, modernity and cosmopolitanism, while others resent having definitions of what is 'ethical' imposed upon them, either in general discourse or more concretely through certification standards. In the two countries, the discursive space of 'ethical consumption' and 'sustainable consumption' is often taken up by terms such as 'conscious consumption' and 'responsible consumption'. In terms of practices, traditional food markets, a longstanding moral discourse condemning wastefulness, the significant informal and solidarity economy spaces, and progressive legislation for sustainable public procurement, in particular in Brazil, offer some endogenous solutions to ethical consumption dilemmas. Significant challenges remain in both countries, such as value-for-money and concerns over whether local produce can compete with global brands on quality.

[4] www.co-operative.coop/PageFiles/416561607/Ethical-Consumer-Markets-Report-2012.pdf.

The investigators on the project all come from different theoretical perspectives and bring different approaches to bear on the data. From one angle, we interpret ethical consumption choices through the lens of Amartya Sen's version of the capabilities approach, which focuses on expanding the real freedoms that people enjoy to lead the lives they have reason to value and to enhance the real choices they have (Sen, 1999). We use a translational tool, the Choice Framework (Kleine, 2013), to operationalise this approach, and to frame reasoned choices within ecological limits.

There is a longstanding debate whether ethical consumption is firstly, too aligned with a view of society as merely a collection of atomised individuals and secondly, a distraction in times when decisive structural, including legislative, changes are needed. In a pioneering move, the Choices Project explored the possibilities for scaling up ethical consumption choices from individual to the collective level by studying citizen-consumers views both on their own consumption and on the procurement decisions the state makes on their behalf. The state is a key buyer in both the Chilean and Brazilian economy. The research findings are nuanced but show a strong citizen support for ecological and social criteria to be used when the state buys goods and services. However, there is a significant trust deficit with citizens flagging concerns over corruption and favouritism. Looking ahead, this will mean that procedures will have to be found which ensure fair competition on these new criteria and transparent decision making processes. This in turn may well lead, somewhat ironically, to a further emphasis on the frequently criticised certification systems. These will have to be improved to be easier to navigate by time-starved consumers and state buyers and to be acceptable to a more culturally diverse audience of ethical consumers.

References

Kleine, D. (2013). Technologies of choice? ICTs, development and the capabilities approach. MIT Press, Cambridge, MA, USA.

Sen, A. (1999). Development as Freedom. Oxford University Press, Oxford, UK.

Intellectual property rights and food security: the role of external relations

S. Hongladarom
Department of Philosophy and Center for Ethics of Science and Technology, Chulalongkorn University, Bangkok 10330, Thailand; hsoraj@chula.ac.th

Abstract

Intellectual property rights (IPR's) have become an important tool in ensuring food security; however, if used inappropriately, it could well create the reverse. This paper looks at the concept of IPR's in order to find a way to harness their use so as food security is ensured. A tentative argument proposed here is that IPR's do not exist in a metaphysical or epistemological vacuum; on the contrary, research and development leading up to patentable products is often related to social, economic or political contexts in such a way that the relation is constitutive. Thus, it is appropriate that claims to IPR's should acknowledge these relations through a scheme of benefit sharing that is fair to all parties. In the course of the paper I will discuss the four major theories of IPR's according to Fisher – the consequentialist theory, the Lockean theory, the Kantian/Hegelian theory, and the democratic order theory. The aim is to criticize each of them very briefly in terms of the constitutive external relations. If it is the case that IPR's are even partly constituted by relations to outside contexts, then elements of these contexts should have a share in the benefits that accrue through the use of IPR's also.

Keywords: intellectual property rights, ethics, food security, external relations, benefit sharing

Introduction

The main concern of this paper is to sketch a possible answer to the question whether, and if so how, food security can be obtained through the use of some kind of intellectual property rights (IPR's). Many scholars have tried to link up the two concepts. For example, Blakeney (2009) devotes a whole book to the topic. Cullet (2004) looks at the issue from the perspective of the developing South. Chapman (2002) links up IPR's and human rights and argues that the former cannot proceed without due recognition of the latter. Since access to food is considered a right, IPR's are linked to food security along this channel. In his book, Dutfield (2004) discusses the use of IPR's, more specifically, protection of plant varieties, and how they could ensure food security in the developing world. The question is important because, on the one hand, it seems that IPR's could foster food security in some way. For instance, technological inventions that are designed to solve food security problems could be protected with IPR's so as to, according to the view of some of their proponents, make it possible for the investor in the research and development of the technologies to recoup their investment and to provide incentives for further development. On the other hand, however, uses of IPR's have been accused of preventing local farmers from maintaining their traditional ways of life and independence, as is the case when they come to rely on some forms of these technological inventions in their farming practices. What I would like to do in this paper, then, is to have a closer look at the whole concept of IPR's and the theory behind it, with an aim toward the problem of food security. What kind of theory and what type of possible modification of how IPR's should be understood is perhaps most amenable to preventing and solving the problem of food security?

I would like to argue that the use of IPR's should be more open. What I have in mind is that, instead of restricting the claim of ownership of an intellectual property to the traditional owners according to most available laws, namely to the investors and the firms that employ those inventors, the rights to intellectual property should be expanded, in some form to be sketched out in this paper, to a wider circle which includes the beneficiaries of the invention, the state, and the public as a whole. This by no means implies that the whole concept of IPR's is being destroyed. On the contrary, inventors still retain a right to ownership of their intellectual creation, but they have to realize that they alone are not the ones who have a stake in those creations. The fact that they are around implies that they owe the fact of their existence to the wider circle. Since nothing and nobody exists in a vacuum, what happens anywhere tends to ripple across all the space everywhere. Thus the wider circle indeed has a say and a share in the intellectual creation that the inventor comes up with. Hence it seems fair that the wider circle should take part in the rights to the intellectual property also.

My approach, then, is similar in spirit to the proposal by Posey and Dutfield (1996: 3), when they aim at refocusing on 'Traditional Resource Rights' (TRR's) rather than IPR's (See also Posey, 1990). However, Posey and Dutfield do not aim at analyzing the concept of IPR's nor criticizing the underlying theories as I intend to do in the paper. In fact the argument I am proposing looks rather simple; yet it is surprising that not many works in the literature deal directly with it. According to Fisher, there are four major theories of intellectual property rights: one that deals with consequentialist reasons – that IPR's bring about desired consequences and provide incentives for innovation, with the right to property arising from one's effort – that one has a right to a piece of property, intellectual or not, if one exerts one's effort and labor in producing or acquiring it, expression of one's personality – that IPR's are an extension of one's own creativity and personality, or a democratic social order – that IPR's are consistent with the kind of social order that is democratic and respectful of individuals' role in it (Fisher, 2013). However, none of these four major theories take into account the rather obvious fact that in order to produce the kind of innovation that merits being granted IP protection, the context is necessarily involved in such a way that to ignore it entirely in claiming the rights to IP would seem to be incoherent. As a result, claims to IPR's have to acknowledge these external relations and the benefits that accrue due to the claim should be shared accordingly.

Role of external relations

Let us look at each of these four major theories in turn. The first one is the most familiar one and perhaps the most cited by corporations benefiting from IP protection as well as by legal scholars and the court in general. IPR's are needed as a guarantee, so to speak, that investments on research and development leading up to the patented product produce adequate return. Furthermore, they are also necessary as an incentive for further effort in innovating. However, effort in innovating, research and development seldom, if ever, exists in a vacuum such that no wider social or political contexts are involved. A firm that develops a technology that would ensure food security, such as a hybrid seed that is resistant to drought and has high yield exists as a node in a complex web of social, economic, cultural relations to other firms, other agencies, as well as the public. These relations do not obtain only at the obvious level of the firm's usual dealings with outside agencies, e.g. paying taxes to the community, buying stuff from suppliers, selling products to consumers, sharing profits to their shareholders, and so on, but the very activity of research and development for the kind of technology that is going to be patented is constituted by these relations to the wider context too. It is highly implausible nowadays that any kind of sophisticated research and development of this kind can start entirely from the ground up. Researchers cannot shut themselves up in their laboratories and can still produce any kind of technology that works in the real world. In order to produce the seed, at least researchers have to rely on past studies, then the resulting seed has to be extensively tested in the field; unless the corporation owns a large tract of land testing the seeds would have to be done in open fields and in order to do that permissions from the relevant authorities have to be obtained. Furthermore, the laboratories have to employ a number of people and engage in various economic and other kind of relationships with other agencies outside of the corporation to which the laboratory belongs. Perhaps the corporation may have had performed a survey of need of the farmers in order to ascertain exactly what kind of seed would be the most preferable to them. All these mean that input from the outside in fact constitutes the very activity of research and development as well as the patentable product that comes up afterwards. The consequence is clear. Claims to IPR's would have to be in some way shared among these wider circles also; if it were possible for the laboratory to shut itself up entirely depending absolutely on nothing from the outside world, then it might be possible for them to claim exclusive rights to their IP. But since the world seldom works that way, the usual claim to IPR's, which is almost always exclusive, would have to be modified.

An obvious rejoinder to this argument is, of course, that to come up with innovation that is patentable might depend in some way on these outside factors, but the product itself must be shown to be sufficiently innovative in order to be able to be patented. The fact that activities leading up to the innovated product require a number of links to the outside world alone is not, so the rejoinder goes, sufficient in guaranteeing that the links and the external relations do have a share in the IPR's that result from the work of the laboratory. However, that is a rather narrow look at how innovation comes about. Even a lone thinker who shuts himself up in a room and thinks up a new idea have at least to base her thinking on some prior ideas that are around at the time which form, among other things, an input to the problem that she has set out to solve through the innovation in the first place. It is commonly acknowledged that Descartes's Cogito Argument is the epitome of an original argument in the sense that, according to the content of the argument, Descartes or the cogito thinker does not need any external relations in order to let the argument go through toward its famed conclusion (Descartes, 1996). The fact that Descartes himself has to eat to survive and is situated in a room in a house, which presupposes that he either owns it or is allowed to remain there, does not imply that the farmer who produces his food, or the relation he has with the authorities to prove his ownership of the house, or the owner of the house who allows him to remain to think, has any role to play in the Cogito Argument. However, one of the familiar objections to Descartes' argument is precisely that the Cogito itself presupposes these very external relations for it to go through. Even if we allow that the house and the food might have been cooked up for Descartes by the Evil Demon, the very fact that Descartes thinks in a language, which presupposes that he has to have learned it through speaking it with others, show that external relations are constitutive of the Cogito from the beginning. According to Wittgenstein, private language, namely a kind of language that in principle only the one who speaks it knows it and no other, is not possible, so Descartes' language is not private either (Descartes, 1996). This means that other people are necessary in the content of the Cogito in the first place. Thus, if the requirement for constitutive external relations is necessary for Descartes' Cogito, then it is obviously the case for a much less stringent argument and practical development that takes place in the laboratory.

The second major theory of IPR's, the Lockean one, states, roughly, that IPR's are justified as a rightful fruit of labor that should belong to the one who has expended it in order to arrive at the intellectual property. This argument is rather similar to the consequentialist one that we have just considered. And as in the case of the former argument, it is rarely the case that one alone or even one corporation, without any relations or any help from the outside, could secure any kind of sophisticated intellectual property that abounds today. If it is the case that any attempt at research and development for patentable product has to rely on a number of contexts and external relations, then it means that it is not the labor or an effort on oneself alone (or that of one group alone, for that matter) that is responsible for the success of finished product. Hence the benefits that accrue through the use of the claim of IPR's should be fairly shared among those who are involved, both directly and indirectly.

The development of technologically advanced hybrid seed that can grow in unfavorable conditions is a case in point. The common assumption is that, since the corporation has invested a sum of money into the research and development for the seed, they are entitled for a period of patent protection where the IPR to the seed is respected. However, in order for the seed to be made meaningful to the majority of the world's farmers who stand to benefit from the seed, the price of the seed needs to be low enough to be affordable. Furthermore, the farmers should be able to save some seeds so that they can grow them in the next season. The practice of some corporations of engineering the seed so as to become sterile is thus not in line with the argument adopted here. The Lockean position would view the research and development for the seed as an investment, an exertion of labor to stake a claim in a piece of property. But in order for the practice to get off the ground, the firm and the team of scientists who do the actual work need to interact with the outside world in one way or the other. Most of all, if the idea is to develop a kind of seed that would help the majority of the farmers, most of whom are poor and live in the tropical countries, then prior research on how the seed would respond to these particular climate conditions is absolutely crucial. Recognizing that the resulting technology is a result not of the work of the scientists alone, but that other factors are critically involved would mean that fair sharing of the benefits should be an important factor in deciding who gains what in the use of the technology in question. Since the scientists do not, and cannot, do their work alone without input and all kinds of relations obtaining between them and the outside world, including the farmers themselves who are on the receiving end, ways need to be found in order to acknowledge the roles that these external relations play in the process of research and development.

The third major theory states that IPR's are justified as an extension, or an expression, of the personality of the creator. Usually this view is used more to justify copyrights than patents, but it has also been used by some scholars to justify patents too. Here one needs to recognize, again, that an entity, be it an individual, a firm, or a scientific laboratory, does not exist in vacuum in total independence from all other factors. Hegel is usually cited as a source of this third major type of IPR's theory, but it is Hegel himself who, in the *Phenomenology of Spirit* (1977), famously presents an analysis of an individual in such a way that an individual, to be the individual he or she actually is, has to be defined through relations with other individuals. In the case of objects this means that an object can only be what it is through whatever lying outside of it, so that a boundary between the object and what is outside of it is defined. Something that has no externally defining boundary would not be an individual object at all. The same analysis also goes for individual persons. A person is the person he or she is only through *recognition* that other people have toward him or her. Without the recognition, there would be no person since there would be no consciousness that this is a particular person with whom one can deal with. This is just another way of saying that a human person cannot be understood as such without reference to his or her community, family or group. The relation of recognition in this sense is a typically Hegelian notion (cf. Taylor, 1977).

The upshot is that firms cannot consistently hold the claim to IPR's justified through the notion of extension of their 'personality' and claim that no external relations are involved. The result then is structurally the same as what I have been trying to say so far regarding the previous two theories. In order consistently to claim IPR's to a product, external relations need to be factored in, and in the next section I will sketch a very rough form of what the factoring in of these external relations consist in.

The last major theory of IPR's is perhaps the most intriguing. The idea is that IPR's are needed to ensure that we live in a fully democratic society where the rights of individuals and presumably corporations are respected. This argument thus recognizes from the first moment the constitutive relation that IPR claims have to their external factors. In order to say that IPR's are necessary for a democratic society, one obviously needs a society to begin with; hence IPR's according to this theory are justified through their relations to the outside factor. That is why it is intriguing, because the previous three theories do not, *prima facie,* seem to admit the relations to external factors as does the fourth one here. Thus, a discussion of this theory has to be a little different from that of the previous three. Here the focus is on whether and how respect of IPR's contributes to a democratic society. However, it is quite clear that a narrow interpretation of IPR's in such a way that the majority of the world's farmers would be worse off because they have to buy expensive drought resistant hybrid seeds, the kind that may one day be needed in order to stave off food insecurity, would be a decidedly less palatable option than a broader one which, as I have been trying to say, includes the role of external factors including stakeholders such as the farmers themselves and other organizations and agencies in the process of developing a patentable product. A fair benefit sharing scheme where the farmers, the community, the local and national authorities, the firms, and the wider public, are all stakeholders are needed in order to ensure food security in the longer run. The scheme is not a radical one where the firms are disincentivized from developing new products, but as it ensures survival of all factors the scheme is in the long-term interest of the firms themselves.

Conclusion

What I have been trying to argue in the short space provided for this paper is that external relations play a constitutive role in IPR's claims. Any attempt to ensure food security through a reliance on IPR's does not even get off the ground, I believe, without the kind of benefit sharing scheme that is based on the recognition of the role external factors play. An upshot of my argument is that one cannot consistently claim a right to an intellectual property and at the same time hold that one has ultimately an exclusive ownership of the property in question with no need to regard any external factors that are obviously involved. It is true, nonetheless, that in a genuine case of innovation, the product has to be proven to be sufficiently distinct from all others that have been invented before. This is the linchpin of the whole idea of intellectual property rights. However, being able to claim innovation is not the same as depending on all others in such a process, and this means that, though one has a right to one's own invention, one does not have an obligation to the world or the context in which such process is possible in the first place. This necessitates a kind of fair benefit sharing scheme that all stakeholders should have a part, including the inventors themselves, the firms that employ the inventors, the farmers, the local and national authorities, and the wider general public. Recognizing a list of stakeholders this wide does not disincentivize the firms from developing future patentable products. Firms still hold on to their IPR's, but they have to realize that their long-term interests do depend ultimately on sharing of benefits of the technology to the wider circles. Holding on to IPR's in order to drive up the price and create a vicious monopoly could only benefit the firms in the short term, as the instability that ensues would create an environment in which further research and business dealings become increasingly difficult.

So what does a fair benefit sharing scheme look like? For one thing, the price of the hybrid seed (or any other agricultural technologies for that matter) should be made affordable to the farmers who need them. Since farmers are at the forefront of the fight to ensure food security, they need to be more fully supported than they are now. The whole idea is that we are living on the same planet earth. As of now there is no possibility of packing up and moving to other planets yet. So firms should lower their profit expectations and look at a broader notion of profit where well-being of people who are not stock holders be taken into account. It certainly requires a tremendous amount of work to sort out in detail who should get what and in what proportion in the fair benefit sharing scheme sketched out in this paper. But if we are to think about how IPR's should play a role in ensuring food security, I believe that this is about the only way to go.

Acknowledgments

I would like to thank Helena Röcklinsberg and Per Sandin for their comments on an earlier draft of this paper, which led to many improvements. I would also like to thank the executive committee of the EurSAFE for inviting me to the conference in Uppsala. Research for this paper has been partially supported by a grant from the Thailand Research Fund and the National Research University Project, grant number BRG5380009 and AS569A, respectively.

References

Blakeney, M. (2009). Intellectual property rights and food security. CABI, Oxfordshire, UK.

Chapman, A.R. (2002). The human rights implications of intellectual property protection. Journal of International Economic Law 5(4): 861-882.

Cullet, P. (2004). Intellectual property rights and food security in the South. Journal of World Intellectual Property 7(3): 261-286.

Descartes, R. (1996). Meditations on first philosophy. J. Cottingham, transl. Cambridge University Press, Cambridge, UK.

Dutfield, G. (2004). Intellectual property, biogenetic resources and traditional knowledge. Earthscan, London, UK.

Fisher, W. (2013). Theories of intellectual property. Available at http://cyber.law.harvard.edu/people/tfisher/iptheory.pdf.

Hegel, G.W.F. (1977). Phenomenology of spirit. A.V. Miller, transl. Oxford University Press, Oxford, UK.

Posey, D. (1990). Intellectual property rights and just compensation for indigenous knowledge. Anthropology Today 6(4): 13-16.

Posey, D. and Dutfield, G. (1996). Beyond intellectual property: toward traditional resource rights for indigenous peoples and local communities. International Development Research Centre, Ottawa, Canada.

Taylor, C. (1977). Hegel. Cambridge University Press, Cambridge, UK.

Section 1. Market and policy

Impacts of sustainability labels on consumers' purchasing decisions for fish

D. Buergelt, I.B. Christoph-Schulz, P. Salamon and D. Weible
Thünen-Institute of Market Analysis, Bundesallee 50, 38116 Braunschweig, Germany; doreenbuergelt@ti.bund.de

Abstract

In Germany, fish products rely on sustainable production and convey it by labels. Currently, fish products with and without sustainability labels can be found. The objective of the study was to elicit consumers' opinions on sustainable fishery in focus groups. Five sustainability labels on captured fish products were presented in focus groups to reveal consumers' perception, labels' credibility and labels' relevance in purchase decisions. In total, 12 focus groups were carried out in four German cities between April and May 2012. Concerning their purchase criteria participants shopping fish products rarely stated sustainability or corresponding labels. In general, participants' answers indicate low familiarity with sustainable fishing. Participants' knowledge on sustainability labels was low. Participants assessed the credibility of the labels as low, were unsure about the meaning of the certification and argued about missing controls. The low perception of the labels and the importance of other issues like time constrain and taste result in a low relevance for consumer's purchase decisions.

Introduction

In many European countries concepts such as sustainable production and consumption have widely evoked public interest (Verbeke, 2007). In Germany, fish products are applying among other products the concept of sustainable production and conveying the message by labels. Currently, fish products with and without a variety of different sustainability labels can be found in shelves or in the counter fridge.

Labels indicating sustainability are supposed to convey information concerning sustainability from production across processing to consumers. The main purpose of labels is to concentrate and communicate product information in order to support consumers in making their purchase choices (BMELV, 2011). Consumers can take advantage of labels as these reduce, at least to a certain extent, their information asymmetry about products' characteristics, thus, adjust their decisions better to their preferences (Teisl, 1998)

However, one has to consider that consumers perceive and process label and their conveyed information differently. Socio-demographics, such as gender, education, presence of children and age influence the perception of labels and the information communicated in labels (Verbeke, 2006). Additional influence have individual characteristics, such as product knowledge, awareness, familiarity, scepticism, motivation, and health status (Wansink, 2004) Most probably, these facts reduce the general benefit of labels and shows the need of personalized information for different consumer groups.

Furthermore, consumers in developed countries such as Germany are surrounded by complex food products; with sustainability as one product characteristic among plenty others. Such product characteristics can be mainly separated into six categories: (1) characteristics related to the process, e.g. sustainability; (2) brand and price; (3) environmental characteristics; (4) health related characteristics; (5) nutritional characteristics; and (6) food safety characteristics (Banterle, 2012). The vast of labels on food products could arouse an information overload (Hwang, 1999) and a resulting tiredness of consumers to consider labels. Sustainability is an intangible credence characteristic not verifiable by consumers even after consumption (Nelson, 1970). Thus, consumers have to rely on labels and to believe in the credibility of their statements when they are interested in purchasing sustainable fish products.

The objective of the study was to elicit consumers' perceived advantage of sustainability labels on fish products in focus groups. The five sustainability labels on captured fish products were presented to reveal consumers' perception, labels' credibility and labels' relevance in purchase decisions.

Research approach

The paper draws on focus groups that were carried out between April and May 2012 in four German cities. Focus groups are carefully planned and moderated discussions and are limited to a defined area of interest (Kahan, 2001) such as sustainable labels. It is 'a method for eliciting respondents' perceptions, attitudes and opinions' (Wilson, 1997) and takes advantage of group interactions to determine participant's motives, which cause their behaviour. Focus groups are not aiming to collect representative data on a certain topic (Padel, 2005). Participants are confronted with other participants' opinions, attitudes or perceptions, and may have to justify their own opinion, attitude or perception. Hence, 'individual response becomes sharpened and refined, and moves to a deeper and more considered level' (Finch *et al.*, 2003).

In total 12 focus groups were carries out in Hamburg, Cologne, Leipzig and, Munich, three groups in each city. In the first group were only men, in the second only women and in the third men and women were mixed. Each group involved seven to 12 participants. For the recruitment of the participants it was most important that they were buying fish at least once a month. Further, the groups were put together based on the demographic characteristics age, occupation and if necessary gender. A guideline was written in advance to structure the discussion. All groups were moderated by one of the authors and were recorded as well as fully transcribed.

Results

Concerning their purchase criteria when shopping fish products participants predominantly named criteria fitting into the categories: taste, intended use, quality and price. Sustainability or corresponding labels was rarely stated in this context. In general, participants' answers indicate low familiarity with sustainable fishing. Participants' knowledge on sustainability labels was low. Thus, we assume that the majority of participants could be characterized as unfamiliar with and not interested in sustainability.

In the following paragraphs the term fish always refers to captured fish products excluding aquaculture. Before we started the discussion about labelling of sustainable fish we asked for participants' understanding of sustainability in regard to fishery. The statements were grouped in following categories: preservation of populations: 'that the fish populations are not overfished' (L2), protection of juveniles and reproduction: 'that the really young fishes are not caught' (K1), fishing methods: 'no endless trawl nets' (L1), fishing quota: 'that only a certain amount of fish can be caught'(L3), aquaculture and fish farming: 'I rather relate it to fish farming because I think if you catch fish sustainable then you farm it' (M1), discard: 'that there is not something else in the nets'(K2), protection of aquatic environment: 'that the aquatic environment is not contaminated' (L2), and preservation of the ecological equilibrium: 'that the ecosystem is sound' (M2). Further, participants expressed their resentment about the inflationary use of the term sustainability: 'Sustainability is a word that was used a lot in the last two to three years, especially in the organic sector. I cannot hear it anymore'(M1).

Participants' knowledge on examples of sustainability labels for captured fish was, in general, low. Only the label of Marine Stewardship Council (MSC) was known to more participants, and sometimes only referred to as 'yes, something oval' (Male, K1) or 'a blue label with letters' (L3).

Afterwards, five labels that can be found on fish products in German supermarkets were presented. These were: (1) Naturland Wildfisch; (2) Friend of the Sea; (3) Dolphin Safe; (4) Iceland Responsible Fisheries; (5) Marine Stewardship Council. Table 1 shows that the participants were only familiar with the Marine Stewardship Council label. Very few had seen the labels of Naturland Wildfisch and Dolphin safe and Friend of the Sea as well as Iceland Responsible Fisheries were completely unknown. Thus from the five labels the Marine Stewardship Council label was best known. Naturland Wildfisch and Dolphin Safe were also recognized by some. The labels of Friend of the Sea and Iceland Responsible Fisheries were not recognized in any focus groups.

The statements of the discussions of each label were classified in categories. All categories were labelled according to the tendency of their statements either as positive or negative. All categories are shown label-wise in Table 2 and Table 3. The Marine Stewardship Council (MSC) label can be classified as the most positive perceived label because it has the most positive categories and the fewest negative ones (Table 2). A positive category is the familiarity of the label. Further, participants saw a relation to certified sustainable fishery which seems to go along with a good conscience when buying fish. Participants rated positively that there seems to be an organisation or entity behind MSC and the presence of an internet address. Thus, some participants rated MSC as their favourite label from the five presented ones. Some participants rated negative that the provided information in the MSC label is insufficient and some stated a general mistrust in labels. The most negative perceived label is Dolphin Safe (Table 2). No positive categories could be compiled but several negative categories. First the label produced a misleading relation to diving clubs, swimming badges and some thought it is against consumption of dolphin meat. Further, participants criticized that it is not self-explaining and it provides not enough information. It was perceived as unserious and incredible and participants were missing a name or an internet address. The visual appearance was rated negatively and the Dolphin Safe label seems not to communicate sustainability. Some participants thought that it is not enough to save dolphins only because there are also other endangered species in the sea and some even stated that the label would prevent them from buying fish. Although, the labels of Friend of the Sea and Iceland Responsible Fisheries were unknown to all participants one positive category could be complied for both (Table 3). The website link shown in the label of Friend of the Sea was rated positive and some participants stated that Iceland Responsible Fisheries is their favourite one from the five presented labels. Participants criticised that Friend of the Sea has more a relation to clothes and sailing than to fish, that it appears incredible, a missing relation to fish, the usage of the English language and a contradiction that was triggered by the picture of a ship in the label. Negative categories of Iceland Responsible Fisheries were the provision of insufficient information, again the usage of the English language and the missing relation to fish due to the abstracted presentation.

Table 1. Label's level of familiarity.

Label	Known	Hardly known	Unknown
Naturland Wildfisch		x	
Friend of the Sea			x
Dolphin Safe		x	
Iceland Responsible Fisheries			x
Marine Stewardship Council	x		

Table 2. Addressed characteristic of known and hardly known labels.

	Positive categories	Negative categories
Marine Stewardship Council	• Familiar ('meanwhile, I have seen it several times that I think it is the right one', L3) • Relation to certified sustainable fishery ('the terms were purposely chosen, certified and that shows a certain..., yes you believe it', M2) • Entity behind ('that there is a kind of organisation, public authority that controls it', K1) • Website link: ('I like the internet address. It is possible to be informed.', M2) • One of the favourites ('evokes the most credence, has somehow a clear message or in inverted commas compared to the others a clear message', M2)	• Insufficient information ('from the label I cannot see if the quota is big or little. Whether the fish is protected', H1) • General mistrust in label ('certificates always discourage me because I think they want to cheat', M2)
Dolphin Safe		• Misleading relation to diving club, consumption of dolphins • Not self-explanatory/ insufficient information (' I never ate a dolphin. The Logo's meaning is not comprehensible', M1) • Unserious/ incredible ('I cannot take this serious', M1) • No name nor website link ('no name, no internet address, no organisation name too', L1) • Visual appearance ('black means death for me', M1) • Communicates no sustainability ('I would not think of sustainable fishing, that this is a sign for that', K1) • Protecting dolphins is not enough (it communicates that dolphins are safe but all other fishes not', M3) • No purchase (when I see this picture, I would think, ok tonight we rather eat chicken', K3)
Naturland Wildfisch	• Relation to good quality ('that is a fish probably not that much contaminated with plumb', K1) • German ('a label for something German...I have more trust in it', M1) • Eye catching ('but it is eye catching...not lost in the melee', H1)	• Barely maritime/ no relation to fish ('no explicit sign for fish', L2) • Misleading relation to plants ('rather looks like something for vegetable', H2) • Company label ('looks less like a seal more like a manufacturer label', M3) • Term: Wildfisch ('a contradiction, because wild fish is not coming from land but from the sea. I think it is a bit misleading', M2) • Meaning unclear ('I really don't know what it should express', L2)

Table 3. Addressed characteristic of unknown labels.

	Positive categories	Negative categories
Friend of the Sea	• Website link ('chance to read what it is all about', K3)	• Misleading relation to clothes/sailing ('rather a cloth logo', H2) • Incredible/ arbitrary ('no relation to something serious', L2) • English ('if one is not able to understand English, then he/ she cannot understand the message', M1) • No relation to fish ('I would not associate it with fish', K2) • Contradiction with ship ('but only saves ships', H1)
Iceland Responsible Fisheries	• One of the favourites ('it is the most clear one for me ...and it says something', K,2)	• Insufficient information ('we don't know what is behind it', H1) • English ('Why does everything has to be in English?', H2) • No relation to fish ('very high quality but I think too much abstracted. The relation to fish doesn't come automatically', M1)

During the whole discussions overall utility of food labels was addressed but intensified when we asked for labels that communicate sustainability, and when participants received the overview with all five labels on it. They mainly criticized: (1) the multitude of labels; (2) their unknown meaning, endorsing entities as well as terms of references; and (3) participants had a general distrust in the controls and validity of the labels. This resulted in a low credibility of the labels. Labels with links to websites such as Friend of the Sea and MSC were rated as more credible. The low perception of the labels and the importance of other issues like time constrain and taste result in a low relevance for consumer's purchase decisions, although some participants stated that labels salve their conscience.

Participants were also asked for their ideas on sustainability labels which they would trust and which would have higher effectiveness regarding their purchase decisions. In regard to provided information the answers were two folded. Some wanted more information and others were already 'overloaded' by the provided information. We found that many participants showed some kind of 'tiredness' due to the rising number of labels or they resign because they do not trust in any label. A reliable label in form of a code number or a seal was requested supported by a governmental institution or an acknowledged environmental organization, such as Greenpeace.

The results of the focus groups show that the endorsing entity behind labels was very important because if insufficient information was criticized participants often asked for the entity behind it and participants rated positively that behind the MSC label seems to be an entity. Familiarity was found to raise the credibility of labels because familiarity evokes trust (Teisl, 2002). Familiarity of participants with the MSC label seemed to influence the perception positively. But it is surprising that Dolphin Safe was rated worse than Friend of the Sea and Iceland Responsible Fisheries because the latter two were completely unfamiliar to participants and Dolphin Safe was known by some participants.

Conclusion

Qualitative studies of this kind provide important insights into the way consumers perceive and evaluate labels. From the results presented in this paper we conclude that sustainability in fisheries is of minor relevance for German consumers' purchase decision. Other criteria like taste, freshness, availability at the point of sale and habits proved to be more important. In addition, consumers' knowledge of sustainable fishery is low. Present labels of sustainability in the market are not well known by many consumers and are, therefore, supposed to have only limited influence on consumers' purchase decisions. Nevertheless, the results support the fact that familiarity with labels such as the MSC is important to evoke trust in the label. But, this preliminary research also indicates that labels on sustainability like any other label require more intensive promotion than currently provided to reach consumers' attention. However, there appears to be a very delicate balance between promotion and not to lose credibility. To underpin the point of equilibrium further research should explore on the reasons for the low level of consumer perception of sustainability labels: is it lack of interest, is it abundance of labels, the emerging confusion, or is it 'tiredness'. To overcome 'tiredness' a vantage point could be pronouncing even more a standardized appearance, perhaps to use a fixed size relation to the package, same placing on the package (e.g. extreme bottom right) in combination with promotion campaigns.

Especially repeatedly mentioned weariness towards labels in general conveys also another message: industries needs to be very careful to implement new labels. Often it seems to be more reasonable to convey the sustainability or other messages via one label than several slightly different ones. Also colour and form of labels need to fit to their message (in our case blue), further hints via a website should be provided and, at least in Germany, the use of the German language is indicated.

References

Banterle, A., Cavaliere, A. and Ricci, E.C. (2012). Food labelled information: an empirical analysis of consumer preferences. International Journal of Food System Dynamics 3: 156-170.

Bundesministerium für Ernährung, Landwirtschaft und Verbraucherschutz (BMELV) (2011). Gemeinsame Stellungnahme der wissenschaftlichen Beiräte für Verbraucher- und Ernährungspolitik, Agrarpolitik: Politikstrategie Food Labelling. Berlin, Germany. www.bmelv.de/SharedDocs/Downloads/Ministerium/Beiraete/Verbraucherpolitik/2011_10_PolitikstrategieFoodLabelling.pdf?__blob=publicationFile.

Finch, H. and Lewis, J. (2003). Focus groups. In: Ritchie, J. and Lewis, J. (eds.) Qualitative research practice. Sage Publications Ltd., London, UK, pp. 170-197.

Hwang, M.I. and Lin, J.W. (1999). Information dimension, information overload and decision quality. Journal of Information Science 25: 213-218.

Kahan, J.P. (2001). Focus groups as a tool for policy analysis. Analyses of Social Issues and Public Policy 1: 129-145.

Nelson, P. (1970). Information and consumer research. Journal of Political Economy, 78: 311-329.

Padel, S. and Foster, C. (2005). Exploiting the gap between attitudes and behaviour. British Food Journal 107: 606-625.

Teisl, M.F. and Roe, B. (1998). The economics of labelling: an overview of issues for health and environmental disclosure. Agricultural and Resource Economics Review 28: 140-150.

Teisl, M.F., Peavey, S., Newman, F., Buono, J. and Hermann, M. (2002). Consumers reaction to environmental labels for forest products: a preliminary look. Forest Products Journal 52: 44-50.

Verbeke, W. and Ward, R.W. (2006). Consumer interest in information cues denoting quality, traceability and origin: an application of ordered probit models to beef labels. Food Quality and Preference 17: 453-467.

Verbeke, W., Vanhonacker, F., Sioen,I., Van Camp, J. and De Henauw, S. (2007). Perceived importance of sustainability and ethics related to fish: a consumer behavior perspective. AMBIO: A Journal of the Human Environment 36: 580-585.

Wansink, B., Sonka, S.T. and Hasler, C.M. (2004). Front-label health claims: when less is more. Food Policy 29: 659-667.

Wilson, V. (1997). Focus Groups: a useful qualitative method for educational research? British Educational Journal 23: 209-224.

Food for good: social movement organizations making sustainable markets for 'good food'

G. Kallio
Department of Organization and International Business, Aalto University School of Business, P.O. Box 11000, 00076 Aalto, Finland; galina.kallio@aalto.fi

Abstract

An ethnographic research approach is used to study how food collective movement organizations organize and create markets for locally produced and organic food in Finland. Increased evidence related to safety, sustainability and justness of contemporary food markets has resulted in the mobilization of various food movements over the world in order to challenge the current systems of food provisioning. These movements are striving towards creating more sustainable markets – the practices of food production and consumption. This study elaborates on how social movement organizations engage in market creation by establishing the market exchange of local and/or organic food.

Keywords: food movements, food collectives, market exchange, exchange practice, local food

Background

Food provisioning is a field rife with concerns over the health, justice and sustainability of current production and consumption patterns. The sources and supply chains of foodstuff are increasingly global and subject to fierce competition. In the era of modern agriculture, industry representatives, advocacy groups and a great variety of expert organizations have strongly addressed food related questions. However, general awareness about the lack of safety of food and scandals related to food production place the reliability of food science and the expert systems that govern market transactions under suspicion.

One of the most striking changes resulting from modern agriculture has been the exponential rise and influence of supermarkets (Ilbery and Maye, 2006), which has resulted in many soci(et)al changes (Colquhoun and Lyon, 2001; Kloppenburg *et al.*, 1996; Pollan, 2008; Seyfang, 2007). At the same time, people's wishes to obtain 'food from the backyard' manifest themselves through a willingness to buy local and organic food. Hence new food products, new diets and new moralities are frequently introduced and new market niches explored. Increasingly, also, farmers and consumers together with the media participate in food debates and various food movements have mobilized themselves in order to provide alternatives to the dominant existing food system.

Many food movements, such as food justice and sovereignty movements (Holt-Giménez, 2009; Levkoe, 2005), organic movement (Raynolds, 2000), the fair trade movement (Wilkinson, 2007), the slow food movement (Van Bommel and Spicer, 2011), and local or alternative food movements (Allen *et al.*, 2003; Delind, 2010; Starr, 2010), among others (Weber *et al.*, 2008), operate actively in promoting both ideological agendas and practical solutions for creating more just and sustainable food markets. The Food Collective movement in Finland provides a good example of an ideological movement in a Western context, consisting of several social movement organizations (SMOs) promoting the movement's aims to bring 'good food' to citizen and engaging in the active construction of more sustainable food markets.

Social movements as market makers

Increasing amount of literature reports that social movements participate in market making (Beckert, 2010; Davis *et al.*, 2008; Fligstein, 1996, 2001; Starr, 2010; Weber *et al.*, 2008; Wilkinson, 2007). Although we can observe that new markets are one of the outcomes of social movements, there are only a few studies that provide insight into the very early stages of collective action processes that shape markets and through which new markets are created (Sine and Lee, 2009; Weber *et al.*, 2008). With such a starting observation, this paper elaborates the everyday running and practice(s) of food collective movement organizations' engagement in the market exchange of local and/or organic food.

Market construction entails new cultural codes, new producer identities and new infrastructure for repeated exchanges (Weber *et al.*, 2008). For their existence, markets require tradable goods or services, need buyers and sellers, and exchange infrastructures (Callon and Muniesa, 2005; Weber *et al.*, 2008). According to Beckert (2010), markets can be understood as 'social arenas for the exchange of goods and services'. More specifically, markets are structures of exchange defining a price for a product traded between producers and consumers, and 'at their core, markets are concrete exchange structures between producers and consumers' (Weber *et al.*, 2008).

In contrast to neoclassical economic views on the functioning of markets, markets can be conceptualized as collective, political and moral projects comprising socially structured institutions and embedded relationships that carry cultural norms and meanings (Beckert, 2010; Fligstein, 1996; Fligstein and Dauter, 2007; Fourcade and Healy, 2007; Granovetter, 1985). In other words, markets are infused with culturally produced meanings and moralities of the market actors (buyers and sellers among others) that influence rules and practice(s) of economic exchange.

The practice of market making

One way of better understanding the process of market making is to look at markets as practices. Besides the importance of ideas and meanings in market making, the practice-based perspective on markets emphasizes the role of materiality, which becomes evident in both enabling and shaping the practices of market making (Kjellberg and Helgesson, 2006, 2007). Market practices are defined broadly as 'all activities that contribute to shape markets' (Kjellberg and Helgesson, 2006) and can be categorized into exchange practices, normalizing practices, and representational practices.

Practices form structures of action (Reckwitz, 2002) and entail various activities that form a particular practice; like cooking, lecturing, or negotiating they all indicate material, contextual, cognitive, and routinized sets of activities constituting the particular practice. The practice-theoretical approach assumes that although individuals are the carriers of practices, practices are collectively created and socially and contextually maintained (Reckwitz, 2002; Warde, 2005). *A practice*, then, following Warde (2005), entails tacit and discursive, mental and bodily, and material and immaterial elements. A practice of exchange – or, exchange (as a) practice – can thus be understood as 'contextually and culturally bound doing and saying consisting of sets of actions and understandings engaging the participants in the exchange of goods and services.'

Research context and method

Owing to oligopolistic competition, in the Finnish food markets it has been hard for small-scale farmers to get their products on the shelves of the chain shops, whose codes of conduct are based on centralized purchases, processing systems, and logistics that require large volumes of regular and standardized food inputs. As a result, the availability of local and organic food in Finland is relatively poor and access to it is difficult. Hence, people have started to mobilize themselves towards finding ways to purchase 'good food'. Finland is a great example of a Western context where the rise of a (globally rooted) local food movement has been mobilizing not only various grass-root actors, but also the media, policy makers and industry representatives, to participate in the (re-)creation of local food markets.

Food collective movement

The food collective (FC) movement comprises active organizations promoting local and organic food at the grass-root level. The appearance of first food collectives in Finland can be traced to the turn of the 1980s to the 1990s. Initially, the great difficulty of sourcing organic food motivated people to seek options for finding it through direct connections with organic producers. These days, purchasing is organized by forming food collectives that engage in direct exchange relationships between farmers and citizens, which enables collective purchases of locally produced and/or organic food. During the past few years, founding of new FCs has accelerated to encompass nowadays over 70 FCs operating in Finland.

Method

A combination of qualitative methods was used for data generation, in order to get a richer picture of the studied phenomenon and its context. The core of the data consists of participatory observation field notes over a three-year time period and 17 interviews. In addition, this data is supplemented with netnographic data (Kozinets, 2002), FC webpages, and secondary source surveys studying FC activity. Data used in the analysis was generated between February 2009 and May 2013.

The analysis was carried out in three stages. First, preliminary analysis was conducted after which more focused thematic analysis was employed based on interviews, observation notes, and secondary data. In the third phase analysing followed the research questions and themes formulated in the previous stages of analysis with additional interview, observation and netnographic data. The focus was primarily on identifying how resources, practices, and knowledge are shared and produced between the food collectives and producers as well as within the FC community/national network. At this stage narrative analysis (Riessman, 2008) was used to study what people say about their practice(s) of exchange.

Food collectives making the markets for good food

Food collectives influence market creation by creating exchange infrastructures between consumers and producers. This entails establishing exchange relations, creating and maintaining the practice(s) of exchange and having the capabilities for constant (re-)negotiation of exchange relations. Establishing exchange relations requires the organization of FCs around a major task of forming a collective of members who agree to common rules of how to run the circle. This is initiated by an active founder-coordinator and involves questions related to the formal organizing of the FC, coordination and management of the activities, and organization of the exchange infrastructure. Equally crucial for the food circle is the establishment of relations with producers. Once exchange relations and infrastructure are established, they appear to be durable. Food collectives cooperate widely with each other in order to exchange information and various management tools for managing the exchange relations.

Exchange is patterned, on the one hand, according to the seasonality of production, harvesting times, and yield, and on the other hand, according to the FC's preferences and resources. In order to run their operations, coordination of the orders between FC members and producers, the exchange of information with the producers regarding the selection of foodstuff, and the organization of the exchange of food all need to take place in a particular place at a particular time and need to be managed. This is made possible with the work input of volunteers.

Despite the aim for durable relations, there is still a (re-)negotiation of exchange relations going on between the FCs and the producers. For example, procedures for the situations of under- and over supply allow that prices remain stable even in such situations. This practice is supported by the interaction whereby FC members learn more about the work of the farmers, which enables making the expenses of production and distribution transparent and understandable to all members. The FC coordinator actively communicates with farmers about harvests in upcoming weeks and uses this information to match the supply and demand for food. Demand is supported by modifying the selection, distributing recipes to make the foodstuff more useable, and creating practices for extending the shelf life of products.

Negotiation practice also relates to the quality of food, where the members of a food collective verify the criteria for 'good food'. Interestingly, it seems that the lack of packaging further amplifies the potential to talk about the qualities of food. Mutual embedding and personal encounters are a prerequisite for trust in food collectives as well as between FCs and the producers: by allowing for flexible payment periods, adjusting prices according to the financial situation of the customers, volunteering, and even accepting labour as a form of currency, trust is (re-)negotiated within the relationships.

There are diverse motivations for participating in food collectives. This reflects the fact that food issues mix many different agendas that range from health and safety concerns to issues of global justice. Hence, rather than for a single purpose, food collectives are brought together on more practical bases. No matter the ideological commitments or reasons for being engaged, FC members depend on others in their attempt to organise markets for good food.

The somewhat an-orthodox result is that demand and supply are actively negotiated and markets are guided without reference to prices. Trust is an outcome of mutual embedding, but it is also explicitly performed within FCs in the form of voluntarism and flexibility regarding compensation for food. As a result of such arrangements and understandings of good food, the notion of quality and fair prices as well as the perception of value all depend on direct contact between suppliers and buyers rather than on global standards. As a result, these practices set the food collectives apart from conventional food markets.

Conclusions

These aspects of 'good food markets' are context specific in at least two different ways. Firstly, the ideology of 'good food' entails broad criticism of capitalism and, in particular, the dominant models of competitive markets. Hence, the way that the markets for local food are organized mirrors the hegemonic order of agro-industrial food production and highly centralised international logistics. In other words, it may be health, safety, or justice that the participants are after, but in all of these cases, a good proxy towards achieving them seems to be providing alternatives to the dominant practices of food production and distribution. Moreover, mutual support and trust between the actors of local food markets is certainly, in part, a phenomenon of emerging markets. In such a context, individual actors benefit from others through increased visibility and institutional support, which is available through the continue increase in the size of the market.

References

Allen, P., FitzSimmons, M., Goodman, M. and Warner, K. (2003). Shifting plates in the agrifood landscape: the tectonics of alternative agrifood initiatives in California. Journal of Rural Studies 19: 61-75.

Beckert, J. (2010). How do fields change? Organization Studies 31(5): 605-627.

Callon, M. and Muniesa, F. (2005). Economic markets as calculative collective devices. Organization Studies 26(8): 1229-1250.

Colquhoun, A. and Lyon, P. (2001). To everything there was a season: deconstructing UK food availability. Food Service Technology 1: 93-102.

Davis, G.F., Morrill, C.M., Rao, H. and Soule, S.A. (2008). Introduction: social movements in organizations and markets. Administrative Science Quarterly 53: 389-394.

Delind, L. (2010). Are local food and the local food movement taking us where we want to go? Or are we hitching our wagons to the wrong stars? Agriculture and Human Values 28(2): 273-283.

Fligstein, N. (1996). Markets as politics: a political-cultural approach to market institutions. American Sociological Review 61(4): 656-673.

Fligstein, N. (2001). The architecture of markets. an economic sociology of twenty-first-century capitalist societies.

Fligstein, N. and Dauter, L. (2007). The sociology of markets. Annual Review of Sociology 33: 105-128.

Fourcade, M. and Healy, M. (2007). Moral views of market society. Annual Review of Sociology 33: 285-311.

Granovetter, M. (1985). Economic action and social structure: the problem of embeddedness. American Journal of Sociology 91: 481-510.

Holt-Giménez, E. (2009). From food crisis to food sovereignty: the challenge of social movements. Monthly Review 61(3): 33-42.

Ilbery, B. and Maye, D. (2006). Retailing local food in the Scottish-English borders: a supply chain perspective. Geoforum 37: 352-367.

Kjellberg, H. and Helgesson, C.-F (2007). On the nature of markets and their practices. Marketing theory 7(2): 137-162.

Kjellberg, H. and Helgesson, C.-F. (2006). Multiple versions of markets: multiplicity and performativity in market practice. Industrial Marketing Management 35(7): 839-855.

Kloppenburg, J., Hendrickson, J. and Stevenson, G.W. (1996). Coming into the Foodshed. Agriculture and Human Values 13: 33-42

Kozinets, R. (2002). The field behind the screen: using netnography for marketing research in online communities. Journal of Marketing Research 39(1): 61-72.

Levkoe, C. (2006). Learning democracy through food justice movements. Agriculture and Human Values 23(1): 89-98.

Pollan, M. (2008). In defence of food. an eater's manifesto. 2nd edition, Atena Kustannus Oy, Jyväskylä, Finland.

Raynolds, L. (2000). Re-embedding global agriculture: the international organic and fair trade movements. Agriculture and Human Value 17: 297-309.

Reckwitz, A. (2002). Toward a theory of social practices. A development in culturalist theorizing. European Journal of Social Theory 5(2): 245-265.

Riessman, C. (2008). Narrative methods for the human sciences. Sage Publications Inc., Thousand Oaks, CA, USA.

Seyfang, G. (2007). Growing sustainable consumption communities: the case of local organic food networks. International Journal of Sociology and Social Policy 27(3/4): 120-134.

Sine, W. and Lee, B. (2009). Tilting at windmills? The environmental movements and the emergence of the U.S. wind energy sector. Administrative Science Quarterly 54: 123-155.

Starr, A. (2010). Local food: a social movement? Cultural Studies – Critical Methodologies 10(6): 479-490.

Van Bommel, K. and Spicer, A. (2011). Hail the snail: hegemonic struggles in the slow food movement. Organization Studies 32(12): 1717-1744.

Warde, A. (2005). Consumption and theories of practice. Journal of Consumer Culture 5(2): 131-153.

Weber K., Heinze, K. and DeSoucey, M. (2008). Forage for thought: mobilizing codes in the movement for grass-fed meat and dairy products. Administrative Science Quarterly 53, 529-567.

Wilkinson, J. (2007). Fair trade: dynamic and dilemmas of a market oriented global social movement. Journal of Consumer Policy 30: 219-239.

Who owns hazard? The role of ownership in the GM social experiment

Z. Robaey
Delft University of Technology, Technology, Policy and Management, Jaffalaan 5, 2628 BX Delft, the Netherlands; z.h.robaey@tudelft.nl

This extended abstract aims to conceptualize the introduction of genetically modified (GM) crops as a social experiment, in order to provide for a new ethical lens on forward looking responsibility. In a first step, I explain what the notion of social experiment allows us to explore with respect to forward-looking responsibility, namely: (1) the care for human lives; and (2) the multitude of actors who qualify as experimenters. In a second step, I look at how moral responsibility is currently distributed in the case of GM crops. The issues around GM crops, however, seem to be more about liability linked to property rights rather than forward looking moral responsibility, or linked to issues of justice in the context of international development. These observations then provide for a basis to explore philosophical notions of ownership and justice from Locke's theory of property and Nozick's entitlement theory, and investigate what is missing from these notions to construct a coherent framework for distributing forward looking responsibility in the case of GM crops. Finally, I suggest a notion of ownership that is more fluid to allow formulating conditions for a fair distribution of forward looking moral responsibility in the GM social experiment.

While a social experiment remains a very open notion[5], the main ideas underlying it are: (1) that a social experiment is unlike a laboratory experiment, i.e. it is not controllable, it does not test a hypothesis, and it does not have clear boundaries therefore impacts directly the *outside* world; and (2) since the experiment is *open*, there is not one single experimenter. All involved party may influence the course of the experiment (Van de Poel, 2011). In the case of GM crops, some actors groups themselves refer as the introduction of these biotechnologies as an experiment, calling them *Frankenfoods* and the likes (Van den Belt, 2009). In those cases, the term social experiment is, however, used with a negative connotation to emphasize the hazards this new technology may bring about. Nevertheless, using the notion of social experiments allows looking at the introduction of technologies in society with a new ethical lens (Martin and Schirzinger, 1983). Indeed, scholars have suggested applying insights of biomedical ethics – with regards to experimentation with humans – to the introduction of technologies in society that have high benefits but also high potential side effects (Van de Poel, 2011; Koepsell, 2010). This would indeed involve the care required when laying out an experiment involving humans (cf Beauchamp and Childress, 2001), and it opens the door for learning about uncertainties, ignorance, and problems emerging from the *actual* use of a technology. This learning component allows fine-tuning on how to best introduce a technology in society. Ultimately, using the social experiment framework allows for specifying the unspecified and allowing adequate action to be taken with the goal of minimizing negative effects on societies that could also benefit from it, in other words to be morally responsible.

The other aspect of the social experiment that is particularly relevant here is the multitude of actors involved in it, unlike in a traditional experiment where one experimenter makes all decisions on the set-up. Indeed, one could think of the journey of a GM crops as a continuous experiment with several experimenters that will influence the set-up and the outcome of the experiment. From the scientists who develop it, the company who markets it, to the regulators who evaluate it, to activists or lobbyist that

[5] Without carrying an extensive literature review, it is important to note that the term social experiments has a long history going back to the Chicago School of Sociology, where at the beginning of the 20th century, sociologists noted that experiments were taking place in society constantly without needing to be set-up as such (Schwarz and Krohn, 2012).

oppose or defend it, and to the farmers who harvest it. The GM crop therefore travels from actor to actor. When it comes to questions of forward looking responsibility, it is important to clearly identify to whom and to what extent moral responsibility may be attributed, and for what, so the question becomes not only a question of being morally responsible, but rather under which conditions can we fairly distribute moral responsibility to the different experimenters/actors involved.

The discourse and literature around GM crops shows the strong polarization of the issue. Indeed, at the center of the debate lies the question whether GM crops are good or bad, from a risk and uncertainty perspective (cf. Levidow, 2001; Levidow and Carr, 2007; Van Asselt and Vos, 2010) – the less scientific certainty we have about a technology, the more it is likely to be introduced with difficulty in society – or from a global justice perspective (Høyer Toft, 2012) – is there a duty towards developing countries to support crops that may potentially be better and why. This paper aims to take a pragmatic approach and look at GM crops as a *fait accompli* that needs to be managed in the best possible way given the possible hazards. Indeed, it is important to discuss why one should or should not plant GM crops, but meanwhile, they are being planted at an exponential rate, as the International Service for the Acquisition of Agri-biotech Applications (ISAAA) points out in their executive summary, '2012 marked an unprecedented 100-fold increase in biotech crop hectares from 1.7 million hectares in 1996 to 170 million hectares in 2012 – this makes biotech crops the fastest adopted crop technology in recent history' (James, 2012).

In these accounts, there is no question of forward-looking responsibility regarding the technology itself. Who is entitled to praise or blame for the results of this rapid expansion? Many issues have arisen regarding blame, but only in a legal sense. Indeed, patents only last a certain number of years, generally 15-20. So in terms of blame-worthiness, one owner of the patent, or licenser, could be blamed for damages only for the extent of the time of the patent or the licence. Also, issues of field contamination are heavily debated, of seed saving, a current farming practice, or of royalties to be paid (Van den Belt, 2009). The legal system has clear limits not only domestically but also internationally. Moreover, companies developing GM crops are not accountable to anyone. But again, all these examples point to backward looking responsibility. And, in order to think of GM crops as a morally responsible social experiment, one must think of forward looking responsibility as well.

One interesting observation emerges from the previous paragraph: backward looking responsibility seems to be mostly linked to questions of ownership, or property. Why would ownership matter in the case of the social experiment? It does matter when it comes to who is liable for hazards in case the experiment goes wrong, in a legal sense. But this alone is not sufficient to make the social experiment morally responsible. In terms of managing the introduction of GM crops, it then appears useful to conceive of a forward looking distribution of responsibility based on ownership. However, the current legal notion of ownership is very limited. Locke and Nozick have suggested two theories that could enlighten the discussion towards distributing moral responsibility in a forward looking way. Indeed, Locke's theory of property claims, in its simplest understanding, that, 'a person's productive work is the basis for a property claim' (Thompson, 2007: 242). In that sense the scientist, regulator, activist, lobbyist, farmer all contribute in a *broad* sense to the way a GM crop is introduced in society. If we then also look at Nozick's entitlement theory which bases a just distribution of holdings (property) on the just acquisition and transfer of holdings (Enbar, 1983), one notices short comings for the case of GM crops. Indeed, a GM crop is an artifact that self-reproduces, mixes with other crops, disseminates in unpredictable and uncontrollable ways, but that can also be voluntarily stolen. The idea of a fair transfer is also present in Locke, and also poses problem as to what can be a fair transfer with a living technology. Also, currently this transfer is based on legal notions of ownership such as patents and licenses and as mentioned earlier, questions of justice are highly debated in the legal sphere. In Nozick's entitlement theory, a provision stands for rectification of injustice as well. What if the current system is unjust because notions of ownership just do not fit the technology in its context? In that sense, providing a better notion of ownership can help addressing the issue of justice. Also, a notion of ownership better fitted to the GM case will help thinking of moral responsibility in a forward looking way.

This paper will suggest that if responsibility may be distributed, so will ownership of the GM crop. Also, inspired by Locke and Nozick, this paper wants to suggest a fluid understanding of ownership. Ownership emerges when one uses an artifact, I will talk of 'a sense of ownership' that develops through 'use' – here use will also be further defined in a broad conceptualization. In that sense, a regulator evaluating a crop for several months will feel a certain sense of ownership over the crop and its evaluation, and in the same way would a farmer feel ownership for a crop even if a license has expired, or the GM crop was acquired illegally. This allows accounting for different scenarios possible with a GM crop, involving all experimenters in the social experiment and assigning them forward-looking responsibility based on the extent of their 'sense of ownership'. In this paper, questions of fairness in the transfer remain to be addressed, and the definition of ownership will be fine-tuned. Linking this form of ownership to the distribution of forward looking moral responsibility will allow establishing further conditions for moral responsibility in the GM social experiment.

References

Beauchamp, T.L. and Childress, J.F. (2001). Principles of biomedical ethics. 5th ed. Oxford University Press, New York, NY, USA.

Enbar, M. (1983). Chapter 1: Equity in the social sciences. In: Kasperson, R.E. (ed.) Equity issues in radioactive waste management., Oelgeschlager, Gunn & Hain, Publishers, Inc. Cambridge, MA, USA, p. 3-23.

Høyer Toft, K. (2012). GMOs and global justice: applying global justice theory to the case of genetically modified crops and food. Journal of Agricultural and Environmental Ethics 25: 223-237.

James, C. (2012). Global status of commercialized biotech/GM crops: 2012. ISAAA Brief No. 44-2012. ISAAA: Ithaca, NY, USA.

Koepsell, D. (2010). On genies and bottles: scientists' moral responsibility and dangerous technology R&D. Science and Engineering Ethics 16(1): 119-133.

Levidow, L. (2001). Precautionary uncertainty: regulating GM crops in Europe. Social Studies of Science December 31(6): 842-874.

Levidow, L. and Carr, S. (2007). GM crops on trial: technological development as a real-world experiment. Futures 39(4): 408-431.

Martin, M.W. and Schinzinger, R. (1983). Ethics in engineering. McGraw-Hill, Boston, MA, USA.

Schwarz, A. and Krohn, W. (2011). Experimenting with the concept of experiment: probing the epochal break. In: Nordmann, A., Radder, H. and Schiemann, G. (eds.) Science transformed? Debating claims of an epochal break. University of Pittsburgh Press, Pittsburgh PA, USA, p. 119-134.

Thompson, P.B. (2007). Chapter 9: Conceptions of property. In: Food biotechnology in Ethical Perspective, 2nd edition. Springer, Dordrecht, the Netherlands, p. 233-259.

Van Asselt, M.B.A. and Vos, E.. 2008. Wrestling with uncertain risks: EU regulation of GMOs and the uncertainty paradox. Journal of Risk Research 11(1-2): 281-300.

Van de Poel, I. (2011). Nuclear energy as a social experiment. Ethics, Policy and Environment 14(3): 285-290.

Van den Belt, H. (2009). Philosophy of biotechnology. In: Meijers, A. (ed.) Handbook of the philosophy of science, Volume 9: Philosophy of Technology and Engineering Sciences. Elsevier, Amsterdam, the Netherlands, p. 1301-1340.

Category management in Swedish food retail: challenges in ethical sourcing

J.P. Rotter, P.-E. Airike and C. Mark-Herbert
Department of Economics, Swedish University of Agricultural Sciences, Box 7013, 750 07 Uppsala, Sweden; Julia.rotter@slu.se

Abstract

Ethical aspects of consumption of fast moving consumer goods, have received increased attention during the last decade. In part, this can be explained by an increasing awareness among consumers at large of environmental and social issues that are embedded in the production of fast moving consumer goods, such as food products. Standards and eco-labels serve as a guiding support for consumer choices, but the role of the food retailer, above and beyond providing services and information for the consumers, is a matter of making corporate decisions in line with responsible corporate governance. This paper aims to explore corporate responsibility, CR, in terms of ethical sourcing in category management, which selection reflects the overall corporate value grounds and create basis for differentiation. Categories are seen as single business units, where considering ethical aspects in business decisions might be conflicting with strategic goals such as profit maximization on a corporate level. Therefore, in a practical setting, category management concerns making choices regarding in which way and what products should be offered in the food retail stores. Key research questions target the conditions for making strategic choices; who are the stakeholders? And what values are given priority? The paper is based on a qualitative research design where two food retail case studies are presented as illustrations of conditions and challenges in addressing complex corporate responsibility issues. These case studies adopt a corporate perspective of key food retailers in Sweden and their approach to managing social and environmental issues in supply chains. Research findings illustrate the central position for food retailers in the supply chain. Yet, it is unlikely that solutions to complex problems, such as those embedded in supply chains, can be achieved by a single actor since businesses that respond individually to complex challenges often find the competitive environment restricting their activities. An extended stakeholder dialogue, in terms of industrial networks and multi-stakeholder dialogue, offers valuable exchange of knowledge and an opportunity to balance a wide set of values and interests. Corporate challenges of creating legitimacy are associated with the selection of dialogue partners, perceptions of created shared values, communicational efforts made to establish trust, and, the underlying query of corporate *raison d'être*.

Keywords: accountability, corporate responsibility, industrial network, legitimacy, product portfolio management

Introduction – perspectives on ethical consumption in food retail

Ethical aspects of consumption of fast moving consumer goods, FMCG, such as food products, have received increased attention in the last decade (Rotter *et al.*, 2012). In part, this is explained by an augmented awareness among consumers at large of environmental and social issues that are embedded in the production and sales of fast moving consumer goods. Ethical issues of particular interest for the consumer may relate to, for example, the production method, human rights issues in the supply chain and the contractual agreement that sets the conditions in the procurement process. Each of these issues are complex and the global value chain for fast moving consumer goods offers numerous challenges in sourcing where ethical aspects are considered throughout the value chain (Hartmann, 2011). For the consumer, ethical sourcing may be validated on a corporate level, as certifications according to standards, as well on a product levels through ethical labels as in a co-branding procedure for a product. Standards and eco-labels sounds like a trustworthy communicational basis for supporting consumers through educational efforts and guidance regarding ethical issues related to food products – but is this really all the food retail can do?

This paper investigates how Corporate Responsibility (CR[6]), in terms of two kinds of category management situations may offer additional ethical consumption values for a wide set of stakeholders. The role of the food retail is given additional responsibilities above and beyond that of education and provider of differentiated products (with ethical labels) for the consumers. This added dimension refers to product portfolio management where the retailer decides which products to offer and which not to offer. It may appear like a simple binary decision, but the background for these decisions is complex and ethically laden (Rotter *et al.,* 2012). This paper is based on two simplified cases to illustrate current conditions for ethical sourcing in food retail. Before presenting the two cases we offer a brief review of CR that leads the way to a selected conceptual framework. The paper does not give any easy answers or general conclusions. It merely offers a context bound understanding of illustrations of two different ways of managing ethical sourcing as a part of product portfolio management and corporate responsibility in a Swedish food retail setting.

Corporate responsibility as category management – a conceptual framework

CR builds on the idea of a 'triple bottom line' (Elkington, 1998) where a private organization is not merely defined by its profitability but by its sense of commitment and responsibility toward its internal and external stakeholders. The concept is closely tied to external pressures as 'CSR is about adapting to the ever-changing social reality and about making oneself fit to take societal demands seriously' (Pompe and Korthals, 2010, 370). Businesses, regardless of industry and size, are active members of society, which is reflected by context bound visions, social realities and operations in accordance with regulations and other societal expectations.

The academic field of CR remains fragmented and contested (Garriga and Melé, 2004), which leaves room for continued epistemological dialogues. A contemporary notion drives CR toward a more *political* perspective (Scherer and Palazzo, 2011) while a more *instrumental* perspective pulls it toward a strategic management standpoint (Porter and Kramer, 2011). Porter and Kramer suggest that CR is about creating shared value for society in order to legitimize businesses and create 'a new way to achieve economic success' (*Ibid.,* 64). A shared value lens can be applied to every business decision as it is seen as an integrated way to create policies and operating practices, which enhance the competitive edge of a company and at the same time affecting social, environmental and economic conditions for the communities they operate in (Figure 1).

A notion of shared value points to needs to rethink models for stakeholders, questioning towards whom the retailers are responsible and accountable. Rainey (2006: 711) defines a stakeholder as 'any individual or group that is directly or indirectly affected by the products, programs, processes, and/or systems, but does not directly benefit as an economic participant such as a customer or supplier'. Stakeholder theory aims at identifying such groups and individuals that are connected to a firm's environment with the intention to not only 'broaden management's vision of its roles and responsibilities beyond the profit maximization function to include interests and claims of non-stockholding groups' (Mitchell *et al.,* 1997: 855).

[6] Corporate Responsibility is referred to as CR in this paper. In much of the contemporary literature synonyms like Corporate Social Responsibility (CSR), corporate citizenship (CC) and corporate accountability are used with reference to a similar phenomenon.

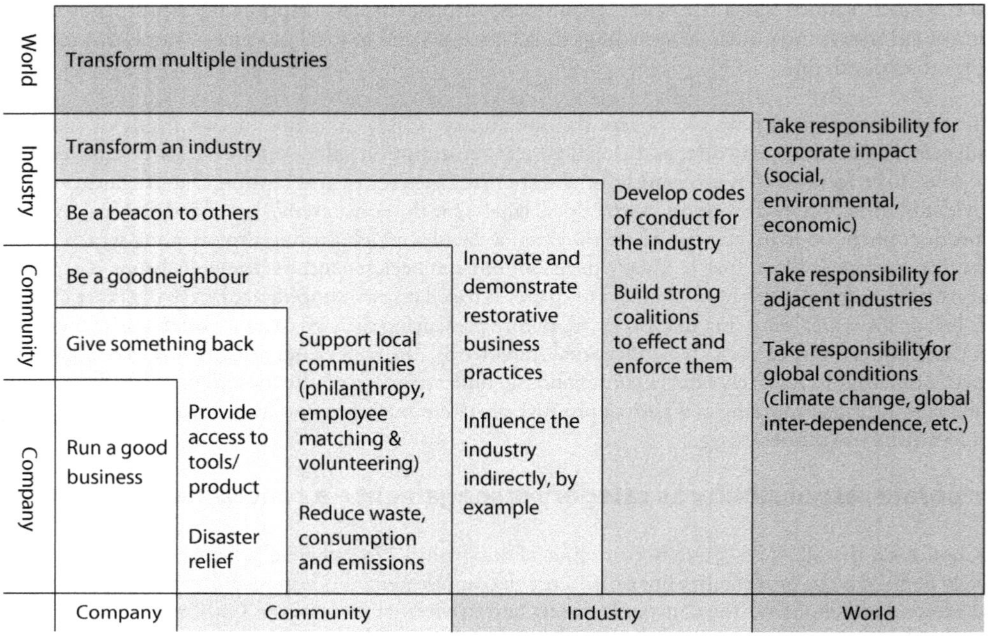

Figure 1. A corporate responsibility landscape (McElhaney, 2008: 230).

Assuming that CR applies to all industries, one such industry that interacts with the vast majority of individuals in the world in its daily operations is the food retailing. Food retailers are central in the value chain as a link between farmers, processors and consumers, which enables them to benefit from a powerful position in decisions not just regarding category management (Tansey and Worsley, 1995). Marketing activities related to category management may refer to decisions to stock or not to stock a particular product. Shelf space is a limited resource where product ranges can be seen as strategic business units (e.g. Corstjens and Lai, 2012).

Central for category management are the industry specific conditions related to boundaries regarding resources dependency, the central role of food as well as characteristics of the food chain (Hartmann, 2011). The food value chain is characterized by a globalized network of different actors that require a high degree of trust for fulfilment of implicit and explicit agreements, expectations and standards. The activities are coordinated in various kinds of dialogues and systems such as labelling, tracking and agricultural practises. Being positioned at one end of the food value chain, food retailers have to ensure and communicate best practices in line with their corporate ambitions throughout the entire value chain, while holding a superior influence on consumers and their choices. Consequently, values for multiple stakeholders can be created in the retailers' daily operations (Figure 1).

Responsible category management, as a part of corporate responsibility, may imply decisions of offering or not offering a product or a product category as well as the selection of suppliers and promotional activities related to a certain category (Ottman, 2011, 168). These decisions are of strategic nature, strongly tied to the corporate identity, and they are expressed in local daily operations related to running a good business as well as in more long-term effects in transforming an industry or a market (Figure 1). The corporate responsibility landscape is further explored in light of two illustrative food-related cases.

Two cases of category management – instrumental and political CR

Given the complex structure of a food chain, the idea of making responsible category management decisions is reflected in needs for developing a dialogue with partners in order to be accountable towards a wide set of stakeholders. These collaboration arrangements are illustrated in the two cases (Table 1) – both of them aimed at efficient use of resources and shared value creation. Category management decisions refer to balancing short-term corporate interests with long-term benefits for stakeholders and society at large.

The scene at which these cases take place is the Swedish food retail industry. It is characterized by three national dominant retail chains, which have well established market positions. These actors jointly control over 80% of the market (Axfood, 2010). ICA controls the majority of the total market share with 45.9%, followed by Axfood (19.3%) and KF-Coop with 18.5% (*Ibid.*). Smaller retail chains, such as Bergendahls, including Vi-stores have a 5.3% market share, while Lidl holds 3.2%, Netto 2.1% and others 5.7% (*Ibid.*). Case A in the table below concerns Axfood, the second largest retail actor, and Case B concerns the four largest retailers in Sweden, i.e. ICA, Axfood, Coop and the Bergendahls.

Exploring CR through the lens of ethical sourcing illustrates how private and public dialogue may offer guidance in category management (in the Tiger shrimp case) and how an industrial network may

Table 1. Two illustrative cases of category management dealing with ethical sourcing.

	A. Tiger shrimp (Rotter *et al.*, 2012)	B. Western Sahara goods (Herrlund, 2011)
The ethical challenge	Tiger shrimp production is related to a number of environmental and social problems (decline of biodiversity, water quality, degradation of mangroves, bycatch and pollution + quality of life for the local communities).	Morocco is occupying Western Sahara, inhabited by the Saharawi people claiming the rights to their land and the produce. The Saharawi people are subjected to severe human rights violations (WSRW, 2013).
Actions taken to ensure wise decisions	A Swedish retail chain, Axfood, chooses to engage in a *partnership/stakeholder dialogue* in which they are advised by World Wide Fund for Nature (WWF) to develop a fish product policy that prohibits the sales of Tiger shrimp.	Four Swedish retail chains, Axfood, Coop, ICA and Bergendahls group signed a declaration of intent that states that food products (mainly fish oil and tomatoes) from Western Sahara will not be sold.
Current outcome/ situation on the market	The sales of Tiger shrimp is discontinued in Axfood (leading to a profit loss) but the total volume sold by the major food retailers in Sweden remains almost the same (consumers buy their tiger shrimp at alternative retailers). Axfood is perceived as having a pro-active ethical sourcing strategy.	The Saharawi people are still not being respected by the kingdom of Morocco. Swedish retail chains have banned products that have unclear origin from what possibly can be Western Sahara. The *industrial ban* is openly communicated.
Corporate responsibility landscape position (in Figure 1)	Effects on company and local communities (it influences the *local production and sales on the Swedish market).*	An *industrial network* that focuses on the ethical challenge with a joint effort that *transforms the industry*, with potential to influence international markets.

coordinate an entire industry in collective action (in the Western Sahara goods' case). Sustainability issues in the food production are of general concern, yet the debate tends to focus on individual products and stakeholders (retailers). *Instrumental CR*, as in the case of Tiger shrimp, may be executed by a single corporation that takes a proactive stand in the case of ethical sourcing (in this case leading to discontinued sales of the product). A more transforming initiative, however, may be seen in a *political CR* initiative in the case Western Sahara goods where an industrial network is formed, influencing a market locally and, perhaps international markets as well in a longer time perspective.

Both of these cases offer an understanding of what Swedish food retail stakeholders are capable of doing in terms of CR above and beyond offering labelled products and educating consumers. Their motives for engaging, being accountable and responsible, are going to determine how far in the value chain the ethical sourcing can be taken – and more importantly, how far in the CR landscape political CR can be developed (Figure 1). It is safe to say that the Swedish food retail industry is gradually adapting to a more political role that is associated with a discourse in which ethical legitimacy will continue to influence ethics in consumption.

References

Axfood (2010). Financial Report 2010. Available at: http://ir.myreport.se/show/axfood/show.asp?pid=135333910008&page=18.

Corstjens, M. and Lal, R. (2012). Retail doesn't cross borders. Harvard Business Review, April, 104-111.

Elkington, J. (1998). Cannibals with forks: the triple bottom line of 21st century business. 2nd ed. Capstone Publishing Ltd., UK.

Garriga, E. and Merlé, E. (2004). Corporate social responsibility theories. mapping the territory. Journal of Business Ethics 53:1/2: 51-71.

Hartmann, M. (2011). Corporate social responsibility in the food sector. European Review of Agricultural Economics 38(3): 297-324.

Herrlund, J. (2011). Västsaharas legitimitet som stat. En undersökande fallstudie om Västsaharas möjligeter att bli en stark stat. Lunds Universitet, Statsvetenskapliga institutionen.

McElhaney, K. (2008). Just good business: the strategic guide to aligning corporate responsibility and brand. Berrett-Koehler Publishers, Inc., San Francisco, CA, USA.

Mitchell, R.K., Agle, B.R. and Wood, D.J. (1997). Toward a theory of stakeholder identification and salience: defining principle of who and what really counts. Academy of Management Review 22(4): 853-886.

Ottman, J.A. (2011). The new rules of green marketing. strategies, tools and inspiration for sustainable branding. Greenleaf Publishing Ltd., Sheffield, UK.

Pompe, V. and Korthals, M. (2010). Ethical room for manoeuvre: playground for the food business. Business and Society Review 115(3): 367-391.

Porter, M. and Kramer, M. (2011). Creating shared value. Harvard Business Review, January-February, 62-77.

Rainey, D.L. (2006). Sustainable business development, 1st ed., Cambridge University Press, Cambridge, UK.

Rotter, J., Özbek, N. and Mark-Herbert, C. (2012). Exploring corporate social responsibility in food retail category management. Social business 2(3): 223-241.

Scherer, A.G. and Palazzo, G. (2011). The new political role of business in a globalized world: a review of a new perspective on CSR and its implications for the firm, governance, and democracy. Journal of Management Studies 48(4): 899-931.

Tansey, G. and Worsley, T. (1995). The food system – a guide. Earthscan Publication Ltd., London, UK.

WSRW, Western Sahara Recourse Watch (2013). WSRW calls for halt of EU trade talks with Morocco. Available at: www.wsrw.org/a105x2528.

Consumer perspectives on ethics in garment consumption: perceptions of purchases and disposal

J. Jönsson, T. Wätthammar and C. Mark-Herbert
Department of Economics, Swedish University of Agricultural Sciences, Box 7013, 750 07 Uppsala, Sweden; cecilia.mark-herbert@slu.se

Abstract

Ethical aspects of consumption of products in the fast moving consumer goods category, such as food and fibre products, have received increased attention in the last decade. In part, this is explained by an increased awareness among consumers at large of environmental and social issues that are embedded in the production of food as well as fibre products. But, to what degree is this awareness expressed in attitudes and consumption behaviour? This project aims to describe consumer behaviour regarding garment products. The focus is put on purchasing and donating garments. As a fibre product, that requires a lot of resources in the production and extensive supply chain is well known for its social challenges, it is assumed that extending the use of a garment is associated with improved sustainable development. Research questions of particular interest are therefore related to consumer attitudes and behaviour in terms of second hand options. The empirical study was executed in two parts; as self-completed survey among 170 consumers and a focus group, targeted in a city environment (in Uppsala) in 2012. The research design was inspired by a recent consumer study that was carried out in a small town setting (Ekström *et al.,* 2012). A tentative hypothesis, that garment consumption patterns may differ in a small town *versus* a city environment, was explored. The preliminary results show that the following factors influence the attitude towards recycling garments: age, gender, where the respondent is living and where he or she has been raised. Given no official recycling program, as is the case for paper, glass, metal, batteries and plastic materials in Sweden there is a need for a program to guide consumers at large in their ethical choices.

Introduction – perspectives on consumption of garments

Consumption of fast moving consumer goods (food and fibre products) has received a lot of attention in the last decades (Gardetti and Torres, 2013). The complicated value chain in most garment supply chains makes it difficult for an average consumer to make decisions with ethical awareness of all aspects of production, sales and use of the product. This project addresses consumer consumption, in other words decisions regarding purchases and disposal of garment products in spite of consumer awareness of the major environmental impact of these products while they are being used. 'The biggest impacts of textiles and garments occur when they are being used by the consumer (estimated at 75-95% of the total environmental impact) and is mainly explained by the use of electricity, hot water and washing and drying processes. This contributes to the generation of greenhouse gases and global warming' (Gardetti and Torres, 2013: 8).

Arguments for studying consumption decisions are related to the importance of these decisions in a global triple bottom line perspective. In the year of 2000 the world's consumers spent roughly US$1 trillion on garments (Gardetti and Torres, 2013: 1). Of the total world export 7% are garments and textiles. The sector is dominated by developing countries, with China in the forefront. Worldwide some 26.5 million people work in the clothing and textile industry, where about 70% of the industrial workers are women. The textile industry is significant to our global economy, being such a large and important industry worldwide. Using a triple bottom line framework the challenges to the current consumption of garments are summarized in Figure 1.

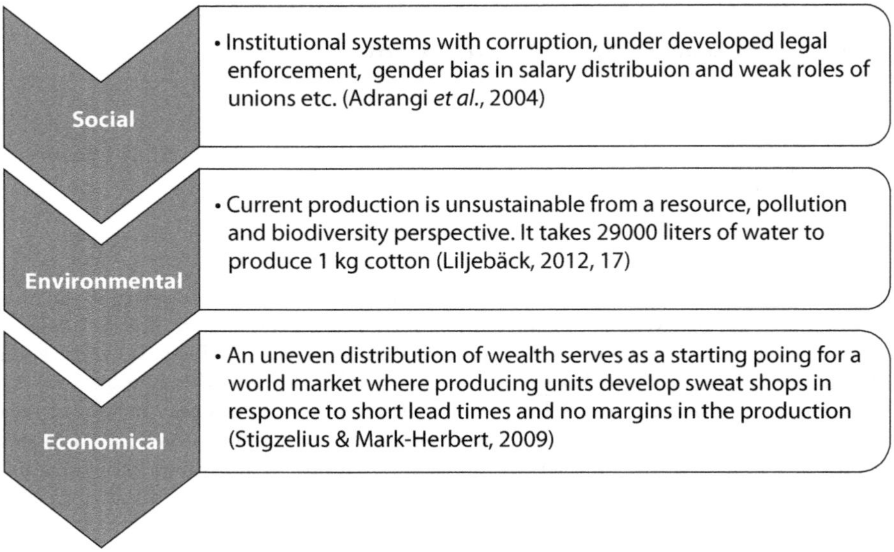

Figure 1. Challenges related to garment production in a triple bottom line framework.

Awareness of these challenges in the garment value chain (figure above) may serve as ethical grounds for making decisions in consumption. Consumption behaviour is described as 'acquisition, consumption, and disposition of goods, services, time and ideas by decision making units' (Jacoby *et al.,* 1977: 22). Consumer disposition behaviour has although a natural part of the consumption cycle received limited research attention and is a relatively new area in research (Albinsson and Perera, 2009; Birtwistle and Moore, 2007). Acquisition and disposal is managed by consumers by engaging in five types of disposition, for example sharing, donating, exchanging, ridding and recycling. The day-to-day behaviour of individuals is one of the most difficult behaviours to change (McDonald *et al.,* 2012). Habits are important when it comes to consumer behaviour and consumer choices can sometimes be regarded as irrational poorly connection to consumer values (Niinimäki, 2010) when consumer fulfils unconscious needs and inner motivations by consuming.

Traditional consumer theory was developed because marketers wanted to know 'why people buy', with intentions to create strategies in how to influence consumers (Blackwell *et al.,* 2006: 4). Without a deeper understanding of consumer attitudes and needs it is difficult for companies to meet their expectations (Kardes *et al.,* 2011). The understanding includes the whole spectrum, from the activities prior to the purchase, during and after consumption that consumers engage in (Figure 2). A more current understanding among marketers recognizes needs to understand consumer behaviour to establish relationships in order to satisfy future and current needs (Solomon, 2006).

In the figure Purchase activities include everything ex ante to purchasing (Kardes *et al.,* 2011). Use activities refers to how, when and where the consumption takes place. Consumption can for example take place immediately after purchase or it can also be delayed. Disposal activities concerns ways in how the consumer gets rid of packaging and products, this includes for example recycle, reuse, giving to charity and resale. Emotional response reflects the emotions, feelings and mood of the consumer. Mental response is for example about the thoughts, beliefs and attitudes the consumer has regarding

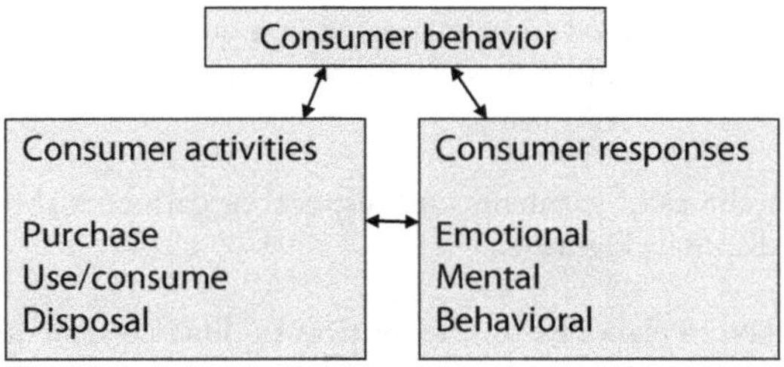

Figure 2. Model: overview of consumer behavior inspiration from Kardes et al. (2011: 8).

the product or service. Behavioural response includes the actions and overt decisions throughout the purchase, use and disposal.

Aim, objectives and approach

The aim of this study is to explain conditions for consumer behaviour related to the purchasing and donating of second-hand clothes. We are assuming an extended use of clothes is associated with sustainable development. The overall objective is to increase the understanding of how consumers *dispose unwanted clothes* and their *attitudes to purchasing second-hand clothes*. The increased understanding may be useful to actors like authorities, second-hand businesses, environmental- and charity organizations, and to the textile commerce and their work on developing sustainable solutions on how to re-use and re-cycle clothes. Key research questions are related to attitudes of second hand clothes and disposal behaviour.

A mixed method approach (Robson, 2011) allows for a combination of data collection methods, in this case an in person survey and a focus group. Our survey study is inspired by a recent study (Ekström *et al.*, 2012) which was carried out in a small town setting. Minor modifications of the survey and the data collection technique make a comparison of the two studies possible. The survey consisted of 25 questions, divided in four themes relating to: waste sorting, buying & disposal of garments, buying second-hand and background questions. In addition to a literature review and survey study (of 170 respondents, in a city environment) a focus group interview (seven persons) was carried out to understand more reasoning behind expressed consumer attitudes and actions.

An empirical study of consumer behaviour – to buy and dispose of garments

The empirical study is presented in detail (theory, approach and results) in the data base Epsilon (open source). Some key findings (after some brief background facts) concerning recycling behavior, awareness of garment options and disposal behaviour are presented below and briefly discussed from a consumer ethics point of view.

A non-randomized survey is technically not suitable for making generalizations. We therefore regard the answers in the study as indicative of views of 'a young generation living in a Swedish city context 2013'. Of the 170 respondents the vast majority (129) were women with an average age of 30 years. Their income was limited (0-2,000 Euro/ month). Approximately one third grew up in the three settings – rural area, small town and big city. A majority of the respondents answered 'Swedish ethnicity' but it was also worth noting that a relatively high proportion of the respondents did not give an answer to this background question.

Environmental awareness was assumed to be expressed in a question relating to what kind of waste the respondents sorted regularly – and it was clear that garments are lagging behind other resources (Figure 3).

Questions concerning purchases of garments and disposal of garments show a frequent purchasing pattern and less frequent disposal (Figure 4).

When respondents are asked to elaborate on the motives to donate garments to charity organization as a way to dispose garments the most common motives related to environmental and social effects. More half of the respondents had thrown away garments in the household garbage the past year. A question of what the possible areas of application that a used garment may have a majority saw all of the following options as possible: sold in a second-hand store, given to a charity organization, re-sewn to new garments, used in the production of new cloths or other textile products, energy production, insulation or as filler in furniture.

The question of buying second hand for own use is answered with a yes by slightly more than half of the respondents (54%) and women are more positive towards that alternative than men. The motives among the positive respondents are related to finding unique and stylish garments, price, environmental awareness and a perception that it is trendy to do so. Motives among respondents who do not see second hand as an alternative are associated with lacking quality aspects, hygienic expectations, a perception of that it is out of style to do so or perceived embarrassment to wear or shop in a second hand shop.

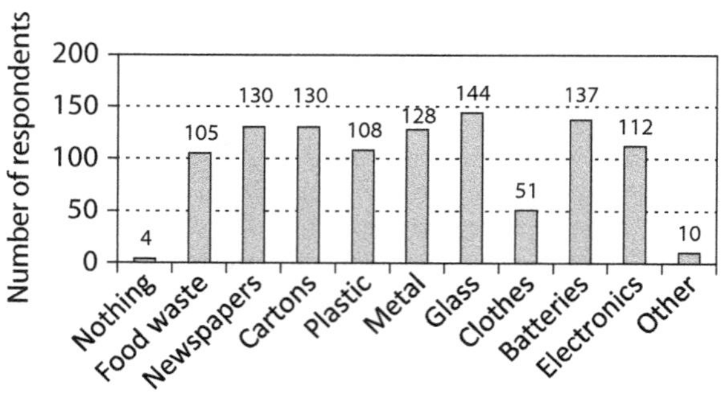

Figure 3. Respondents answer to a question relating to what kind of waste they sort regularly.

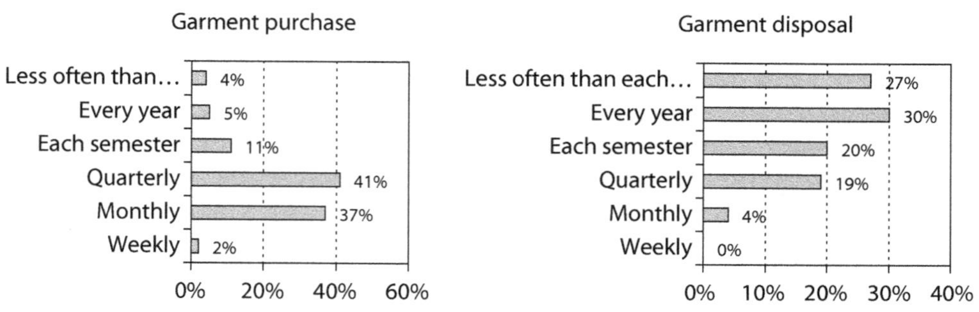

Figure 4. Frequency of garment purchases respective frequency of garment disposal.

Previous research suggests that consumers that recycle and sort garbage (plastic, paper etc) are more likely to recycle clothes (Morgan and Birtwistle, 2009). In our study, the recycling frequency is rather high. The attitude towards recycling garments, within the home (sharing clothes or using old clothes as rags) as well as selling to a second hand shop or donating to a charity organization is positive. Reasons to donate clothes to charity are mainly related to social reasons, to help those less fortunate, than a desire to limit environmental effects, which supports the findings in Ekström's *et al.* (2012) study.

Suggestions for continued research relate to monitoring awareness of ethical aspects of garment consumption and attitudes towards alternatives for purchasing as well as disposing of used garments. This includes policy evaluations of programs that are geared towards information and creation of disposal systems as a part of societal efforts to promote sustainable development.

References

Adrangi, B.K., Dhanda, K.K. and Hill, R.P. (2004). A model of consumption and environmental degradation: making the case for sustainable consumer behavior. Journal of Human Development 5(3): 417-432.

Albinsson, P.A. and Perera, B.Y. (2009). From trash to treasure and beyond: the meaning of voluntary disposition. Journal of Consumer Behaviour 8: 340-353.

Birtwistle, G. and Moore, C.M. (2007). Fashion clothing – where does it all end up. International Journal of Retail & Distribution Management 35(3): 210-216.

Ekström, K., Gustafsson, E., Hjelmgren, D. and Salomonson, N. (2012). Mot en mer hållbar konsumtion – en studie om konusmenters anskaffning och avyttring av kläder. Vetenskap för profession. Rapport 20.

Epsilon Computer data base at the Swedish University of Agricultural Sciences, where the full report of this research study can be found (www.slu.se/sv/bibliotek/publicera/sok-epsilon/).

Gardetti, M.A. and Torres, L.A. (2013). Sustainability in fashion and textiles: values, design, production and consumption. Greenleaf publishing limited, UK.

Jacoby, J., Berning, C.K. and Dietvorst, T.F. (1977). What about disposition – what do consumers do with products once they have outlived their usefulness, and how does this relate to the purchase of replacement products? Journal of Marketing 44(2): 22-28.

Kardes, F., Cline, M. and Cronley, T. (2011). Consumer behavior: science and practice. Cengage Learning, Andover, UK.

Liljebäck, L.-E. (2012). Billigare kläder, dyrt för miljön. Naturvetare 6: 12-17.

McDonald, S., Oates, C.J. and Alevizou, P.J. (2012). Individual strategies for sustainable consumption. Journal of Marketing Management, 28(3-4): 445-468.

Morgan, L.R. and Birtwistle, G. (2009). An investigation of young fashion consumers' disposal habits. International Journal of Consumer Studies 33(2): 190-198.

Niinimäki, K. (2010). Eco-clothing, consumer identity and ideology. Sustainable Development 18: 150-162.

Robson, C. (2011). Real world research. 3rd edition. John Wiley and Sons, Ltd, Chichester, United Kingdom.

Solomon, M.R. (2006). Consumer behaviour – buying, having, and being. 7th edition. Pearson Education, Inc., Upper Saddle River, NJ, USA.

Stigzelius, I. and Mark-Herbert, C. (2009). Tailoring corporate responsibility to suppliers: managing SA 8000 in Indian garment manufacturing. Scandinavian Journal of Management 25(1): 46-56.

The international regulation of the food market: precedents and challenges

D.S. Calley
The School of Law, The University of Essex, United Kingdom; dscall@essex.ac.uk

Abstract

This paper rests on the simple premise that agri-businesses exists to make money. Like any other commercial sector, agri-businesses make money by maximising their share of the market whilst minimising their outlay. Minimising outlay can, and often does, mean ensuring that animal welfare standards are applied at the most minimal level allowed under the laws of the State in which the 'products' are 'produced'.[7] Hence if the agri-sector of State 'A' has very few regulatory requirements of animal welfare then this will place State 'A' at a competitive advantage in terms of its exports to State 'B' (which might adopt and implement higher animal welfare standards for its own agri-sector). This is clearly a problem for State B (from a financial and ethical point of view), and to correct this imbalance State B might wish to invoke trade restrictive measures ('TRMs') against imports from State A. By closing off the markets of State B to a business entity based in State A, this clearly has an impact on that entity's ability to maximise their share of the market and, consequently, *might* compel them to consider modifying their practices. Unfortunately whilst this logic is simple to understand, in practice it becomes very difficult to apply due to the historic reluctance of courts to allow such fetters on trade. This paper will demonstrate how, in the sphere of the international trade in fisheries it is lawful for States to implement TRMs against those States whose vessels do not abide by certain conditions (concerning the manner in which the fish were harvested) and effectively prohibit the import of non-compliant fisheries products into their territory. Following this analysis the paper will posit the simple question of what the real differences between constraints on trade for the protection of the marine environment and constraints on trade for the protection of animals actually are.

Keywords: animal welfare, fisheries, trade-restrictive measure, barriers to trade

The orthodox view: trade restrictive measures are an unlawful restraint on free trade

During the 1990's much was made, in legal circles, of a number of high profile cases involving States that sought to unilaterally impose trade restrictive measures (TRMs) when the 'means of production' processes adopted by other States did not conform with the standards imposed domestically. In all cases, whether the measures involved the export or import[8] of 'products' but the relevant tribunals declared the TRMs to be incompatible with international agreements, such as the Treaty of the European Union or the General Agreement on Tariffs and Trade ('GATT').

[7] Throughout this paper terms such as 'product' and 'harvested' are used, simply on the basis that these are the terms employed throughout the legal discourse and the vast difference being living creatures (and not products) that are caught (not harvested) is acknowledged.

[8] Although the distinction between imports and exports is acknowledged, for present purposes and in the context of the hostility shown by tribunals towards TRMs generally, this distinction is not immediately relevant.

The 'EU veal calves' case

For instance, in 'R v Minister of Agriculture, Fisheries and Food ex parte Compassion in World Farming Limited' (European Court of Justice, 1998a) the claimants, CIWF, sought a judicial review of the decision of the Minister of Agriculture not to prohibit the export of veal calves from the United Kingdom to mainland Europe. CIWF claimed that to allow the exports of calves to other States would run counter to the provisions of the European Convention for the Protection of Animals kept for Farming Purposes and Appendix C of the subsequent 'Standing Committee Recommendation Concerning Cattle' (Council of Europe, 1988). The Recommendation was issued under the article 9(1) of the Convention in order to give detailed effect to the 'General Principles' contained in Chapter of the Convention by providing calves with a more comprehensive degree of protection than the relevant European Community laws on the protection of veal calves (European Commission, 1991). Consequently, under the Recommendation calves were afforded the more space and provided with an improved diet when compared to the requirements of the relevant Community law. Thus, said the claimants, the UK (having already banned the use of veal crates in its own territory) should – in order to fully give effect to its obligations under the Convention and its progeny – prohibit the export of calves to territories that merely implemented the lower welfare standards imposed by Community laws.

Upon the referral of the issue of the question from the UK's High Court, the European Court of Justice ('ECJ') was of the opinion that were the UK to, as the claimant CIWF maintained they should, prohibit the export of calves then this would operate as a restriction of the free trade contrary to article 34 of the EC Treaty which could not be justified as necessary as serving one of the legitimate aims listed under article 36 of the Treaty. Whereas article 36 of the Treaty *could* be invoked in circumstances where, for instance, public morality or the life and health of animals was threatened, recourse to article 36 would not be possible, in cases such as the present one, when the Community had specifically enacted harmonising legislation (such as Directive 91/629/EEC) that served the same purpose. In other words, where the Directive had harmonised EC law in such a way as to afford a degree of protection to veal calves it would not be possible for one particular State to invoke article 36 to justify a higher degree of protection, even if such measures could be provisionally justified under the Convention and its associated Recommendation because, said the ECJ, these documents did not contain any legally binding obligations for the contracting parties.

Underpinning this judgment was the territorial scope of any domestic measures adopted: whereas it was perfectly permissible for the UK to adopt higher standards of welfare for calves than the minimum standards imposed by the relevant European Directive, these higher standards must be strictly limited to the UK's territorial boundaries. Consequently, if the export ban that the CIWF sought were to be permitted, this would have the effect of extending this higher degree of protection to calves beyond the UK's territorial boundaries and, as explain by Attorney General Léger, giving his preliminary view on the *Veal Calves* case: 'To allow a Member State to prohibit or restrict the export of animals located on its territory in order to protect them against practices affecting their health or life beyond its own borders would in practice mean giving Member States the right to monitor, and even to influence, the practices or rules applied by the other Member States' (European Court of Justice, 1998b).

Clearly then from the perspective of the 1990's European Community the most important objectives were those of facilitating the free movement of goods, and no TRMs that would unjustifiable distort the operation of the common organisation of the market would be permitted.

If the 'Veal calves' case represented something of a disappointment to animal welfare campaigners, the decision of the ECJ should have come as no surprise: throughout the 1990s the World Trade Organisation's ('WTO') Dispute Settlement Body decision in the *Tuna-Dolphins* litigation (World

Trade Organisation, 1991) had cast severe doubt on the ability of States to impose trade related environmental measures ('TREMs') upon imports, where the importing State objected to the manner in which 'products' were harvested.

The view of the WTO: 'Tuna-Dolphins' and 'Shrimp-Turtle'

In *Tuna-Dolphins* the United States had attempted to prohibit the importation of tuna into its territory on the basis that the means by which the tuna was harvested by certain other States represented a threat to the dolphin population of the Eastern Central Pacific and contravened the US Marine Mammal Protection Act. In a highly symbolic judgment, the tribunal found that the USA's prohibition violated article XI of GATT, which prohibits the use of quantitative restrictions on imports. More legally significant than *Tuna-Dolphins*, however, was a further case brought before the WTO's Appellate Body (the Organisation's senior tribunal) six months after the *Veal Calves* case was heard by the ECJ. In *Shrimp-Turtle* (World Trade Organisation, 1998) the USA had again imposed TREMs against the import of certain shrimp products and cited as a justification the manner in which certain States 'harvested' these shrimp and which, said the USA, caused irreparable damage to populations of sea turtles. Thus, claimed the USA, any restriction on trade (ostensibly contrary to GATT article XI) could be justified by recourse to article XX of GATT which allows for States to invoke, in a non-discriminatory manner, TREMs 'relating to the conservation of exhaustible natural resources.' Whilst the WTO's Appellate Body ultimately found against the USA (because of the discriminatory and arbitrary way in which the TREMs were applied) the decision of the tribunal was a clear endorsement of the ability of States to restrict imports of products when natural resources (including living ones) and humans, animals and plants are threatened under GATT. Furthermore, the judgment of the WTO tribunal gives a number of clues as to how States may invoke GATT article XX in a legitimate and non-discriminatory manner.

The most noteworthy element of *Shrimp-Turtle* was that, unlike their European counter-parts in the *Veal Calves* case, the WTO was clearly not as hostile to the contention that to allow TRMs would be to allow invoking States to extend their jurisdictional capabilities beyond their sovereign territory. Whilst falling slightly short of stating in entirely unequivocal terms that there was, in all cases, no jurisdictional limitation to the effects of TRMs, the tribunal was, in the present case, content that there did, in fact, exist a sufficient nexus between the harms to be avoided (the destruction of migratory sea turtle populations) and the State attempting to invoke the TRMs which could justify the *de facto* application of US domestic laws to foreign-owned fishing vessels, even when the activities that these vessels engaged in (harvest the shrimp in the high seas) fell beyond US territory.

Embracing fisheries-related trade restrictive measures post-millennium

Trade measures as a means of regional fisheries management

Perhaps buoyed by the (partial) success of *Shrimp-Turtle*, since the turn of the 20th century many regional fisheries management organisations ('RFMOs') have embarked upon a sustained programme of utilising trade measures against the perceived ill of Illegal, Unregulated and Unreported ('IUU') fishing. RFMOs are generally groups of States with a common interest in a specific fishery, or geographic location, that are empowered under international law to regulate, manage and conserve fish stocks within their zones of competence (United Nations, 1982; United Nations, 1994). Hence, a typical RFMO might specify how, when and where certain fish stocks are harvested, and when a fishing vessel does not comply with relevant regional rules, that vessel will be barred from landing and transporting its catch within the territory of any RFMO member State. This is routinely accomplished by the use of a 'negative list' whereby the details of individual vessels, known to or having been suspected of engaging in IUU fishing, are passed to State parties to any RFMO with the instruction that no trade with these vessels should be permitted (ICCAT, 2011a; IATTC, 2005; WCPFC, 2010). In other words, RFMOs operate a series of small-scale TRMs targeted at individual vessels, whether that vessel is registered in a State that is a member of the RFMO or not.

Because of the serious problem of name and detail changing by individual fishing vessels – if a vessel finds itself on a negative list in many cases all it must do to avoid any RFMO prohibitions is to change its name and registration details – many RFMOs have more recently begun to adopt a positive list methodology: rather than denying landing rights to known offenders, positive list operate on the basis that unless it can be certified to the contrary, *all* vessels are presumed to be engaged in IUU activity. To be placed on a positive list an individual vessel must demonstrate positive compliance and produce a verified catch document whenever it wishes to land or trans-ship a catch. In practice, the vessels of States that are members of the RFMO, and that have no prior record of engagement with IUU fishing are generally presumed to be of good standing and, as such, are included on any positive list. Vessels registered to States that are not parties to the RFMO or vessels with a notoriously poor record of compliance with measures adopted in the fight against IUU fishing are, conversely, generally barred from entry onto positive lists (ICCAT, 2011b; IATTC, 2006; WCPFC, 2009).

Certain RFMOs have, as well as operating measures against individual vessels, the power to operate wholesale prohibitions of trade against all fisheries products from entire fleets. For instance, the Inter American Tropical Tuna Commission ('IATTC') can recommend to its members (which include the major economies of China, Japan, the USA and the EU) that TRMs are adopted, following a sufficient period of consultation and notice, against States when they have failed to fulfil their obligations under the IATTC Convention (in the case of IATTC member States) or when non-party States have failed to exercise sufficient control over their registered vessels to prevent those vessels undermining IATTC conservation and management measures (IATTC, 2006). Likewise, the International Commission for the Conservation of Atlantic Tuna ('ICCAT') and the Indian Ocean Tuna Commission ('IOTC') adopt exactly the same system to prohibit trade with member States and non-party States whose vessels undermine ICCAT and IOTC conservation measures (ICCAT, 2003; IOTC, 2003). The requirement under all of these regulations that allow for trade measures to be imposed against non-compliant States that the sanctioned States are notified and consulted is an important one in the context of GATT/WTO rules: in *Shrimp-Turtle*, it will be recalled, the United States' measures were deemed by the WTO to have been applied in an arbitrary discriminatory fashion, contrary to the *Chapeau* of GATT article XX because, *inter alia*, the US had failed to enter into dialogue with the complaining States prior to invoking the TRMs against them, but had engaged in dialogue with (and provided assistance to) certain other States whose activities had a similarly detrimental effect on sea turtles. The requirements of the ICCAT, IATTC and IOTC regulations (and those of any other RFMOs adopting similar provisions) of prior consultation and notification, coupled with the explicit prohibition on the discriminatory application of TRMs should ensure that no RFMO measures fall foul of GATT article XX's *Chapeau*.

A further key feature of RFMO TRMs is that they can, if necessary, be applied *whenever* a State's vessels undermine all conservation and management measures adopted within the RFMO's area of competence: there is, for example, no requirement that TRMs are adopted for the singular purpose of preventing deleterious activities that simply affect target species. Consequently, if a particular RFMO has adopted conservation measures, as most have, for the protection of non-target species, such as sea turtles, (ICCAT, 2010; IATTC, 2007; IOTC, 2012) then any State whose vessels pose a threat to those sea turtles, by failing to comply with the RFMO measures, will be deemed to be failing to fulfil their obligations and/or undermining the effectiveness of these conservation measures and thus be potentially subject to TRMs.

Furthermore, whilst RFMO TRMs (despite following a certain template) are individual agreements between member States to adopt trade measures against the effects of destructive fishing practices (and, as a consequence, be criticised as being 'collective unilateralism'), they do have a clear pedigree under international law. It will be recalled from earlier in this paper that the United Nations Convention on the Law of the Sea (UNCLOS) – the widely accepted definitive statement on the laws of the sea – mandates that States co-operate for the conservation and protection of the marine environment (United Nations, 1982). This rather vague obligation is further embellished by the UN Fish Stocks Agreement which empowers States to take necessary 'port State measures' – essentially prohibitions on landings and transhipments – in situations where vessels' activities are deemed to be undermining the effectiveness of sub-regional, regional or global conservation and management measures (United Nations, 1994).

Fisheries trade measures and the European Union: Regulation 1005/2008

In 2008 the European Union adopted Regulation 1005/2008, which had as its avowed aim of preventing, deterring and eliminating illegal, unreported and unregulated fishing. In accomplishing this task Regulation 1005/2008 mirrors many RFMO TRMs by prohibiting the importation of fisheries products into the EU by third party vessels (that is, the vessels of non-EU member States) unless those vessels can demonstrate positive compliance with the 'anti-IUU' provisions of the Regulation. The EU places a requirement on the vessels of non-parties to adhere to the EU's own internal policies and regulations: in other words, the EU assumes extra-territorial jurisdiction over the manner in which fisheries operations are conducted. In the context of the corpus of RFMO TRMs the Regulation is, perhaps, slightly unremarkable: it does nothing more than the numerous measures adopted by many RFMOs in the last decade, but what is remarkable is the fact that the merely a decade on from the *Veal Calves* case – in which the notion that UK might impose welfare standards that might have an extra-territorial application was roundly rejected by the EU's own court – the EU is happy to apply exactly the same form of extra-territorial jurisdiction over the fishing operations of third party States' vessels.

Conclusion

This, therefore, leads to the question that if it is permissible for the EU (and RFMOs) to invoke TRMs that have extra-territorial application in the context of fisheries, aimed at conserving fish stocks and to protect the living creatures of the marine environment, then why would TRMs that have an internal (to the EU) extra-territorial scope – such as those claimed by CIWF in the *Veal Calves* case – be so radically different so as to render these unlawful? If the opposition to the measures that have extra-territorial application is based upon the notion that it is offensive to the principle of State sovereignty for one State, or group of States, to impose their higher standards of animal welfare on other States, then why is this permitted in the case of fisheries? The answer might be, of course, that in the context of fisheries, the TRMs are simply the natural consequence of all States' agreement to co-operate for the protection and conservation of the marine environment, evidenced by their signature of UNCLOS: hence these measures are justified under international law as supporting – and being an integral part of – the commitments that States have made to protect the oceans. In which case one might legitimately ask what the difference in principle between a State's commitment to protect the oceans under UNCLOS and a State's commitment to protect animals under the European Convention for the Protection of Animals kept for Farming Purposes might be. Whilst the ECJ may, in 1998, have thought that the farmed animals Convention contained no binding rules of law, again the question remains to be answered what the point of the Convention actually was. By operating as the justificatory statement of international law – as UNCLOS does – the Convention can underpin, provided that it is done in a non-discriminatory manner, TRMs aimed at the protection of animals used for farming purposes. Within the EU, however, there remains the fundamental question of what the Union's primary purpose is: is it a multi-faceted animal protection Treaty? Clearly not. The EU's primary goal remains the facilitation of trade between member States, and so on this basis the judgment in the *Veal Calves* case remains beyond reproach. Since the judgment was delivered, however, the EU has made strides towards improving the welfare on animals and, in Article 13 of the Treaty on the Functioning of the European Union there rests a reminder to all member States that they should 'pay full regard to the welfare requirements of animals' whilst Union policies are formulated and implemented. Whilst the animal welfare requirements of Art 13 TFEU are clearly subjugated to the workings of the internal market, it is submitted here that rather than operate on a 'race to the bottom' model, whereby high achieving States' ability to impose TRMs on lower achieving States is stymied by the workings of the market, to give full effect to article 13 TFEU TRMs can, and should, be used to raise (and pay full regard to) animal welfare standards. Rather than viewing those States that operate the highest of animal welfare standards as being disruptive to the functioning of the market, TRMs can be a practical way in which lower achieving States are compelled to apply the provisions of international agreements – such as the European Convention for the Protection of Animals kept for Farming Purposes – that those States have already agreed to implement.

References

Council of Europe (1988). Standing committee recommendation concerning cattle. Adopted by the Standing Committee on 21 October 1988; available at http://www.coe.int/t/e/legal_affairs/legal_co-operation/biological_safety_and_use_of_animals/farming/Rec%20cattle%20E.asp.

European Commission (1991). Directive 91/629, laying down the minimum standards for the protection of calves.

European Court of Justice (1998a). Case number C-1/96; reported at [1998] 2 CMLR 661, judgment of the European Court of Justice.

European Court of Justice (1998b). Case number C-1/96; reported at [1998] 2 CMLR 661, opinion of the Attorney General.

IATTC (2005). Inter American tropical tuna Commission Resolution C-05-07 to establish a list of vessels presumed to have carried out illegal, unreported and unregulated fishing activities in the Eastern Pacific ocean.

IATTC (2006). Inter American tropical tuna Commission Resolution C-11-06 on a vessel register.

IATTC (2006). Inter American tropical tuna Commission Resolution C-06-05 on the adoption of trade measures to promote compliance.

IATTC (2007). Inter American tropical tuna Commission Resolution C-07-03 to mitigate the impact of tuna fishing vessels on sea turtles.

ICCAT (2003). International Commission for the conservation of Atlantic tuna Recommendation 03-15 concerning trade measures.

ICCAT (2010). International Commission for the conservation of Atlantic tuna Recommendation 10-09 on the by-catch of sea turtles in ICCAT fisheries.

ICCAT (2011a). International Commission for the conservation of Atlantic tuna Recommendation 11-18 establishing a list of vessels presumed to have carried out illegal, unreported, and unregulated fishing activities in the ICCAT convention area.

ICCAT (2011b). International Commission for the conservation of Atlantic tuna Recommendation 11-12 establishment of an ICCAT record of vessels over 20 m authorized to operate in the convention area.

IOTC (2003). Indian Ocean tuna Commission Recommendation 03-05 concerning trade measures.

IOTC (2012). Indian Ocean tuna Commission Recommendation Resolution 12-04 on the conservation of marine turtles.

United Nations (1982). The United Nations convention on the law of the sea.

United Nations (1994). The United Nations agreement for the implementation of the provisions of the United Nations convention on the law of the sea of 10 December 1982 relating to the conservation and management of straddling fish stocks and highly migratory fish stocks.

WCPFC (2009). Western and Central Pacific Fisheries Commission Conservation Measure 2009-01 record of fishing vessels and authorisation to fish.

WCPFC (2010). Western and Central Pacific Fisheries Commission Conservation Measure 2010-06 to establish a list of vessels presumed to have carried out illegal, unreported and unregulated fishing activities in the WCPO.

World Trade Organisation (1991). United States – Restrictions on the import of tuna (1 and 2). Report of the Panel, 30 ILM 1594 (1991); 33 ILM 839 (1994).
World Trade Organisation (1998). United States – import prohibition of certain shrimp and shrimp products, report of the appellate body. WT/DS58/AB/R (1998).

Precaution or prudent vigilance as guiding the path to global food security?

M. Kaiser
Centre for the Study of the Sciences and the Humanities, University of Bergen, Norway; matthias.kaiser@svt.uib.no

Abstract

Food safety is an obvious ethical demand as seen from the consumer's perspective. It is also, at least in principle, in the best interest of States to protect their citizens from harm through nutrition. Food security is also an ethical interest of people, albeit perhaps not this obvious to all those who presently live in abundance. Sustainable food production can also be seen as basically an ethical task, since it recognizes basic interdependencies between people, environment and economy on a global scale, and aims at a remedy of existing or emerging imbalances. Globalisation is a fact of our times, and in the food sector globalisation can be assumed to increase rather than decrease due to more varied demands that follow socio-economic advances. Thus one may assume that the global trade of food, food technologies, and seeds and animals for food production will become even more important in the near future. The trade of genetically modified (GM) food-products has already triggered a heated debate about the Precautionary Principle, mainly in respect to food safety, but also in respect to other national interests. While some States emphatically endorse the Precautionary Principle as a regulative mechanism (as e.g. the EU countries), other States equally emphatically seem to deny this principle any role in international (or even national) regulations (as e.g. the USA). New food technologies are under way (e.g. functional foods, synthetic biology, in vitro meat or GM animals), and are typically proposed as a way to tackle the challenge of global food security. The trade of these products can be assumed to be even more problematic than what one has experienced so far. The US Presidential Bioethics Commission has recently addressed the regulatory needs of synthetic biology, and concluded that there is no need of the Precautionary Principle. Instead they propose a policy of 'prudent vigilance'. The paper will discuss whether there is a real, substantive difference between these proposals, and whether global food security will need to endorse the one or the other.

Keywords: precautionary principle, ethics, food trade, GM food

Introduction

It is a disturbing fact that malnutrition has not been eradicated on a global scale, and it is even more disturbing that this is happening despite concerted efforts to combat malnutrition, on the background of food abundance in many parts of the world. Furthermore, given realistic assessments of further population growth, the global population is expected to having surpassed the 9 billion mark by 2050, the question of global food security acquires even more urgency to act now in order to secure equal and sufficient access to nutrition. Admittedly, scenarios of how to develop the global production of food in order to reach this goal differ widely. What is taken as a fact and what as a value typically varies between different actors, and even within the sciences. Both factors influence what scenarios we take as realistic options for a sustainable course of action. Thus there is considerable uncertainty surrounding the policies that we might adopt.

Yet, there is also some more or less well established and assumedly robust knowledge which constrains our policies. We observe that the availability of food resources is not necessarily congruent to the size of the populations which live in proximity to the resource. We also observe that increased welfare and wealth typically leads to a preference for greater variety in food, thus stimulating global trade of food products. Furthermore, we face already now a situation where natural food resources, i.e. both arable land and fish stocks, are exploited close to their limit or above. These and other factors may lead us to two conclusions. First, future global food security will need the aid of some form of improved and more efficient technologies in food production, and second, we may safely assume that also future food security will to a significant extent involve the global trade of food products.

As we have seen, the development of genetically modified (GM) food products was originally advanced with great fervour to 'feed the world'. As we also have seen, this development, though by no means at its end, has faced significant difficulties both in terms of technology, economy, regulatory measures, and societal acceptance. Presently the goal of feeding the world by GM food remains utopian at least. About GM food the world has been divided into those who favour a precautionary approach in approaching its trade and development, and those who adopt the more traditional scientific risk-cost-benefit approaches. The division in politics between the EU and the USA can exemplify this.

But as far as we know, GM food exemplifies only the first wave of technological innovation in food production. Other developments are on the way and promise also to provide a way out of the problem of global food security. One works on supplementing GM crops with GM animals for agricultural (or aquacultural) production, in vitro meat is at least on the scientific horizon, and synthetic biology aims at engineering food products from scratch. These developments will again pose the question what kind of regulatory mechanisms we want to employ for their global trade (in addition to the national policies surrounding them).

So far, the regulatory principle that has been advanced to deal with large scale technological and scientific uncertainty has been the so called Precautionary Principle (PP). However, recently the US Presidential Commission for the Study of Biological Issues (PCSBI 2010) launched a report on synthetic biology in which they argued for an allegedly new principle, termed 'prudent vigilance'. This report was well received within the US government and indeed most of the US scientific community apparently. The question we shall therefore pursue in the following is whether prudent vigilance differs significantly from precaution, and if yes, in what respect. The answer to this may be significant for the future of global food policies and trade.

The Precautionary Principle in context

A common reference point when discussing the PP is often §15 of the Rio Declaration (United Nations, 1992) where it reads:

'In order to protect the environment, the precautionary approach shall be widely applied by States according to their capabilities. Where there are threats of serious or irreversible damage, lack of full scientific certainty shall not be used as a reason for postponing cost-effective measures to prevent environmental degradation.'

Several points need to be noted here: First, this quote does not mention a *principle*, but only a precautionary *approach*. Second, the text is far from functioning as a satisfactory definition. It contains a triple negation (i.e.: *lack* of full certainty ... shall *not* be used ... as reason *not* to act) which makes it difficult to see what is actually implied. Third, it contains two important provisos which allow for problematic interpretations, i.e. both the reference to national capabilities, and reference to cost-effective measures. Fourth, from a philosophical point of view one may rightfully question whether science ever achieves full certainty. In spite of this, the mentioning of precaution in this context marked something close to a watershed in international law. The PP has already earlier played a role in some national legislation (e.g. in Germany and Sweden), and, indeed, to some extent in international law, as e.g. in the North Sea Treaties. But most significantly the PP in several varieties of definition has entered many other international agreements in later years, such as e.g. the Cartagena Protocol on Biosafety (2000), the Stockholm Convention on Persistent Organic Pollutants (2001), or the EU Communication on the PP; the EU Nice Treaties (2000).

At the same time, a lively debate about the PP ensued, both politically and academically. This is certainly not surprising at all, since the PP can be seen as touching different contexts at the same time, as e.g. a scientific, a legal, a political, and a cultural and ethical context. Bridging all these areas involves grasping rather complex connotations, traditions and institutions. Furthermore, the fact that the PP in its official documents does not have a unique, clear and precise definition also invites critical comments. Per Sandin (1999) identified not less than 19 definitions, and one might safely assume that others have emerged since. Though the conceptual vagueness of the PP might be a crucial reason for its political success (at least in some states and contexts), it is also raising issues about the implications for its actual application. Thus, several authors make do with delineating between weak and strong versions of the PP. Roughly speaking, weak precaution would simply state that some kind of preventive or monitoring action is not precluded when being in a situation where there is no conclusive evidence that some serious harm actually will occur. This somehow captures the folkloristic wisdom: 'better safe than sorry'. A strong version, on the other hand, would go beyond weak precaution and require some active measure to counteract or delimit some uncertain but scientifically plausible and serious future harm. This typically appeals to institutions to consider regulatory or other kind of controlling measures to accompany a scientific or technological development as long as basic uncertainties prevail. One should note that even a strong version of the PP is not identical to an extreme form of the PP which would require that an activity should not proceed until proven to be safe. This is obviously extreme and unrealistic, since from a scientific point of view zero-risk is an impossibility and proofs of total safety are beyond the possibilities of science.

Another issue of debate has been whether the notion of approach versus principle actually makes much of a difference. Again commentators are divided. Some view the notion of approach as the more flexible term, allowing for contextual adjustments and various operational strategies. This they contrast to the notion of principle which is then viewed as a legally binding concept to apply strict measures to prevent uncertain but possible harm of an activity to occur. Yet, this debate may in the end only reflect the difficulty of different legal systems to accommodate broad principles that allow for some discretion and judgement in their concrete application (Cooney, 2005). Thus, in an international context it makes perfect sense not to differentiate sharply between these two notions.

It is important to understand the PP as a specific response to a specific perception of a societal (and by extension: environmental) threat. De Sadeleer (2003) reminds us that there is an evolution in perceived environmental threats and their corresponding legal measure. The Polluter-Pays-Principle is basically a curative, a posteriori control mechanism, aimed at deterring actors from taking the risk of pollution. The Prevention-Principle, on the other hand, is an anticipatory measure to allow interventions at various thresholds of risk, allowing for the weighing of the goods to be protected. While the Prevention-Principle rests on some reasonably well established and quantified risk assessment, the PP goes beyond this and aims at dealing with uncertain risks, i.e. risks where no reliable quantification of magnitude is at hand, but which still are backed up by some plausible evidence or plausible scientific model. Thus, the PP is about managing uncertainty, and not about risks in general. This is an important distinction which even otherwise well informed commentators sometimes overlook (e.g. Sunstein, 2005). For the rest of the paper, I shall (for obvious reasons) endorse the definition of the PP as it is given in COMEST 2005 (see ibid. p.14 for further explanations):

'When human activities may lead to morally unacceptable harm that is scientifically plausible but uncertain, actions shall be taken to avoid or diminish that harm.'

Prudent vigilance

In response to the announcement of the Craig Venter Institute in 2010 of having created the first synthetic self-replicating genome, Barack Obama asked the Presidential Commission for the Study of Bioethical Issues to review the development of synthetic biology and identify emerging ethical issues in this field. The result was an in many ways impressive document with altogether 18 specific policy recommendations (PCSBI, 2010). The Commission identified five ethical principles as significant in this context: (1) public beneficence; (2) responsible stewardship; (3) intellectual freedom and responsibility; (4) democratic deliberation; and (5) justice and fairness. We shall not enter a detailed discussion of the report or these principles. What is of interest, however, is the fact that the Commission under responsible stewardship claims to identify a principle of *Prudent Vigilance* which some commentators celebrated as a viable alternative to the PP. Here we want to critically assess to what extent prudent vigilance introduces a new management principle for uncertain risks.

> The principle of responsible stewardship rejects two extreme approaches: an extreme action-oriented approach that pursues technological progress without limits or due regard for public or environmental safety, and an extreme precautionary approach that blocks technological progress until all possible risks are known and neutralized. While the action-oriented approach is irresponsibly brazen, the precautionary approach is overly wary (p.26).
>
> Responsible stewardship calls for prudent vigilance, establishing processes for assessing likely benefits along with safety and security risks both before and after projects are undertaken. A responsible process will continue to evaluate safety and security as technologies develop and diffuse into public and private sectors. It will also include mechanisms for limiting their use when indicated (p.27).

Amy Gutman makes it clear that prudent vigilance is viewed by the Commission as being in contrast to the PP:

> Prudent vigilance shares with the precautionary principle a concern for identifying and mitigating risks. However, it advocates continued progress in the pursuit of potential benefits in tandem with that ongoing sensitivity to risks and the development of appropriate responses (Gutman, 2011: 22).

Yet, in view of what has been said about the PP it is far from clear what the substantial difference to the PP is supposed to entail. It is certainly not far removed from any weak version of the PP, but even in regard to stronger versions it is difficult to point to any feature which would clearly delineate the two principles as different in substance. Both are obviously dealing with situations where large uncertainties prevail, bordering on the level of ignorance and precise quantitative risk assessments being illusionary. Both are also calling for some proportionality between potential harms and potential benefits. As COMEST (2005) states: 'Interventions should be proportional to the chosen level of protection and the magnitude of possible harm' (p13). And both are including the consideration of costs of action and opportunity costs of inaction. The PP in its common meanings does not imply a simple choice between yes-or-no (go-or-no-go) decisions about certain technologies, but opens for a repertoire of possible responses, aiming at measures which are either constraining the possibility of harm, or which would contain the harm should it occur. Case-by-case and step-by-step assessments, combined with continuous monitoring would fall under such measures. It is certainly not implied to rule out potential benefits just because some risks might occur and there is some level of uncertainty about the nature and scope of these risks. The EU policy on GM crops, as they are being guided by the PP, may illustrate this.

So, then one may ask what is left of the claimed uniqueness of prudent vigilance, if a clear substantial difference to a policy based on the PP cannot be identified? Perhaps the difference between these principles is merely semantic? This hypothesis would imply that there may be a prevalent reading of the PP in the USA which places it much closer to an extreme and ultimately unrealistic version than the readings which are prevalent in e.g. Europe. To some extent this may indeed be a reasonable conclusion, given that there have been quite a bit of rhetorical exchanges about the PP between the USA and the EU.

But on closer inspection one may doubt that this is in fact the sole reason. There are many areas where the USA may claim to be more precautious than Europe, e.g. in regard to carcinogens in food additives. In political practice, the US decision for a nationwide swine flu vaccination in 1976 can reasonably be said to be based on an implicit endorsement of the PP (President Ford said: 'I think you ought to gamble on the side of caution. I would always rather be ahead of the curve than behind it.'). In this sense, precaution in general is certainly not alien to the US policy context.

In a Master thesis at the University of Utrecht Nick Nieuwenhuijsen (2012), comparing the PCSBI study to a study from the European Group on Ethics (EGE), claims to have identified three morally relevant differences:

> Firstly, the European emphasis on a more principled approach versus the domination of pragmatic arguments on US-side. Secondly, a more conservative attitude from the EGE due to the application of the precautionary principle against a strong emphasis from the PCSBI on public beneficence and thirdly, a mainly governmental responsibility to keep watch (on the benefits and risks) of synthetic biology according [to] the EGE against more trust in individual responsibility by the PCSBI (ibid.: 3).

These observations are well taken. They can be strengthened by looking at one particular area of conflict where reference to the PP has been highly problematic: the area of international trade. The problem is whether the PP may be in conflict with the Sanitary and Phytosanitary Agreement (SPS Agreement, 1994) of the World Trade Organisation (WTO). The WTO is particularly concerned about measures which may occur as protective or precautionary measures, but are in effect measures to restrict free international trade. Therefore restrictions on import need to be based on 'sufficient scientific evidence' and risk assessments. In other words, where precaution in the light of significant uncertainties would potentially mandate some import restrictions, the SPS Agreement would call for scientific proof that risks are real. It is this difference of standards which has been a stumbling block for GM crops.

Now, adding to this the general observation that the US so far have been reluctant to endorse legal principles which could bind their policies in international affairs, or make American actions liable in international law, the endorsement of prudent vigilance rather than precaution seems to make some sense from a US point of view. First, it further separates policies from any governmental oversight responsibilities, where a precautionary approach would call for some active measures. Second, it retains the freedom of American business to enter markets with different risk perceptions. Third, it conforms to an ideological bias to base policy on 'sound science' before interfering with private or market initiatives. And fourth, it refrains from sanctioning a legal reference point from which American (export) activities may become liable to principles of international law.

If this is indeed indicative of future US policies for the trade of novel food technologies to combat food security, the hollowness and weakness of prudent vigilance should be the target of responses from countries which place a high value on environmental integrity and sustainable uses of their resources.

Conclusion

Though a principle of prudent vigilance does not seem to add substantially new considerations about uncertain risks compared to the PP, it does imply a further move to privatize responsibilities and a separation from internationally binding principles of law. Striving for global food security with extensive international trade needs to be based on a stronger basis, such as that provided by the PP.

References

COMEST (2005). The Precautionary Principle. A report of an ad hoc working group, endorsed by COMEST /UNESCO April 2005 (members of the expert group: M. Kaiser (chair), J.v.d. Sluijs (rapporteur), S. Beder, V. Hösle, A. Kemelmajer de Calucci, A. Kinzig; UNESCO secretariat: H. t. Have, S. Scholze, E. Kuok), UNESCO: Paris; also available on: http://unesdoc.unesco.org/images/0013/001395/139578e.pdf (accessed: 31.03.2013)

Cooney, R. (2005). From promise to practicalities: the Precautionary Principle in biodiversity conservation and sustainable use. In: Cooney, R. and Dickson, B. (eds.) Biodiversity and the Precautionary Principle – risk and uncertainty in conservation and sustainable use. Earthscan, London, Sterling, VA, USA, pp. 3-17.

De Sadeleer, N. (2003). Environmental principles: from political slogans to legal rules. Oxford University Press, Oxford, UK.

EGE (European Group on Ethics in Science and New Technologies) (2009). Opinions on the ethics of synthetic biology. Luxembourg: Publications Office of the European Union. Available at: http://ec.europa.eu/bepa/european-group-ethics/docs/opinion25_en.pdf.

Gutman, A. (2011). The ethics of synthetic biology: guiding principles for emerging technologies. Hastings Center Report 41, no.4 (2011), pp. 17-22.

Nieuwenhuijsen, N. (2012). Synthetic biology – an atlantic expedition. Master thesis, Dept. of Philosophy, Applied Ethics, University of Utrecht, Utrecht, the Netherlands.

PCSBI (2010). The Presidential Commission for the study of bioethical issues, 'New directions: the ethics of synthetic biology and emerging technologies'. Washington DC, USA. Available at: http://bioethics.gov/cms/sites/default/files/PCSBI-Synthetic-Biology-Report-12.16.10_0.pdf.

Sandin, P. (1999). Dimensions of the Precautionary Principle. Human and Ecological Risk Assessment 5: 889-907.

Sunstein, C.R. (2005). Laws of fear – beyond the Precautionary Principle, Cambridge University Press, Cambridge, UK.

Certification for sustainable biofuels

A.J.K. Pols
Department of Philosophy & Ethics, IPO 1.09, School of Innovation Sciences, Eindhoven University of Technology, 5600 MB Eindhoven, the Netherlands; A.J.K.Pols@tue.nl

Abstract

Certification for sustainable biofuels has been developed to ensure that biofuel production methods adhere to social and environmental sustainability standards. As such, requiring biofuels to be certified has become part of policy documents such as the EU Renewable Energy Directive (RED) that aim to promote energy security, reduce emissions and promote rural development. According to the EU RED, in 2020 10% of our transport energy should come from renewable sources, including biofuels. Only certified biofuels may count towards this target. In this paper I examine what biofuel certificates are, what they can do and what their weak points are. I argue that the EU RED makes an important but unjustified assumption in demanding certified biofuels for its target: that if biofuel *production* is sustainable, then biofuel *use* is too. Applying the use plan approach from the philosophy of technology to biofuel certification, I show why this assumption is unjustified and why the EU is in fact making 'improper use' of biofuel certification. Finally, I discuss ways in which biofuel certification could be used in working towards the EU RED's goals.

Keywords: agrofuels, bio-energy, EU Renewable Energy Directive, stakeholder participation, use plan approach

Introduction

Biofuels have been hailed by the EU as a potential step towards energy security, reducing greenhouse gas emissions and stimulating rural development (EC, 2009; statement 1). According to the EU's 2009 Renewable Energy Directive (RED; EC, 2009), in 2020 10% of the EU's transport fuel and energy consumption should come from renewable sources: Bowyer (2011) anticipates that 92% of this target will be attained by using biofuels.

However, biofuels have been strongly criticised for increasing food insecurity through using food crops for fuel, and arable land for fuel crops; actually contributing to climate change, as direct and indirect land use change for fuel crop cultivation can release large amounts of greenhouse gases into the atmosphere (Gomiero *et al.*, 2010); and increasing rural poverty (Levidow, 2013). The EU RED proposes several strategies to deal with these problems, including increasing efficiency in energy use, developing second-generation biofuels and developing sustainable production criteria for biofuels, laid down in certificates, where only certified liquid biomass may count towards the 10% target.

Much like the Forest Stewardship Council or Fairtrade certificates, biofuel certification is meant to guarantee that fuel crop cultivation and biofuel production adhere to certain sustainability standards, e.g. with regard to labour conditions, protection of the local environment and projected greenhouse gas emissions. Much discussion in the last years has focused on the content of those certificates and whether they should be 'universal' or adapted to specific regions (e.g. Guariguata *et al.*, 2011). Less attention, however, has been given to certificate *use*, and whether it manages to accomplish the EU's goals: delivering the supposed biofuel benefits of energy security and emission reductions while minimising social and environmental costs and stimulating rural development. In this paper I argue that the EU does not and indeed cannot achieve these goals, given how biofuel certificates are currently used. I argue that this is because of an assumption implicitly present in the EU RED, that if biofuel *production* is sustainable, then biofuel *use* as prescribed in the EU RED is sustainable, but this assumption is currently not justifiable. I show what measures could be taken in order to make this assumption justified by applying the 'use plan account' for technical artefacts (Houkes and Vermaas, 2004, 2010) to biofuel certification. Specifically, I argue that capping or taxing consumption or emissions is a necessary part of those measures.

Biofuel certificates

The development of biofuel certification serves two purposes. First, it is a way of identifying and establishing general *principles*, that can be agreed upon by all stakeholders involved in the development process, which biofuel production has to adhere to in order to be classified as 'sustainable'. Second, it establishes how these principles should be translated into practically measurable *criteria*, in order to prescribe biofuel producers what they actually should do in order to produce biofuels sustainably. For example, a principle might describe the general importance of protecting human and labour rights of workers, while its related criteria prescribe specific regulations concerning unionisation, against child labour, etc. Principles and criteria can be established together by one party (e.g. as by the Roundtable on Sustainable Biofuels (RSB) 2010), but they can also be established in different ways by different parties. For example, the Dutch Cramer Criteria (Cramer, 2007) are a set of principles established by a committee, which was headed by a government minister and included experts from both industry and civil society. These principles were translated into criteria, however, by the Netherlands Standardization Institute (as the NTA 8080/81).

Whether certificates actually work and can be considered legitimate depends on the processes by which principles and criteria are arrived at. Some certificates are generally considered to be effective in dealing with direct social and environmental impacts, such as the FSC sustainable forestry certificate (Guariguata *et al.*, 2011). Much of their success, however, depends on the degree of stakeholder involvement. For example, if the principles are not discussed by a comprehensive set of stakeholders, as is often the case with voluntary industry standards such as the RSPO, legitimacy of the certificate is threatened, as different stakeholders have different conceptions of what counts as 'sustainable' production (Partzsch, 2011). Similarly, if the principles are formulated by a comprehensive set of stakeholders, but the criteria are not, the resulting criteria may not be implementable 'in the field' (this charge is levelled by Romijn *et al.*, 2013 against the NTA 8080/81). Even well-working certificates have inherent weak points, however, that fall into two classes. First, the scope of certificates is usually limited, excluding, for example, considerations of indirect land use change and social and environmental impacts above farm or plantation level (Guariguata *et al.*, 2011). Second, certification as an instrument tends to favour powerful stakeholders in the global biofuel trade. This is partly inherent to certification – companies and plantations have more administrative resources to spend on the certification process than smallholder farmers, even though the latter tend to have less environmental impact (Romijn *et al.*, 2013). Furthermore, in practice the process of drawing up principles and criteria is often dominated by powerful stakeholders in developed countries, sometimes not even involving stakeholders from the global south (Partzsch, 2011).

The sustainable production assumption

In this section I examine another factor that falls outside the scope of biofuel certification, namely how sustainably produced biofuels are used. In particular, I argue that sustainable biofuel *production* does not entail sustainable biofuel *use* in the way envisaged by the EU RED. I do so by examining certificate use through the lens of the 'use plan approach' for technical artefacts, as developed by Houkes and Vermaas (2004, 2010).

The use plan approach

The use plan approach for technical artefacts (Houkes and Vermaas, 2004, 2010) is a rational reconstruction of the product design process that sets standards for the design of functional technical objects, or technical artefacts. Central to the approach is the idea that product design is not primarily about *artefact* design, but about *use plan* design. According to Houkes and Vermaas, a use plan describes a series of actions leading to a goal, where the actions involve the use of one or more technical artefacts. For example, according to this approach designing a jatropha seed press should not start with designing the physical press, but rather with designing a plan for extracting jatropha oil from seeds. This plan, in turn, informs the physical artefact design (e.g. that the artefact needs a funnel to put seeds in, a mechanism to press them, etc.). This is not to say that physical design questions are not important, but rather that they are subservient to the overall use plan: the physical design depends on considerations of what the artefact is to be used for, and how.

Usually a use plan is developed by the designer and then communicated to the user: the user's manual is one way by which this can be done, but there are others, including labels, symbols and affordances. Communication of the use plan is not only important in that it enables the user to use the artefact for its intended purpose; it also helps transfer responsibility for the consequences of artefact use from designer to user (Pols, 2010). Users may create their own use plan for existing artefacts as well, such as when using a jatropha seed press as a doorstopper. Designers have only limited responsibility for the consequences of such 'improper use', however, especially when they warn against doing so in the use plan. This is not to say that improper use can never be successful, just that users have no ground to complain to the designers if it is not, and experiment at their own risk.

Use plans and certificates

Though biofuel certificates are social rather than technical artefacts, there is reason to assume that the use plan approach can help us evaluate their conditions for 'proper use'. As mentioned, the use plan approach evaluates *plan* design, according to values of practical rationality, rather than specific *technical artefact* design. This implies that biofuel certificates can be evaluated with this approach, insofar as they are designed for one or more specific goals, and specify means for reaching these goals.

The goal of biofuel certificates, or expected result of implementing certification criteria, is usually defined as *the sustainable production of biofuels* or *biomass* (e.g. Cramer, 2007; RSB, 2010). Applying the concrete criteria is then performing the 'series of actions' that should lead to that goal. However, much like the manual of a microwave that tells one *not* to put whole eggs or metal bakeware in it, several certification systems come with added disclaimers of what would constitute *improper* use. The RSB Principles and Criteria document, for example, states that '...the Principles & Criteria do not attempt to quantify an amount of biofuels which could be sustainably produced, or whether, as a whole, biofuels are sustainable. Biofuels cannot replace all of our fuel consumption and must be accompanied by significant changes in lifestyle and efficiency of use' (2010: 3). In a similar vein, the Nuffield Council on Bioethics (NCB) proposes six ethical principles for biofuel production and development, but states that we only have a *moral duty* to produce biofuels if additional considerations are met, including considerations of alternative energy sources, alternative applications of biomass and stakeholder participation (NCB, 2011: 78).

If we compare these 'instructions for use' with the way the EU is actually using certification, as described in the EU RED, several discrepancies emerge. Most salient among them is that the EU takes (rising) levels of transport fuel consumption as a given, rather than as a habit to be changed (Levidow, 2013). Though it suggests 'soft' incentives to reduce energy consumption, such as transport planning and supporting public transport (EC, 2009, statement 28), it is not clear whether these can bring about the 'significant changes in lifestyle' that the RSB prescribes. And while the EU acknowledges the importance of increasing efficiency of use, doing so without unambiguously limiting use could lead to a rebound effect where efficiency gains are at least partly offset by increased use (Alcott, 2010). Finally, by setting a quantitative target of 10% renewable transport energy in 2020, the EU sets a goal that is not guaranteed to be achievable by biofuel production according to certificate criteria, as the RSB has noted (cf. Bindraban *et al.*, 2009; Sues, 2011). In terms of the use plan approach, the EU is making *improper use* of certification as an instrument to achieve a goal, which might not be achievable in that way. Moreover, if transport fuel consumption were to keep growing, possibly even stimulated by the large-scale production of biofuels, increased biofuel production might not even help to bring the target within reach.

From a use plan point of view, what is worse is the EU RED statement that 'The *main purpose* of mandatory national targets is to provide certainty for investors and to encourage continuous development of technologies which generate energy from all types of renewable sources' (EC, 2009, statement 14, author's italics). This suggests that the goals of the EU RED regarding energy, emissions and rural development are not similar to the purpose of one of its key parts, the 10% target. Indeed, as we have seen, both goals may well conflict, and this might create pressure to lighten certifications' sustainability requirements.

Can proper use of certification fulfil the EU Renewable Energy Directive goals?

I have argued that, according to the use plan theory, the EU's reliance on certified biofuels to fulfil its 10% target constitutes *improper use* of certification. In this section I examine which of the EU RED's goals of energy security, lowering emissions and rural development proper use could facilitate.

First, as the RSB makes clear, proper use of biofuel certification is intended to yield an unspecified quantity of sustainably produced biofuels. Biofuel production certificates cannot guarantee any contribution to the EU's *energy security* as they cannot put limits on the EU's transport fuel *use*. Worse, the rebound effect suggests that increasing biofuel production without limiting transport fuel consumption could increase rather than offset transport fuel consumption, leaving us no closer to this goal than before (Alcott, 2010). For the same reason, biofuel certification cannot guarantee an actual reduction in *greenhouse gas emissions*, even if we disregard indirect emissions, such as those resulting from indirect land use change, and the fact that many biofuel crops have a relatively modest potential for reducing emissions. Certificates cannot guarantee this actual reduction in greenhouse gas emissions even despite the fact that explicit emission criteria are currently part of systems like the RSB and the Cramer Criteria and the EU RED itself (EC, 2009, criterion 2). For example, the EU RED demands that the greenhouse gas emission savings from biofuel use should be at least 35% now, compared to fossil fuel emissions, rising to 60% in 2018. Again, however, biofuel production certificates cannot put limits on EU transport fuel use or total level of transport emissions. Therefore, if the rebound effect were to lead to more fuel consumption, the *absolute* emissions could rise, even if biofuels emit relatively less than fossil fuels, if emissions are not capped or taxed in some other way. Certificates have more promise to contribute to *rural development*, as they explicitly concern conditions regarding biofuel production. As mentioned in section two, though, special attention should be given to ways in which certification could be made to work for smallholder farmers and poor stakeholders. Certificates can establish a procedure for sustainable biofuel production, but they do not guarantee that smallholder farmers benefit from them, or can even implement them in practice (cf. Romijn *et al.*, 2013). Also, considering the extreme poverty and food insecurity many smallholder farmers in developing countries face, it could be argued that we should help them meet their basic needs with more tried-and-tested means *before* we ask them to experiment with biofuel schemes.

Conclusion

Biofuel certification has been used by the EU as a means towards energy security, reduction of greenhouse gas emissions and rural development, while avoiding the social and environmental costs associated with large-scale biofuel developments. In this paper I have used the use plan approach to argue that biofuel certificates can be used for producing biofuels sustainably, but that there is no guarantee that proper certificate use will yield the quantity of biofuels required to meet the EU RED's target. Nor can certificates guarantee that using them to produce sustainable biofuels will ultimately help the EU achieve its goals of energy security and greenhouse gas emissions. Finally, both certificate structure and use practice place restrictions on the ways certificates can be used for rural development.

If biofuel certification cannot guarantee a contribution towards at least two of these three goals, and assuming that abandoning those goals is not an option (though their relative priorities merit discussion), how could we embed biofuel certification in a use plan that *does* contribute to those goals? Two approaches have been put forward in this paper: combining certification with other measures, and improving the process of establishing certification principles and criteria.

With regard to other measures, fuel consumption or emission caps or taxes have been suggested to be the only way to guarantee achievement of set emission or consumption goals (Alcott, 2010). Combining these measures with biofuel certification could help prevent the rebound effect and thus help the EU towards energy security and emission reductions. These measures would also create a more fair competition of biofuels with other liquid fuels such as fossil fuels, and other renewable energy sources (cf. NCB, 2011, Chapter 6). Furthermore, they would remove the need for separate greenhouse gas emission reduction targets in certification. Finally, policies using biofuel certification should be aware of certification's 'weak spots' and compensate for them, e.g. through the inclusion of indirect land use change factors such as discussed for the EU RED (Levidow, 2013) and considerations of other potential biomass applications.

Improving the certification process would be more useful for achieving the goals of rural development (and poverty alleviation) through sustainable biofuel production. As mentioned in section two, the certification process should then be tailored to better fit business models that favour smallholder farmers, such as cooperatives (as Fairtrade is aimed at) or outgrower schemes. Also, the formulation of principles and criteria should involve stakeholders from both business and civil society from consuming *and* producing countries. Extra efforts should be made to involve those stakeholders who have urgent and legitimate moral claims, but little power or opportunity to defend their interests, such as smallholder farmers. 'Experiments' with the most vulnerable should be avoided. In these ways, use plans could be developed that allow linking the use of biofuel sustainability certification to effectively achieving societally desired goals.

References

Alcott, B. (2010). Impact caps: why population, affluence and technology strategies should be abandoned. Journal of Cleaner Production 18: 552-560.

Bindraban, P.S., Bulte, E.H. and Conijn, S.G. (2009). Can large-scale biofuels production be sustainable by 2020? Agricultural Systems 101: 197-199.

Bowyer, C. (2011). Anticipated indirect land use change associated with expanded use of biofuels and bioliquids in the EU – an analysis of the national renewable energy action plans. Institute for European Environmental Policy, London, UK.

Cramer, J. (2007). Testing framework for sustainable biomass. Available at: www.agentschapnl.nl/en/node/93396.

European Commission. (2009). Directive 2009/28/EC of the European Parliament and of the Council of 23 April 2009 on the promotion of the use of energy from renewable sources and amending and subsequently repealing Directives 2001/77/EC and 2003/30/EC. Official Journal of the European Union L 140. June 5: 16-62.

Gomiero, T., Paoletti, M.G. and Pimentel, D. (2010). Biofuels: efficiency, ethics and limits to human appropriation of ecosystem services. Journal of Agricultural and Environmental Ethics 23: 403-434.

Guariguata, M.R., Masera, O.R., Johnson, F.X., Maltitz, G. von, Bird, N., Tella, P. and Martinez-Bravo, R. (2011). A review of environmental issues in the context of biofuel sustainability frameworks. Occasional paper 69. CIFOR, Bogor, Indonesia.

Houkes, W.N. and Vermaas, P.E. (2004). Actions versus functions: a plea for an alternative metaphysics of artifacts. The Monist 87(1): 52-71.

Houkes, W.N. and Vermaas, P.E. (2010). Technical functions: on the use and design of artefacts. Springer, Dordrecht, the Netherlands.

Levidow, L. (2013). EU criteria for sustainable biofuels: accounting for carbon, depoliticising plunder. Geoforum 44: 211-223.

Nuffield Council on Bioethics. (2011). Biofuels: ethical issues. Nuffield Press, Abingdon, UK.

Partzsch, L. (2011). The legitimacy of biofuel certification. Agriculture and Human Values 28: 413-425.

Pols, A.J.K. (2010). Transferring responsibility through use plans. In: Van de Poel, I. and Goldberg, D. (eds.) Philosophy and engineering: an emerging agenda. Dordrecht, Springer, the Netherlands, pp. 189-203.

Romijn, H.A., Heijnen, S. and Arora, S. (2013, in press). Standardizing sustainability: certification of Tanzanian biofuel smallholders in a global value chain. In: Lindgreen, A., Sen, S., Maon, F. and Vanhamme, J. (eds.) Sustainable value chain management. Gower, Surrey, UK.

Roundtable on Sustainable Biofuels. (2010). RSB principles & criteria for sustainable biofuel production. Reference code: [RSB-STD-01-001 (Version 2.0)].

Sues, A. (2011). Are European bioenergy targets achievable? An evaluation based on thermoeconomic and environmental indicators. Eindhoven University of Technology, Eindhoven, the Netherlands. Available at http://alexandria.tue.nl/extra2/695295.pdf.

Employing the capability approach in assigning individual responsibility for sustainable development

L. Voget-Kleschin
Kiel University, Institute of Philosophy, Leibnizstr. 6, 24118 Kiel, Germany; voget-kleschin@philsem.uni-kiel.de

Abstract

The capability approach (CA) gives a static picture of how individual capabilities come about. Sustainable development (SD) deals with societal development, that is, a collective and temporal process. Employing the CA in re-conceptualizing SD thus necessitates accommodating that individual behavior individually and collectively creates feedbacks on natural and social conditions. Acknowledging negative feedbacks of individual behavior on natural conditions allows reasoning the necessity for changing individual behavior. However, since SD constitutes a societal challenge, assigning responsibility for SD to individuals alone overburdens individuals. This contribution argues that employing the CA in re-conceptualizing SD allows explaining the occurrence of such overburden and demonstrating how it can be alleviated.

Keywords: sustainable lifestyles, collective responsibility, collective capabilities, recursiveness

Introduction

The capability approach (CA) gives a static picture of how individual capabilities come about. Sustainable development (SD) deals with societal development, that is, a collective and temporal process. Employing the CA in re-conceptualizing SD thus necessitates accommodating that individual behavior individually and collectively creates effects on natural and social conditions (Schultz *et al.*, 2013). Acknowledging negative feedbacks of individual behavior on natural conditions allows reasoning the necessity for changing individual behavior. However, since SD constitutes a societal challenge, assigning responsibility for SD to individuals alone overburdens individuals. This contribution analyzes if and in how far employing the CA in re-conceptualizing SD allows assigning individual responsibility for collective sustainable development. Furthermore, the paper argues that in addressing the capability approach, sustainability and (more) sustainable lifestyles respectively, it is fruitful to distinguish between conceptual and normative issues.

Distinguishing conceptual and normative issues

Prescriptive arguments can be depicted in the form of a practical syllogism, that is, as consisting of prescriptive and descriptive premises and a conclusion. This is why ethical decision making is not only about reviewing and reasoning norms and values, but also about reviewing and reasoning descriptive or conceptual information (cf. Dietrich, 2009). In the following contribution, I will use this distinction as an analytical lens for discussing if and in how far the CA can be employed in assigning individual responsibility for collective sustainable development.

The capability approach

Normative issues

The CA primarily constitutes a metric of justice. It holds that what makes human life valuable are the beings and doings human beings achieve. However, the CA does not claim that evaluating humans' quality of life should focus on functionings. Rather, it argues for a focus on capabilities, that is, freedoms to achieve certain functionings. Thus, the CA holds that for human beings to be able to *choose* constitutes an essential aspect of what it means to live a good human life. Nevertheless, such choice must not draw on unscrutinized conviction. Rather, Amartya Sen persistently emphasizes the importance of reasoning with oneself and with others. For him, capabilities are about what people 'value and *have reason to value*' (Sen, 2007: 18). Finally, it is important to note that the CA is firmly committed to normative individualism.

Conceptual issues

In arguing in favor of capabilities, the CA draws on an account of what it means to be a human being. That is, the CA conceptualizes human beings as being able to convert resources into functionings. Furthermore, the CA argues that humans' ability to convert resources into functionings differs, depending on personal heterogeneities, social and environmental conditions (Sen, 2009: 254-256).

In regard to environmental conditions, Sen points out that these 'need not be unalterable – they could be improved with communal efforts, or worsened [...] But an isolated individual may have to take much of [these conditions] as given [...]' (Sen, 2009: 255). I agree but would like to argue that this statement does not only hold in regard to environmental but also in regard to social conditions.

The CA acknowledges two pathways via which social conditions influence human beings: First, social conditions act as *conversion factor*. That is, the CA assumes that social norms, expectations, structures or recognition and the like affect an individual's ability to convert resources into functionings. Second, the CA conceptualizes the conversion of capabilities into functionings as necessitating an act of choice and acknowledges that *social conditions* also *influence decision making*.

Sustainable development

Normative issues

In this paper, I understand sustainability broadly as the claim that all contemporary and future human beings should be able to live a decent or good (in the philosophical term) human life (cf. Christen and Schmidt, 2011). In regard to the normative dimension, the concept of SD is quite unspecific. Nevertheless, there is one way in which the normative dimension of SD can be said to transcend the normative dimension of the CA: While the CA can be read as dealing primarily with the *claims* human beings can make, SD seems to place equal importance on the *duties* (or *responsibilities*) that correspond to these claims. This is not to say that the concept of SD encompasses a specific elaboration of these duties. Rather, responsibility is broadly assigned to 'us' or 'society'. This is also why SD can be described as a (primarily) societal and collective concept.

Conceptual issues

It is constitutive of SD that it conceptualizes the claim that all contemporary and future human beings should be able to live a decent human life as a 'dilemma': the aim to achieve a decent quality of life for contemporary humans potentially threatens the possibilities to achieve a decent quality of life for future humans (Christen and Schmidt, 2011). SD can therefore be characterized by the following descriptive SD-assumptions (cf. Schultz *et al.*, 2013: 118):

- society (and thus individuals as members of society) is/are dependent on the natural dimension;
- the relationship between society and the natural dimension is recursive, that is, human activity partially determines how the state of nature occurs and vice versa; and
- the natural dimension can be characterized as fragile and (leastwise moderately) scarce.

Employing the CA for conceptualizing SD – existing proposals

Recently, several scholars have analyzed in how far SD can be conceptualized within the CA (cf. the contributions in the Journal of Human Development and Capabilities Vol. 14, No. 1 2013). These works show that the CA is able to depict human *dependence* on certain environmental conditions. However, *recursiveness*, i.e. that human use of the environment results in feedbacks onto the environment, is so far only partly accommodated by the CA. Accordingly, some authors proposed to integrate 'feedback-loops' of human behavior on the (natural) base of livelihood into the CA (ibid.). These feedbacks can be positive as well as negative. The notion of feedback-loops refer to the conceptual rather than the normative dimension of the CA. It allows developing an argument in favor of (more) sustainable lifestyles: *if* an individual's behavior in terms of achieved functionings results in negative feedbacks on natural capital, thus diminishing the individual's own as well as contemporary and future other's abilities to lead a decent human life, *then* this individual has a reason for not behaving in this way. Subsequently, I will employ the distinction between normative and conceptual issues for addressing some objections to this claim.

However, before doing so I would like to point to an important caveat: in discussing the notion of sustainable lifestyles, I do not hold that it is possible to define a (group of) lifestyle(s) as sustainable. What I do hold is that it is possible to ordinally rank different lifestyles according to their sustainability. Again, I do not hold that it is possible to give a complete ordering. The less demanding thesis I presuppose here is that it is (sometimes) possible to identify different choices as well as different lifestyles – understood as the individuals decisions as to what functionings to achieve and how to achieve them – as more and less sustainable respectively.

Assigning individual responsibility for SD? The importance of structures as a conceptual issue

An important objection against claims for (more) sustainable lifestyles points out that it is dangerous to conceive of (more) sustainable lifestyles as a substitute or alternative of political commitment to and promotion of SD. This argument builds on the conviction that SD is constitutively and necessarily a political issue and cannot be brought forward without political measures (Grunwald, 2010). I propose that in terms of employing the CA for conceptualizing SD this argument points to the necessity of further elaboration of dependence and recursiveness between an individual's capabilities and achieved functionings on the one hand and social conditions on the other. I will coin this as the importance of structures, using the term in a broad way, encompassing not only formal institutions but also social norms, expectations and structures of recognition.

Dependence: social conditions as conversion factor and social influence on decision making

Social structures effectuate that capabilities for social functionings such as the ability to appear in public without shame or to take part in the life of the community, very much depend on lifestyles of others. Thus, in a society with a very high average standard of living and in a culture where the standard of living determines the individual status to a large degree, a quite high standard of living could be described as subsistence – even though certain goods are not necessary for the material wellbeing of the individual they may be conceived of as merely satisfying its psychosocial needs in regard to recognition and status (cf. Sen, 2007: 71, 2009: 255). This means that interventions aiming at (more) sustainable lifestyles:

- have to take into account the relation of standard of living to social functionings and the cultural and symbolic meaning(s) of owning certain goods; and
- should proceed in a way that allows and possibly even advances accompanying change in the social context.

Recursiveness: collective (cap)abilities for change

As in regard to environmental conditions dependency on social conditions needs to be complemented by the notion of *recursiveness*. While in regard to the environment the focus is primarily on *negative* feedbacks of achieved functionings on environmental conditions, the notion of collective (cap)abilities points towards the possibility of *positive* feedbacks of individual behavior on social conditions. One example would be individual behavior that contributes towards changing structures in a direction that enables or even pushes individuals to live (more) sustainable lifestyles. A more specific concept of how this works is (to my knowledge) lacking in the CA-discourse but might be found in sociology or political science. However, rather than of collective *capabilities* I would speak of collective *abilities*. This is the case because in my view in the terminology of the CA the term capability is inevitably (to be) understood as something which is intrinsically valuable. Meanwhile, in the view that I have proposed here, the ability to collectively contribute towards SD *is* valuable. However, rather than being valuable as such, it is instrumentally valuable in that it is a necessary (though not sufficient) ingredient in bringing about SD. To emphasize this difference, I propose to speak of collective abilities rather than collective capabilities.

Assigning individual responsibility for SD? Normative issues

The normative significance of aggregate results regarding individual responsibility

It will seldom be the decision of a specific individual to live her life according to her ideas of what constitutes a good life that result in harmful consequences. In contrast, harmful consequences will mostly occur as aggregate consequences of the behavior of many individuals. However, if it is not possible to link specific lifestyle choices with specific harmful consequences, how can individuals be hold accountable for these consequences? This issue is linked to the philosophical discourse about collective responsibility (cf. inter alia Parfit, 1984, pp. chapter III and Isaacs, 2011, for a discussion of individual responsibility regarding (mitigation of) climate change cf. the contributions in Ethics, Policy and Environment, Vol. 14, No.1, 2011).

I have cited Sen in arguing that while no individual can on its own change social and structural constraints, every individual can contribute towards changing them. Drawing on this insight, I propose that individual responsibility for SD encompasses three tasks:

- First, individuals are responsible for trying to and engaging in changing the structural constraints in a direction that enables SD (cf. Cripps, 2011).
- Second, individuals are responsible to accept changes in structures and the necessity of lifestyle-changes implied by them (similarly Hourdequin, 2011).
- Third, in this perspective, changing ones individual behavior in the absence of or prior to structural changes nevertheless has its place: In so far as individuals are able to identify and influence the (significant) ecological effects of their actions, they are responsible for limiting these effects. Furthermore, taking up (more) sustainable lifestyles can be conceived of as advancing the possibility of implementing political change – by sending political signals as well as by demonstrating that such lifestyles are possible, acceptable and (hopefully) enjoyable (cf. ibid.).

The danger of overburdening the individual

The CA places high importance on individual's ability to choose a life she has reason to value. Accordingly, one could argue that changing social conditions with the aim of pushing individuals towards (more) sustainable lifestyles illegitimately constrains the ability of individuals targeted by these constraints to choose a way of life they value and have reason to value.

However, Sen understands capabilities not as individuals' opportunities to live a life they value but rather to live a life they value and *have reason* to value. Sen argues that what *we* as a society have reason to value should be decided by democratic deliberation. Applying this general argument to the issue of distinguishing legitimate and illegitimate constraints yields that such a distinction should be drawn in deliberative discourse: Decisions regarding governmental intervention in individuals' lifestyles are decisions that have to be taken by a democratic government. Ultimately, it is citizens themselves who need to impose limits on their own behavior. But why should individuals do so?

One achievement of a conception of SD is that it allows explaining that and how certain human actions harm contemporary and future others. According to the liberal harm principle, people should be able to choose whatever they want only as long as these choices do not harm others. Thus, *if* a conception of SD manages to convince individuals that certain lifestyle choices do result in consequences that harm contemporary and future others, *then* they have a normative reason to politically constrain these choices. In a deliberative process about which capabilities individuals value and have reason to value, the argument that the results of certain choices (will) harm others therefore constitutes a strong reason why not to value the capability to make these choices.

This means first, that reasoning the critique that claims for (more) sustainable lifestyles overburden the individual in that they illegitimately constrain an individual's ability to live a good human life needs to draw on a *shared, societal notion* of what it means to live a good human life – and thus what qualifies as *illegitimate* constraints. Second, if a shared notion of the good is employed in the context of one question (i.e. distinguishing legitimate from illegitimate measures) it is leastwise counter-intuitive to not presuppose the same notion in the context of other questions (i.e. what kind of life qualifies as 'decent' in terms of that we owe to other contemporary and future human beings as well as what we expect tother contemporaries to owe to us). Accepting the need to employ one notion of the good in regard to different questions might drastically decrease the range of notions of the good that can be reasonably argued for in deliberative discourse. Against the background of contemporary political discourse which predominantly conceives of (any) measures constraining individual choice as illegitimate, this argument opens up space for publicly discussing what kind of constraints citizens are prepared to impose on their own behavior.

Conclusion

In this paper, I have analytically distinguished normative and conceptual issues in analyzing if and in how far employing the CA in re-conceptualizing SD allows assigning individual responsibility for collective sustainable development. In conclusion, the CA can be employed in specifying the concept of SD. Current work on these issues focuses on dependence and recursiveness regarding environmental conditions. However, assigning individual responsibility for collective/aggregate results of individual behavior points to the necessity for further elaborations. First, in regard to the conceptual dimension this is linked to the relation between social conditions on the one hand and capabilities and achieved functionings on the other. The CA encompasses ample conceptual considerations regarding individual's *dependence* on social conditions. However, it does not yet sufficiently allow for depicting *recursiveness*. Acknowledging that achieved functionings can (positively or negatively) feedback on social conditions allows for depicting the notion of collective abilities for SD. Second, in regard to the normative dimension I discussed two issues: While the question of overburdening the individual can be solved by pointing to deliberative discourse as a means of distinguishing legitimate from illegitimate measures, the theoretical question if and in how far a notion of collective responsibility exists and how it relates to (corresponding, derivative or complementary) individual duties is much less clear. I have thus demonstrated that beside the necessity to complement the conceptual dimension of the CA so as to accommodate issues of dynamics and the system level, making sense of the notion of individual responsibility also necessitates investing further normative premises regarding the notion of collective and individual responsibility respectively.

References

Christen, M. and Schmidt, S. (2011). A formal framework for theories of sustainability. Sustainable Development 20(6): 400-410.

Cripps, E. (2011). Climate change, collective harm and legitimate coercion. Critical Review of International Social and Political Philosophy 14(2): 171-193.

Dietrich, J. (2009). Zum Verhältnis von Ethik und Empirie: ein Überblick am Beispiel der Schmerzmedizin. In: Vollmann, J. (ed.) Klinische Ethik. Aktuelle Entwicklungen in Theorie und Praxis. Campus, Frankfurt, Germany, pp. 225-239.

Grunwald, A. (2010). Wider die Privatisierung der Nachhaltigkeit. Warum ökologisch korrekter Konsum die Umwelt nicht retten kann. Gaia 19(3): 178-182.

Hourdequin, M. (2011). Climate change and individual responsibility: a reply to Johnson. Environmental Values 20(2): 157-162.

Isaacs, T. (2011). Moral responsibility in collective contexts. Oxford University Press, Oxford, UK.

Parfit, D. (1984). Reasons and persons. Clarendon Press, Oxford, UK.

Schultz, E., Christen, M., Voget-Kleschin, L. and Burger, P. (2013). A sustainability-fitting interpretation of the capability approach: integrating the natural dimension by employing feedback loops. Journal of Human Development and Capabilities 14(1): 115-133.

Sen, A.K. (2007). Development as freedom. 13th ed. Oxford University Press, Oxford, UK.

Sen, A.K. (2009). The idea of justice. Allen Lane, London, UK.

Crop protection in horticulture: how to rescue growers from punishment for shortfall of control agents?

J.S. Buurma and V. Beekman
LEI Wageningen UR, P.O. Box 29703, 2502 LS The Hague, the Netherlands; jan.buurma@wur.nl

Abstract

The EU and its Member States have been working on reducing the use and risk of pesticides for decades. This has largely been achieved by regulating the authorisation of pesticides under Directive 91/414/EEC and Regulation (EC) 1107/2009. Consequently, the number of active ingredients authorised for pest and disease control in ornamental crops (e.g. flowers) has dramatically decreased. Now, growers face a shortfall of control options, and there are growing numbers of reports on unlawful use of pesticides in ornamental crops. The Food Inspection Authorities try to change this by imposing fines on these growers. Furthermore, retail companies start to impose restrictions on pesticide residues on ornamental products. In case of exceeding they reject the products supplied. On the one hand, the growers are thus justifiably punished for unlawful use of pesticides. On the other hand, their violations indicate that growers find themselves in a desperate position. The question is how this unsatisfactory situation can be solved. The objective of this paper is to improve understanding of the positions and interests of the involved parties in relation to pesticides and pest control. We therefore study how pesticide use in ornamental crops is framed by the various parties involved. Furthermore, power relations in both the knowledge and value chain are studied. Examples of framing pesticide use are: my pests are difficult to control, surface water quality is below standards, working in greenhouses should be safe, authorisation of control agents is too expensive, ornamental crops should be safe for consumers, chemical pesticides close the door to biological control agents, growers should apply decision support systems. These examples illustrate the frictions among the parties on the playing field of crop protection. The aim is to explore some options for sustainable development of crop protection in floriculture. Our suggestion is that new interactions and initiatives have to be developed between flower growers, value chain partners and/or knowledge partners. Bringing partners together for collective action under a national agreement or in a public-private partnership for plant health research are considered to be the most promising options. The lesson learned is that social innovation needs special attention in governance of sustainable crop production.

Keywords: pesticides, value chain, framing, power, knowledge system

Introduction

The EU and its Member States have been working on reducing the use and risk of pesticides for decades. This has been largely achieved by regulating the authorisation of pesticides under Directive 91/414/EEC and Regulation (EC) 1107/2009. Furthermore, the Water Framework Directive (2000/60/EC) adds urgency to taking pesticide risk reduction seriously, especially in the Netherlands with its many ditches and canals. The number of active ingredients and plant protection products authorised in the Netherlands strongly decreased with the introduction of Directive 91/414/EEC. According to De Snoo and Vijver (2012) the number of authorisations dropped from 1,700 in 1991 to 700 in 2005.

A complicating factor for horticultural crops is the shift from broad spectrum pesticides (controlling several pests simultaneously) to biological control and selective pesticides (controlling specific pests only). As a result relatively unknown pests emerge, which are no longer controlled by broad spectrum pesticides (Boertjes *et al.*, 2003). Examples of such pests are woolly aphids and scale aphids in roses. A recent survey (LTO Groeiservice, 2011) among rose growers in the Netherlands revealed severe bottlenecks in the control of woolly aphids and scale aphids. Both biological control agents (according to 96% of the respondents) and chemical control agents (according to 91% of the respondents) were found inadequate to control the aphids.

As a result the rose growers are inclined to seek refuge in unlawful use of pesticides. An analysis of inspection reports of the Food Inspection Authorities over the period 2006-2009 (Smit, 2011) revealed that 86% of the ornamental growers had unauthorised pesticides in stock or in use. In the other sectors (outdoor crops) 'just' 40% of the growers committed this offence. Unlawful use of pesticides -to an increasing degree- has consequences on the market. Retailers want to reduce the risks of social unrest about pesticide residues on horticultural products, including flowers and plants. Therefore they monitor fruit, vegetables and also flowers and plants on compliance with maximum residue limits (MRLs). In case of exceeding these requirements the products are rejected (E. Klein, personal communication).

In summary, inspection authorities and value chain partners justifiably punish flower growers for unlawful use of pesticides. The flower growers in turn face outbreaks of emerging pests for which no acceptable control options are available yet. Somewhere in the innovation system of crop protection there seems to be a mechanism that causes this unsatisfactory situation. In the next sections of this paper we will try to improve understanding of this mechanism and to explore some options for improvement.

Methodology

The first author of this paper is involved in action research on the societal aspects of innovation in crop protection in the Netherlands. He was involved in working groups for the National Action Plan (2013-2018) for sustainable crop protection in the Netherlands. Furthermore, he was involved in the project Sustainable Crop Protection Roses 2020 of the Horticultural Marketing Board in the Netherlands. In the action research he collected reports, presentations and contributions of the parties involved. Through content analysis of these documents he improved understanding of the positions and interests of the various parties involved in the aforementioned activities (see Buurma and Beekman (2012) for more methodological details).

Power relations

In the period 1995-2008 the Netherlands experienced a public debate on pesticides and crop protection. This public debate was described and analysed in Buurma (2011). Food safety and registration policy were the main issues in the debate. The debate on food safety resulted in retail companies imposing residue requirements on trade companies, farmers and growers. In fact, the retailers took over the inspection on compliance with MRLs from the government. The debate on pesticide registration focused on the compliance with environmental criteria in the authorisation of pesticides. This debate resulted in the notion of economic essentiality, implying a trade-off between environmental and economic interests. Under the banner of economic essentiality specific applications of pesticides are accepted in crops, which otherwise would have become economically infeasible. In a further reflection Buurma (ibid.) related the success of the debate to the diminishing power position of the agricultural sector in the Netherlands. Due to saturation of agricultural and horticultural markets the value chain partners (retailers and wholesalers) gained a power position in which they could impose requirements on growers at the beginning of the value chain. The power positions and interests of the partners in the value chain of flowers are depicted in Figure 1.

The power in the value chain runs from right to left (grey chevrons) in Figure 1. The interests of the distinct value chain partners are specified in the green rectangles. In a saturated market the consumer can choose according to preferences. In case of flowers colour and ripeness, long vase life, product safety

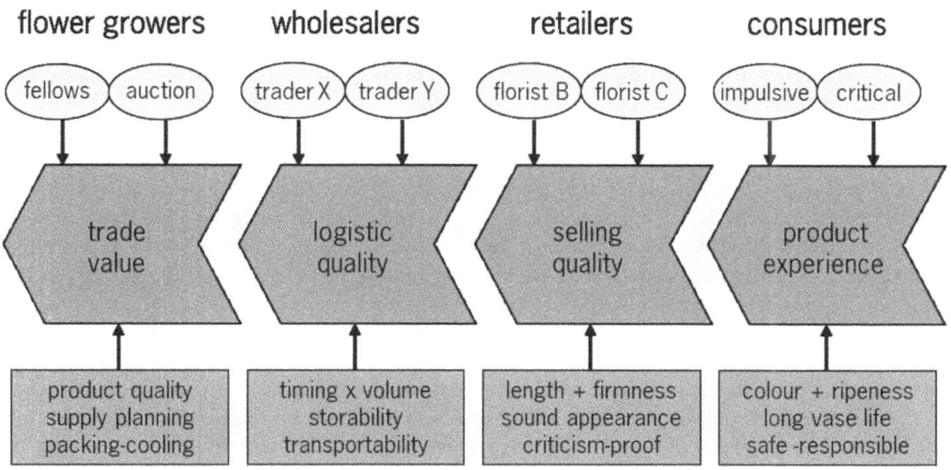

Figure 1. Power relations and interests of partners involved in the value chain of flowers.

and social responsibility are relevant factors (Benninga, 1998). For the florist strong flowers, sound appearance and product safety are important factors, defining the selling quality (Marissen en Benninga, 1999). For wholesalers the timing and volume of the supply are crucial, in addition to storability and transportability. All these factors together are shaping the license to deliver for the flower grower and his colleagues.

Figure 1 displays the position of the grower as a trade partner in the value chain. In order to produce the required product the grower depends on support from the knowledge chain. This knowledge chain is depicted in Figure 2.

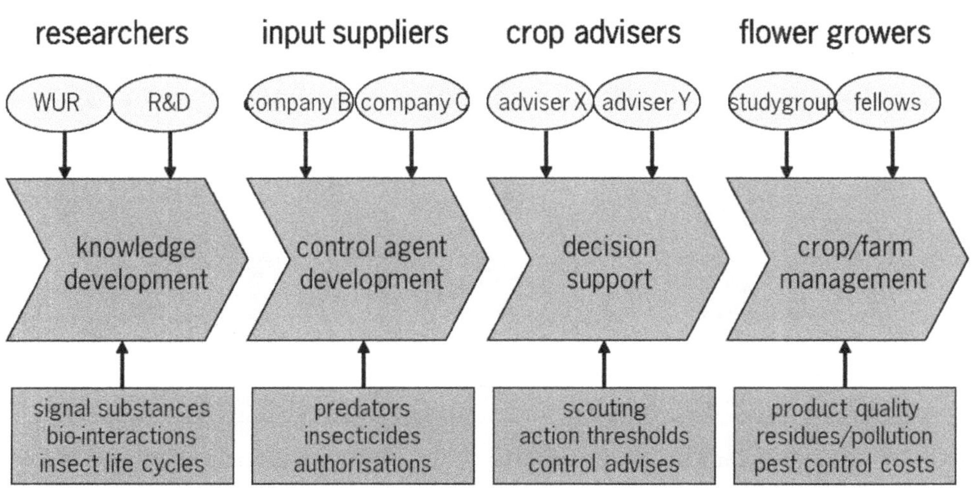

Figure 2. Power relations and interests of partners in the knowledge chain of crop protection.

The power in the knowledge chain runs from left to right (grey chevrons). The interests of the distinct knowledge chain partners are specified in the green rectangles. Biological control is an important issue in crop protection research for horticulture in the Netherlands. Crop protection scientists conduct research on signal substances and bio-interactions and improve knowledge of insect life cycles. Manufacturers of biological and chemical control agents focus on the development of predators and pesticides. They invest time and effort in getting their control agents authorised for application in practice. Crop advisers provide services in monitoring pest/disease levels (scouting) in greenhouses, comparing them with action thresholds and advising control measures when needed. The challenge for growers is achieving the required product quality, meeting the residue requirements of value chain partners, meeting the water quality requirements of water boards, and keeping pest control cost effective.

The directions of the grey chevrons in Figure 1 and Figure 2 indicate a two-sided dependant position of the flower grower. In Figure 1 he has to achieve a license to deliver from the value chain partners. In Figure 2 he has to wait for the ability and willingness of the knowledge chain partners to provide new control options. Our appraisal of the current situation is that the technology developments in the knowledge chain of crop protection do not keep up with the product quality requirements of the value chain. As a result the flower grower are unable to control new, emerging pests in a socially responsible way.

Framing of pesticides and crop protection

The challenge in this unbalanced situation is to better align the dynamics of the value chain and the knowledge chain. The dynamics in interactions between parties in social processes are strongly determined by the way in which a situation is framed (Schnabel, 1992). In his work for the National Action Plan the first author identified several combinations of parties and framings (Figure 3).

Obviously the involved parties have quite different opinions on what is problematic with pesticides and crop protection. Starting from their frames the parties involved try to bend the dynamics of the public debate in their own direction and according to their own interests. That is normal behaviour in an individualised society. The challenge is the lack of coordination among the parties. The representatives of the general public want to reduce the use and risks of pesticides. The knowledge chain partners committed themselves to provide new control options but they just partly did their job. The flower growers still struggle with woolly aphids and scale aphids (LTO Groeiservice, 2011). In fact the flower growers experience that pests are running out of control and are thus in urgent need of new control options.

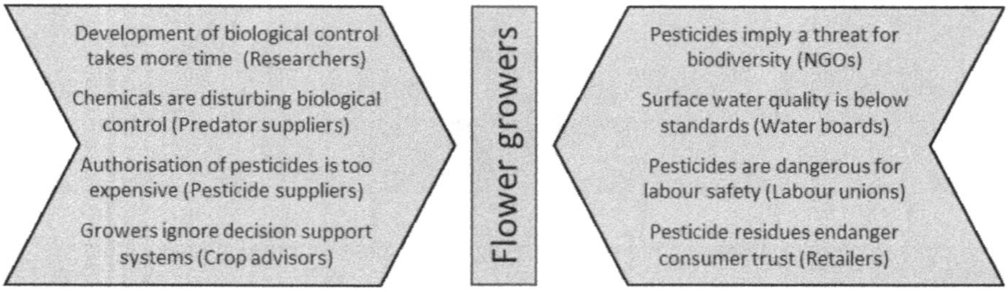

Figure 3. Framings of pesticides and pest control in knowledge chain (left) and value chain (right).

Reflections on improvement options

In the previous sections we saw flower growers in a dependent position between value chain partners and knowledge chain partners. The growers are waiting for the knowledge chain partners to provide new control options for woolly aphids and scale aphids. A complicating factor is that the various parties in value chain and knowledge chain are framing the situation according to their own interests. This implies that new interactions and initiatives have to be developed between the various parties. Starting from the main keywords in Figure 3, we identify the following options:

- Bringing retailers and growers together on the common interest of gaining access to higher market segments by supplying flowers with low pesticide residue levels. This may improve the return on investments in new (biological) control options.
- Bringing NGOs and growers together on the common interest of biodiversity. This may result in awarding good farming certificates for distinct levels of biological control and positive price effects for the flower grower.
- Bringing growers' unions, value chain partners and knowledge partners together for collective action under a national agreement. This may result in a collective sense of urgency to serve the common interest of sustainable crop protection.
- Bringing knowledge chain partners together in a public-private partnership for research and development on plant health in floriculture, under coordination of the national growers' organisation. This may improve the consistency in the knowledge chain.

Now, the question is which options provide the best chances for sustainable development of crop protection in floriculture. We consider the options (3) and (4) as most promising on the short term, since options (1) and (2) depend on hitherto unavailable control options. We learned from this exercise that social innovation needs attention in governance of sustainable crop production.

References

Benninga, J. (1998). Causes of differences in auction prices of roses. Report 141. Experimental Station for Floriculture and Greenhouse Vegetables, Aalsmeer, the Netherlands (in Dutch).

Boertjes, B.C., Fransen, J.J. and Marissen, N. (2003). Inventory of problems with woolly aphids and scale aphids in floriculture crops. Applied Plant Research, Wageningen, the Netherlands (in Dutch).

Buurma, J.S. (2011). Changing the crop protection or pesticide use regime in the Netherlands: an analysis of public debate. In: Vellema, S. (ed.) Transformation and sustainability in agriculture; connecting practice with social theory. Wageningen Academic Publishers, Wageningen, the Netherlands.

Buurma, J.S. and Beekman, V. (2012). Adressing farmers or traders: socio-ethical issues in developing a national action plan for sustainable crop protection. In: Potthast, T. and Meisch, S. (eds.) Climate change and sustainable development: ethical perspectives on land use and food production. Wageningen Academic Publishers, Wageningen, the Netherlands, p. 280-286.

De Snoo, G.R. and Vijver, M.G. (2012). Pesticides and water quality. Leiden University, Centre for Environmental Sciences, Leiden, the Netherlands (in Dutch).

Janssens, S.R.M., Stokreef, J.W. Smit, A.B. and Prins, H. (2012). Evaluation of the policy document on sustainable crop protection. Sub-report compliance. Report 2011-091. LEI Wageningen UR, the Hague, the Netherlands (in Dutch).

LTO groeiservice (2011). Results of survey on integrated crop protection in roses. Memorandum April 2011. LTO, Bleiswijk, the Netherlands (in Dutch).

Marissen, N. and Benninga, J. (1999). Comparative studie on storability of roses. Report 193. Experimental station for floriculture and Greenhouse Vegetables, Aalsmeer, the Netherlands (in Dutch).

Schnabel, P. (1992). Erving Goffman. In: Rademaker, L. (ed.) Hoofdfiguren uit de sociologie 3. Het Spectrum, Utrecht/ Antwerpen, the Netherlands/Belgium, pp. 26-44.

'Oil versus fish' in northern Norway: perspectives of the market, the law, and the citizen

A. Blanchard
University of Bergen, Centre for the Study of the Sciences and the Humanities, Allegaten, 34, Postboks 7805, 5020 Bergen, Norway; anne.blanchard@svt.uib.no

Abstract

The coastal marine area constitutes one of humanity's most important food baskets. However, food production must compete with other uses of scarce coastal resources, demanding high-stakes trade-offs. This paper focuses on one particularly heated trade-off: that of 'oil versus fish' in the Lofoten area of northern Norway. Lofoten is an important spawning ground for numerous fish species, but also contains petroleum reservoirs estimated at 1,300 billion barrels. In order to highlight the complexity of this trade-off this paper examines it from three broad perspectives: the market, the law, and the citizen. From a market perspective, the 'oil versus fish' trade-off is relatively simple, and can be reduced to a single monetary measure. Petroleum resources are today worth more than fish, and have brought 2,800 billion NOK to the Norwegian state since 1970. From this market perspective the trade-off is framed as a technical issue: the best decision is to allow oil production in Lofoten, and trust in the ability of current and future technologies to deal with an oil spill. From the perspective of the law, the trade-off is governed by the 'Barents Sea – Lofoten Management Plan' of 2006, and its updated version of 2011. This policy document recommends a precautionary approach to petroleum activity in the Lofoten area, which extends to a moratorium on petroleum activities, and suggests standards for minimum environmental harm, such as the 'zero discharge' of chemical substances to sea. Finally, from the citizens' perspectives, the trade-off is more complex. Indeed, it does not implicate only one measure, but as many measures as there are citizens concerned; ranging from culture, to identity, socio-economic development or the protection of nature's intrinsic value. The best political decision is less straightforward as it needs to take into account these various arguments and measures. These three perspectives show the 'oil versus fish' trade-off as multifaceted, and in making a decision, all of these perspectives need to be taken into account.

Keywords: petroleum, fish, multifaceted trade-off, Lofoten

Setting the scene: Lofoten as an area with multiple competing resources

The coastal marine area constitutes one of humanity's most important food baskets (Costanza *et al.*, 1997), with one billion people relying on seafood as their primary source of protein (Gutiérrez *et al.*, 2011). However, food production must compete with other uses of scarce coastal resources, demanding high-stakes trade-offs. This is the case in the Lofoten area of northern Norway, where fish has to compete with petroleum production.

Lofoten is an archipelago of about 1,200 km^2, which hosts a population of 24,500 inhabitants. It is described by the 'Norwegian integrated management plan for the Barents Sea – Lofoten' as both a very valuable and vulnerable region (NME, 2006), owing to its ecological and cultural diversity. Indeed, the area is a key spawning ground for Northeast Arctic cod and Northeast Arctic haddock, the eggs and larvae of which then drift north on a narrow advection route along the coast of Lofoten to join the Barents Sea (Hjermann *et al.*, 2007). Lofoten is also a nursery for numerous fish species including cod, herring and haddock (Olsen *et al.*, 2010). The area thus supports a successful traditional fishing industry. Since the Viking Age, cod from Lofoten has been traded and has made Norway an important trading partner for Europe (Jentoft and Kristoffersen, 1989). Northeast Arctic cod is economically the most important fish stock in Norway: 340,000 tons were caught in 2011, with a value of ca. 4 billion NOK (NDF, 2011); ranking Norway as the third largest exporter of fish worldwide (Hjermann *et al.*, 2007).

Beyond fish, Lofoten also hosts a very high density of migratory seabirds (Barrett *et al.*, 2006), more than 20 species of marine mammals (Larsen *et al.*, 2001), and the world's largest deep-sea coral reef, of about 100 km^2 (Forsgren *et al.*, 2009). In addition, Lofoten is known for its unspoiled landscapes that figure on the UNESCO list for the 'Protection of the World Natural and Cultural Heritage'. This natural diversity supports a lucrative tourism sector, evaluated as worth 2.75 billion NOK for the period from 2005 to 2011 (NME, 2011).

On the other hand, exploratory drillings in Lofoten have indicated the presence of important petroleum reservoirs amounting to 1,300 billion barrels of oil equivalent. The petroleum sector has since the 1960s been the most important economic activity in Norway, and has brought about 2,800 billion NOK to the Norwegian state (Dale, 2011). Due to its ecological value, Lofoten has not been opened yet to petroleum production. However, petroleum activity in this area brings the prospect of socio-economic development, with new infrastructure and the creation of between 720 and 1,340 jobs, depending on the scenario chosen (Dale, 2011).

We can see that this 'oil versus fish' trade-off is very complex and is surrounded by uncertainties regarding the impacts of petroleum activity in Lofoten. This complexity and uncertainty legitimates a broad range of perspectives, which all focus on different aspects of the trade-off. In this paper we will look at three such perspectives: that of the market, the law, and the citizen. We will explore how these three perspectives frame the trade-off in very different ways, before discussing in the conclusion the need to look at all three perspectives to achieve a holistic understanding of the 'oil versus fish' debate.

The market perspective: a mere technical issue

From a market perspective, endorsed by the petroleum industry, the 'oil versus fish' trade-off is relatively simple, and can be reduced to a single monetary measure. The petroleum industry develops two main arguments using this measure, for encouraging petroleum activities in Lofoten.

First, as shown by the numbers above, petroleum resources are today worth more than fish and tourism put together. In political debates, the petroleum industry puts forward what Dale calls the 'tale of prosperity and progress' (Dale, 2011: 95). Petroleum activities will continue to provide further development and wealth to Norway, and the projected benefits of petroleum activity in the area of Lofoten could amount to 105 billion NOK (NME, 2011).

The second argument is related to a 4% discount rate, which indicates that a barrel of oil today is slightly more valuable than a barrel of oil in the future. Petroleum should therefore be extracted as soon as possible. However, this discount rate is fairly low. This suggests that Norway recognises petroleum as an important resource for the country, both for present and future generations. Indeed, when the oil runs out, a 'poorer future for Norway' is feared (Gerhardsen, 2012).

When faced by critiques on potentially severe consequences of petroleum activity on the vulnerable ecosystems of Lofoten, the petroleum industry displays two arguments in which they frame the 'oil versus fish' trade-off as a technical issue.

First, they emphasise the extremely low probability of a 'worst-case scenario' oil spill: about once chance in 10 million (Bergsli *et al.* 2009). Accordingly, the risk of impacts on marine ecosystems should therefore be considered as acceptable, and the co-existence of petroleum activity and fisheries is seen as possible. However, we can challenge whether the use of statistics in cases where the probability of an accident is very low but the potential impacts are very serious, is relevant. Indeed, some have framed accidents as 'normal' (Perrow, 1999) and the history of petroleum activity worldwide has illustrated this. From this angle, the trade-off is less a matter of balancing how much oil extraction and fisheries can co-exist in Lofoten, but rather a choice between one or the other: fish or oil. Another critique is that it is advantageous for the petroleum industry to talk in terms of worst-case scenario, which has a very low probability; rather than in terms of everyday activity. Indeed, routine petroleum activity may also impact on the marine environment, and disagreements arise over whether fish are affected by the discharge of chemicals, whether marine mammals are perturbed by sound surveys, and whether benthic communities living on the seabed are disturbed by drillings (see e.g. Meier *et al.*, 2008; Popper and Hastings, 2009).

The second argument put forward by the petroleum industry is that of reliable technologies and 'oil spill preparedness', which allow them to 'detect anomalies at an early stage, and [...] prevent or limit the impact of accidents' (NMFA, 2012: 106). Technology is the toolbox for progress according to a representative of the Norwegian Oil Industry Association: 'Some people argue that these sea areas should be protected for eternity. We feel this is to close the door to the future. The world does move forward' (Dale, 2011: 96). However, we can here reassert the 'normality' of accidents, sometimes caused by interactions that are 'not only unexpected, but are incomprehensible for some critical period of time' (Perrow, 1999: 9). This was the case with the Ekofisk field accident in 1972, which released up to 20,000 m^3 of petroleum into the North Sea. 105 days passed before the origin of the blow-out was determined and fixed, making the Ekofisk accident the biggest blow-out in Norway.

Another shortfall of the market perspective is that decision-making processes are mostly steered by a focus on economically important fish stocks, and less on other environmental and non-monetary aspects. We see below that the citizens' arguments go beyond economic aspects, and frame the Lofoten question not as a mere technical issue, but as a complex, multifaceted trade-off.

The law perspective: precaution… but how?

As seen above, decision-making around the 'oil versus fish' trade-off are focussed on a worst-case scenario oil spill on the marine environment, and precisely on the impacts of such oil spill on the economically most important fish stock: Northeast Arctic cod. Petroleum activities present risks to fish eggs and larvae which are the most sensitive to oil contamination. While laboratory studies have shown the capacity of adult fish to detect low concentrations of petroleum and avoid them (Farr *et al.*, 1995), fish eggs and larvae drift with currents in the upper layers of the sea and are not able to escape polluted water (Hjermann *et al.*, 2007). These risks are compounded in Lofoten given that Northeast Arctic cod have short, intensive spawning seasons, and the larvae gather in large concentrations in the very localised advection route along the coasts (Brude, 2005). The worst-case scenario would be to have a major oil spill during the spawning season, with the potential result of wiping out a whole year-class (all fish born in the same year) of cod.

Given these risks, petroleum activities in the area surrounding Lofoten are managed by the precautionary principle. This principle is mentioned in the integrated management plan for the Barents Sea – Lofoten (NME, 2006) and defined as such: 'When a decision is made in the absence of adequate information on the impacts it may have on the natural environment, the aim shall be to avoid possible significant damage to biological, geological or landscape diversity' (p. 17).

However, implementing the precautionary principle is problematic given that the 'oil versus fish' trade-off is situated in the 'high impacts, low probability' category. This opens an arena for negotiation of how the precautionary principle should be applied over time and across space. An example of such negotiation is found in the integrated management plan, where the precautionary principle finds expression in a policy measure of 'zero discharge to sea'. Discharges of chemicals used and/or created during petroleum production (such as produced water or drilling muds) are hereby prohibited. In 2006, when this measure was adopted, petroleum companies claimed that they had technologies to re-inject these chemicals back into the reservoirs. As a result, the 'zero discharge' principle was rephrased in the updated plan of 2011 (NME, 2011), to 'zero *harmful* discharge': 'As a rule, oil and substances that may be environmentally hazardous may not be discharged to the sea' (p. 64).

In Norway, there is a broad acceptability of the precautionary principle for managing petroleum activities. However, since its application is adapted to each petroleum field or area, there is a risk of distorting it from what it initially means. Therefore, there is a need in each decision-making process to make explicit why the precautionary principle is applied in a particular way, and highlight the values and interests of the various actors concerned by the 'oil versus fish' trade-off.

The citizens' perspective: a complex constellation of interests and values

The question of whether to allow petroleum development in Lofoten has been at the forefront of political debates since 2005 (Dale, 2011). The uncertainties linked to the impacts of a worst-case scenario oil spill and everyday petroleum activity, as well as the high stakes that are linked to this 'oil versus fish' trade-off, give rise to very different interpretations of the scientific 'facts' by the various groups of citizens concerned. In 2010, a scientific report (von Quillfeldt, 2010) carried out by more than 20 research institutions was issued to serve as a knowledge base to update the integrated management plan for Lofoten of 2006. The various actors, including the petroleum industry, fishery sector, the tourism industry, environmental NGOs and local citizens all interpreted the results of this report in such ways that they would support their own perspectives and concerns (Dale, 2011). These various interpretations are steered by different interests and values; we roughly map some of these, keeping in mind that the reality is far more complex.

For understanding local citizens' perspectives, it is important to remember that power in Norway is centralised to Oslo. Therefore, in northern Norway, and particularly in Lofoten, there is some resistance to accept political decisions designed by the 'south'; i.e. by people who do not know and experience Lofoten as much as the local citizens. Particularly, a large part of Lofoten citizens strongly value the identity of Lofoten as a fishing culture: 'we don't see ourselves as a petroleum nation, but as a fishery nation' (Dale, 2011: 98).

Environmental NGOs such as the 'Bellona Foundation' put forward arguments for the conservation of nature. On their website (bellona.org), the intrinsic value of nature is emphasised: 'the polar bear can feel the ice melting underneath its paws', with the objective of 'preventing petroleum activities in the High North'. Similarly, the 'People's movement for an oil free Lofoten' insist on the intrinsic value of nature that could not be replaced by any monetary or technological alternative: 'Lofoten, Vesterålen and Senja have a fantastic natural environment and ecosystems of unimaginable importance' (folkeaksjonen.no). This social movement claims that petroleum activities in Lofoten could not coexist with tourism and fishing; and supports sustainable fisheries.

Fishermen in Lofoten, represented by the Norwegian Coastal Fishermen's Union, voice their fear of an oil spill jeopardising their fish stocks. This would have negative consequences both in the short-term, with an immediate loss of income, and in the long-term, with a declining reputation of Lofoten's fish. They also argue that fisheries are an important source of work, income and settlement along the

Norwegian coast, especially in the north, which is increasingly deserted by younger generations. For them as well, 'coexistence with the petroleum sector is utopia' (Hamnes, 2009).

The tourism industry, represented by the region's joint promotional body 'Destination Lofoten', stresses its valuable contribution to the employment of young people and particularly women, and its input in the maintenance of a decentralised industrial and population structure. Allowing petroleum activities in Lofoten would break the pristine image of the area, by the simple fact of introducing petroleum platforms on the horizon.

The citizens' arguments go beyond the mere economic development or technological progress put forward by the petroleum industry, to reach dimensions like cultural identity, traditions or the intrinsic value of nature. The different interests and values of the various groups of actors are all legitimate, and should be discussed in decision-making processes to achieve a sustainable future for Lofoten.

Conclusion

Exploring the 'oil versus fish' trade-off through the perspectives of the market, the law and the citizens, allows us to highlight its complexity, and the many, often intertwined, dimensions that it implies. Today, political discussions around this trade-off are dominated by a market perspective, which leads decision-makers to argue for safer technologies, and focus on only few aspects of the trade-off: those which are economically relevant (i.e. Northeast Arctic cod). This paper constitutes a critical challenge, as it argues that in order to make sustainable decisions regarding the future of Lofoten, all three broad perspectives described in this paper should be included in political decisions. Indeed, we have seen that these three perspectives of the market, the law and the citizens frame the issue in their own particular way; narrowly seeing it as a technical matter, a case for precaution, or a value conflict. In the past we have seen that decision-making processes have been biased towards a technical approach, supported by science and economics. This paper argues that no one perspective is sufficient on its own as this trade-off gathers all three aspects.

In order to include all three perspectives, participatory mechanisms such as hearings are already in place, and fishermen, Lofoten citizens, environmental NGOs and the tourism industry often voice their perspectives in the media. What is lacking, however, is transparency in the decision-making process. With so many uncertainties, on what basis are decisions made? Are they based in science, on economic projections or with attention to the citizen's interests? What is the role played by power in the policy processes? Is it to the most powerful actor (the petroleum industry in our case) that the benefit of the doubt is given when there is uncertainty? Attention to these key questions of power, uncertainty and the plurality of values and interests should be increased and made more transparent in the 'oil versus fish' trade-off.

References

Barrett, R.T., Lorentsen, S.H. and Anker-Nilssen, T. (2006). The status of breeding seabirds in mainland Norway. Atlantic Seabirds 8(3): 97-126.

Bergsli, A., Rudberg, A. and Kruuse-Meyer, R. (2009). Simulering av oljeutblåsning utenfor Lofoten og Vesterålen (No. 125POAF-6). Høvik, Norway: Det Norske Veritas (DNV).

Brude, O.W. (2005). Areas vulnerable to acute oil pollution in the Norwegian Barents Sea (No. 168567). Norway: Det Norske Veritas (DNV).

Costanza, R., d'Arge, R., De Groot, R., Farber, S., Grasso, M., Hannon, B., Limburg, K., Naeem, S., O'Neill, R.V., Paruelo, J., Raskin, R.G., Sutton, P. and Van den Belt, M. (1997). The value of the world's ecosystem services and natural capital. Nature 387: 253-260.

Dale, B. (2011). Securing a contingent future: how threats, risks and identity matter in the debate over petroleum development in Lofoten, Norway. University of Tromsø (UiT), Tromsø.

Farr, A.J., Chabot, C.C. and Taylor, D.H. (1995). Behavioral avoidance of fluoranthene by fathead minnows (Pimephales promelas). Neurotoxicology and Teratology 17(3): 265-271.

Forsgren, E., Christensen-Dalsgaard, S., Fauchald, P., Järnegren, J. and Næsje, T.F. (2009). Norwegian marine ecosystems – are northern ones more vulnerable to pollution from oil than southern ones? Norwegian Institute for Nature Research (NINA), Trondheim, Norway.

Gerhardsen, W. (2012). En fattigere fremtid for Norge? Aftenposten. Retrieved from www.aftenposteninnsikt.no/1-januar-2012/en-fattigere-fremtid-norge.

Gutiérrez, N.L., Hilborn, R. and Defeo, O. (2011). Leadership, social capital and incentives to promote successful fisheries. Nature 470(7334): 386-398.

Hamnes L. (2009). Varsler aksjoner mot seismikkskipene. Teknisk Ukeblad Online.

Hjermann, D.Ø., Melsom, A., Dingsør, G.E., Durant, J.M., Eikeset, A.M., Røed, L.P., Ottersen, G., Storvik, G. and Stenseth, N.C. (2007). Fish and oil in the Lofoten–Barents Sea system: synoptic review of the effect of oil spills on fish populations. Marine Ecology Progress Series 339: 283-299.

Jentoft, S. and Kristoffersen, T. (1989). Fishermen's co-management: The case of the Lofoten fishery. Human Organization 48(4): 355-365.

Larsen, T., Boltunov, A., Denisenko, N., Denisenko, S., Gavrilo, M., Mokievsky, V., Nagoda, D., Spiridonov, V. and Von Quillfeldt C. (2001). The Barents Sea ecoregion: a biodiversity assessment. WWF, Olso, Norway.

Meier, S., Morton, H.C. and Misund, O.A. (2008). Does the discharge of alkylphenols in produced water cause oestrogenic endocrine disruption among fish in the North Sea? Exploration & Production – Oil & Gas Review 6: 64-67.

NDF (2011). Economic and biological figures from Norwegian fisheries. Norwegian Directorate of Fisheries, Bergen, Norway.

NME (2006). Integrated management plan of the marine environment of the Barents Sea and the sea areas off the Lofoten Islands. Report No. 8 to the Storting. Norwegian Ministry of the Environment, Oslo, Norway.

NME (2011). First update of the integrated management plan for the marine environment of the Barents Sea – Lofoten area. Report No. 10 to the Storting. Norwegian Ministry of the Environment, Oslo, Norway.

NMFA (2012). The high north: visions and strategies. Report No.7 to the Storting. (2011-2012). Norwegian Ministry of Foreign Affairs, Oslo, Norway.

Olsen, E., Aanes, S., Mehl, S., Holst, J.C., Aglen, A. and Gjøsæter, H. (2010). Cod, haddock, saithe, herring, and capelin in the Barents Sea and adjacent waters: a review of the biological value of the area. ICES Journal of Marine Science 67(1): 87-101.

Perrow, C. (1999). Normal accidents: living with high risk technologies. (2nd ed.). Princeton University Press, Princeton, NJ, USA.

Popper A.N. and Hastings M.C. (2009). The effects of anthropogenic sources of sound on fishes. Journal of Fish Biology 75(3): 455-489.

Von Quillfeldt, C.H. (2010). Det faglige grunnlaget for oppdateringen av forvaltningsplanen for Barentshavet og havområdene utenfor Lofoten. Bergen, Norway.

Food consumerism in today's China: towards a more experience-oriented economy?

S. Andersen Øyen
Centre for the Study of Sciences and the Humanities, University of Bergen, Allegaten 34, 5020 Bergen, Norway; simen.oyen@svt.uib.no

Abstract

In this article I will analyse food as a stratification marker in the increasingly more dominant consumer and experience economy of China. In what way constitute food and the consumption of food important markers in China? How is this related to a gradually more dominating Chinese consumerism? Culturally speaking food is generally used as a common marker for social identity but can therefore also constitute a marker serving to highlight differences and individual identity. I will in the following address how this takes place and acquires cultural expression in the Chinese late-modern experience economy.

The experience economy

Through consumerism we construct ourselves through the pragmatic choices we make, through the jobs we take, who we socialize with, what we eat, where we travel. This very aspect of the consumer culture can be acknowledged as, but also analysed in terms of a broader symbolic economy or an experience-based economy (Pine and Gilmore, 2011). It is an economy that is based on value production through the production of signs. But the experience economy is also about something more. It is about the feeling of being on the way to a better life. Within the experience-based economy, the aesthetic experience is the key to how we construct ourselves. In the perception of social mobility and the understanding that this is something we can signify through the choices we make, products, services, and experiences that primarily offer us something other than the product itself are central. The feeling of adventure is an example of one aspect used to sell products: It is the experience of the feeling rather than the product in itself that is the most important. Simultaneously, this type of consumption also exhibits an ethical manner of being and a demonstration of a better moral self. One accumulates experiences and acquires the moral characteristics that can be attributed to these experiences, which then in turn demonstrates the type of person one is (Hammer, 2010). This is predominantly a late-modern phenomenon, but the French 19th century economist Emile Dupuit described already in his time a similar phenomenon – of how when the railroad system emerged in France in the beginning there was a third class service for which the cars had no roofs, even though it did not actually cost any more to construct roofs. The point of this was to induce people to pay more for the second class cars because they did not want to be the type of poor people who were willing to be shipped like cattle simply to save a little money. An example of the same type of strategy as employed today is found in the pricing policy of airline companies (Hammer, 2010).

The new economy requires a consumer to continuously acquire new skills and new forms of capital. In an experience-based economy, in which the consumerism represents, the value of cultural capital becomes increasingly more important; In order to acquire value, one must invest in oneself and its own cultural qualifications, as products of larger social processes. The investment is about acquisition of the 'legitimate' culture and the legitimization of the value landscape as a whole. Here the working class represents the middle class's constitutive outside, and the difference is no longer exclusively economic. It is not, in the first case, a matter of a lack of material resources, but rather about a lack of 'taste', knowledge and the 'right way of being and behaving'. The working class and those who constitute the lower social strata in the new Chinese Economy imply the worst sort of conservative, backward world, which includes 'bad food' and 'bad taste'. One need not be well-off to be middle class, but one must have progressive opinions, the right aesthetic judgment, and the right kind of taste (Lawler, 2007). So-called civilized people demonstrate cuisine, manners, and etiquette to distinguish themselves from other groups. This is also the case in China where migrant workers, the farmers, and the generations that grew up under Mao differs in their food preference to the new, urban middle class. Food thereby entails a variety of interests, moral universes, temporal horizons, and action repertoires. But first and foremost, food consumption might represent *the* exemplifying shift towards a new consumerism in the modern Chinese modernity.

The rise of the new Chinese consumer culture

In China a consumer- and experience-oriented culture has emerged in tandem with the socio-historical individualisation processes that have occurred, especially following the opening and reform policy launched in 1978. The consumption-based individualism has not to the same extent as in the West been connected to the development of a rights-based individualism, through the welfare state or the emergence of a state of law, but it has acquired many of the same consequences and cultural expressions with respect to social differentiation: 'Along with the flow of material goods and cultural products, consumerism has in China changed consumers' spending patterns and served as a new mechanism of social differentiation' (Yan, 2009: 236). As an example, proper and stylish consumption has become so important that knowledge about consumption has become a commodity in its own right. A number of magazines and newspapers are dedicated to consumption-related topics and one can find daily consumption-related news items, columns, and debates in all media (Yan, 2009). In contradiction to the social egalitarianism that reigned under Mao, social differentiation, distinction, and hierarchies have been re-established and glorified both through the mass media and the emergence of consumerism. Luxury products, exclusive shopping centres, exclusive clubs and food consumption are frequently used to create new differences between social classes and to provide social status (Kleinman *et al.*, 2011).

The first businessmen after 1978 and the reform and opening politics, called 'getihu', were eager to legitimate and demonstrate their newly acquired economic status (Yan, 2010). They did so through luxury consumption and the purchase of expensive clothing, luxurious banquets, gambling, etc. Social anthropologist Yunxiang Yan (2009) tells of a man who spent 188,000 Yuan (approximately EUR 24,500) on a dinner, while another wallpapered his house with 10-yuan bills. With the second wave of self-employed business people who predominantly came from the public sector – a phenomenon that was referred to as 'jumping into the ocean', *xiahai feng*, at the beginning of the 1990s, the character of consumption changed. These were described as 'private entrepreneurs', *siying qiyejia*, who wanted to emphasise their dissociation from the *getihu*. While in the 1980s a getihu could easily spend 188,000 Yuan on a banquet, the qualifications and status markers in the 1990s would be, for example, a membership in an exclusive golf club with a price tag of more than $100,000. This has also been a defining factor for the consumption-based individualism that blossomed, connected to Chinese cities where a consumer economy is emerging (Yan, 2009). Along with this development a shift in resources and capital for social differentiation has occurred since the time under Mao up to the post-Mao period and the transition from a plan economy to a late-modern capitalist experience economy:

> The real significance here is that the symbolic meaning of luxury goods now has much more to do with one's purchasing power than one's political power [...] As a result of these changes, a dual structure of social stratification is emerging in the village, in which the socialist-bureaucratic rank or der co-exists with a market-based economic class order (Yan, 2009: 14-15).

This development also finds expression in the Chinese consumption of food. In this context, the tensions that have arisen between the migrant workers, the farmers, and the generations that grew up under Mao on the one hand, and the new, urban middle class on the other, are characteristic indications of how new cultural and normative discourses, new disciplinary processes, and new distinction markers are undergoing development.

The elevated taste: cultural capital in the social space of China's new consumerism

The French sociologist Pierre Bourdieu (1984, 1996) has investigated taste as a principle of social differentiation. In analyses of taste in the different social classes of France, he shows how the experience and classification of aesthetics distinguishes social classes from one another. Taste is a part of the daily practice that produces class relations. We position people among the dominating or dominated by affirming or looking down on the way they walk or express themselves – what Bourdieu calls *patronising strategies*. Taste is included here among the mechanisms that produce social inequality and establish relations of power in society. The distinction between highbrow and lowbrow culture becomes thereby extremely strategic for the function of the entire society. This is connected with Bourdieu's view of how relations of superiority and inferiority are created and upheld. It is not simply that we recognise and rank styles. We also assign them value. This occurs through a continuous evaluation of the way that people express themselves through objects, the body, and language. For example, bodies – how they look, how they are carried, how they are ornamented, and the preferences they exhibit – are central in the representations of different classes. This can be described through what Bourdieu calls *symbolic* and *cultural capital*, which in simple terms must be understood as cultural resources and as a practical kind of knowledge about how society is constructed (Brekke *et al.*, 2003).

One factor that Bourdieu maintains characterises this entire social space, is a division between an economic pole and a cultural pole. These are the main dimensions according to which the individual and institutions are valued. This valuation at the economic pole is based on access to money, property, or control of property, while the valuation at the cultural pole is assessed according to access to cultural commodities, such as education, art, and literary works (Johansen, 2011). The amount and the quality of these different social resources or capital are included in the determination of the individual's social position. Taste is thereby included as a reproduction strategy. Taste results in all social actors bearing particular signs, where the sign for being distinguished is just a subgroup – and these signs unite and separate with the same effectiveness as barriers and explicit prohibitions. Attitude, manners, the way one eats, food, manner of speaking, and taste are embodied signs and production principles for all activities that signify and display social position through a play with specific distinguishing features. For Bourdieu, taste comprises the ability to master aesthetics, not only in narrow arts-related contexts, but just as much in the education system, working life, politics, the media, the family, and all types of informal meetings between individuals where something is at stake. This ability finds expression in the expertise required in discerning and knowing what has a high or low value in society and deposits itself in the habitus of the dominant class fractions. Mastering the distinctions gives people an exclusive status and tells us that they belong to the dominant class fractions. It distinguishes or elevates some to the detriment of others. This is an effect of social factors that give the middle class and the elite the power to assert their definitions (Bourdieu, 1984). What is considered 'tasteful' is determined by those who have power in society to define the tasteful, as we shall also see when we move on to address China.

Food as a social distinction

In China as in all cultures, food has a central position. One way of saying 'How are you' in Chinese, for example, can be directly translated as 'have you eaten' (*shilefan meiyou*). The anthropologist Wu Fei (2011) has looked at food's social significance and in particular at the connection between symbols connected to food and suicide. He tells of one case in Jianli where an elderly man hung himself because his son and daughter-in-law had hidden bread from him and instead given him paltry and rancid food. Food has also had a political dimension in China. Gong bao chicken, which in the West is known as *Kung Pao chicken* was politically incorrect during the Cultural Revolution because it was associated with imperial cuisine and was named after a governor of a province during the Qing Dynasty (Jackson, 2006).

In China the dichotomy of 'cold' versus 'warm' is central. This is not a matter of the temperature of the various dishes, but of whether the food is fundamentally yin or yang. While yin food is mild and cooling, yang food is spicy and warming. Chinese food and the mealtime have symbolic functions reflecting a context comprising a world view and social factors. For example, long rice noodles (*shoumian*) express a wish for a long life and good health. Coconut meat symbolizes togetherness, dried longan shall give many sons, and lotus seeds shall in general produce many children (Jackson, 2006). Those wishing to cultivate good relationships must 'give face to' (gei mianzi) others and this finds expression at the table by taking into consideration social status, such as through a meal that has so many courses, or is so expensive that it highlights the importance of the guest or the occasion (Strandenæs, 2011). Here the rice is served last and not eaten because it is a sign that more than enough of the other food has been served. In Chinese, incidentally, 'to eat food' is called to eat rice (chi fan).

In China as in the West, food does not exclusively represent communities of shared taste but also signals social mobility through price and availability. For example, shark fins are considered exclusive and in some districts are requisite for a New Year's dinner, large celebrations, and important business dinners. The seasonal, hairy freshwater crab in Shanghai is another popular and extremely expensive food. Freshwater shrimp are far more exclusive than saltwater shrimp because they are more rare. Freshwater fish from the rivers is generally more in demand than saltwater fish from the sea. Here another component is also important and that is that the fish must be as fresh as possible. The fish is often displayed first alive before it is prepared. The livelier the fish, the better. The fish is served whole – complete with head and tail as a symbol of a good beginning and a good end of the year (Jackson, 2006). Foreign and imported food is also in certain cases considered exclusive and a luxury – particularly by the new urban middle class. In China, for example, chicken feet are used as an ingredient in soups. Norwegian chicken feet are here considered especially exclusive, with a good flavour and guaranteed to be free of salmonella and bird flu. In China we also see a tendency to classify high-quality products according to the production site in much the same way as how champagne is defined that which is exclusively produced in the French town of Champagne. The most famous cured ham in China is produced in the Yunnan province (Jackson, 2006). Consumption of organic food has, as in the West, become a marker of social status, but in China has been dominated by the focus in recent years on food safety and the related scandals. Organic and safe food has therefore become a resource, predominantly for the country's economic and political elites. This finds expression through an exclusive access to specially selected food manufacturers who supply organically grown and meticulously tested food products, such as the well-known Jushan farm outside of Beijing which for many years has cultivated organic food solely for the political elite. At the same time, organic food is as in many other locations more expensive than other food and has also attracted the attention of China's growing middle class.

What characterises this development is that the food consumption of the new economic and political elites as well as the new, emerging, urban middle class and their preferences plays a defining role for lifestyle and normative standards when it comes to food. This is not something that is solely regulated by availability, demand, and price, but also on the basis of disciplinary processes and the means by which one acquires cultural capital. Here taste as an aesthetic experience is wholly central to how we construct ourselves. In China we find these tensions between those who master this aesthetic assessment and those who are marginalised along a number of axes where the lines of conflict run between urban and rural areas, migrant workers and those who reside permanently (*hukou*) in the large cities and classes, but also between the Mao and post-Mao generations. In China the migrant workers, the farmers, and the generation that grew up under Mao are those who make up the growing middle class' constitutive outside.

The Chinese food and mealtime produce thereby different moral universes and action repertoires which the emerging economic and political elite employ as differentiation and stratification markers. This highlights not only social differences, but also contributes to the construction of individual identity in contrast to the identity of the collective moral ethos under Mao, and which now provides greater opportunities to display and differentiate improved moral selves. Through their normative dominance these elites are seeking here to incorporate weaker ideological and social strata on their own terms. This finds expression in different online debates where migrant workers and farmers are depicted as uncouth and uncivilised and where the debate is about everything from slurping to eating dog meat. The members of the new, young, and urban middle class do not eat dog meat, and neither do they slurp. In Shanghai this was particularly evident in connection with the preparations for EXPO (the World's Fair in 2010). Here the customs of the average Shanghai resident and the migrant workers were to be cultivated through publicly initiated, planned management campaigns. These are processes which are also the result of an increase in interaction and new connections with the West, but which have a distinctly cultural dimension and can be seen as an expression of deep-running modernisation and civilisation processes. They thereby contribute to understanding cultural change in China.

Refernces

Bourdieu, P. (1984). Distinction. A social critique of the judgement of taste. Routledge, London, UK.

Bourdieu, P. (1996). Symbolsk makt – artikler i utvalg. Pax, Oslo, Norway.

Brekke, O.A., Høstaker, R. and Sirnes, T. (2003). Dimensjonar i moderne sosialteori. Samlaget, Oslo, Norway.

Fei, W. (2011). Suicide, a modern problem in China. In: Kleinman, A., Yunxiang, Y., Jing, J., Sing, L. and Zhang, E. (eds.) Deep China: the moral life of the person. University of California Press, Berkeley, CA, USA, p. 213-236.

Johansen, J.C. (2011). Sosial kapital og multiposisjonalitet blant kvinner i eliteposisjoner. Master thesis. University of Bergen, Sociologic institute, Bergen, Norway.

Hammer, E. (2010). Velkommen til opplevelsesøkonomien. In: Madsen, O.J. and Øyen, S. (eds). Markedets fremtid, Cappelen Akademiske Forlag, Oslo, Norway.

Jackson, A. (2006). Kina – mat og livsstil. Spektrum, Oslo, Norway.

Kleinman, A., Yunxiang, Y., Jing, J., Sing, L., and Zhang, E. (2011). Deep China: the moral life of the person. University of California Press, Berkeley, CA, USA.

Lawler, S. (2007). Att känna avsmak: så bildas medelklassens identitet. In Fronesis 24: 131-148.

Pine, J. and Gilmore, J. (2011). The experience economy. Harvard Business School Press, Boston, MA, USA.

Strandenæs, T. (2011). Når maten snakker – det kinesiske måltidet som veiviser til kulturelle verdier. In: Drønen, T., Fretheim, K. and Skjortnes, M. (eds.) Forståelsens gylne øyeblikk – Festskrift til Øyvind Dahl. Tapir Akademisk Forlag, Trondheim, Norway.

Yan, Y. (2009). The individualization of Chinese society. Berg, Oxford, UK.

Yan, Y. (2010). The Chinese path to individualization. British Journal of Sociology 61(3): 489-512.

Section 2. Labelling and market forces

The relevance of sustainability for the consumer in a food context: a segmentation analysis

F. Vanhonacker, E.J. Van Loo, X. Gellynck and W. Verbeke
Department of Agricultural Economics, Ghent University, Coupure links 653, 9000 Ghent, Belgium; filiep.vanhonacker@ugent.be

Abstract

Current western food consumption is associated with a high ecological impact. A way to reduce this impact is to shift to more sustainable food choices. This study investigates consumer attitudes towards more sustainable food choices. The alternatives under study range from well-known meat substitutes to alternatives which are more radical or innovative and that require an adaptation of food habits and cultural patterns. Findings are based on responses of 221 Flemish consumers to a survey conducted in Spring 2011. Results revealed an underestimation of the ecological impact of animal production. Further consumers accepted well-known alternatives such as organic meat, moderation of meat consumption and sustainable fish, although their willingness to pay was pronouncedly lower than their willingness to consume. Consumers were more reluctant to alternatives that (partly) ban or replace meat. Opportunities of introducing insects were non-existent. A segmentation analysis based on self-evaluated ecological footprint and personal relevance of the ecological footprint revealed five consumer segments, termed Conscious, Active, Unwilling, Ignorant and Uncertain. Each segment is defined in terms of demographics, attitudinal and behavioral characteristics. Opportunities for sustainable food choices in each segment are discussed.

Keywords: consumers, ecological footprint, meat substitutes, segmentation, sustainability

Introduction

The concept of ecological footprint has come to the forefront, along with the emerging societal significance of sustainability. Food consumption accounts for a significant proportion of the ecological footprint of individuals with a carnivorous diet (FAO, 2006b). As global meat consumption is expected to increase further (FAO, 2006a), innovative technologies and increased efficiency of animal production alone will not suffice to solve the environmental problems related to the growing meat demand (Garnett, 2011). To maintain a sustainable environment, a better balance between meat consumption and livestock production's environmental impact will be essential, and a change in meat consumption will be inevitable to lessen food-related GHG emissions (Garnett, 2011). This study investigates opportunities of more sustainable food choices from a consumer perspective. In order to acknowledge differences in society, a segmentation analysis is included. The alternative food choices are based on de Bakker and Dagevos (2010), and include hybrid meat products (meat variants in which part of the meat is substituted by plant-based ingredients), meat types with lower environmental impact (e.g. chicken, game), plant-based meat substitutes or replacements (foods containing proteins from plants such as soy and cereal protein, e.g. tofu, seitan, vegetarian burger), sustainable farmed fish, organic meat, proteins from insects and a moderated meat consumption (less meat per meal). Insights in consumers' opinions about ecological food choices and meat consumption alternatives are of paramount importance to better position these products in the market.

Method

Study design and subjects

A web-based survey was carried out in Flanders during Spring 2011. A valid sample of 221 participants was obtained. The sample was biased to a higher share of females (65%), a higher share of 18-30 year old people (36%), higher educated people (77%), participants with the main responsibility for food purchases (54%) and participants with a self-reported above average financial situation (65%).

Questionnaire and scales

The questionnaire consisted of three parts. First awareness and concern in relation to the concept of ecological footprint was measured. Participants were further asked to self-evaluate their personal footprint on a scale from 'very small' (score 1) to 'much too high' (score 5). Next, participants were presented with a list of sectors, among them livestock production. For each sector, they were asked to score the contribution to GHG emissions on a scale from 'no contribution' (score 1) to 'very strong contribution' (score 5). Concern about the ecological footprint was measured through two statements on a five-point scale, where a higher score indicated a higher concern. Finally, personal relevance was measured through the question '*To what extent is your ecological footprint important in your consumer choices?*'. Response categories ranged from 'not important at all' (score 1) to 'very important' (score 5).

The second part dealt with sustainable food choices. Participants were informed about the contribution of animal production to CO_2-emissions. Then, they were asked how aware they were of the extent of this contribution. After that, they were introduced to the alternative food choices. Five-point semantic differential scales were used to obtain an evaluation score. Additionally, participants were probed for their willingness to consume and willingness to pay for each alternative, both registered on a five-point scale. The third part registered demographics, consumption frequency of different meat types and enjoyment from eating meat. These variables were used to profile different consumer segments and to verify relationships previously reported in literature. Data were analyzed using SPSS 19.0.

Results

Awareness and concern

A majority (68.6%) claimed to be aware of the concept 'ecological footprint', but under-estimated the ecological impact of animal production and meat consumption. When self-evaluating their personal ecological footprint, some perceived their footprint to be too high or much too high (42.3%), all right (42.7%) and small or very small (15%). Most participants (63.3%) were concerned about the climate change and the ecological footprint. Only a small group (8.7%) indicated not to be concerned and 28% were neutral. Similarly, 65.4% disagreed that the issue of climate change and ecological footprint is overstated while 10.4% agreed.

Alternatives for a more sustainable meat consumption

Reduced meat consumption (amount of meat per meal) was the most preferred alternative, followed by sustainable farmed fish, meat types with lower environmental impact and organic meat (results not shown). These alternatives all received distinct positive evaluation scores. Hybrid meat products and plant-based meat substitutes were evaluated neutral to slightly positive. The consumption of protein from insects was rejected. Regarding willingness to pay, lower scores were found. Only for organic meat, a positive willingness to pay was registered (wtp was not registered for a reduced meat consumption).

Participants were mainly willing to adjust the meat quantity in their traditional meal, followed by changing to organic meat and meat types with a lower impact. Participants reported a higher frequency for replacing their traditional meal by hybrid meat products compared to plant-based meat substitutes.

Cluster analysis

Five consumer segments were defined following the cluster analysis (Table 1). Segment 1 ('Conscious') reports a too high ecological footprint and a moderate personal relevance for the ecological footprint concept. Segment 2 ('Active') indicates a high personal relevance, combined with a small personal footprint. Segment 3 ('Unwilling'), opposite to segment 2, combines a high personal footprint with a low personal relevance. Segment 4 ('Ignorant') scores low on both variables. Finally segment 5 ('Uncertain') is composed of respondents that have selected the mid-point of the scale on both questions.

Profile of the segments

Segments did not differ on most of the socio-demographic characteristics, including gender, financial situation and living environment (rural or urban). Significant differences existed in mean age and education (Table 1). The Unwilling were the youngest on average and significantly younger than the Active and the Uncertain (the oldest). The Conscious were significantly younger than the Uncertain. The large majority of the Conscious, the Active and the Unwilling were higher educated (education beyond the age of 18), while a more balanced distribution in education level was found among the Ignorant and the Uncertain.

The Conscious and the Unwilling reported the highest meat consumption frequency (in terms of beef, pork and poultry); the Active the lowest. The Active consumed most frequently meat substitutes, and were the least positive towards the enjoyment of eating meat and the perceived necessity of meat in a healthy diet, opposite to the Unwilling (results not shown).

The Conscious, the Active and the Unwilling were more aware of the ecological footprint concept as compared with the Ignorant and the Uncertain. Despite the high awareness, the Unwilling do not take it into account in their purchase decisions. This differentiates them from the Conscious and the Active, who were also most aware of the negative environmental impact of animal production. A higher level of concern towards ecological footprint among the Conscious and the Active suggests a positive association between concern and personal relevance. Finally the Conscious, the Active and the Unwilling were

Table 1. Profile of the segments on the segmentation variables and demographics[1].

	Conscious	Active	Unwilling	Ignorant	Uncertain	*P*-value
Segment size (%)	25	19	16	18	22	
Segmentation variables						
Self-evaluation (mean)	4.22^{a}	$2.74^{b,c}$	4.19^{a}	2.38^{c}	3.00^{b}	<0.001
Personal relevance (mean)	3.31^{b}	4.12^{a}	1.86^{e}	2.23^{d}	3.00^{c}	<0.001
Socio-demographic profile						
Age (mean)	$36.3^{a,b}$	$44.4^{b,c}$	33.0^{a}	$43.2^{a,b,c}$	48.1^{c}	<0.001
Higher education (%)	94	81	94	58	56	<0.001

[1] Values with the same letter as subscript indicate not statistically significant differences.

more strongly opposed to the statement that CO_2-emissions, climate change and ecological footprint are overstated.

Differences in terms of the evaluation of sustainable food choices were mainly present for alternatives that (partly) ban the meat component from the meal (reduced meat consumption, hybrid meat types and plant-based meat substitutes), mainly due to a difference between the Active (most positive) and the Unwilling (least positive) (Table 2). The evaluation of protein from insects was negative in general. Overall, the Active were most positive towards the alternatives, followed by the Conscious. The Unwilling were most averse to alternatives that replaced meat and relatively more open to alternatives that still contained the entire meat component (e.g. organic meat). The Uncertain seemed to be rather averse or skeptic towards more innovative alternatives such as hybrid meat types. Comparable between-group findings were found regarding willingness to consume and willingness to pay (Table 2). Also, the willingness to pay was clearly lower than the willingness to consume, for each alternative and within each segment. Within each group, willingness to pay was the highest for organic meat.

Discussion and conclusion

In general, findings showed an imbalance between reported awareness of the ecological footprint concept (rather high) and perceived ecological impact of animal production (underestimation). Similar

Table 2. Evaluation of the different alternatives: cross-segment comparison (n=219).[1]

	Conscious	Active	Unwilling	Ignorant	Uncertain	*P*-value
Evaluation of more sustainable meat alternatives (scale: bad (1) – good (5))						
Reduced meat consumption	4.33^{ab}	4.57^{b}	3.81^{a}	3.84^{a}	3.85^{a}	0.001
Sustainably farmed fish	4.28	4.29	3.92	4.16	4.02	0.237
Meat types with lower impact	4.39	4.31	4.17	3.87	4.17	0.067
Organic meat	4.20	4.33	4.22	3.95	3.92	0.218
Hybrid meat types	3.44^{ab}	3.79^{a}	3.34^{ab}	3.33^{ab}	2.96^{a}	0.039
Plant-based meat substitutes	3.48^{ab}	3.98^{b}	2.72^{a}	2.92^{a}	2.72^{a}	<0.001
Protein from insects	2.54^{a}	2.55^{ab}	2.06^{ab}	2.05^{ab}	1.81^{b}	0.012
Willingness to consume						
Reduced meat consumption	4.15^{ab}	4.40^{b}	3.58^{a}	3.74^{ab}	3.77^{ab}	0.010
Sustainably farmed fish	3.87^{ab}	4.07^{b}	3.25^{a}	3.64^{ab}	3.75^{ab}	0.009
Meat types with lower impact	4.22^{a}	4.07^{ab}	4.00^{abc}	3.41^{c}	3.67^{bc}	0.001
Organic meat	4.13	4.26	3.78	3.59	3.69	0.010
Hybrid meat types	3.57^{b}	3.67^{b}	2.92^{ab}	3.05^{ab}	2.60^{a}	<0.001
Plant-based meat substitutes	3.26^{bc}	3.51^{c}	2.28^{a}	2.70^{abc}	2.56^{ab}	<0.001
Protein from insects	1.74	1.93	1.58	1.55	1.43	0.113
Willingness to pay						
Reduced meat consumption	3.09^{ab}	3.43^{b}	2.47^{a}	2.85^{ab}	2.71^{a}	0.002
Sustainably farmed fish	3.00^{b}	3.10^{b}	2.31^{a}	2.72^{ab}	2.62^{ab}	0.011
Meat types with lower impact	3.43^{ab}	3.61^{a}	3.06^{ab}	2.77^{b}	3.00^{b}	0.015
Organic meat	2.72^{b}	2.93^{b}	1.97^{a}	2.58^{ab}	1.94^{a}	<0.001
Hybrid meat types	2.52^{ab}	3.10^{b}	1.81^{a}	2.14^{a}	2.10^{a}	<0.001
Plant-based meat substitutes	1.39	1.64	1.25	1.31	1.33	0.107

[1] Values with the same letter as subscript indicate not statistically significant differences.

underestimations were found in other studies (De Boer *et al.*, 2013; Latvala *et al.*, 2012; Tobler *et al.*, 2011) and could explain why environmental concerns are rarely dominant in changing meat consumption behavior (Fox and Ward, 2008; Ruby and Heine, 2011). These findings imply a need for better informing consumers and citizens on the actual impact of livestock production on the environment, and on the influence of present (meat) food choices on the ecological condition of the planet (Tobler *et al.*, 2011).

It will be a challenge to change meat consumption behavior as numerous barriers exist including preconceptions towards vegetarian diets, habits and prices. Different findings in this study confirm that meat is still considered the centerpiece of the traditional meal in the Western food culture. Alternatives that partly (hybrid meat products) or fully (plant-based products) substitute meat were clearly less appreciated. This relates to societal preconceptions that vegetarian diets have a lower sensory and health quality (Lea and Worsley, 2003; Povey *et al.*, 2001). There will be a need to readdress the negative image of meat substitutes. Currently, the low penetration (not many consumers try out meat substitutes) makes it difficult to change the perceived lower quality and sensory properties of meat substitutes. As such, it could be worth to promote trials and expose consumers repeatedly to these products to positively affect their acceptance (Hoek *et al.*, 2011). Also many consumers feel uncertain about where to find and how to prepare meals with meat substitutes (Lea and Worsley, 2003; Schösler *et al.*, 2012).

The findings of the present study reveal some burdens for alternatives such as hybrid meat, meat substitutes and insect protein to gain presence in the Western diet. Alternatives where meat is not substituted were evaluated more positively. The highest acceptance was found for meat moderation, i.e. consuming less meat per portion. Many studies have focused on meat substitution and several concluded that the enjoyment people experience from eating meat strongly hampered the potential success of meat alternatives (e.g. Schösler *et al.*, 2012). The highest willingness to pay among all alternatives was reported for organic meat. This could be explained by the familiarity with the price premium for organic products, and by consumers' beliefs about both environmental and personal health benefits from organic food consumption (Hoefkens *et al.*, 2009; Van Loo *et al.*, 2010).

Our findings support a targeted approach. Five consumer segments were identified based on personal relevance of the ecological footprint concept and self-evaluated ecological footprint. The cluster analysis resulted in two opposite segments – termed the Active and the Unwilling – and three intermediate segments (the Conscious, the Ignorant, the Uncertain). The Active attribute high importance to the ecological footprint when making consumption decisions and evaluate their personal footprint to be small. This segment is highly aware of and concerned about ecological issues. They were most open towards alternative sustainable food choices, in terms of evaluation, willingness to consume and willingness to pay. This segment is probably more involved and more actively searching for information. To reach this segment, an appealing and diversified product offer seems most appropriate, including visible and transparent information at the point of purchase, eventually complemented with additional information (e.g. leaflet). This segment consumed less meat and expressed higher preferences for alternative food choices without meat. This is consistent with de Boer *et al.* (2012) who reported more positive attitudes towards meat-free meals among consumers caring more about the environment. On top they were the least favorable towards meat in terms of necessity and taste, two important barriers for meat reduction (Lea and Worsley, 2002; Schösler *et al.*, 2012). This segment is largely composed of vegetarians, flexitarians (consumers with a mainly vegetarian diet, who consume meat from time to time) or flexivores (consumers with a varied diet, who alternate meat with fish, vegetarian meals and other alternatives). The Unwilling are the opposite of the Active. They realize that their footprint is too high, yet do not include it in their food choices. They are mainly interested in their personal life quality and can be considered 'meat lovers'. They like meat very much, consume a lot of meat and are not willing to lessen this consumption. This was illustrated in a low rank of a reduced meat consumption (compared with other consumer segments) and their preference for (or lower rejection of) hybrid meat products over plant-based meat substitutes (due to the absence of meat in plant-based substitutes). Market opportunities for alternatives that are harmful to the perceived taste and quality of meat and that reduce the quantity of the meat component in the meal are probably non-existent. For example, organic meat which is not associated with these barriers was clearly better evaluated within this segment. The Conscious reported to have a too high ecological footprint, despite an acknowledged importance of the concept in their consumer choices. This segment thus expresses a gap between their attitudes (and intentions) and their actual behavior. This segment enjoys eating meat and expresses the highest meat consumption frequency together with the Unwilling. However they are not averse towards meat consumption moderation and more sustainable alternatives. Alternatives that occasionally or not entirely abandon meat, but that focus on moderation and more sustainable food choices seem most appropriate. Appealing meat substitutes for this segment could be products that resemble meat in terms of taste, appearance and taxonomy (Hoek *et al.*, 2011). The Ignorant score low on both segmentation variables. They do not take ecological issues into account in their food choices. On top they are poorly aware of the ecological footprint and the ecological impact of animal production. This seems mainly related to disinterest and/or an indifferent attitude towards the theme. As such opportunities to involve this segment seem more limited. Finally, the Uncertain are respondents that have selected the mid-point of the scale on both segmentation variables. This specific answering behavior could be due to their low awareness of the ecological footprint concept. Still they engaged more in environmental friendly behaviors compared with the Unwilling and the Ignorant. As such this segment offer some opportunities, based also on the evaluation scores and their willingness to consume certain alternatives. This segment was older than the other segments and seemed more averse towards the more innovative or technology-driven alternatives (e.g. hybrid meat). It thus seems more appropriate to change the composition of the meal with existing more sustainable substitutes rather than with innovative components.

References

De Bakker, E. and Dagevos, H. (2010). Vleesminnaars, vleesminderaars en vleesmijders: duurzame eitwitconsumptie in een carnivore eetcultuur. LEI-report 2010-003. LEI, Den Haag, the Netherlands, 207 pp.

De Boer, J., Schösler, H. and Boersema, J.J. (2013). Climate change and meat eating: an inconvenient couple? Journal of Environmental Psychology 33: 1-8.

Food and Agriculture Organization of the United Nations (FAO) (2006a). World agriculture: towards 2030/2050, Interim Report. FAO, Rome, Italy, 71 pp.

Food and Agriculture Organization of the United Nations (FAO) (2006b). Livestock's long shadow: Environmental issues and options. FAO, Rome, Italy, 390 pp.

Fox, N. and Ward, K.J. (2008). Health, ethics and environment: a qualitative study of vegetarian motivations. Appetite 50: 422-429.

Garnett, T. (2011). Where are the best opportunities for reducing greenhouse gas emissions in the food system (including the food chain). Food Policy 36: S23-S32.

Hoefkens, C., Verbeke, W., Aertsens, J., Mondelaers, K. and Van Camp, J. (2009). The nutritional and toxicological value of organic vegetables: Consumer perception versus scientific evidence. British Food Journal 111: 1062-1077.

Hoek, A.C, Luning, P.A, Weijzen, P., Engels, W., Kok, F.J. and de Graaf, C. (2011). Replacement of meat by meat substitutes. A survey on person- and product-related factors in consumer acceptance. Appetite 56: 662-673.

Latvala, T., Niva, M., Mäkelä, J., Pouta, E., Heikkilä, J., Kotro, J., and Forsman-Hugg, S. (2012). Diversifying meat consumption patterns: consumers' self-reported past behavior and intentions for change. Meat Science 92: 71-77.

Lea, E. and Worsley A. (2002). The cognitive contexts of beliefs about the healthiness of meat. Public Health Nutrition 5: 37-45.

Lea, E. and Worsley, A. (2003). Benefits and barriers to the consumption of a vegetarian diet in Australia. Public Health Nutrition 6: 505-511.

Povey, R., Wellens, B. and Conner, M. (2001). Attitudes towards following meat, vegetarian and vegan diets: an examination of the role of ambivalence. Appetite 37: 15-26.

Ruby, M.B. and Heine, S.J. (2011). Meat, morals, and masculinity. Appetite 56: 447-450.
Schösler, H., De Boer, J. and Boersema, J.J. (2012). Can we cut out the meat of the dish? Constructing consumer-oriented pathways towards meat substitution. Appetite 58: 39-47.
Tobler, C., Visschers, V.H. and Siegrist, M. (2011). Eating green. Consumers' willingness to adopt ecological food consumption behaviors. Appetite 57: 674-82.
Van Loo, E.J., Caputo, V., Nayga, R.M, Meullenet, J., Crandall, P.G. and Ricke, S.C. (2010). The effect of organic poultry purchase frequency on consumer attitudes toward organic poultry meat. Journal of Food Science 7: S384-S397.

'Unnecessary suffering' as a concept in animal welfare legislation and standards

F. Lundmark[1], C. Berg[1] and H. Röcklinsberg[2]
[1]Department of Animal Environment and Health, Swedish University of Agricultural Sciences, Box 234, 53223 Skara, Sweden; [2]Department of Animal Environment and Health, Swedish University of Agricultural Sciences, Box 7068, 75007 Uppsala, Sweden; frida.lundmark@slu.se

Abstract

The project 'Legal systems and ethical values behind the official and the stakeholder-based animal welfare control' covers studies of farm animal welfare legislation and ten private animal welfare standards from four European countries. Their expressed aim of prevention of suffering is in focus. Many pieces of legislation stating the aim of preventing 'unnecessary suffering' lack a clear distinction between 'unnecessary' and 'necessary' calling for definitions, as it is difficult to decide where to draw the line in actual farming, or for a citizen to interpret what is regarded acceptable by the legislator. Several interpretations are possible, e.g. the intensity and duration of the suffering, the intention behind the act, the fulfillment of human interests and the animals' interests. Furthermore, countries differ regarding what species are legally protected and at what level. We will further discuss ethical values behind such differences. Painful management procedures are legal in many countries, and hence regarded as 'necessary suffering' in some – but not all – countries. As private standards are developed to meet consumer demands for a stricter interpretation of 'unnecessary' it is important to clarify inherent values. We tentatively argue that besides utilitarian ethical thinking duty, contractarian and/or virtue ethical thinking can be found in both legislation and private standards. If so, this mirrors consumer interest in an integrated and complex reasoning related to the concept of 'unnecessary suffering', a complexity that needs to be considered in forthcoming legislation.

Keywords: ethics, contractarianism, consumer, virtue ethics, farm animal welfare, law

Introduction

From early to mid 20th century a number of countries developed animal welfare (AW) legislation covering the prevention of cruelty and unnecessary suffering (Vapnek and Chapman, 2010). Since the 1970's EU has also taken initiatives to protect animals by legislation (Veissier *et al.*, 2008) and in 2009 the Lisbon Treaty entered into force recognizing animals as 'sentient beings' and stating that full regard shall be paid by the member states to the welfare of animals. However, exceptions are allowed when it comes to 'religious rites, cultural traditions and regional heritage' (European Commission, 2007). Today, animal keeping producers do not only have the legislation to comply with, but also different kinds of private standards (Veissier *et al.*, 2008). Most of these standards have been established during the last decade (Bayvel, 2004).

It seems rather clear that scientists agree that birds and mammals can feel for example pain and distress and hence are capable of suffering (Gregory, 2004; Underwood, 2002). There has been more uncertainty as to whether invertebrates and fish are capable of suffering, but research over the last ten years has showed that investigated fish species have the capacity to feel pain and fear and adapt their behavior to current situation (Braithwaite and Boulcott, 2007; Chandroo *et al.*, 2004; EFSA, 2009), and Elwood *et al.* (2009) concluded that crustaceans can feel pain. Notable is that in the new directive for the protection of animals used for scientific purposes (2010/63/EU) cyclostomes and cephalopods are also included in addition to vertebrates, with the motivation that there is scientific evidence of their ability to experience pain, suffering, distress and lasting harm (see preamble 8 in 2010/63/EU). Given that there is evidence enough for answering 'yes' to the question whether non-human animals can suffer the next question is how we should act upon that knowledge. According to Lund *et al.* (2007) the capacity of having sensations such as pain and pleasure is perhaps the single most important criteria for moral status of an animal, at least in the Western society.

In this study the following national legislation; Sweden, UK, Spain and Germany, and the following private standards; KRAV, Arlagården, Swedish Seal of Quality; the Laying hen programme, the Broiler programme, Soil Association, Freedom Food, Marks & Spencer, Neuland and Carnes Valles del Esla, were included.

The aim of this study is to discuss how 'unnecessary suffering' is defined and used in legislation and standards studied in this project. We want to investigate if there is a difference between different pieces of legislation and standards for different species as to what is considered necessary or not.

Preventing 'unnecessary suffering'

The prevention of animal suffering was a central concept in all four countries' national legislation. 'Unnecessary suffering' should be avoided in Sweden, UK and Spain, whereas German legislation use 'avoidable suffering' or suffering 'caused without a good reason'. Some of the legislation/standards clarified that both physical and mental suffering is included. As there is unnecessary suffering, there is also suffering that is necessary or unavoidable and therefor legal to expose animals to. Our interpretation is that the private standards also strive to prevent suffering but they often used other words than 'suffering'. The organic standards for example more often use the concept 'stress'; focusing on minimizing the stress for the animals.

Studying preambles and similar texts we found that what policymakers actually define as 'unnecessary' or 'necessary' were not well explained. In the Swedish government bill (1987/88:93) it is claimed that it is unavoidable to sometimes cause animal suffering and that this kind of suffering should be seen as legal. Two examples are mentioned; procedures for the treatment of sick animals and the use of animals for scientific purposes. In the UK AW Act Section 4 (3) there is a list of considerations that should guide the courts when determine whether suffering is unnecessary or not. The 'considerations focus on the necessity, proportionally, humanity and competence of the conduct' (Explanatory notes, 2006). Also others have noticed the insufficient explanations of 'unnecessary suffering' in the legal context (Forsberg, 2011; Wahlberg, 2011). Hurnik and Lehman (1982) wrote the following summary about unnecessary suffering; 'Suffering of animals is unnecessary suffering if it is not essential for purposes of sufficient importance or if it could be avoided by adopting alternative practices that would achieve the same important purposes, but would result in less suffering, providing that such alternative practices were not too expensive for the community in question to bear.' Also Landera-Luri (2010) mentioned that both morally and economical aspects are taken under considerations when it comes to 'unnecessary'. Wahlberg (2011) drew the conclusion that according to the legislators, acceptable suffering is suffering that is in practice unavoidable, impossible to prevent or prohibit, but possible to legally justify and hence an acceptance of all current farming practices.

Does species matter?

The private standards chosen for this study concerned one or several species of production animals. What species AW legislation included differed between countries. It seems like the animals which ought to be protected by the legislations in the UK and Spain were the ones capable of suffering. The UK AW Act covered animals that were commonly domestic in the British Islands, were under control of man and not living in a wild state. An 'animal' in the UK Act meant a vertebrate (other than man), but the authorities could include invertebrates if there is scientific evidence that these animals were capable of experiencing pain or suffering. The Spanish legislation covered 'all animals (including fish, reptiles and amphibians) kept for production of food, wool, fur or other agricultural ends', e.g. no invertebrates.

In Sweden and Germany it did not seem to be the capability of suffering alone that determined what animal species should be protected, as also all invertebrates were included. The Swedish AW Act protected animals when kept by man in captivity, while the German legislation also included wild animals. Notable is, however, that there are no regulations about invertebrates and there have almost never been any legal cases concerning these animals in Sweden. In one case regarding invertebrates the court questioned if centipedes should be protected by the AW Act at all, but they self-evidently included the scorpions (Striwing and Åslund, 2005).

The countries did not only differ regarding what species were legally protected but also regarding on what level, e.g. what procedures and farming systems were necessary or unavoidable. We know for example that castration of piglets without analgesia or anesthesia, beak trimming of birds and hot branding are painful procedures (Underwood, 2002). These procedures are not allowed in every country so what is necessary suffering obviously differ between countries. There are also differences between legislation and standards in the same country. A procedure that can be allowed in the legislation can be banned according to a certain private standard. Furthermore, we noticed that the same species can be treated differently within the same legislation depending on the purpose of keeping the animal (pet, farm, zoo, laboratory animals).

The consumers' interests and views

The majority of industry, retailer and farmer initiated standards had taken consumers' attitudes into account, beside other factors which also the legislators had considered; such as animal behaviour and animal health. As private standards are developed to meet consumer demands for a stricter interpretation of 'unnecessary' it is important to clarify inherent values. The ideals expressed in the legislations and standards serve as a starting point. The justification for letting animals suffer because it is 'necessary' will diverge between different normative ethical views. The idea of maximizing welfare/pleasure and preventing suffering for sentient individuals but accepting some suffering if needed to maximize total good, is a hedonistic utilitarian thought (Bentham, 1789). According to Peter Singer's utilitarian preference theory acceptable or necessary suffering is when suffering in one individual or smaller group is a prerequisite for the good of a larger group. This is often the central normative theory expressed in the animal welfare legislation (Behdadi, 2012; Sandøe *et al.*, 2003; Würbel, 2009). It is important to keep this question of normative ethical theory apart from the question of whom or what is an object of moral concern (the 'center position') (Röcklinsberg, 2001). According to the anthropocentric position humans have a higher moral standing than all other animals, whereas a sentientistic view takes the stance that all sentient beings are morally significant in themselves and a biocentric view include all living objects, i.e. also the species level (Curry, 2011; Röcklinsberg, 2001). So when justifying if suffering is necessary or not in a specific case the decision will differ with regard to who is at all taken into consideration – humans, the individual animal or a species? In an animal welfare standard, animals are self-evidently included but also other concerns are taken into account.

The private standards more explicitly stated that they cared both about the consumers' demands and producers' prerequisites. Some of the standards were clear about the fact that they included rules believed to be the most important for the consumers. This indicates that also a contractarian ethical view can be seen in the policy work. According to a contractarian not the consequence is important but the mutual agreements (explicit or implicit) within the moral community. As animals cannot enter into agreements they are not a part of the moral community (cf. Sandøe and Christiansen, 2008). Therefor it is, according to this standpoint, not wrong *per se* to let animals suffer. But suffering can be unnecessary if an act or procedure is generally unaccepted by the society. Hence; if the consumers would not accept castration of piglets this would then be an action causing unnecessary pain. It all depends on what is in the 'contract' between producers and retailers/consumers; high quality meat, cheap products, safe products, animal friendly husbandry, environmental friendly production, etc. Or what is in the 'contract' between policy makers and producers; a guarantee to financially manage the competition that exists on the international market? Or even between producers and animals and the environment (Lund *et al.*, 2004).

There are signs of virtue ethics in legislation (Behdadi, 2012), and also in studied standards, such as emphasis on stock keeper's attitudes and actions in UK Freedom Food standard. The virtue ethics propose one should develop certain favorable character traits; be generous, courageous, just, self-controlled, sympathetic, loyal, patient and honest (Rachels, 2007). Focus is not on the amount of suffering and welfare of the animal, or on who has the right to be a part of the moral community, but rather on what kind of person I want to be in relation to the animal – how do I want to act in relation to other beings? (Hursthouse, 2006). It is not a virtue to cause suffering and you are not a good person if you do, and hence quite a lot of suffering is unnecessary for a virtue ethicist. Virtue ethicists such as Nussbaum and Hursthouse question the intensive animal husbandry systems of today were the animals have small opportunities to develop and 'flourish' as the animals/species they are (Hursthouse, 2006; Nussbaum, 2006).

An example of duty ethics is the obligation on the moral duty to treat the animals well for producers certified by Swedish Seal of Quality. Another example of signs of duty ethics is the German standard Neuland stating that production animals are to be treated with respect as ends and not as means.

What '(un)necessary' suffering refers to differs between normative ethical theories and consequently it differs between people, views and countries (EURObarometer, 2005). This is probably one of the reasons why some people think that the intention with the legislation is not fulfilled when certain procedures and actions are allowed, and why others think that the legislation sometimes is too strict. Also the definition of animal welfare is not at all solely dependent on science, particularly in relation to legislation and standards, but depends also on philosophy, ethics and economics. Policymakers have to balance the interest of different stakeholders (Croney and Millman, 2011) but do they know what the consumers want? And do consumers have enough knowledge about animal production to know what they buy? Algers (2011) pointed out that the consumers of today are lacking knowledge about animal welfare and husbandry to make good choices from an animal welfare point of view.

Other studies (Lassen *et al.*, 2006) however show that allowing natural behavior and outdoor access seem to be what concerned consumers ask for. To have 'natural behaviour' as a claim is probably strategically wise when appealing to consumers and citizens. It can be hypothesized that this is why some legislation and standards have a focus on the animals' possibilities to 'behave naturally' as a way to promote non-suffering. According to Lassen *et al.* (2006) 'living a natural life' is of importance for laymen when it comes to what good animal welfare is, while the absence of suffering and frustration are central in the experts (farmers, technicians, scientists) approach.

Conclusion

There are several definitions of the concept 'unnecessary suffering', e.g. legally acceptable level of suffering, in relation to AW legislation and standards. These different definitions can depend on different factors; the intensity and duration of the suffering, the intention behind an act that caused the suffering, the fulfillment of human interests and the animals' interest, but also aspects related to consumer expectations.

It was clear that different initiators build on different definitions of welfare and hence make different assessments as to whether an action or situation results in a suffering for the animals or not, and have different views on what *kind* of suffering is unnecessary. The legislation/standards allowed and forbid different actions/situations. These differences could be seen on four levels: (1) between countries; (2) within a country but between different legislation/standards; (3) between different species of animals covered by the same legislation/standard; and (4) between individuals of the same species covered by a legislation/standard.

The differences between legislation and standards in different countries indicate that several factors, besides animal welfare science, affect the outcome of policy work. Above cultural views, traditions and financial calculations ethical theories accept different levels of suffering as unnecessary, which is relevant for both policymakers and consumers when making decisions. The fact that indicators of different ethical views can be detected in legislation and standards is not unexpected, since this mirrors the complexity of the society. When using specific concepts in legislation (like 'unnecessary suffering' or 'natural behaviour') that can be interpreted in different ways depending on the individuals' own ethical view one must be prepared for comments about an inconsistent framework. We believe that one way to prevent and handle such opinions and assist the consumers when making choices is to support transparency as regards levels of acceptable/unacceptable suffering, both within the food chain and the policy processes. If legislation and especially standards are supposed to mirror consumers' will and intentions, they must have a chance to know what reality looks like and based on this inform the policy makers about what they want.

References

Algers, B. (2011). Animal welfare – recent developments in the field. Perspectives in Agriculture, Veterinary Science, Nutrition and Natural Resources. 6. Available at: www.cabi.org/cabreviews.

Bayvel, A.C.D. (2004). Science-based animal welfare standards: the international role of the Office International des Epizooties. Animal Welfare 13: 163-S169.

Behdadi, D. (2012). 'The compassionate stock-keeper' and other virtous ideals. Values and definitions in the animal welfare legislations of the United Kingdom, Spain and Argentina. Report 32. Department of Animal Environment and Health. Swedish University of Agricultural Sciences, Skara, Sweden.

Bentham, J. (1789). A utilitarian view. In: Regan, T. and Singer, P. (eds.) Animal rights and human obligations (1989), Prentice Hall, Englewood Cliffs, NJ, USA, pp. 25-26.

Braithwaite V.A. and Boulcott, P. (2007). Pain perception, aversion and fear in fish. Diseases of Aquatic Organisms 75: 131-138.

Chandroo, K.P., Duncan, I.J.H. and Moccia R.D. (2004). Can fish suffer?: perspectives on sentience, pain, fear and stress. Applied Animal Behaviour Science 86: 225-250.

Croney, C.C. and Millman, S.T. (2007). Board-invited review: The ethical and behavioral bases for farm animal welfare legislation. Journal of Animal Science 85: 556-565.

Curry, P. (2011). Ecological ethics – an introduction. Polity Press, Cambridge, UK.

EFSA (2009). Scientific Opinion of the Panel on Animal Health and Welfare on a request from European Commission on General approach to fish welfare and to the concept of sentience in fish. The EFSA Journal 954: 1-26.

Elwood, R.W., Barr, S. and Patterson, L. (2009). Pain and stress in crustaceans? Applied Animal Behaviour Science 118: 128-136.

Eurobarometer. (2005). Attitudes of consumers towards the welfare of farmed animals.

European Commission (2007). Treaty of Lisbon.

Forsberg, E-M. (2011). Inspiring respect for animals through the law? Current development in the Norwegian animal welfare legislation. Journal of Agricultural & Environmental Ethics 24: 351-366.

Gregory, N.G. (2004). Physiology and behavior of animal suffering. Blackwell Publishing, Oxford, UK.

Hurnik, F. and Lehman, H. (1982). Unnecessary suffering: definition and evidence. International Journal for the Study of Animal 3(2): 131-137.

Hursthouse, R. (2006). Applying virtue ethics to our treatment of the other animals. In: Welchman, J. (ed.) The practice of virtue. Hackett Publishing Company, Cambridge, MA, USA.

Landera-Luri, M. (2010). Animal welfare regulations in the EU: a matter of ethics or a matter of human economic interest? In: Casabona, R. (ed.) Global food security: ethical and legal challenges. Wageningen Academic Publishers, Wageningen, the Netherlands, p. 354-359.

Lassen, J., Sandoe, P. and Forkman, B. (2006). Happy pigs are dirty! Conflicting perspectives on animal welfare. Livestock Science 103: 221-230.

Lund, V., Anthony, R. and Röcklinsberg, H. 2004. The ethical contract as a tool in organic animal husbandry. Journal of Agricultural and Environmental Ethics 17(1): 23-49.

Lund, V., Mejdell, C.M., Röcklinsberg, H., Anthony, R. and Håstein, T. (2007). Expanding the moral circle: farmed fish as objects of moral concern. Diseases of Aquatic Organisms 75: 109-118.

Nussbaum M. (2006). Frontiers of justice: disability, nationality, species membership (The Tanner Lectures on Human Values). The Belknap Press of Harvard University Press, Cambridge, MA, USA.

Rachels, J. (2007). The elements of moral philosophy. 5th edition. McGraw-Hill Higher Education, New York, NY, USA.

Röcklinsberg, H. (2001). Das Seufzende Schwein. Zur Theorie und Praxis in deutschen Modellen zur Tierethik. Monograph, doctoral thesis. Harald Fischer Verlag, Erlangen, Germany.

Sandøe, P. and Christiansen, S.B. (2008). Ethics of animal use. Blackwell Publishing, Oxford, UK.

Sandøe, P., Christiansen, S.B. and Appelby, M.C. (2003). Farm animal welfare: the interaction of ethical questions and animal welfare science. Animal Welfare 12: 469-478.

Striwing, H. and Åslund, M. (2005). Djurskydd – fakta och tips. Striwing, Falun, Sweden.

Underwood, W.J. (2002). Pain and distress in agricultural animals. Journal of the American Veterinary Medical Association (JAVMA) 2: 208-211.

Vapnek, J. and Chapman, M. (2010). Legislative and regulatory options for animal welfare. FAO, Rome, Italy.

Veissier, I., Butterworth, A., Bock, B. and Roe, E. (2008). European approaches to ensure good animal welfare. Applied Animal Behaviour Science 113: 279-297.

Wahlberg, B. (2011). Reglering och förvaltning av produktions- och slaktdjurs välbefinnande – En offentligrättslig undersökning. Doctoral thesis. Åbo Akademi. Turku, Finland.

Würbel, H. (2009). Ethology applied to animal ethics. Applied Animal Behaviour Science 118: 118-127.

Animal welfare labelling: is the market the right governance structure to meet people's moral concerns?

F. Pirscher
Institute of Agricultural and Nutritional Sciences, Martin Luther University Halle-Wittenberg, 06099 Halle, Germany; frauke.pirscher@landw.uni-halle.de

Abstract

The introduction of an animal welfare label is intensively discussed in Europe. The political debate is dominated by the technical question how to develop the label, while the fundamental question whether a market based approach is the right governance structure to deal with public moral concerns is nearly neglected. However, the decision to turn states of different animal welfare level via labelling into a process quality of meat is based on normative assumptions that are not without controversy. Following the school of social constructivism one can assume value articulating institutions as a social construct that determines what kind of values can be expressed and also influences the formation of values. Markets as institutional settings imply assumptions on the understanding of rationality, preferences formation and making choices. Therefore, it is the aim of this paper to disclose the normative implications associated with commodifying moral concerns regarding animal welfare and to induce a discussion whether markets are the appropriate institutional frame for animal welfare problems. The paper focus' on the following questions: Who's moral concerns can be articulated and who's are excluded on a market? What kind of ethical value system regarding animal welfare is implicitly assumed by a market approach? What does the interpretation of moral concerns as individual preference for process quality imply for dealing with ethical conflicts? It can be shown that only the moral concerns of compassionate carnivore consumers are taken into consideration. This group must also have a utilitarian belief systems that allows for trade-offs between animal well-being and other material goods. Deontological systems are not compatible with the assumptions of neoclassical economics. Lexicographical ordering of preferences is here not assumed as rational. Finally, by interpreting animal suffering as a problem of market failure, the normative discussion of how should we deal with animals is reduced to a technical question of making individual preferences explicit. Preferences themselves are not under discussion.

Keywords: animal ethics, institutions, value articulation, social constructivism

Introduction

A growing part of Europeans worries about the wellbeing of farm animals in standard production systems. They are concerned about the living conditions and treatment of animals at farm level, their transport from farm to slaughterhouse and the process of slaughtering itself. For many people the current legal standards do not fulfil their conception of animal welfare. Therefore, they have being calling for more animal friendly production systems since many years. As response a multitude of production lines with higher welfare standards has been developed in many countries. All of them created an individual animal welfare label. To increase consumers' transparency and trust in labels, some countries developed a nation-wide label with an external certification system. Since several years an European wide label system is under discussion. The political debate is dominated by the technical question how to develop the label, while the fundamental question whether a market based approach is the right governance structure to deal with public moral concerns is nearly neglected. However, the decision to turn states of different animal welfare level via labelling into a process quality of meat is based on normative assumptions that are not without controversy. Markets as institutional settings imply assumptions on the understanding of rationality, preferences formation and making choices (Vatn, 2005). Therefore, it is the aim of this paper to disclose the normative implications that are associated with commodifying moral concerns regarding animal welfare via labelling and by this to induce a discussion whether markets are an appropriate institutional frame to treat animal welfare problems.

Labelling as a market based instrument to increase animal welfare

To understand the functioning of a labelling systems it is important to first look at key assumptions of neoclassical economics that serves as normative basis of the market approach for decision making. Relaying on an utilitarian approach individual preferences are at the core of decision making in neoclassical economics. They are viewed as given and stable for each individual, but might differ between them. Under the assumption of full information, zero transaction costs and individual property rights, an individual can satisfy her preferences on a market, where she articulates them via her willingness to pay for certain products. Here, individuals are interpreted as consumers and preferences exist only for commercial goods. Preferences are value neutral and their creation is seen as independent of any cultural or social influence. The limited budget, however, does not allow the realisation of all individual preferences. Therefore, the individual has to select from all possible options that mix of goods that maximize her utility under the given restrictions. Within the market model this behaviour is viewed as rational choice. It is assumed that individual utility maximizing leads also to the societal welfare optimum. Interpreting values as preferences and transforming them into willingness to pay allows to make them commensurable and exchangeable. Losses in one sphere can be compensated by benefits in another. Goods can be substituted and consequently a trade-off between them regarding their ability to satisfy individual needs exists.

Modern consumer theory assumes that not the product as such but a bundle of quality attributes determines its value for consumers (Lancaster, 1966). However, not all attributes of a good can be observed before the buying decision. Some, like the taste or the storage life, can be evaluated afterwards, some, like environmental impact of production or animal welfare, cannot be evaluated by the consumer at all or only with prohibitive high transaction costs. In this case information asymmetries between consumers and producers about the product or process quality can lead to market failure. As a consequence uncertainty and positive transaction costs for consumers make the maximization of utility costly or even impossible. In such a situation labels serve as an instrument to overcome this information asymmetry and reduce transaction costs. If consumers trust this instrument, whether public or private, it will help to overcome market failure.

In the economic interpretation animal welfare is either assumed to be a process quality of meat with credence character, because it cannot directly be observed by consumers (Kehlerbach *et al.*, 2012; Liljenstolpe, 2008) or an external effect that is not reflected in the meat price (Lusk, 2011). Both cases lead to market failure. A label for animal welfare can be interpreted as an instrument to allow consumers to articulate their preference for animal welfare beyond the current legal minimum standard via their willingness to pay on a market. Animal welfare is viewed as part of their individual utility function. While some economists (Carlsson *et al.*, 2007) argue that beside this private valuation of animal welfare there exists also a public one that is not exclusively linked to own consumption and thus should not be regulated by labelling, they still view monetary bits as the adequate value articulating institution to elicit the societal attitude towards animal welfare. Against this background I would like to reflect the decision to opt for a market based approach for regulation and elicitation. In the following I will do this from the perspective of social constructivsm.

Social constructivism

Following Vatn (2005) markets can be understood as one of several possible institutions to organize choices. In contrasts to the view of neoclassical economics, choices are not understood as independent from the institutional context, but reflect norms, rules and expectations, people have into the institutions of a society. What is viewed as rational cannot be generalised for all institutional settings. To interpret rational choice as individual utility maximizing is appropriate when goods are exchanged on markets. For institutional structures like a family, a club or a community more social or cooperative behaviour may fit better. Thus, rationality is institutionally dependent. Institutions themselves are a social construct that determines how a society interprets the world and creates common values about what people view as normative appropriate behaviour. This means that institutions have a cognitive and a normative aspect. Choosing one institutional setting implies assumptions on the understanding of rationality, preferences formation and making choices.

Who's moral concerns can be articulated and who's are excluded on a market?

Choosing markets as the appropriate institutional arrangement means that improving animal welfare is viewed as a consumer and not as a citizen issue. 'We focus on transparency and the power of consumers' said the German Agricultural Minister, when she introduced the new animal welfare label in January 2013 (Agrarheute, 2013). One can argue that consumer behaviour is not independent of citizen norms and that it is the central idea of the label to allow people to articulate their moral attitude towards animal welfare via their buying decision and thus act in accordance to their normative rules. However, a market based approach discloses all those people, who do not buy meat products or animal products at all because of their different moral understanding of animal welfare, like vegetarians and vegans. Thus, the market allows only one group, the compassionate carnivores, to contribute to an improvement of animal welfare and, as Lusk (2011) points out, they have to bear all costs of improvement. Vegetarians and Vegans can only act on a market via boycott and by this indicate a reduced demand for animal products. Less demand means keeping less animals under conditions that are not acceptable for vegetarians or vegans, but it does not allow to influence the living conditions for those animals that are still demanded by the carnivores. Thus, the moral concerns of vegetarians and vegans are disclosed by the selection of the institutional setting.

What kind of ethical value system is implicitly assumed by a market approach?

Empirical studies show that for a great part of the society high animal welfare standards are an important attribute of meat (European Commission, 2005). The reasons for this are manifold. Some view the products as qualitatively better, e.g. more tasty or healthier, or they feel better when animals had a better live before consumption. The introduction of a label allows them to identify those products with a higher standard and in buying them consumers' wellbeing rises by increasing animal wellbeing. The higher the aggregate willingness to pay the more animals are better of. How many animals can profit from the label is determined by the individual preference structure of the carnivore consumers expressed by their willingness to pay, the production costs for increased welfare standards and the relative prices of all other products. Changes in each factor leads to changes in the demand for animal welfare labelled products. Animal wellbeing is assumed to be fully commensurable with other goods or product qualities. The development of a label with different levels of animal welfare (legal minimum standard for animal welfare, increased level, premium level) implies that wellbeing can be gradated and that it exists in discret units. If the motivation for improved animal welfare is to increase ones own individual welfare, the market is an adequate value articulating institution. Although utilitarism was one of the first ethical systems where sentient animals were considered intrinsically valuable (Bentham, 1789, Singer, 1975), in it's neoclassical reduction animal welfare cannot directly be taken into account. An anthropocentric welfare function is always assumed.

However, individual utility maximisation is not the only reason why people care about animal welfare. Some people want a higher welfare standards because they view sentient animals as intrinsically valuable like in pathocentric but also in biocentric and holistic ethical concepts. In this understanding the life of each individual, or at least the life of those who feel pain and pleasure counts, whether other individuals value that life or not. When intrinsic values are interpreted as rights, wellbeing exists independently of any utility considerations (Regan, 1983). Not only within the animal welfare movement but within the whole society a growing percentage argues for animal rights. These people do not follow an utilitarian but a deontological ethical value system. Having a right means that all entities, who fulfil the criteria for assigning the right, posses it equally. If farm animals as sentient beings have a right to live a life appropriate to their species, all farm animals have the same right. A benefit of an action cannot justify any violation of the right. Such a deontological ethical position does not follow a compensatory decision making process, but either assumes that some values are totally incommensurable or it decides on the basis of hierarchal ordering. Incommensurability means that no appropriate value exists at all that allows to compare different options as 'better', 'worse' or 'equal' (Holland, 2002). Vatn and Bromley (1994) call it the incongruity problem. Hierarchal ordering or lexicographical choices only limit the scope for trade-offs. Before a certain threshold of a good and services has not been realised no other choices for goods are made (Rosenberger, 2003). Incommensurability and lexicographical ordering violate the exchange value assumption in neoclassical economics (Anderson, 1993). Thus, assigning animal rights means rejecting the ability to make trade-offs. Here, not the rationality of individual utility maximisation but the social rationality of formulating norms about how to treat animals rightly is applied by those who value animals intrinsically. Price bids over a product or a product attribute that people do not view as a marketable good will lead to randomness of the result rather than exactly eliciting their attitudinal believes towards animal welfare.

What does the interpretation of moral concerns as individual preference for process quality imply for dealing with ethical conflicts?

Most contemporary societies do not give a homogeneous answer to the question how humans should treat farm animals. Neither do philosophers. Value pluralism is a typical characteristic of secular societies not only in animal ethics. Accepting this situation does not mean that any kind of animal treatment is morally acceptable but as Beekman and Brom (2007) pointed out that for pluralistic democracies there is a need to design ethics as a platform for value debates. Values and beliefs of oneself and others have to be critically reflected and justified. This moral inquiry can only succeed by collective engagement of individuals (Kupper and De Cock Buning, 2011). The decision for or against a value articulating institution influences itself the value debate and its outcome. Each institutional frame has its advantages and disadvantage. The merits of a market based approach that aims increasing animal welfare via labelling are its universal applicability, its low transaction cost for implementation and control, its voluntary character and its easy political implementation. However, a market approach implicitly assumes several value judgements and excludes them from a societal discussion. As already mentioned it is not compatible with alternative value concepts besides anthropocentric utilitarism. Furthermore, there is no need to justify ones desires or preferences, because they are viewed as given and stable. Therefore, a market approach cannot be open for deliberation and reasoning if we understand deliberation not as process of discovering preferences, but as 'a process reflecting upon what there is most reason to want' (Holland, 2002: 23). A moral discourse will also be suppressed when animal welfare becomes a process quality. Here, the decision to choose or not to choose animal friendly products is transformed to a question of liking or disliking one special aspect of food, namely the animal friendliness. The normative question will be reduced to a question of exchange. By this it looses its moral content. Finally, the responsibility for animal welfare is automatically located at the level of the individual consumer, not at that of e. g. citizens. Here does not the communicative rationality – importance of an argument – count, but the calculative one – monetary bits (Vatn, 2005). Thus, it does not support a societal debate how to solve ethical conflicts in a value plural society. Nevertheless, if the majority of the society interprets improving animal welfare as an individual option of consumers to increase their individual utility, the market will be the right institutional setting and a trustworthy labelling systems will help to overcome information asymmetries between producers and consumers. However, if people do not view a market as the best fitting frame for the problems at stake it will lead to confusing results regarding value articulation. Hypothetical valuation approaches for environmental goods show that people react with protest bids in refusing to pay or with considerable uncertainty about the hypothetical value when asked for their willingness to pay for issues that include moral judgement like the existence value of species (Clark *et al.*, 2000; Spash and Haneley, 1995; Stevens *et al.*, 1991). By this they reject the framework chosen to deal with these ethical questions. We also can observe that the behaviour of consumers on markets for animal products is inconsistent with their articulated attitude towards animal welfare. While asked as citizens people stress the importance of animal welfare and often indicate a high willingness to pay for it. However, their real consumption patter does not show a corresponding consumer behaviour. The common interpretation of this phenomenon is a lack of acceptance of the price discrepancy between conventional and animal friendly products because of budget restriction, a tendency to free riding behaviour or a lack of transparency and trust in the existing labels. Sometimes we find as explanation that people do not feel responsible for animal welfare as consumers but want to shift the responsibility to others. From the perspective of neoclassical economics this behaviour can only interpreted as irrational. However, from the perspective of social constructivism it can also be seen as a rejection of the institutional framework.

Conclusion

If we do not question the institutional framework and automatically accept the market and its neoclassical assumption as the right governance structure to solve animal welfare problems a value debate within the society will not come up. Preferences and desires of consumers remain without any need for justification and reasoning. However, as Holland (2002) put it: 'desires function as reason only if there are reasons for a desire'. There is evidence that a considerable part of the European society do not share the value assumptions of this institutional framework and reacts with not acting on the market for animal products at all or with buying decisions not consistent with their preference structure.

References

Agrarheute (2013). Neues Tierschutzlabel: Viel Kritik und wenig Zustimmung. Available at: www.agrarheute.com/reaktionen-tierschutzlabel.

Anderson, E. (1993). Values in ethics and economics. Harvard University Press, Cambridge, MA, USA, 245 pp.

Beekman, V. and Brom, F.W.A. (2007). Ethical tools to support systematic public deliberation about the ethical aspect of agricultural biotechnologies. Journal of Agricultural and Environmental Ethics 20: 3-12.

Bentham J. (1789). Introduction to the principles of morals and legislation. Payne, London, UK, 347 pp.

Carlsson, F., Frykblom, P. and Lagerkvist, C.J. (2007). Consumer benefits of labels and bans on GM foods – choice experiments with Swedish Consumers. American Journal of Agricultural Economics 8981: 152-161.

Clark, J. Burgess, J. Harrison, C.M. (2000). 'I struggled with this money business': respondents' perspectives on contingent valuation. Ecological Economics 33: 45-62.

European Commission (2005). Special Eurobarometer 229. Attitudes of consumers towards the welfare of farmed animals. Available at: http://ec.europa.eu/food/animal/welfare/euro_barometer25_en.pdf.

Holland, A. (2002). Are choices tradeoffs? In: Bromley, D.W. and Paavola, J. (eds.) Economics, ethics and environmental policy. Contested choices. Blackwell Publishing, Oxford, Uk, pp. 17-34.

Kehlbacher, A., Bennett, R. and Balcombe K. (2012). Measuring the consumer benefit of improving farm animal welfare to inform welfare labelling. Food Policy 37: 627-633.

Kupper, F. and De Cock Buning, T. (2011). Deliberating animal values: a pragmatic-pluralistic approach to animal ethics. Journal of Agriculture and Environmental Ethics 24: 431-450.

Lancaster, K. (1966). A new approach to consumer theory. Journal of Political Economy 74: 132-157.

Liljenstolpe, C. (2008). Evaluating animal welfare with choice experiments: an application to swedish pig production. Agribusiness 24: 67-84.

Lusk, J.L. (2011). The market for animal welfare. Agricultural and Human Values 28: 561-575.

Regan, T. (1983). The case for animal rights. University of California Press, Berkeley, CA, USA, 425 pp.

Rosenberger, R. S., Peterson G.L., Clarke, A. and Brown, T.C. (2002). Measuring dispositons for lexicographic preferences of environmental goods: integrating economics, psychology and ethics. Ecological Economics 44: 63-76.

Singer, P. (1975). Animal liberation. A new ethics for our treatment of animals. New York Review, New York, NY, USA, 301 pp.

Spash, C.L. and Hanley, N. (1995). Preferences, information and biodiversity preservation. Ecological Economics 12: 191-208.

Stevens, T.H., Echeverria, J., Glass, R.J. Hager, T. and Moore, T.A. (1991). Measuring the existence value of wildlife: what do CVM estimates really show. Land Economics 67: 390-400.

Vatn, A. (2005). Rationality, institutions and environmental policy. Ecological Economics 55: 203-217.

Vatn, A. and Bromley, D.W. (1994). Choices without prices without Apologies. Journal of Environmental Economics and Management 26: 129-148.

Why pure procedural justice doesn't remove the individual responsibility to make right economic judgments

A. el-Wakil
University of Geneva, Av. Des Cavaliers 19, 1224 Chêne-Bougeries, Switzerland; elwakil0@etu.unige.ch

Abstract

In his *Theory of justice*, John Rawls argues that we as individuals can decide self-interestedly how we are to behave in the markets. In the economic system, justice doesn't require us to consider anything more than whether we can afford what we want or not. Distribution of wealth in society is just insofar as the basic structure of that society respects the two principles of justice and the rules of the market – a pure procedural justice system. However, our personal economic choices do modify the distribution of wealth in relevant ways; according to Gerald A. Cohen, Rawls' focus on institutions prevents him from achieving properly his project of a just society. His reliance on invisible hand mechanisms cannot guarantee the just character of the distribution; our individual choices, being an object of justice just as much as the institutions of the basic structure, should be guided by an ethos of justice. Having their 'ideal theory' considerations in mind, I follow Cohen by arguing that, contrary to what the Rawlsian model of pure proceduralism claims, our everyday choices in the current economic sphere should be informed by considerations of justice. I contend that the ideal notion of ethos can participate in the development of our capacity to carefully read prominent speeches, characteristic of the neoliberal age, that encourage uncritical, and self-interested consumption, if it is complemented by the practice of reflective judgment in the economic sphere. Should we achieve this, we will be better equipped to acknowledge and act upon the implications of our consumption on human and non-human life and development of the planet. The relevance of acknowledging this type of overlooked individual responsibility is exacerbated by the current context of increasing inequalities and decreasing amounts of resources, a situation of seemingly unrelenting crisis. By exercising a pressure upon how and which goods are to be produced and how wealth is to be distributed among people, our reflective consumption choices have the capacity, at the citizens' level, to make things change towards a fairer world.

Keywords: consumer, responsibility, pure procedural justice, reflective judgment, ethos

Introduction

As the behaviour of a great number of persons in the current Western liberal democratic societies shows, it seems to be commonly thought that the way we[9] spend our money should be a purely personal choice. My generation and the younger ones were raised in a context in which, as long as we manage to pay, we can decide what we want to buy according to our own preferences; it is the government's job, not ours, to organise things so that the production of food products and its distribution be fair to all. However, our everyday choices' impact is consequential: depending on how we spend our income, we can participate in wasting alimentary goods or make efforts to reduce the amounts thrown away; we can encourage the development of sustainable agriculture or slow it down; we can lower the economic power of farms that exploit animals or endorse their actions. In the current market economies, we as simple individuals have power in deciding what is to be produced, how it will be produced and who will receive money for it through the way we spend the money we have. Having that in mind, should we still consider that we can spend our money self-interestedly, in the way we want? Or should we use this power in particular, more responsible ways for the economic system to be just?

[9] Considering the nature of the project that should include the most people possible, as well as the message of personal engagement it promotes, I will make use of the 'we' form most of the time.

John Rawls and Gerald A. Cohen offer two different ways to consider this question and to think about the role individuals should play for the economic distribution to be just from an ideal theory perspective.[10] In his *Theory of justice*, Rawls (1999) argues that the distribution of income and wealth is just independently from individual preferences and economic behaviour. He defends what he calls 'pure procedural justice' mechanisms: if the background institutions of the society are just, and that if the procedure they set is respected, the distribution of income and wealth will always be just. The just character of the distribution is totally independent from individual choices, which can thus be completely self-interested.

In *Rescuing equality and justice*, Cohen (2008) criticizes this conception of a just society that, according to him, misses something essential. He claims that by focusing only on institutions, Rawls actually ignores the impact that individual choices have on the distribution of income and wealth. Rawls' reliance on invisible hand mechanisms prevents him from achieving his own project of reducing the impact of unequally distributed, morally arbitrary features on our life opportunities. In Cohen's view, the just character of the distribution cannot be guaranteed if we can behave in a self-interested way; our individual choices, guided by an ethos of justice, should be an object of justice just as much as the institutions of the basic structure.

Because it promotes the idea that distributive justice doesn't need to consider individual actions, the Rawlsian pure procedural justice cannot guarantee the existence of a just society. I thus argue here with Cohen that our everyday choices within the just basic structure should also be informed by considerations of justice. In the political, social, as well as in the economic spheres, we have responsibilities towards our fellow-citizens, and, even in the limits of the just institutions' regulations, we are thus not free to make completely self-interested decisions; we should collectively be guided by an ethos of justice promoting responsible and accountable consumption when we make economic choices. This guiding ethos should be complemented by the individual responsibility to practice reflective judgment in the economic sphere. As I show, these elements don't restrict freedom or impose a too heavy burden on individuals. On the contrary, the social pressure for adopting a new and just ethos of consumption could be a way to fight the current prominent speeches, characteristic of the neoliberal age, that encourage uncritical, unreflective and self-interested consumption. Because our choices have a big impact on other human and non-human beings and on the planet, we should develop our capacity to make accountable decisions regarding our consumption.

Rawls' pure procedural justice and individual responsibility in the markets

While I take the main arguments of *A theory of justice* to be well known, I will briefly present Rawls' view of the role played by individuals in the just society and the conditions for a just distribution of income and wealth. The Rawlsian just society follows two principles of justice: the first one sets the right to equal liberties for every individual, while the second one allows for conditional social and economic inequalities once the first principle is secured. These inequalities are justified if two conditions are met (Rawls, 1999: 72, 266): first, the structure must make it possible for everyone to access all 'offices and positions' (Rawls, 1999: 72) in the society (*fair equality of opportunity*), and second, while they can't reward morally arbitrary individual features as talent or beauty, all the inequalities should benefit the worst-off (*difference principle*).

These two principles apply to the basic structure of the just society, namely the institutions that have the most profound impact on individuals' lives. They are the 'primary subject of justice' (Rawls, 1999:

[10] They both consider only one hypothetical and closed society. Even if that seems very limitative in today's world economy, I will adopt their scope and stay at the level of one society in which only citizens reside.

6) because they regulate the distribution of 'primary goods', namely liberties, rights, opportunities, and income and wealth – defined as the goods all individuals are supposed to want in order to pursue their plan of life (Rawls, 1999: 6, 79). The society is thus just and avoids moral arbitrariness by distributing the primary goods in a just way if the background institutions respect the two principles of justice.

Whereas the justice for institutions is essential for the just society, 'a complete theory of right includes principles for individuals as well' (Rawls, 1999: 93). Rawls thus argues that individuals would unconditionally have 'natural duties' (Rawls, 1999: 100). The most important of them is the duty 'to support and to comply with just institutions that exist and apply to us' (Rawls, 1999: 99), in the limits of what is acceptable 'with little cost to ourselves' (Rawls, 1999: 294).

Called the 'duty of justice' (Rawls, 1999: 99, 293), this duty is essential to Rawls' concept of pure procedural justice, which he describes as follow: 'pure procedural justice obtains when there is no independent criterion for the right result: instead there is a correct of fair procedure such that *the outcome is likewise correct or fair, whatever it is, provided that the procedure has been properly followed*' (Rawls, 1999: 75, emphasis added). The duty of justice ensures that individuals will properly follow the procedure set by the just institutions, thus carrying out the fairness of the procedure to its outcome (Rawls, 1999: 75).

However, all the procedures are not equivalent to ensure that the outcome they produce will be just. For example, the majority rule is an imperfect procedure regarding efficiency. While its objective is to set a just constitution and just regulations, it is not possible to claim that any outcome of a vote will be a just one: 'this question is one of political judgment' (Rawls, 1999: 313-314). The procedure will not be just no matter what the individuals do; 'a just constitution must rely to some extent on citizens and legislators adopting a wider view and exercising good judgment in applying the principles of justice' (Rawls, 1999: 317). While the rules voted must be accepted as such, individuals should keep their own judgments regarding their just character (Rawls, 1999: 314).

On the contrary, structures as the structure of gambling or the market one, which Rawls adopts as the exemplary system to distribute income and wealth (Rawls, 1999: 239-242), are perfect procedures regarding efficiency. 'The market achieves an efficient outcome even if everyone pursues his own advantage' (Rawls, 1999: 316). Once the rules aiming at limiting the 'market failures and imperfections' (Rawls, 1999: 240) are set by the just institutions, this perfect procedure implements pure procedural justice (Rawls, 1999: 242). Starting from a position of absolute equality between the people (Rawls, 1999: 69), economic actors and individuals following the procedure only have to pursue their own, self-interested ends (they actually are 'not prepared to abandon their self-interest' (Rawls, 1999: 248)) and to believe that the outcome they will collectively produce will be a just distribution. 'It is a mistake [...] to consider that every change, considered as a single transaction viewed in isolation, be in itself just' (Rawls, 1999: 76). Individuals don't need to have an opinion about 'what is from a social point of view the most efficient economic configuration' or even 'to know what it means' (Rawls, 1999: 316) for the outcome to be efficient. The distribution of income and wealth is just *a priori*, independently from individual decisions and choices, as long as other institutions that aim at implementing the two principles of justice set its rules and boundaries.

Hence, a pure procedural justice system allows us to behave in the way we want in the economic sphere as long as we stay in the limits of the just institutions' rules and of our natural duties of justice (Rawls, 1999: 248). Moreover, we have no duties at all in noninstitutional situations (Rawls, 1999: 295). Despite the huge impact that the economic system has on our lives (Rawls, 1999: 229), we are not required to have opinions about the way we behave in it; 'any judgment [we] make is from [our] own point of view' (Rawls, 1999: 316). Applied to the issue of consumption, these ideal theory considerations mean that we can consume how and what we desire in the limits of the rules set by the just institutions. We couldn't buy, for example, products produced by exploitation of others, because equal freedom of all would be protected; but it would not be required from us to understand and evaluate the meaning of our consumption, or to think about its impact on other beings or on the planet. Since the institutions would guarantee the just character of the distribution, it will be equally just if we spend our income self-interestedly or altruistically, or if we buy organic cheese at a local farm or industrial one at the biggest supermarket. Justice in the Rawlsian sense doesn't require us to consider anything but our own interest when it comes to economic decisions; all we have to do is to comply with the rules.

Cohen and the necessity to consider individual behaviours

Cohen calls the Rawlsian distinction between the basic institutions, to which the principles of justice apply, and the individuals, who don't have to respect these principles in their everyday life, the 'moral division of labor' (Cohen, 2008: 8). In his egalitarian critique of the *Theory of justice*, he adopts the same starting point as Rawls: the inequalities of life opportunities as well as of income or wealth are unjust if they are based on morally arbitrary criteria. However, Cohen claims that the just basic structure alone cannot guarantee that the system of distribution will eliminate these unjust inequalities (Cohen, 2008: 123).

Actually, he argues that the difference principle, which states that economic inequalities are permitted only if they enhance – or at least do not deteriorate – the situation of the worst-off (Cohen, 2008: 29), leaves a possibility for individuals to justify a higher income on an unjust ground (Cohen, 2008: 119): while they should only ask for a higher income if that incentive was literally necessary for them to be as productive as they can be (Cohen, 2008: 29), the talented people might demand a higher income even if they actually could already produce more for the society without this inequality. Implemented by the institutions of the basic structure and not by the individuals themselves, the difference principle thus makes it possible to reward talent, a morally arbitrary feature, despite the fact that such an inequality is denounced as unjust in Rawls' theory.

According to Cohen, the cause of this inconsistency is to be found in an ambiguity in Rawls' imprecise definition of the basic structure: it is indeed unclear whether Rawls includes or not the family in the social institutions (Cohen, 2008: 132). In Cohen's view, it is clear that *A theory of justice* must include the family in the basic structure because of the profundity-of-effect criterion that justifies taking the basic institutions of the society as primary objects of justice in the first place (Cohen, 2008: 136). And if this informal structure is comprehended, so must be they 'day-to-day choices within it' (Cohen, 2008: 135), because it is impossible to distinguish them from this structure. The Rawlsian theory thus implicitly requires the informal institution of the family to be included in the basic structure – and individual choices must thus be understood as subjects of justice as well.

A just distribution requires more from individuals than pure self-interestedness. In Cohen's view, 'principles of distributive justice, [...] that is, about the just distribution of benefits and burdens in society apply, wherever else they do, to people's legally unconstrained choices' (Cohen, 2008, 116). Even if Rawls repeats several times that it would be good if the citizens had a desire to act in a virtuous way (Rawls, 1999: e.g. 49) and accumulates the mechanisms to ensure their compliance with the institutions (hypothesis of full-compliance, natural duty of justice, assurance problem), there cannot be a rule for each of our actions: that is not only an impossible task, but also an undesirable one with respect to freedom (Rawls, 1999: 496). It is thus 'not possible to achieve distributive justice by purely structural means' (Cohen, 2008: 127).

Then how is it possible to obtain that individuals make just choices? Cohen claims that, for the Rawlsian model to make justice possible, 'not only just coercive *rules*, but also an *ethos* of justice that informs individual choices is needed' (Cohen, 2008: 16). He defines this ethos as a 'structure of response lodged in the motivations that inform everyday life' (Cohen, 2008: 123). Such an ethos is necessary for the Rawlsian difference principle to set the basis for a just distribution (Cohen, 2008: 16); it is only if all the individuals actually believe in the righteousness of the principle that they will want to produce more for a normal income (Cohen, 2008: 101) and 'to share equally the greater product produced' (Cohen, 2008: 102). This ethos of justice should thus be shared broadly in the society, as it 'promotes a distribution more just than what the rules of the economic game by themselves can secure' (Cohen, 2008: 123).

Because the Rawlsian pure procedural justice doesn't evaluate individual behaviours, it cannot guarantee alone the achievement of a just distribution. As Cohen shows, 'the individual must be as dedicated to such justice as the state is' (Cohen, 2008: 10); because we have a responsibility to make choices in accordance with our conception of justice in order to make a just distribution of income and wealth possible, we, as individuals, are accountable for the way we decide to work, to produce something or to spend our money. 'Justice evaluates everyday economic choice' (Cohen, 2008: 139-140). This also means that, contrary to what Rawls claims, we should have opinions in the economic sphere as well as in the political one; we shouldn't stop 'adopting a wider view and exercising good judgments in applying the principles of justice' (Rawls, 1999: 317) when it comes to economic decisions.

The ethos and freedom

Cohen's critique to Rawls shows that our individual choices matter for the distribution of social wealth to be just. But imposing a specific ethos and responsibilities to individuals might trouble those concerned with freedom – a value emphasized in the first and lexically prior Rawlsian principle of justice.

However, the Rawlsian idea of freedom already limits the extent of individual freedom in several ways. First, it restricts the possible scope of pluralism by adopting the very exclusionary model of individuals as rational *homini-oeconomici*, capable of planning their lives on the long-term, not envious of others and so forth (Rawls, 1999: e.g. 109, 360, 465); by emphasizing the need for them to share a 'thin morality' (Rawls, 1999: 348-349); and by claiming that, once the just basic structure is set, a broader consensus, a 'thick morality,' should exist among the individual (Rawls, 1999: 349). Second, as I discussed in the first section, Rawls claims that we have natural unconditional duties. He argues that these duties are not in conflict with freedom because equal freedom would already be guaranteed by the implementation of the first principle of justice by the institutions (Rawls, 1999: 295). Moreover, 'Rawls claims that in a well-ordered society, people will internalize the natural duty to support just institutions and acquire a sense of justice that guides and motivates them' (Carens, 1989: 42), thus transforming these coercive duties into voluntary behaviours. Finally, the third and most obvious restriction to freedom is the existence of coercive social institutions that constrain our choices and force us to behave in a certain way. In Rawls' view, these coercive interventions on how people live their lives aren't violating freedom as long as they respect the two principles of justice. But then, Cohen asks, 'why might an egalitarian ethos be thought to be unacceptably demanding, when maximizing legislation is not thought to be?' (Cohen, 2008: 203).

These elements show that Rawls' concept of freedom is far from being absolute. It is thus difficult to object that a guiding ethos of justice inspiring our choices violates it. What Cohen actually does is to remind us that, outside of the Rawlsian frame of coercive institutions and self-interested individuals, we have plenty of other reasons to act in a certain way: 'moral convictions and fellow-feeling' (Cohen, 2008: 197), developed in relation to institutional (formal and informal) regulations and rules (Cohen, 2008: 144-146), are motivations as well. 'Morally inspired motivations' (Cohen, 2008: 198) can even do more than coercive institutions without being coercive, for example to fight racist attitudes (Cohen, 2008: 199).

In Cohen's view, social pressure can make an ethos change: as in his examples of sexism and ecological consciousness, 'a path becomes easier and easier to follow as more and more people follow it' (Cohen,

2008: 142). I believe this element to be very relevant for the issue of consumption in the current, non-ideal world. As a matter of fact, since the end of the 1980's, most of the people think about their consumption in the way the dominant neoliberal ethos has set. Encouraging uncritical and self-interested consumption, this ethos has certainly made it possible to sell much more goods than before; but, in the alimentation domain, it has also prevented new models of agriculture to develop and participated in wasting food products, as in Europe where, according to the European commission, 90 million tonnes of food is wasted annually. A new Cohenian ethos of more responsible and altruistic consumption should thus replace the current one and guide our economic choices, which are relevant to the way our society develops itself, as well as the way in which wealth is distributed.

Accountable, reflective choices

However, I would like to express a reservation concerning the notion of ethos. Actually, Cohen doesn't define its nature very clearly; he gives no answer regarding how an ethos should be chosen, defined, and implemented in the society. Would there be a unified ethos or various ones? Would we choose to adopt it or would it be imposed to us? And how far would we be allowed to criticize it?

Knowing that, as Cohen himself acknowledges (Cohen, 2008: 141), some senses of justice can contain unjust guiding lines or be unfair, taking an ethos as granted can be dangerous. Whereas the virtues of an ethos to engage a collective shift towards a new way of behaving and to guide people's free choices must be adequately appreciated, we should do more than just follow a rule or a guide. As I attempted to show, against Rawls, that judgment is necessary in the economic sphere, I argue that it is necessary to keep practising reflective judgment in the context of an ethos.

Reflective judgment requires us to go beyond our own point of view and to widen our perspective before making a choice. In Rawls' view, (political) good judgment demands that we adopt a 'wider view'; this new perspective can be obtained through discussion with others. 'In everyday life the exchange of opinion with others checks our partiality and widens our perspective; we are made to see things from their standpoint and the limits of our vision are brought home to us' (Rawls, 1999: 315). This way of making judgment should apply to all the choices we face, the issue of consumption included. If the ethos encouraging us to consume in a more altruistic, justice- and impact-oriented way can help us changing our behaviour by the means of social pressure, the practice of judgment plays a part in choosing which new ethos to adopt. As a long term process, it allows us to become aware of and to detach ourselves form our own social conditioning and biases, by making it possible for us to reflect upon our behaviour (Cohen, 2008: 142) and, thus, to be responsible for the choices we make.

It could be argued that the burden imposed on individuals by this responsibility is too heavy. Taking time and energy, would the practice of judgment in our economic choices be a threat to the 'legitimate personal prerogative [...] to be something other than an engine for the welfare of other people' (Cohen, 2008: 10)?[11] However, on the one hand, while we should be conscious that we are accountable for our choices, the process of making reflective judgments is a long term one. Rather than reflecting for all our everyday decisions, we could reflect upon one choice at a time, in an on going process of judgment and actualisation of these judgments. Moreover, as Cohen and Rawls both emphasize, what might be felt as a burden in the first place often becomes normal and easy after some time (Carens, 1989: 42; Cohen, 2008: 142).

On the other hand, there would be no sanctions, formal or informal, imposed to the persons who would fail to make reflective judgments, or even to follow the new ethos. Even if we do have choices, there also exists a 'heavy social conditioning' (Cohen, 2008: 141) that might influence our choices in a wrong way,

[11] I would like here to thank Professor Joseph Carens for pushing me on that question.

or make us unaware that we are making a choice, so that either we are not conscious of the injustice we cause or we think that departing from the right behaviour is allowed (Cohen, 2008: 141). This is why our responsibilities are not absolute; as Cohen puts it, there should be a limited blamability (Cohen, 2008: 140-143). We should condemn the injustice, but not persecute its perpetrator.

Individual responsibility to follow an ethos and to make choices in accordance with our idea of justice in the economic sphere thus doesn't violate the liberal-egalitarian conception of freedom nor imposes a too heavy burden on people. It is thus our responsibility to make accountable choices that participate in the just distribution of wealth in the society and in its development towards a fairer one.

Conclusion

The currently dominant neoliberal discourse that encourages the de-responsibilisation of the citizens-consumers, promoting the primacy of their desires and preferences over their responsibility, has had disastrous impacts on the globalized and poorly regulated world economy as well as at the nation-state scale. The way my generation has been raised, in a context of abundance and diversification of consumption goods, promoted the newest as the best, the older as something to throw away, and an only question: 'what do I want to buy?'

I contend that the ideal considerations offered by Rawls and Cohen I presented here help us to think a way out of the problems we face in the current, non-ideal world. Cohen's critique to Rawls shows, a pure procedural justice mechanism cannot guarantee a just distribution of income and wealth in the society because it allows individuals to behave self-interestedly. Justice requires individuals to behave in a way that reflects their conception of justice in the economic sphere as well as in the political one. With his notion of ethos, Cohen offers a precious tool to think of what could collectively be done to engage a change in mentalities. However, I emphasized the need to encourage individual reflective judgment as well, as both a way for us to reflect upon our own biases, to question our own behaviour, the things we want to buy, and to consider the implications of our consumption for others and for the planet, and to become responsible and accountable for the choices justice requires us to make.

It is now the time to acknowledge the type of overlooked individual responsibility. The question of how to make the essential step of changing the current dominant and domineering ethos, sustained by so many lobbies, economic interests and domineering discourses, will here remain open. It will certainly be necessary to act upon the different possible levels: the institutional, formal level, which can create new regulations necessary to the exercise of individual reflective judgment, as the rules of transparency, while the informal level, the medias or opinion leaders for example, could promote the new ethos in a more bottom-up way. Education should also certainly play a role in developing one's capacity for the practice of reflective judgment, promoting the development of the ability to live in uncertainty, to widen our perspective, and to valuate the impacts of our choices. I firmly believe that the solutions we need can only be designed through the collaboration of political theorists, empirical scientists, and practitioners.

References

Carens, J.H. (1989). Rights and duties in an egalitarian society. Political Theory 14(1): 31-49.

Cohen, G.A. (2008). Rescuing justice and equality. Harvard University Press, Cambridge, MA, USA, 430 pp.

European Commission, Health and Consumer Division, http://ec.europa.eu/food/food/sustainability.

Rawls, J. (1999). A theory of justice (revised edition). Belknap Press of Harvard University Press, Cambridge, MA, USA, 538 pp.

Section 3. Legal issues in food production and consumption

The gullible consumer in EU food law

W.W. Huizing Edinger
Department of Food and Resource Economics, Science, University of Copenhagen, Rolighedsvej 25, 1958 Frederiksberg C, Denmark; whe@foi.ku.dk

Introduction

EU consumers are free to choose how they live their lives and what they eat. This right of consumer autonomy can be seen as the passive counterpart of the market access rights that are at the very heart of European integration (Micklitz *et al.*, 2009: 48). Although the free movement of goods essentially addresses the 'active market participant' (Micklitz *et al.*, 2009: 9), its effective functioning depends to a large extent on whether consumers actively engage in cross-border trade. Therefore, the internal market concept simultaneously presupposes that consumers have access and are at liberty to choose between what is at offer.

EU law prescribes, at the same time, that consumers must be offered a high level of protection of their health, safety and other interests in relation to their choice of consumption (Article 12 and 169 of the Treaty on the Functioning of the European Union ('TFEU')). From a legal point of view, the idea of free choice may seem difficult to reconcile with the principle of consumer protection, since freedom presupposes absence of intervention, while protection calls for regulation, for limits to freedom.

This paper looks into how this seeming conflict between freedom and protection is dealt with at a European level, with a special focus on the consumption of food. Is it, indeed, possible to adequately protect consumers while, at the same time, guaranteeing them a genuinely free choice of what they eat, and *vice versa*? In case of conflict, which interest should prevail: freedom or protection? And what is the position of the weaker, credulous and gullible consumer: he who has difficulties managing his freedom and making the 'right' choices?

The Court of Justice of the European Union ('CJEU' or 'Court'), seeking to strike a fair balance the various interests at stake, has developed a protective standard, which is based on an objectified image of the EU consumer and his needs. Usually, the impetus for legal protection is the will to balance an unequal relationship in favour of the weaker party. Therefore, one would expect the objectified food consumer to be relatively weak and unable to protect himself. However, as will be discussed below, this is not necessarily the consumer image that prevails throughout EU food law.

The EU consumer, consumer protection and consumer rights

The words *consumer*, *consumption* and *consume* are readily associated with food. Somehow, one is just more likely to imagine someone consuming an *apple* than a *contract*. Nevertheless, in a legal sense, the overall consumer concept comprises both of these activities – and many more.

Arguably, the everyday consumer concept is very broad: we all seem the fall within its scope one way or another. This may explain why it is so hard to find a single legal description of the EU consumer. The EU Treaties do not contain such definition and, although Article 169 TFEU holds a reference to consumers' basics rights and interests, it does not specify by whom and under what conditions these can be enjoyed. The consumer notion remains implied in the Treaty and must be further delimited by exploring its historical and practical application.

The consumer movement emerged as a response to the growing complexity of the market. At the same time, however, it is essential to keep in mind that consumer protection policy developed as a corollary of the internal market before it was accepted as an independent task of the EU (Unberath and Johnston, 2007). The result, as identified by Unberath and Johnston, is a 'double-headed approach' to consumer protection within the EU. Where national rules aimed at protecting consumer interests directly compete with free market interests protected by the Treaties, a liberal, market-oriented approach is adopted. This 'negative harmonisation' (harmonisation through the suppression of national rules that violate the Treaties) favours deregulation and a relatively low common denominator for protection of consumers' rights and interests.

For the purpose of 'positive harmonisation' (harmonisation through the adoption of positive EU legislation), Unberath and Johnston have identified a protective standard that is, as a rule, based on a fairly high common denominator for protection, generally resulting in a rather consumer-friendly, trade-restricting interpretation of the rules in question (Unberath and Johnston, 2007). The result is an intriguing blend of 'consumerism', where the EU legislator has adopted harmonising legislation, and 'trade-ism', where this is not the case.

Considering the above schism, it is not surprising that the consumer concept remains vague and consumer rights have not been made very explicit in the EU Treaties. In fact, only few of these rights, i.e. the rights to information, education and association, were actually formalised therein.

In the following section, we will establish how the consumer concept has further developed within positive EU food law.

The EU food consumer and his protectable interests

As suggested in the above section, the EU legislator has been quite active in the field of consumer protection in recent years. Contemporary secondary EU law comprises an extensive set of harmonised rules aimed at offering a high level of protection to consumers. These various laws operate with different definitions of 'the consumer', depending on the consumptive activity covered by the law in question.

Also within the area of food law, consumer protection legislation has been established. Article 3(18) of Regulation 178/2002 ('General Food Law' or 'GFL') defines the 'final consumer' as 'the ultimate consumer of a foodstuff, who will not use the food as part of any food business operation or activity'. Accordingly, consumers are the end-users of the food: those who actually *eat* it – alone or shared with others in a private atmosphere.

The Preamble to the General Food Law identifies two preconditions for the well-functioning of the internal market: consumer safety and consumer confidence. These basic concerns have been balanced and translated into two general principles of EU food law: the principle of food safety (Article 14 GFL) and the principle of informed choice (Article 8 GFL), which can be seen as a context-specific application of the consumer right of information. Apparently, despite the fact that consumer choice and safety are not formalised consumer rights in the context of the EU Treaties, both have gained unambiguous status as guiding values within the area of EU food law.

On the basis of the principle of food safety, a dense net of strict safety rules has been developed, aiming to protect the health and safety of food(Article 5(1) GFL), and placing the full responsibility for safe food production and marketing on the food producer (Article 17 GFL). Effectively, this system establishes a horizontal consumer right to food safety, in the sense that consumers may expect the food they purchase not to be harmful to their health, and they have a claim on the producer if this is proven not to be the case.

The principle of informed choice is reflected in mandatory rules that complement the general principles on unfair commercial practices with specific requirements for the provision of food information to consumers. The main labelling rules are contained in Regulation (EU) 1169/2011 on the provision of food information to consumers ('FIR').

According to Article 3(1) FIR, food information rules aim to provide a basis for consumers to make an informed choice and to make safe use of food, with particular regard to health, economic, environmental, social and ethical considerations. Although the provision of adequate food information is a responsibility of the food producer (Article 8 FIR), we will see that consumer autonomy is maintained as a guiding principle and the responsibility for reading, understanding and applying the supplied information is left with the consumer.

The consumer right to information is thus converted into a subjective entitlement to be offered insight in an objectified series of mandatory particulars about any food product. The result is, effectively, the acceptance of a consumer right to *informed choice*, its protective element contained in an objectified consumer need for data in order to enable him to fully understand his choice and its consequences – and not necessarily the wish to guide consumer choice in a certain direction – e.g. towards what objectively seems to be the better option.

EU food law thus comprises two designated areas of consumer protection legislation, which differ considerably in character. The overall legislative approach to food safety has resulted in a protective system which is aimed at effectively ruling out any chemical, biological or physical hazard in relation to food. Apparently, the EU legislator has concluded that the consumer, in general, is unable to protect himself from such health threats, resulting in solid harmonisation based on a relatively high protective level. Only incidentally we see that a responsibility for dealing with food safety risks is placed with the consumer. For instance, salmonella-infected raw poultry meat may be placed on the market, provided that adequate instructions for use are placed on the product's packaging, even though inadequate compliance may induce health risks (Article 9(1)(g) FIR).

Where mandatory information requirements are preferred over stringent food safety rules, the conclusion seems justified that the EU legislator has deemed it disproportional to impose additional requirements on the food producer and concluded, instead, that the consumer must be able to protect himself, provided he be offered objectively necessary information. Thus, albeit it implicitly, certain objective qualities are attributed to the consumer, resulting in a lower common denominator for protection.

As a result, the objectified EU food consumer can be regarded as having two faces. In areas with well-identified health and safety challenges, he is, in principle, perceived as weak and vulnerable, and in need of protection. Where such a direct risk is not deemed present, however, the consumer is expected to take care of himself by making a subjective choice that will maximise his personal benefit.

This is all relatively unproblematic for as long as health and safety risks are clearly delimited and predictable and fall under the seemingly narrow definition of food safety in Article 14 GFL (Szajkowska, 2012: 18). In reality, however, it is not always easy to foresee the longer or short-term effects of a food on a person's health and well-being, especially if the manifestation of such effects depends on the overall pattern of consumption of the consumer in question, which seem to make 'health' an issue of personal choice rather than a food safety-related problem.

Where uncertainty as to health risks exists, the precautionary principle contained in Article 7 GFL may be helpful. According to this principle, provisional risk management decisions may be taken if there is a possibility of harmful effects, but scientific uncertainty persists with regard to the reality and seriousness of the risks. However, the application of the precautionary principle is restricted to risks as defined by natural science, which limits its scope to food safety in the above narrow sense. Despite the existence of uncertainties with regard to the health effects of a food in view of the overall diet, the final responsibility for the consumptive decision in such cases remains with the consumer.

Hereafter, I will look further into how the two-faced food consumer came into being, what characteristics he possesses and what this means for his protection in the field of food law.

The food consumer according to the Court of Justice of the European Union

The two-faced food consumer can be seen, primarily, as a product of the case law of the CJEU in the field of negative harmonisation, which, as mentioned above, seems to favour trade over other legitimate interests.

In Case 120/78 Rewe-Central ('*Cassis de Dijon*') and subsequent rulings, the CJEU has established the basic rule that national measures that have the potential to hinder intra-community trade can be permitted only in so far as 'necessary on grounds of public interest listed in Article 36 [of the EEC Treaty, now Article 36 TFEU], such as the effective protection of human health, or in order to satisfy mandatory requirements relating *inter alia* to consumer protection' (Case 216/84 Commission v. France, para.7). This applies for any national rules that lay down product requirements relating to the product's name, form, size, weight, composition, presentation, labelling and packaging, even if those rules apply to national and imported products alike (joined Cases C-267 and 268/91 Keck and Mithouard, para.15; Case C-470/93 Mars, para.12). Furthermore, a Member State may, on the basis of the proportionality principle, not invoke the grounds of public interest or mandatory requirements in order to justify a measure restricting imports unless the same objective cannot be achieved by any other measure which restricts the free movement of goods less (Case 216/84 Commission v. France, para.7).

Where the protection of health and life of humans is concerned, the Court has consistently held that 'in so far as there are uncertainties at the present state of scientific research it is for the Member States, in absence of harmonisation, to decide what degree of protection they intend to assure, having regards however to the requirements of the free movement of goods within the Community' (Case 178/84 Commission v Germany, para.41).

As far as health protection is concerned, the Court has further delimited the application of the proportionality principle underlying Article 36 TFEU to the existence of a 'real threat to human health' (Case 216/84, para.15). Safe provisional measures based on the precautionary principle Member States may only go so far as to effectively ban a certain product from their territory 'if the real risk alleged for public health appears sufficiently established on the basis of the latest scientific data available' may (Case C-192/01 Commission v Denmark, para. 48-49).

This strict application and interpretation of the proportionality principle has interesting implications for the part of food law that is concerned with food quality and nutritional value. When keeping in mind that it is very difficult for a Member State to prove that a food is unsafe beyond the requirements of positive EU food law, it may be clear that it is, indeed, practically impossible to intervene in the circulation of foodstuffs that are principally safe, but that are qualitatively or nutritionally inferior to other, comparable products. And in view of the limitation of the application of the precautionary principle to scientific criteria, it cannot be of help in this respect.

The underlying thought is that consumers are best served with genuine choice and competition – even if such implies that increased choice is, at least partly, created by allowing products of lesser quality or with inferior nutritional characteristics to enter the market. In these cases, mandatory labelling requirements may demonstrate to be the last resort as a protective means that least restricts the free movement within the EU (Case 216/84, paras. 12, 15). Although this standpoint is understandable from a commercial point of view, it may prove to be particularly problematic where the gullible consumer is concerned: those consumers that find themselves in – socially, intellectually or otherwise – disadvantageous positions.

The protective effect of food information depends heavily on how the consumer understands and processes the provided particulars, as well as how he reacts thereupon. If, for example, information is provided on the expiry date of a foodstuff, it is important that the consumer understands this information as imperative – and acts accordingly by not consuming the food after this date.

The situation gets more complicated when the provided information does not relate to an immediate health threat, but to (nutritional) characteristics that may induce negative health consequences depending on the consumptive pattern adopted by the consumer in question. This type of information presupposes that the consumer is capable not only of deciphering often quite technical messages, but also to act rationally upon them. The consumer is, in other words, expected to fully comprehend the effects of food on his health and well-being in the shorter or longer run and in view of his overall diet and lifestyle. It may be clear that this is a mouthful for *any* consumer, and in particular for the weaker kind.

The Court, confronted with the dilemma of what information is required in order to enable consumers to adequately protect themselves, has adopted an objectified consumer image generally known as the 'average consumer' (Case C-210/96, opinion of Advocate General Mischo, para.55). The average consumer is understood to be reasonably well-informed, reasonably observant and circumspect (Regulation (EC) 1924/2006, Preamble 16; FIR, Preamble 41), thus representing quite an optimistic standard of consumers' general insight and cognitive skills.

When provided with product information, the average consumer is expected to be 'reasonably circumspect' (Case C-470/93, para.24), implying that he is attentive to the overall characterization of the product, as opposed to the consumer who regards the provided information only casually and uncritically, and who is more likely to be influenced by the colour of the pack, as well as by designs and slogans on the pack (Case C-210/96, opinion of Advocate General Mischo, para.97). Consumers are assumed to be aware of the implications of their purchasing decisions and, therefore, expected to read, understand and act rationally upon information particulars provided to them, such as the ingredients list provided for foodstuffs (Case C-51/94 Commission v Germany, para.34). At the same time, consumers are deemed able to distinguish between mandatory particulars and commercial information, and just as they are expected to fully rationalise consumer essentials, they are presumed to take commercial statements with a grain of salt (Case C-470/93, para.24).

The adoption of harmonised rules for food safety and food information legislation has not changed the Court's practise with respect to consumer (health) protection. In fact, the adoption of these rules positively harmonises a 'negative' interpretation of consumer's rights and interests. Could it be that the Court reasons that where no harmonised rules exist, they are, indeed, unnecessary?

As a consequence, where a food's quality and nutritional characteristics are concerned, the lower protective standard adopted in the field of consumer information becomes decisive also in questions where, essentially, the safety of a food may be at stake. Because of the difficulty to prove the existence of a real, immediate threat to consumers' health from the consumption of a food with inferior (nutritional) characteristics, consumer protection from the potential negative effects on their health resulting from the consumption of these products is not considered a food safety, but rather a consumer information issue. The consumer is expected to protect himself from the materialisation of such negative effects by rationally applying his free choice.

Conclusion

EU consumer protection is not only 'double-headed' in the sense that, generally speaking, the protection of the consumer in the field of negative harmonisation is based on a lower common denominator than in the field of positive harmonisation (Unberath and Johnston, 2007).

When we look into the merits of positive EU food law, the consumer also appears double-faced: The overall characterisation of the consumer depends on which of his interests are at stake, and how they are categorised.

The seemingly sharp distinction between, on the one hand, food safety related issues and, on the other, food information matters, has given way to a split consumer personality. The consumer is viewed as essentially helpless in relation to health risks that fall under the scope of Article 14 GFL and which can be nominated as 'scientific'. Where potential health risks are more diffuse and unpredictable – for instance under influence of other than purely scientific considerations, such as economic and behavioural reflections – the consumer is viewed as fundamentally capable and in charge, as long as he is provided with minimum information particulars to guide his choice.

The guiding principle behind this split is the freedom to choose, which remains, although not explicitly recognised as a consumer right within an EU context, one of the core values of the internal market. In practise, free choice can only be limited in case of clear and present danger. If this is not the case, freedom rules out protection, not matter how difficult the choice and how multifaceted its impact.

References

Case 120/78 Rewe-Zentral ('Cassis de Dijon)' [1979], ECR 649.

Case 178/84 Commission v Germany ('Beer purity') [1987], ECR 1227.

Case 216/84 Commission v. France [1988], ECR 793.

Case C-192/01 Commission v Denmark [2003], ECR I-9693.

Case C-210/96 Gut Springenheide and Tusky [1998], ECR I-4657.

Case C-210/96, opinion of Advocate General Mischo of 12 March 1998, ECR I-4659.

Case C-470/93 'Mars' [1995], ECR I-1923.

Joined Cases C-267 and 268/91 'Keck and Mithouard' [1993], ECR I-6097.

Micklitz, H.-W. Reich, N. and Rott, P. (2009). Understanding EU consumer law. Intersentia, Antwerp, Belgium.

Regulation (EC) No 178/2002 of the European Parliament and of the Council of 28 January 2002 laying down the general principles and requirements of food law, establishing the European Food Safety Authority and laying down procedures in matters of food safety.

Regulation (EU) 1169/2011 of the European Parliament and of the Council of 25 October 2011 on the provision of food information to consumers, amending Regulations (EC) No 1924/2006 and (EC) No 1925/2006 of the European Parliament and of the Council, and repealing Commission Directive 87/250/EEC, Council Directive 90/496/EEC, Commission Directive 1999/10/EC, Directive 2000/13/EC of the European Parliament and of the Council, Commission Directives 2002/67/EC and 2008/5/EC and Commission Regulation (EC) No 608/2004.

Szajkowska, A., (2012). Regulating food law: risk analysis and the precautionary principle as general principles of EU food law. Wageningen Academic Publishers, Wageningen, the Netherlands.

Unberath, H. and Johnston A., 2007. The double-headed approach of the ECJ concerning consumer protection. Common Market Law Review 44: 1244.

Ethics and consumerism: legal promotion of ethical consumption?

F. Dias Simões
Faculty of Law of the University of Macau, Av. Padre Tomás Pereira, Taipa, Macau, China P.R.; fernandodsimoes@umac.mo

Abstract

Consumption patterns influence our quality of life in deep ways, having a substantial effect on both the individual and the society as a whole. Most consumption decisions we make on a daily basis may have an ethical content. Normally our ethical concerns result from personal beliefs and societal influences, not from the Law. The question then becomes whether public entities have legitimacy to impose certain dimensions of ethical behaviour and whether Consumer Law is the appropriate mechanism to achieve such purposes. Government's role in moulding our consumption habits is doubtful as there is no consensus about the role that Law should play in the promotion of ethical conducts. Traditionally Consumer Law was aimed at protecting the consumer himself. Should legal intervention be expanded to guide consumer behaviour even when the protection of consumers as such is not the goal? In some cases public intervention may be justified with the impact of individual decisions on the community as a whole. In fact, the State is interested in controlling the social costs of individuals' behaviours that are borne by society at large. However, going beyond the frontier of public interest may be troublesome. This new field of regulatory intervention touches upon some of the values that define us as a society: personal freedom, choice and liberty. Any intervention in this regard should consider the balance between individual freedom and public interest. Citizens should not be treated as creatures deprived of any moral or ethical character.

Keywords: freedom of choice, consumer law, ethical consumers

Consumer law and freedom of choice

Consumption habits and patterns influence our quality of life in profound ways, having a considerable effect on both the individual and society at large. The quantity and quality of products and services we acquire is an important aspect of our lifestyle. Freedom, namely freedom of choice, is an eternal topic of philosophical interest. In the realm of Consumer Law freedom of choice is viewed as priceless, a right that should be defended whenever possible and only subject to very narrow exceptions. Consumer Law consists of legal mechanisms aimed at the general protection and promotion of consumer interests. It is composed, first of all, of mandatory rules that guarantee that parties will not depart from the legislative rules to the detriment of the consumer. It further comprises the obligation of information disclosure as information plays a vital role in consumers' decisions. In fact, the relationship between professionals and consumers is often characterized by information asymmetry and consumers depicted as weak contracting parties that lack bargaining power. Only well informed consumers can choose freely and (hopefully) make pondered decisions.

The emergence of consumer ethics

Ethical consumer behaviour can be broadly defined as the 'decision-making, purchases and other consumption experiences that are affected by the consumer's ethical concerns' (Cooper-Martin and Holbrook, 1993: 113). This notion has arisen around the turn of the twenty-first century and refers to consumers that have political, religious, spiritual, environmental, social or other motives for choosing one product or service over another. It can be seen as an evolution of green consumerism that incorporates concerns with aspects other than just the price or quality of products and services. Ethical consumers are concerned with the effects that a purchasing choice has, not only on themselves, but also on the external world around them. They ponder on the external consequences of their choices instead of focusing merely on its internal impacts. Obviously, consumers will not choose some types of products if they cost half a month's salary of if they taste bad. Ethical consumers are not, therefore, ignoring price and quality, but applying some additional (and sometimes prior) criteria in the decision-making process (Harrison *et al.*, 2005: 2).

The current global recession, on the other hand, presents new challenges for those consumers who wish to consume ethically. In times of economic hardship many consumers will struggle to accommodate their ethical convictions with the need to rationalise consumption – it is necessary to satisfy basic needs like food and clothing at the lowest cost possible. Consumers will probably be more price-sensitive and will not gladly accept to be determined by criteria other than price and quality.

Should consumer law be inspired by ethics?

Public entities worldwide are incorporating ethical concerns in their policies. This is clear in the case of environmental protection. Legislation has been in place to incite citizens to adopt a sustainable lifestyle (e.g. incentives to buy eco-friendly vehicles) or to punish behaviours that have a negative impact on the environment (e.g. eco-taxes). This type of public policy can be easily understood as the State has the legal obligation to protect the environment and, as a result, the quality of life of all citizens. Most consumption decisions we make on a daily basis may have an ethical content: buying a newspaper or accessing its online version; buying normal coffee or fair-trade coffee; having beef or a salad for lunch, *etc.*; each of these options carries a variety of ethical stances that range from environmentalism to concerns with animal welfare. Government's role in moulding our consumption habits is disputed as there is no consensus about whether the Law should promote ethical standards of consumption. The question then becomes whether public entities are legitimized to promote and regulate certain dimensions of ethical consumption and whether Consumer Law is the appropriate mechanism to achieve such purposes. Should the State guide our decisions by providing incentives or creating barriers to certain options? Should Consumer Law affirm a certain morality?

Any Law student starts his adventure in the legal world by learning about the difference between Morality and the Law. Morality is concerned with the norms, values and beliefs entrenched in a given community which define right and wrong. The question arises: should consumer ethics be written into the law? Should we move from codes of ethics to legal codes with an ethical content? Should concepts such as 'fair trade', 'ethical consumption' or 'sustainable' be used as legal standards of behaviour? Should we move from 'right and wrong' to 'legal and illegal'?

The traditional perspective that aims to promote 'consumer sovereignty' may be moderated by the existence of limits to the freedom of choice. This is not exactly a novelty. Freedom of contract is becoming increasingly circumscribed by other social concerns and purposes. Statutes have fragmented the Law of Contracts into legal subsets such as Labour Law, Insurance Law and Consumer Law, and in so doing they have placed significant limits on freedom of contract (Pettit, Jr., 1999: 266). Therefore, we may argue that the very existence of Consumer Law imposes limits on the freedom of consumers. These limits are justified as a means to protect consumers themselves. Legislation has played an important part in raising consumer expectations of marketing behaviour, and regulation has also helped us move from the traditional *caveat emptor* rule to a more socially responsible era. Producers and advertisers have had to face responsibility for their role in promoting certain products. However, this type of legal intervention was aimed at protecting the consumer as such. Should legal intervention be expanded to guide the consumer behaviour even when his protection is not the goal? Should the legislator promote vegetarianism, for instance, because it improves animal welfare? Should the legislator restrict meat products advertising in order to reduce its consumption? Should Law inspire the creation of an ethical marketplace? Should governmental entities regulate what people buy? What sort of criteria may be used to justify possible restrictions to purchasing choices? If certain products are deemed to be unethical and thus unacceptable, should citizens be denied the right to freely keep on buying them?

This 'new frontier' of public intervention is troublesome. Some may argue that regulatory measures amount to a limitation to the freedoms of individuals, namely, their freedom of choice, freedom of enterprise and even freedom of contract. State's role in the promotion of ethical lifestyles is a hot topic in modern societies. How should we find an appropriate balance between private liberty and morality? Does morality have a political meaning? Is there a right-wing morality and left-wing ethics? The quantity of products that we purchase is an important aspect of our lifestyle. To a large extent, and not overlooking other relevant factors, we can say that *we are what we buy, and how much we buy*. The State should play an active role in the promotion of an ethical society, but to what extent? Is there any 'official ethics'? Which one? Should Law even promote any lifestyle?

An example: gastronomical freedom vs. animal welfare

Food choices may be particularly contentious. While all food has ethical implications and significance, some food has taken on the connotation of being, in particular ways, more 'ethical' (Goodman *et al.*, 2010: 3). Take the example of foie gras, the popular delicacy in French cuisine. Gavage-based foie gras production is controversial due to the force feeding procedure used: birds are fed with more food than they would eat in the wild and more than they would voluntarily eat domestically. Animal rights and welfare activist groups contend that foie gras production methods, especially forced feeding, consist of cruel and inhumane treatment to animals. Some argue that foie gras should at least come with a warning so that 'people know what's being done to the animals'. The force feeding of animals, essential to current foie gras production practices, is explicitly prohibited by specific laws in several countries. However, foie gras can still be imported into and purchased in most countries. Should States prohibit the importation and purchase of foie gras? Or should they just advertise, on the product, the ethical concerns associated with its method of production, thus trying to curb its consumption?

It cannot but cause some surprise to observe the legislator aspiring to regulate such apparently ordinary and intimate behaviours as buying decisions. From a traditional viewpoint, purchasing options are inappropriate subjects for government regulation, as these are taken as private matters for the individual – and only the individual – to ponder and decide on. However, with the evolution of consumer ethics there is increasing social awareness that some purchasing decisions may have ethical consequences or a considerable effect on the community as a whole. What used to be a private matter is now a question of public interest. What once was a pure private behaviour is now seen as a public concern. The biggest problem is that this new field of public regulation touches upon some of the values that define us as a society: personal freedom, choice and liberty. The exploration of the new frontiers of public intervention also illustrates the perennial tension between the Law and the citizen.

There are different perceptions on how to address different ethical concerns. The perspectives on ethical concerns surrounding animal consumption, for instance, differ between 'rightists' and 'welfarists'. The difference in perspective between is clarified by Payne (2002: 597): for the rightist, nothing short of total cessation of animal consumption will suffice; consumers have a moral obligation to avoid the consumption of meat, and producers have a parallel obligation not to sell it. Welfarists, by contrast, while also encouraging vegetarianism, accept the reality that animals are raised for human consumption and should be afforded humane treatment throughout the entire process of breeding, rearing, transporting, and processing. Consumers have a corresponding ethical obligation to avoid consuming meat that is not produced in a humane, natural fashion, and food retailers have a parallel obligation to provide patrons with information that facilitates ethical consumer choice.

Where should the Law stand? Should the State devise a legal framework that makes clear ethically driven choices, for instance, regarding animal welfare? Should the legislator regulate the production of specific categories of products and prohibit some practices, even proscribe certain industries? Legal rules that eliminate or restrict purchasing options are not a novelty. Most countries have enacted laws prohibiting or limiting the acquisition of some products. These measures are generally justified by different reasons, *inter alia* the protection of consumers from products which can endanger their health and safety. The question arises: should governments go even farther, limiting options or directing them on the grounds of ethical concerns, instead of merely basing such legal rules on the protection of consumers themselves?

According to the classic libertarian view, minimal state intervention is the only way to ensure the protection of individual freedom. This doctrine postulates that governments should not regulate private behaviour that does not directly cause any harm to a third party. How individuals behave, particularly their purchasing options, belongs to their private sphere. According to this worldview, the state should only regulate actions that directly cause, or have a very high probability of causing, unacceptable harm to others. Therefore, governments should only intervene in individuals' lives when their behaviour is harming other persons. The promotion of animal welfare, for instance, is not based on the protection of 'other persons'. Should the State restrict the freedom of citizens to acquire meat due to ethical concerns on its consumption? Is Consumer Law running the risk of becoming an ideology, a purchasing deontology? Should Consumer Law, instead of promoting the freedom of consumers, orientate them towards certain options that are assumed to have an ethical content, thus limiting their freedom to the 'right' (ethical based) choices?

Ethical elements are a particularly challenging matter to regulate. Historically, prohibitory legislation most often occurs when a social majority objects morally to the specific conduct, value system, or culture of others and imposes regulation upon them (Hunt, 1999: 1). The core ethical code accepted in each society is likely to be incorporated in Law. Differently, ethical issues outside the core are difficult to control formally as the borderline between 'right' and 'wrong' can be subjective and vary from person to person, making mandatory regulations difficult to design and enforce. A more practical alternative seems to be to complete the consumer's information set by introducing guidelines, rules, decrees with voluntary compliance supplemented by policies to ensure traceability and market segregation (Tothova, 2009: 12).

Different methods of legal intervention and different degrees of intrusiveness

In some cases public intervention may be justified with the impact of individual decisions on the community as a whole. This may be true, for instance, in the environmental field. This takes the problem beyond the individual and makes it a legitimate focal point of public concern. The State has a reasonable interest in controlling the social costs of individuals' behaviours that are borne by society at large, for instance, the excessive use of plastic bags. According to this perspective, there may be valid justifications for intervention in the market beyond a minimalistic libertarian level. When private purchasing decisions have high socioeconomic costs, there seems to be a natural point for government intervention, since government officials ought to take action for the public good. The question has to do with achieving the appropriate balance between private liberty and public interests.

Public information campaigns are the preferred policy approach to change the behaviour of consumers. If the State considers that the consumption of some products takes a toll which is ethically unacceptable (for instance, because it has an impact of animal welfare) it may decide to promote public awareness campaigns about the problem. Public awareness is a powerful mechanism as it can encourage citizens to consciously change their behaviour. Education and information are the cornerstones of many public initiatives, generally supported by media communication and other social marketing tools. Well-informed and educated individuals will probably decide to take the necessary changes for a different lifestyle without the need to resort to more 'intrusive' measures. Of course consumers are not necessarily willing to be informed and – more importantly – to act properly upon that information. However, this is a problem of efficacy, not of legitimacy. Individuals are free to refuse such information, to ignore it or not to change their behaviour. In any case, their options remain untouched. It is obvious that measures that only aim at educating or informing consumers do not 'curtail the liberty of individuals'. In fact, legislation that only aims to inform and educate consumers does not interfere with individual liberty. Rather, it promotes the exercise of individual freedom by enabling informed decision-making. Such measures do not limit the freedom of citizens to buy whatever they choose, so freedom is not constrained. All rational consumers can be safely assumed to be interested in knowing material information with respect to the products that they consume. Hence, there is a good reason to presume that such policies have the consent of the people.

Frequently acting ethically has a price. Consumers will often pay more for humanely produced goods because those goods often cost more to produce. For instance, meat which is produced according to standards that respect animal welfare might be more expensive than meat produced without such concerns. It is easier to justify legal intervention if the government decides to influence the market by adopting a positive perspective. For instance, instead of limiting or forbidding the access to certain types of meat, due to concerns with the production methods used, it will probably receive more popular acceptance to subsidize certain types of meat that are not subject to the same concerns. Faced with two products (one type of meat produced respecting animal welfare and another one which was not) with equal prices, consumers would not have to be concerned about the price, and would be free to consciously think about the ethical implications of their choice. This type of intervention would not raise controversy as it would reduce the price of some products instead of making it harder to have access to them. People are always in favour of lowering prices, regardless of the specific policy behind the decision. Governments might have a hard time trying to justify that some types of food are going to be harder to acquire (or even proscribed) because they raise concerns about ethical standards when citizens are concerned with putting food on the table in the first place. Do not expect people to contemplate ethical dilemmas on an empty stomach.

Of course there are other options available, namely prohibiting the purchase of meat that is produced with resort to practices that cause animal suffering. This type of legal tool is far more intrusive, as it denies freedom of choice instead of just limiting it. Banning the production or sale of certain types of products is the most controversial form of legal intervention. This type of legal tool should be used carefully, based not only on ethical grounds but also on a comprehensive vision capable of justifying legal intervention, taking proper account of the balance between individual freedom and public interest. Citizens are especially sensitive to laws that limit their freedom, and it is particularly hard to persuade them that they should refrain from certain purchasing options. What we buy, and in what amount, is still considered as a space of privacy and freedom. Any intervention in this regard has to be especially supported and justified.

The most powerful and disruptive tool of public regulation is a complete prohibition of certain products thought to be produced according to ethically questionable procedures. Food prohibitions are often perceived as intrusive or a coercive form of intervention as they can result in the elimination of ingredients or products. Some argue that this type of measures constitute an unjustifiable restriction on the freedom to decide what one eats. This type of measure curtails the gastronomical liberty of citizens, forcing upon them an ethical lifestyle without their consent and probably against their will. As in other situations, the temptation to just prohibit something is sometimes too big, as if that would just erase the problem. We should address ethical concerns globally, not by erasing parts of the problem (eliminating products from the market) but rather by widening the options that citizens face and enhancing their knowledge about such options. Product prohibition measures should be the *ultima ratio*: in fact, the ability to decide what one buys, though not important as freedom of speech or religion, is an important freedom nonetheless. All of the other legislative measures should be considered and discussed before deciding to engage in such a radical tool. This type of intervention calls for an enhanced legitimacy and justification. Again, the question has to do with legitimacy. The legitimacy of public regulation tools depends on individuals' acceptance of such policies. Citizens should not be treated as creatures deprived of any moral or ethical character that require constant guidance. Public policy should nurture ethical behaviours without turning itself into a sort of official ethos. Citizens (and therefore, consumers) do not enjoy receiving sermons, much less from the State.

References

Cooper-Martin, E. and Holbrook, M.B. (1993). Ethical consumptions experiences and ethical space. Advances in Consumer Research 20: 113-118.

Goodman, M.K., Maye, D. and Holloway, L. (2010). Ethical foodscapes?: premises, promises and possibilities. Environment, Politics and Development Working Paper Series, paper no. 29. Department of Geography, King's College London, London, UK.

Harrison, R., Newholm, T. and Shaw, D. (2005). The ethical consumer, SAGE Publications, London, UK, 280 pp.

Hunt, A. (1999). Governing morals: a social history of moral regulation. Cambridge University Press, Cambridge, UK, 273 pp.

Payne, R. (2002). Animal welfare, animal rights, and the path to social reform: one movement's struggle for coherency in the quest for change. Virginia Journal of Social Policy & the Law 9: 587-633.

Pettit, Jr., M. (1999). Freedom, freedom of contract, and the 'rise and fall'. Boston University Law Review 79: 263-354.

Tothova, M. (2009). The trade and trade policy implications of different policy responses to societal concerns. OECD Food, Agriculture and Fisheries Working Papers, no. 20, OECD Publishing, Paris, France.

Local foods, food quality and agricultural soil consumption: new challenges for the European Union

S. Montaldo
University of Milan Bicocca and University of Turin, Department of law, Italy; stefano.montaldo@unito.it

Abstract

The article explores some of the most debated challenges that EU food and agricultural law is facing. Firstly, the analysis focuses on the newly reformed EU rules on the protection of geographic terms, as a powerful means to safeguard cultural and historical heritage of local and rural communities. Secondly, the paper investigates the role played by private and public technical standards on the global foodstuff market. Finally, the problem of agricultural soil consumption is addressed, being it a threat to EU as a whole. The three selected topics combine to sketch a promising scenario for the future of food policies in the EU, even if threats to important non-market values are still far from being properly addressed.

Keywords: food, geographic terms, technical standards, agricultural soil

Introductory remarks

European food law has been traditionally inspired by two major and often opposing categories of values: market priorities and non-commercial values, such as consumers' rights, environmental protection, traditional cultures preservation. On the one hand, the provisions of the Treaty on free movement of goods and the participation in the World Trade Organization (WTO) have played a key role in fostering foods import and export. On the other hand, EU food legislation has gradually developed – as it is actually evolving – to the benefit of safety, security and quality of food products. The remarkable body of rules at stake directly or indirectly covers all the main aspects of the subject, from the recently reformed food labeling rules to GMOs use regime to highly debated questions on the Common Agricultural Policy (CAP). Complexity is therefore one of the key features of the EU legal framework on food, since it deeply intertwines political, technical, legal and economic options, in order to pursue the objectives prescribed by the Treaties. Even if such an entangled background raises several debated questions on the influence of legal regimes on the trends of food consumption and on consumers' preferences, a limited number of specific issues is worth being addressed here.

Indeed, the mentioned scenario is the stage on which the EU legislator is trying to face global challenges brought by arm wrestling between market and non-commercial values, further amplified by parallel phenomena such as climate change and the threats to global food security. Under this perspective, three main aspects highlight the advances, expectations and pitfalls EU institutions – and, indirectly, EU citizens and consumers – are coping with. First of all, the EU is striving to increase consumers' awareness on the need to protect local and traditional foods, which are often a remarkable cultural factors and express the ethics and values of a community. Secondly, a significant impact on market and consumers' behaviour derives from the definition of common technical or quality standards for food-related products, which has been fostered at both EU and international levels by public and private actors. Thirdly, as far as shortcomings are concerned, even if massive consumption of agricultural lands is an urgent global concern, the efforts displayed by EU institutions in order to adopt comprehensive rules on soil consumption and the maintenance of agricultural lands in good conditions met on their way stumbling blocks which have not been overtaken yet.

The three challenges offer a clear image of the connection between legal regimes and consumers' attitude and in this particular perspective they will be analyzed.

The protection of local and traditional foods: the case of geographic terms

Origins, purposes and evolution of the geographic terms regime

The protection of geographic terms of foods has acquired an increasing importance in the EU, as evidently testified by the significant reform process the subject has undergone. At the beginning of the '90s, the Commission boosted a comprehensive and innovative strategy for a new era of the CAP, whereby abandoning a subsidy-based form of protection of farmers' activities and moving toward a higher level of competitiveness of EU agriculture. For the first time, this ambitious strategy took into due consideration fundamental non-economic values, namely environmental protection, sustainable use of natural resources and consumers' safety. In particular, one of the cornerstone aspects of the launched reform was a renewed consciousness of the implications of the concept of territory, which – to the purposes of the CAP – had to be considered not only a mere geographical area, but also a set of precious historical, cultural and environmental elements. Agricultural lands could therefore be conceived as powerful sources of cultural expressions, which include both local communities' traditions and the specificities of agricultural practices, methods and products. Such approach suggested a sharp shift from the ordinary approach of food production to a newly balanced relationship with market dynamics, under which the characteristics of a geographical area could become a relevant driver for consumers. These principles soon took the shape of a binding and comprehensive set of rules, thanks to the adoption of Council regulation No 2081/92 on the protection of geographic indications (GIs) and designations of origin (DOs) for agricultural products and foodstuffs. By the means of this act, the Council offered a strong 'European view' of the protection of local traditional foodstuffs, a subject which had already been elaborated by (feeble and ineffective) international treaties since the last decade of the XIX century and was actually under scrutiny and dense (but unfruitful) negotiations within the WTO. Aiming to grant a high-level market protection to geographic indications (and the non-commercial values they expressed), Regulation 2081/92 introduced specific and harmonized forms of intellectual property rights, which identify a product as originating in a particular region or locality and deriving its fundamental qualities from such geographic factor. In fact, as already stated by the European Court of Justice (ECJ) with regard to the international rules on the matter, the essence of a GIs protection system '*fulfills* [its] *specific purpose if the product which they describe does in fact possess qualities and characteristics which are due to the fact that it originated in a specific geographic area*' (ECJ judgment, February 20, 1975, case 12/74, *Commission* v. *Germany*). As a consequence, the definitions themselves of GIs and DOs were intimately structured on the link between the authenticity and reputation of the product and its territorial origins, in terms of historic roots, manufacturing methods and practices, cultural and social heritage, nutritive and organoleptic values (McMaolàin, 2007). More specifically, the notion of DO identified – and still identifies today, under the recently reformed regime – products whose quality and characteristic are essentially or exclusively due to a particular environment, with its inherent natural and human factors, and all the production steps of which take place in the defined territory. On the contrary, the concept of GI meant a less close relationship between the product and its original environment, since at least one of the phases of its manufacturing process must be performed in such area and its qualities are essentially attributable to its geographic origins. Regulation 2081/92 was eventually repealed by Council regulation No 510/2006, which harmonized all the aspects of the system, but did not solve some inconsistencies the former legal regime had shown, also due to highly questioned interpretative solutions given by the ECJ in its case-law. On the one hand, the Court has always stated that EU rules on DOs and GIs are exhaustive in nature: therefore, no further space for Member States to protect geographic products by their own is left (ECJ judgment of 7 November, 2003, case C-216/01, *Bud I*). On the other hand, the Luxembourg Court underlined that Member States are not under the duty to protect *ex officio* the products labelled with DOs and GIs (ECJ judgment of 28 February, 2008, case C-132/05, *Commission* v. *Germany* – '*Parmesan* case'). Thereby, the ECJ risked to undermine the whole system of geographical terms at its roots, denying the fundamental role of national authorities as the actors in the best position to ensure the effective and coherent application of EU law (Ubertazzi and Muniz Espada, 2009). Indeed, the judgment could allow each Member State to set its own limits and conditions for the protection of widespread products such as Parmesan cheese, to the detriment of both the truly local and original producers and the consumers, in terms of loss of competitive power and (with the highest probability) of food quality.

The reform of the Common Agricultural Policy and the consequences on GIs and DOs

Bearing in mind this fragmented situation, the debate on the extension of the safeguards in favour of DOs and GIs was eventually at the core of the recent reform process of the CAP. Indeed, in 2011 the Commission issued a series of communications, upholding a comprehensive renovation of agricultural policies. The set of proposals was framed in the general 'Europe 2020 strategy', through which the Commission identified the political priorities of its activity for the years to come. The strategy is aims to reach a smart, inclusive and sustainable growth, also by means of an increased attention to agricultural products quality policies and the need to provide the producers with the tools to select, promote and protect the products revealing value-adding characteristics. To this regard, two main aspects have to be underlined. Firstly, the second and the third fundamental pillars of the proposed CAP reform show an unpredictable priority for agricultural-led cultural treasures. The second pillar is represented by the support to farming communities, that provide the EU citizens with quality, value and diversity of food produced sustainably. The latter focuses on the efforts to maintain viable rural communities, whose activities deliver multiple social, cultural and environmental benefits, preserving characteristic local landscapes and ensuring wide choice of quality products for consumers. In a few words, even if competition and economic growth still lie at the core of the agricultural policies, the solely market-centered perspective has been abandoned, at last. Agriculture is also a matter of culture, traditions, bond with the land and food quality, for the consumers' benefit. Secondly, the rules on DOs and GIs have been reformed as well, thanks to the adoption of the Parliament and Council Regulation 1151/12, which entered into force in December 2012. The new set of rules updates the registration mechanisms of the geographic terms, but also introduces relevant innovations as to some general aspects concerning the extent of the protected intellectual property rights. To our purposes, it is important to remark that a clear answer to the ECJ case-law has been provided, since now all Member States are bound to take all appropriate administrative or judicial steps to prevent or stop the unlawful use of registered DOs and GIs. Moreover, the regulation establishes a scheme for optional quality terms, in order to facilitate the communication within the internal market of the value-adding characteristics of agricultural products by the producers thereof (Article 27). In particular, the optional terms refer to mountain and island farming products. In such cases, in fact, the direct link between geographic environment, local communities and qualities of the products or their processing is often evident, but at the same time the communication of such added values proves to be extremely difficult, because of the structural obstacles imposed by the environmental context itself.

The definition of international standards: common standards, different values?

The role of technical standards in the agri-food chain

Another aspect directly involving – or, better, influencing – consumers' behaviour is represented by the use of private or public technical standards as powerful marketing tools. To this regard, the EU has become a major actor in promoting voluntary standards for the market of foods. Indeed, the agri-food chain is a battlefield where the need for an adequate protection of opposing values has fostered – together with the adoption of formal legal rules – the establishment of an informal system of governance, where standardization by private or public actors is a widely accepted model of (informal) self-ruling. At European level, a key-role is played by the European Committee for Standardization (ECS), which serves under the auspices and the mandate of the Commission, but whose activity is mainly inspired by the different stakeholders involved. In 1991 this body and the International Standardization Organization (ISO) signed the Vienna Agreement, in order to optimize the organization of standard setting and to reach the adoption of shared technical rules. Food production and supply chain turned out to be one of the most challenging sector for the adoption of shared common standards, but the coordination between the two technical bodies eventually led to remarkable outcomes. In fact, after long negotiations, 'technical diplomats' representing both the ECS and the ISO managed to approve jointly the EN ISO 22000:2005 standard, whose main purpose is the harmonization of food safety requirements. The standard fixes basic requirements for the reduction and detection of food borne hazards and has a direct legitimization by the Commission, since its adoption is grounded on the mandate to the ECS provided by directive 98/34/EC, the so called transparency directive. As a consequence, the overall set of EU rules on food security – namely, for instance, the acts on hygiene and traceability, rapid alert system or official controls on food and feed – have largely benefited from the establishment of widely recognized and shared technical standards on food security. And so it is for European consumers.

The EU legal order and the standard adopted by the Codex Alimentarius

On the contrary, major concerns arise when technical standards specifically entailing the processing or composition of a product are at stake. To this regard, in fact, especially at international level, it is even more difficult to find and adopt shared technical rules, due to the extreme variety of interests that different States and international bodies or organizations respectively uphold. Under this perspective, the relationship between the EU system and specialized international bodies has originated significant frictions. As a matter of fact, the Council Regulation 178/2002 acknowledges the role of the EU as a major global food and feed trader and formally claims for an increasing cooperation with international standard setting organizations. Nonetheless, no clear mention is made to WTO rules and especially to the Agreement on sanitary and phytosanitary measures and the Agreement on technical barriers to trade. Neither it is for the Codex Alimentarius, which specifically focuses on food-related standards (Costato and Albisinni, 2012). Therefore, despite a formal reverence for such international sources of hard law and soft law rules, the EU legislator seems to be willing to maintain a wide discretionary power as to any decision on food safety, in order to provide consumers with the highest level of protection available, especially in case internationally defined standards are not satisfactory. Indeed, Article 5(3) of the regulation manifestly acknowledges the importance of imminent international standards, but also states that the EU legislator – and, we may say, the other institutions and bodies as well – is free to ignore such technical rules in case they would be ineffective or inappropriate means for the fulfillment of legitimate objectives of food law or where they would result in a different level of protection from the one determined as appropriate by the EU itself. This statement is particularly important when dealing with the Codex Alimentarius, a system developed by FAO and WHO which adopts private technical standards on food and feed on a voluntary acceptance basis. Nowadays, the Codex is an essential reference point for food safety and in many cases highly influences national legislations and case-law (Fortin, 2009). On the one hand, the Codex has tried to reach an increasing degree of integration with other international standard setting bodies, for instance by establishing contact points as a direct link with the ISO. On the other hand, the activity of the Codex to a certain extent lacks transparency and democratic legitimization and also reflects the obstacles that multi-stakeholder negotiations on food-related issues suffer from in the WTO and TRIPs contexts. This fact sooner or later erupts in radically different approaches and to the same question. This is particularly true when the EU tries to set up standards ensuring a high level of protection, but which turn out to be stricter than those defined by the Codex. The well-known *beef hormone* (EC Measures Concerning Meat and Meat Products Complaint by the USA – WT/DS26/R/USA) and *sardines* (European Communities – Trade Description of Sardines WT/DS231/R) cases, both treated by the dispute settlement body of the WTO, describe this concern properly, since bans imposed by EU institutions to the importation of foodstuff which under EU standards or scientific elaboration had to be considered unsafe were considered to be unjustified and contrary to the principles international trade is grounded on. At the same time, it is important to remark that the ECJ, in a handful of cases, has expressly mentioned the Codex Alimentarius, describing its soft law standards as a useful interpretative criterion, which has to be taken into consideration when dealing with food-related disputes (Van Gestel and Micklitz, 2013).

Consuming...soil! The (failed) attempts to sketch an EU policy on (agricultural) soil protection

According to the data collected and provided by the EU Commission, soil consumption and agricultural lands quality depletion are increasingly alarming trends. Between 1990 and 2000, an average of 275 ha were lost every day in the EU, amounting to 1,000 Km^2/year. In the last years, such loss increased by 3%, mainly due to urban sprawl or further land threats, such as erosion and salinization. As a consequence, between 1990 and 2006, with large regional variations, EU Member States lost a potential agricultural production capability amounting to 6.1 t of wheat per year. In order to face such situation, in 2002 the European Commission issued a communication to the Council and the Parliament, laying the foundations for a future thematic strategy for soil protection in Europe (COM(2002)179 final, 16 April 2002, not published in OJ). The document underlined the importance of an integrated long-term approach to the subject, urging a wide political consensus and direct commitment of member States. On the basis of that document and of the debate that soon followed, the Commission decided to take a further step, by adopting another communication generally focusing on the matter (Commission communication, COM(2006)231 final, 22 September 2006, not published in OJ). The new document remarked the need to advance practical solutions to address the various threats soil is affected by, expressly connecting ground protection to human health and food safety for the first time. Due to the large-scale questions at stake, the rationale of the proposal was to adopt a multi-stakeholder approach, involving horizontally several European policies and calling the different recipients to their duties. The document lied on four key-pillars, including raising people awareness, integrating European and national measures, closing knowledge gaps through research programmes and data collection, adopting framework legislation on sustainable land use and soil protection. In particular, as to the last point, the Commission proposed the adoption of a framework directive covering the main challenges to soil quality and establishing conditions and procedures for a proper cooperation between European institutions and national authorities. Different EU policies were actually contributing to soil protection, but as these policies pursue other scopes of action, they were not considered sufficient to ensure an adequate level of protection for all soils in Europe. At the same time, in spite of the importance of the matter, the Commission pointed out the lack of a comprehensive and effective set of rules, at both national and European levels. The framework directive was therefore intended to set a non-prescriptive and flexible legal background, in order to allow Member States to set their own levels of ambition on targets and to select the measures under the programmes and remediation strategies which they consider most appropriate and most cost-effective. In 2012, breaking an enduring silence, the Commission issued a report on the implementation of the 2006 thematic strategy, in order to assess results and shortcomings of the embryonic policy on soil protection (Commission communication COM(2012)46 final, of 13 February 2012, not published in OJ). The report presents an updated summary on soil degradation trends, upholding the need for a comprehensive European solution to the problem. Despite some remarkable achievements with regard to the first three pillars of the strategy, the Commission has been forced to raise its hands in front of the stumbling blocks the legislative process for the adoption of the framework directive – the most strategic aspect of the whole plan – has met. The European Parliament adopted its first reading on the proposal in November 2007 by a majority of about two thirds.

Nonetheless, at the Environment Council held in March 2010, a minority of member States blocked further progress on grounds of subsidiarity, excessive cost and administrative burden. In practice, the most relevant aspect of the integrated approach fostered by the Commission is now under the severe scrutiny of the member States and the Council itself: the persisting *statu quo* undermines the true launch of a European policy on soil protection and risks to endanger the effectiveness of the whole strategy. Soil protection, indeed, is now divided up in bits and pieces falling under various EU policies, which often pursue different objectives. Moreover, despite political alarms, no specific rules limiting soil consumption have been adopted: the preservation of quality and quantity of agricultural lands is still left to episodic and fragmented rules on environmental protection and CAP.

Conclusions and perspectives

The three selected topics combine to sketch a promising scenario for the future of food policies in the EU, even if threats to important non-market values are still far from being properly addressed. Indeed, while fostering the free movement of goods within the internal market, EU institutions have tried to set up adequate balances between the neutralization of borders and the risks that this phenomenon could have entailed, in terms of standardization of food production, environmental protection and threats to consumers' safety. Such commitment led to remarkable outcomes from a formal and institutional perspective, thanks to the establishment of strict procedures for controls over each step of the food production and supply chain. In parallel, the EU, under the impulse of the Commission, has been trying to pair the economic and technical advances it reached with the ethical concerns raised by civil society, starting a race to the top toward the definition of increasingly demanding rules binding the actors of the European food market stage. To this regard, the case of the strategy on soil consumption is particularly meaningful, since up to now the EU Commission has been promoting a forward-looking approach to the subject, while member States are still constrained by national interests and priorities.

References

Costato, L. and Albisinni, F. (eds.) (2012). European food law. CEDAM, Padua, Italy, 511 pp.

Fortin, N.D. (2009). Food regulation. Wiley, Hobocken, NJ, USA, 712 pp.

McMaolàin, C. (2007). Food laws of Europe. Hart, Oxford, UK, 285 pp.

Ubertazzi, B. and Muniz Espada. E. (eds.) (2009). Le indicazioni di qualità degli alimenti: diritto internazionale ed europeo. Giuffré, Milano, Italy, 404 pp.

Van Gestel, R. and Micklitz, H.W. (2013). European integration through standardization. Common Market Law Review 50: 145-182.

Section 4. Citizenship and consumers

The choice that disappeared: on the complexity of being a political consumer

M. Gjerris and H. Saxe
Institute of Food and Resource Economics, University of Copenhagen, Rolighedsvej 25, 1958 Frederiksberg C, Denmark; mgj@ifro.ku.dk

Abstract

This article criticise the notion that ethical consumerism can solve the ethical issues related to sustainability and food production through an analysis of the complexity of the concept of sustainability as related to food choices. The current trend of leaving the political discussion and regulation of the food area to the political consumer is shown to be problematic as shopping for sustainability might be much harder than initially believed due to the conflicting considerations entailed in the concept. Thus political consumerism may give way to fatalism as the complexity of choices become apparent and acts of citizenship increasingly are reduced to ethical consumerism supposed to be performed while shopping. The suggested solution is to let food policies be decided to a much higher degree through the political process engaging humans as citizens rather than consumers in the process.

Introduction

The notion of the 'political consumer' and 'ethical consumerism' plays an increasingly important role in discussions on how to bring about social and environmental fairness and health in a market driven economy (Trentmann, 2010). Initially, the idea of creating change by consuming 'correctly' might seem compelling. As the future of companies and products are decided by their ability to satisfy the demands of the consumers all that is needed to create 'a better world' is for consumers to demand the right products. Political consumerism does, however, come at a price – and a whole lot of confusion.

Three areas of critique seem especially important: (1) it threatens to obliterate the public responsibility of citizens and individualize and commercialize it into a market-oriented consumer responsibility (Lockie, 2009); (2) the notion of political consumerism runs the risks of (inadvertently) supporting current consumption through creating commodity fetishism thus not addressing deeper environmental and social issues (Carrier, 2010); and (3) the possibility of the individual for navigating the cacophonic marketplace paved with competing promises of green, health, welfare-oriented, sustainable, organic, and climate friendly products and lifestyles has been questioned as these different social and environmental agendas might not be mutually compatible, if only dealt with through consumerism (Morgan, 2010).

In the following we explore some of the confusion facing consumers when attempting to express their values through consumption in the area of *sustainability* mainly illustrating the third area of critique. Sustainability can be found at many levels: Environmental impact, climate impact, social impact, economic impact, etc. Deciding which parameters are best at measuring the sustainability of a product not only entails the risk of arbitrariness, but also means choosing among different aspects of sustainability that might not always go hand in hand.

The present analysis will demonstrate how specific ethical values sometimes exclude each other. This brings forth a demand of transparency if political consumerism is to be realized – a demand which is hard to realize in a consumerist setting where information and commercials often turn into infomercials. Furthermore it brings forth a demand of time; time to discuss and choose among the different notions of sustainability on the basis of a reflected foundation of values. Something that is more likely to be achieved in a political discussion among citizens than at the meat counter on a Thursday afternoon.

Sustainability

Sustainability has been a political, corporate and social buzz-word since it was picked up by main stream media in the wake of the Brundtland Commission back in 1987. Originally intended to describe long-term management of natural resources such as forestry and fishery, the Brundtland report employed it to put focus on the rights of future generations. Today sustainability has grown into an almost all-encompassing heading under which notions of environmental protection, animal welfare, social justice, and climate responsibilities are gathered (Gamborg and Sandøe, 2005).

Sustainable consumption is therefore one of the key elements of being a political consumer. It has become the Western consumerist's way to 'be the change that we want', as Gandhi said, that we seek to buy products that are sustainable and through this choice seek to change the world for the better. But what do you buy when you want to support sustainability through your choice of food and beverages? Should you opt for local produce, for organic food for low climate impact,or forless animal products? Or maybe you should in general just consume less?

In the following we will illustrate the complexity of answering this question by showing how different values, underlying different aspects of the sustainability concept sometimes conflict, making the choice between products more than difficult for the political consumer.

Going local

Consumers are often recommended to buy local produce since long-distance transport of produce plays a role in the environmental impact of those goods (e.g. Fogelberg, 2008). But comparison between the environmental impact of producing the foods and beverages and the environmental impact of transport shows that transport is typically not very important compared to production methods and types of food (Saxe, 2012). Some products, like e.g. lambs meat may even be better for the environment from a European consumer perspective when produced in New Zealand than when produced in the UK even when deducting the impact by transport (Saunders *et al.*, 2006). One reason is the sheep production system in New Zealand: an extensive system bases on natural grass ecosystems under ideal climate conditions. Another reason is that the environmental impact of animal produce is much higher than for vegetables, since animals consume wheat, barley, oat, maize, rape, sunflower and soy many times their weight before they end up as food on a plate. This makes the relative contribution of transport much higher for plant products, sometimes up to 50%. Buying local products might thus not necessarily be the most climate friendly choice.

Another example is that countries like Greece, Italy and Spain have an ideal climate for growing tomatoes outdoors most of the year, so heated greenhouses are not called for. The transport to consumers in Northern Europe does, however, add to the environmental impact of the Southern tomatoes. During the summer, tomatoes grown outdoors in Northern Europe have the least environmental impact when seen from the perspective of the Northern European consumer as no heated greenhouses are needed, and neither any long-distance transportation. However, most of the year tomatoes in Northern Europe are grown in greenhouses that add substantially to the environmental impact. So which tomatoes should the Northern European consumer-citizen chose?

Life cycle assessment shows that during the winter fresh tomatoes from Southern Europe have the least environmental impact. But when a greenhouse which grows tomatoes in Northern Europe is heated by waste heat from a nearby power plant or from renewable energy sources, then the environmental impact may be smaller than for imported Southern tomatoes. During the summer, locally grown outdoor tomatoes should be preferred, or tomatoes from greenhouses supplied with waste or renewable heat. Canned tomatoes typically have lower environmental impact than fresh tomatoes. This is because they make use of tomatoes that cannot be sold as fresh tomatoes, and because storage and transport is cheaper than for fresh tomatoes (British Tomato Growers' Association, 2012).

If the political consumer interprets 'sustainable' as meaning 'least climate impact', the choice of tomato will depend on obtaining knowledge about the growth season, country of origin, transportation, production form, etc. To many consumers 'sustainable' also concerns the more immediate environmental impact and 'organic' will therefore be seen as part of a sustainability concept. But even though organic tomatoes have a lower environmental impact in the form of pesticides, etc. they do, however, have a higher climate impact because the production per area is less than for convention tomatoes.

But climate impact might not be all that matters from a sustainability perspective. Buying locally produced foods can also be connected with a social attitude regarding small scale production system, creation or sustaining local jobs and a focus on connecting with the place, community or region through meals. This, however, can counteract another social issue in the age of globalization which is the idea that international trade through consumption of labelled products can function as a development tool in poorer countries. The consumer is thus left with the choice of supporting local jobs or economic development in poorer countries. Obviously these goals can be reached through other means than consumerism, but as the notion of the consumer-citizen becomes more and more embedded as the way for individual political action this becomes more and more difficult.

Going organic

As mentioned above, 'organic' is often part of the broader sustainability paradigms. Increased consumption of organics at the expense of reduced intake of conventionally produced foods is viewed as a road to an improved environment, human health and animal welfare. But including more organic food and beverages to a diet may significantly increase the environmental impact per kg food (Saxe *et al.*, 2012) since the yield in organic agriculture is typically lower than in conventional agriculture (Seufert *et al.*, 2012). However, organics are better for the environment when measured per hektare agricultural land, as it does not add pesticides to the soil, and the soil may retain more carbon than in conventional agriculture.

The issue most often discussed with regard to organic agriculture is the presumed lower yields compared to conventional agriculture in a situation where 870 million humans starve. There is huge disagreement about whether organic agriculture actually can 'feed the world' (Halweil, 2006), but assuming that organic agriculture would give a smaller total output, the question is whether this counts against buying organic from the point of view of social sustainability. But other factors could also be included when assessing which products to buy, e.g. the long-term detrimental effects to soil structure and water quality of conventional agriculture.

Yet another issue related with organic foods is whether organic food is better for your health or not. Dangour *et al.* (2010) found no evidence for nutrition-related health effects from choosing organics rather than conventionally produced foods. But the health of organic food is not only about nutritional content; it is also about content of residues of pesticides, hormones and antibiotics (Forman and Silverstein, 2012) and about the indirect health effects coming from living in an environment affected by conventional rather than organic farming, including pesticide accumulation in the food chain, water shed protection and soil quality. Whether organic food is healthier or nor is thus, also, a question of perspective and, to a certain extent, the research that one trusts.

Any such thing as a sustainable diet?

A recent Danish study (unpublished data) analyses the overall effect of the composition of a diet (reducing meat content 35% and increasing wholemeal products, fruit and vegetables relative to the average Danish diet), the long-distance transport (local produce vs. actual imports), and the type of production (80% content of organics vs. the present 8% organics). The results demonstrate that the diet composition can reduce the overall environmental impact by 30%, while buying local produce further reduced the environmental impact to 36% of a normal diet. Including 80% organics in the alternative diet removes most of the environmental advantages, so that it is now only an 8% improvement compared with the average Danish diet.

This study indicates that eating in a way where one takes heed of all the considerations gathered under the broad heading of sustainability might be impossible as the different considerations sometimes collide. In a Western setting a range of environmental considerations can be shown to collide with organic considerations as the intensive farming systems delivers a more efficient production compared to the organic production.

Going vegetarian

There is, however, one method by which the political consumer who wishes to put emphasis on a sustainable development through her or his dietary choices, has an easier time navigating the many considerations gathered in the concept of sustainability. Meat and dairy production typically cause greater greenhouse gas emissions than the production of fruit and vegetables (e.g. Audsley *et al.*, 2009). With a complete diet, animal produce account for more than half the greenhouse gas emissions (Saxe *et al.*, 2012). Reducing meat and increasing fruit and vegetables in the typical Western diet would therefore decrease the global warming potential of food and beverage consumption.

Animal production is responsible for other environmental problems globally than just global warming, they are responsible for a whole range of environmental, health and resource-related problems (unpublished data). According to FAO (2011) there is an overconsumption of meat and dairy products in Europe, North America, Latin America and Oceania. Weidema *et al.* (2008) showed that the environmental improvement potential of meat and dairy products in Europe is only about 20%, so no matter what we do, reducing meat and dairy intake will always be the most efficient way of reducing environmental impact. According to FAO (2011) by 2050 the global demand for meat and dairy products will be respectively 73% and 58% higher than today, so even with all efficiency improvements in the animal husbandry sector the supply of meat and dairy products will be deficient in a decade or two. A report from UNEP (Hertwich, 2010) concludes that a significant shift in diets away from animal based proteins towards more vegetable-based foods is needed in order to significantly reduce the environmental impact of diets.

It should be noted, however, that in some areas animal production based on grazing might be beneficial to biodiversity as they support an environment for many species of flora and fauna (Ejrnæs, 2010) and that there are indications that animal production based on grazing could help prevent desertification in e.g. North America and Africa (Sullivan and Climatewire, 2013).

Conclusion

Supporting a sustainable development regarding climate change, environmental degradation, social fairness, etc. through ethical consumerism is complicated – to say the least. The maze of information and interdependent effects of different production systems, foods and diets in the food system that creates the clash of otherwise often simultaneously held values makes the life of the ethical consumer a tough one. There is no such thing as a perfectly sustainable diet. All human food consumption has some kind of impact. It is possible to step much lighter on the ground than we do today, but not to hover above it. It is necessary to choose between a range of unwanted impacts and figure out which ones can be considered the lesser ones. Doing that in the situation of shopping for groceries where other considerations such as economy, taste, commercial influences, time, tiredness, etc. also play a role, might not be the best time to express the values you as a citizen wish to endorse in the food production system that supports modern Western societies.

Reducing citizenship in the food area to ethical consumption as e.g. suggested by the chairman of the Danish Agricultural and Food Association (Thomsen, 2013) is therefore highly problematic. No doubt that ethical consumerism can to some degree help move things in a more sustainable direction. At the same time though, it is important to remember the very existence of the consumer society itself is a challenge to sustainability and the notion that we can consume our way to a sustainable world to some degree can be viewed as yet another illusion invented and pushed by commercial producers (Barber, 2007).

There are, however, as mentioned in the previous section examples of consumer choices that from almost any sustainability-perspective would improve the state of the world. As pointed out by the United Nations Environmental Programme a reduction in meat production is one of the most efficient steps towards solving the unmet need for the 850 million starving people in the world and at the same time the most efficient step towards solving a range of escalating environmental problems, including global warming (UNEP, 2010). The consumers' diet choice could thus become a healthy and cost-effective instrument in environmental protection (Saxe, 2011, 2012). And as the American Dietetic Association recommends vegetarian diets from the standpoint of health (ADA, 2009) this choice could also alleviate some of the consequences of the obesity epidemics threatening both public health and the economy of public health sectors around the world.

This is not happening though. Ethical consumerism is not solving these issues. There are many reasons why the transition towards a more vegetarian or vegan-based diet is not happening. From our perspective one of them is that as long as the ethical values at stake are left as an individual consumer responsibility in competition with many competing considerations and influences, it becomes hard to realize the values embedded in the sustainability agenda. And when this happens, even in a relatively clear-cut case as reduced consumption of animal products, how much more difficult does it not get in the other cases discussed in this paper?

This study has emphasized and illustrated the classical difference between a the consumer and the citizen perspective, and supplied evidence that choices made through the latter might facilitate a more reflective process where the conflicting values are made clear so that our choices can be based on transparent values. Finally, we suggest that the choice of sustainability made within the citizen perspective might support ethical consumerism through labelling, tax-regulations or out-rights bans of certain products, all in all faciliteting the 'good' choice by consumers – even though it will still often be the choice among the lesser evils.

References

American Dietetic Association (2009). Position of the American dietetic association: vegetarian diets. Journal of the American Dietetic Association 109:1266-1282

Audsley, E., Brander, M., Chatterton, J., Murphy-Bokern, D., Webster, C. and Williams, A. (2009). How low can we go? An assessment of greenhouse gas emissions from the UK food system and the scope to reduce them by 2050. Cranfield University, Cranfield, UK

Barber, B.R. (2007). Consumed. How markets corrupt children, infantilize adults and swallow citizens whole. W.W. Norton & Company, New York, NY, USA.

British Tomato Growers' Association (2012). The greenhouse effect – environmental issues. Available at: www.britishtomatoes.co.uk/environment.

Carrier J (2010). Protecting the environment the natural way: ethical consumption and commodity fetishism. Antipode 42: 672-89.

Dangour, A.D., Lock, K., Hayter, A., Aikenhead, A., Allen, E. and Uauy, R. (2010). Nutrition-related health effects of organic foods: a systematic review. American Journal of Clinical Nutrition 92: 203-10.

Ejrnæs, R.P., Wiberg-Larsen, T.E., Holm, A., Josefson, B., Strandberg, B., Nygaard, L.W., Andersen, A., Winding, M., Termansen, M.D.D., Hansen, M., Søndergaard, A.S., Hansen, S., Lundsteen, A., Baattrup-Pedersen, E., Kristensen, P.H., Krogh, V., Simonsen, B., Hasler, A. and Levin, G. (2011). Danmarks biodiversitet 2010: status, udvikling og trusler [Biodiversity in Denmark 2010: Status, development and threaths]. Report DMU no. 815. University of Aarhus, Aarhus, Denmark.

Fogelberg, C.L. (2008). På väg mot miljøanpassade kostråd. Vetenskapligt underlag inför miljökonsekvensanalysen av Livsmedelverkets kostråd. Livsmedelsverket, National Food Administration, Uppsala, Sweden.

Food and Agricultural Organisation of the United Nations (2011). World livestock 2011 – livestock in food security. FAO, Rome, Italy.

Forman, J. and Silverstein, J. (2012). Organic foods: health and environmental advantages and disadvantages. Pediatrics 130: 1408-1415.

Gamborg, C. and Sandøe, P. (2005). Sustainability in farm animal breeding. A review. Animal Production Science 92(3): 221-231.

Halweil, B. (2006). Can organic farming feed us all? World Watch Magazine 19(3).

Hertwich, E., Van der Voet, E., Suh, S., Tukker, A., Huijbregts, M., Kazmierczyk, P., Lenzen, M., McNeely, J. and Moriguchi, Y. (2010). Assessing the environmental impacts of consumption and production: priority products and materials. United Nations Environment Programme, Nairobi, Kenya.

Lockie, S. (2009). Responsibility and agency within alternative food networks: assembling the 'citizen consumer'. Agriculture and Human Values 26(3): 193-201.

Morgan, K. (2010). Local and green, global and fair: the ethical foodscape and the politics of care. Environment and Planning A 42: 1852-1867.

Saunders, C., Barber, A. and Taylor, G. (2006). Food miles – comparative energy/emissions performance of New Zealand's agriculture Industry. Research Report No. 205.

Saxe, H. (2011). Diet as a healthy and cost-effective instrument in environmental protection. In: Nriagu, J.O. (ed.) Encyclopedia of environmental health, Vol. 2. Elsevier, Burlington, MA, USA, p. 70-82.

Saxe, H., Larsen, T.M. and Mogensen, L. (2012). The global warming potential of two healthy Nordic diets compared with the average Danish Diet. Climatic Change 116: 249-262.

Seufert, V., Ramankutty, N. and Foleey, J.A. (2012). Comparing the yields of organic and conventionsl agriculture. Nature 485: 229-232.

Sullivan, C. and Climatewire (2013). Can livestock grazing stop desertification? Scientific American 05.03.2013. www.scientificamerican.com/article.cfm?id=can-livestock-grazing-stop-desertification.

Thomsen, P.B. (2013). Gade: markedskræfterne skal styre den grønne omstilling [Gade: The Market should control the green conversion]. Altinget 04.03.2013 www.altinget.dk/artikel/gade-markedskraefterne-skal-styre-den-groenne-omstilling.

Trentmann, F. (2010). Citizenship and consumption. Journal of Consumer Culture 7(2): 47-58.

Weidema, B.P., Wesnæs, M., Hermansen, J., Kristensen, T. and Halberg, N. (2008). Environmental improvement potentials of meat and dairy products. JRC Scientific and Technical Report. EUR 23491 EN.

Green food consumption: whose responsibility?

S. Meisch
University of Tuebingen, International Centre for Ethics in the Sciences and Humanities, Research group 'Ethics of sciences in the research for sustainable development', Wilhelmstaße 19, 72074 Tübingen, Germany; simon.meisch@uni-tuebingen.de

Abstract

Starting from the discussion of the concept of responsibility, the paper asks to what extent the issue of counteracting climate change by different food consumption patterns can be broken down to the level of individuals. As no one is obligated beyond what he or she is able to do ('Ought presupposes Can'), one has to ask for the limits of individual responsibility. Green Food Consumption is not limited to consumers changing their shopping lists in replacing conventional by green products. It rather requires more encompassing lifestyle changes in accordance with the concept of sustainable development. If one agrees that consumers with green shopping lists are to be the solution, then one implicitly agrees that markets are the mechanism to solve the problem. An opposite view calls for the state as problem-solving mechanism. However, there are mechanisms beyond market and state. The paper asks to what extent these other institutional mechanisms alter the assessment of individual responsibility.

Keywords: responsibility, green consumerism, social ethics/political philosophy, institutions, sustainable development

Green food consumption: whose responsibility?

Agriculture and food production greatly contribute to total anthropogenic greenhouse gas emissions. These could be significantly reduced if people consumed differently, in particular less animal products (Popp *et al.,* 2010, Voget-Kleschin, 2012). Hence, there are calls for changes in individual food consumption as a means to counteract climate change. They assign responsibility to act against climate change to individuals. Some approaches highlight the power of virtuous consumers on markets. If people consumed 'green', food production would follow by providing 'greener' products (Bilharz *et al.,* 2011). An opposite view rejects individual responsibility and calls for political action to regulate food markets (Grunwald, 2012, Sinnott-Armstrong, 2005). Without denying that consumers have a choice they can take responsibly, it appears way too simple to reduce Green Food Consumption (GFC) to individual consumer choices. Yet, rejecting GFC and calling for state action underestimates the importance of individual action and leaves its possibilities unexploited.

Discussions on green consumerism seem to focus on the questions whether or not: (1) people have a moral obligation to act individually; (2) individuals are overburdened by taking over a responsibility they cannot shoulder (Sinnott-Armstrong, 2005, Sandberg, 2011); and (3) different individual acts of purchase are an effective means to achieve sustainable development (SD) and to counteract climate change (cf. discussions in GAIA 19(3), 20(1), 20(3)). These are relevant ethical questions.

By discussing the concept of responsibility, the paper focuses on institutional aspects. From the view point of social ethics (Mieth, 2011), the paper argues for a wider perspective on institutions that enable people to take responsibility collectively. The paper builds on other works claiming that there is a moral obligation of individuals to act for SD and that different consumption patterns can significantly contribute to that aim (Voget-Kleschin, 2012, 2013). However, as no one is obligated beyond what he or she is able to do ('Ought implies Can'), institutions become important. In the last century, many have contributed to a smaller or greater extent to global warming. Climate change has not been caused by one person. Similarly, it cannot be fought by any individual alone. Yet, the joint effort of many can help to reduce aggregated emissions. Institutions organise collective action and enable individuals to take their responsibility. Institutions link individual effort (consuming differently) and the final result (reducing GHG emission). Commonly, markets and states are seen as the main institutions for collective action to achieve SD. However, market and state are only two ways to solve collective action problems. This paper points to other institutions such as deliberative policy making or Commons (Immergut, 2011, Ostrom, 2010) that allow individuals to consume differently and thereby take responsibility. In conclusion, the paper argues that there are different ways to act for SD: GFC is one of them and it consists of more than just making sure that people are able to buy the right products. Institutions play an important role as they establish different fields of action and enable people to take responsibility.

Responsibility – conceptual considerations

Responsibility is an often employed, but nevertheless complex concept (Meisch, 2012). Within the academic literature, it is understood as a *multi-relational ascription* (Ott, 1997; Ropohl, 2009; Werner, 2011). Evolving in a speech situation, responsibility is ascribed to (natural, legal) persons. As such, it is distinct from traits such as intelligence or body size. Ascription also means that responsibility is not a norm by itself. Rather, responsibility is ascribed against the background of given normative standards: Because of given values and norms, the concept of responsibility is used to hold a natural or legal person accountable to a competent authority (e.g. consciousness, court, society) for an attributable act (or omission) that can be interpreted as value or norm violation (Ott, 1997). Responsibility therefore presupposes a moral Ought. It has also become clear why one can describe it as multi-relational: Responsibility is at least a four-digit relation: Someone (agent) is responsible for something (object) to somebody (addressee, authority) because of certain normative standards (norms, values) (Ropohl, 2009; Werner, 2006).

One can distinguish between a prospective and a retrospective meaning of responsibility. In its prospective sense, responsibility means that someone is responsible for something because of a task or a role he or she took on (role responsibility, *Rollenverantwortung*). Fulfilling the tasks that come with a social role does not say anything about their moral quality. Therefore, moral responsibility is given special importance as it reflects, justifies and circumscribes the validity of other forms of prospective responsibility.

In its retrospective sense, responsibility means that because of given normative standards someone is held responsible for attributable acts and their consequences (responsibility for consequences, *Folgenverantwortung*). However, there are preconditions to this. In order to hold an agent responsible for an act (or omission), it has to be attributed to him or her. This attribution cannot be reduced to a mere empirical causation. Responsibility presupposes an autonomous agent with the ability to direct his or her acts by reasons and reasoning as well as an effort of will (Stoecker, 2007; Tugendhat, 2010). We do not hold children or animals responsible, although we can empirically attribute an act to them, because we do not (yet) allot this ability to them. Moreover, if actions happened accidentally or due to external or internal constraint, we do not hold agents responsible either as they were not able to act differently or they did not influence the actions intentionally (Pauen, 2006; Tugendhat, 2010). An agent needs to have a certain scope of action and the freedom to act differently. To put it the other way round, this means that no one is obligated beyond what he or she is able to do: An Ought presupposes a Can. Taking over responsibility requires the ability and freedom to do so. Therefore, the concept of responsibility protects from excessive claims on an individual agent (Ropohl, 2009; Tugendhat, 2010). Individuals are only held responsible to the degree that they were indeed able to influence an act intentionally.

Stoecker (2007) points to the dynamic side of the concept of responsibility. Moral responsibility depends on an individual's ability to influence a course of action intentionally. Moreover, it is also restricted to specific spheres of responsibility where such an influence can be expected. Those spheres are different for different people and in different situations. Responsibility is understood as dynamic because spheres of responsibility can change. They can be influenced by new external circumstances (such as illness or war), by an individual itself (by an offer of help or an escape) or by third parties (by an election to an office). This dynamics creates a second layer of moral assessment. On the first one, we ask whether or not an individual meets its responsibility and judge whether or not it conformed to its moral duty. On the second layer, we ask how much responsibility an individual takes over and assess it as praiseworthy or disdainful. Taking over more responsibility will not necessarily lead to more praise if it results in self-overburdening, the neglect of other duties or the presumptuous intrusion in somebody else's sphere of responsibility. One cannot expect that individuals become moral saints by being responsible for everything. Choosing and accepting one's spheres of responsibility forms part of an individual's self-constitution as a person (Stoecker, 2007).

Green food consumption

It has been mentioned that we ascribe responsibility because of given normative standards. The normative basis of green consumerism is the theory of SD. In a nutshell, SD is a concept of inter- and intragenerational justice: We are morally obligated to organise ourselves in a way that all human beings have a chance to live a good human life. In this endeavour, we proceed in such a way that the natural basis necessary to live a good human life is at least retained (and in the best case recreated) for all contemporaries and future human beings (Ott and Döring, 2008; Voget-Kleschin, 2012).

Green consumerism answers the observation that our using and consuming resources and ecosystem services and our producing goods destroy humanity's natural livelihood. It ascribes to individual consumers a responsibility to use their purchasing power. In this context, buying different, i.e. more sustainable (food) products is understood as a contribution to protect humanity's natural livelihood and (as part of this) to mitigate climate change. Green consumption then is 'adopting a different perspective on our disposable income. Instead of seeing money as a means to buying us status, luxury goods or an improved quality of life, we also need to consider our money as a vote which we use every time we go shopping' (Ethical consumer *n.y.*: 8). In this understanding, substituting conventional (food) products by 'greener' ones also becomes a political act of using consumer power ('dollar voting').

There are many objections against this understanding of green consumerism. Some refer to institutions and doubt that markets are appropriate and efficient means to achieve SD in general and sustainable agriculture and food production in particular. They call for political action in order to regulate markets (see e.g. Grunwald, 2012). Others criticise that green consumerism reduces individual action for SD to acts of purchase. Voget-Kleschin (2012), for example, adopts another approach by specifying green food consumption as a low share in animal products, a certain share in regional and seasonal products and taking social issues into account. She also takes general consumption patterns into account: shopping, cooking and eating. Changing one's own consumption patterns might require more time and different knowledge and is possibly more expensive (Voget-Kleschin, 2012: 365).

Individual responsibility of consumers

From an individual ethical point of view, an individual can voluntarily take over responsibility for his or her food consumption patterns before his or her conscience because of a given norm (SD). As a consequence, individuals might buy more sustainable products, they might even try to cook differently, avoid food waste, reduce animal products, etc. They might do so because they deem it right and because they want to reduce their impact on climate change, environmental degradation and animal suffering. Virtuous consumers make contributing to a more SD their business. For them, it becomes a question of good life. We can praise them as well as criticise them in case they overburden themselves. This is a rough draft of a virtuous consumer and yet it probably matches the self-image of many 'green' consumers. However, there are many objections against this approach. This paper addresses three of them:

1. A very basic objection argues that achieving SD by building solely on (prudent or) virtuous consumers is not enough to meet the challenge. According to this argument, advancing SD is not a matter of virtues but of duties and rules, that is of deontological ethics (Eser, 2012).
2. Some argue that individual contributions to climate change are so insignificant that it cannot be an individual's duty to change his or her lifestyle. Rather than making the (emotional, motivational, etc.) efforts of acting alone and changing one's own food consumption patterns (and without knowing whether others will do so too), individuals would be better advised to campaign for a political solution that regulates all individuals' food consumption patterns (Sandberg, 2012; Sinnott-Armstrong, 2005).
3. Others doubt that markets are the appropriate means to achieve SD. They stress that individual consumers lack full information about production processes and their (environmental) consequences. In order to function properly, markets presuppose such information. Aggregation of consumers' individual actions via market mechanisms might not yield the (more sustainable) outcomes individuals aim for (cf. e.g. Grunwald, 2012). It is questionable whether one gets a life-cycle assessment that really considers SD indicators. Even if that were possible, it would overburden individual consumers to study those assessments for every single item they want to buy. State regulation of markets is seen to be the best solution to guarantee more sustainably produced food.

With regard to the concept of responsibility, all three objections carry important insights. SD cannot be solely based on virtuousness. It is a justice-based approach that imposes moral obligations and therefore has to be considered as a deontological approach (Ott and Döring, 2008). Yet, SD in general and GFC in particular still remain connected to virtue ethics. Food and eating are essential elements of what people consider as a good life and they are entrenched in cultural images (especially with regard to food consumption). Individuals' ability to achieve the kind of lives they have reason to value is then limited by the moral philosophical boundaries set by the concept of SD (Voget-Kleschin, 2013). Accepting that SD imposes moral obligations plays a role when it comes to ascribing responsibility. Against this background, it becomes relevant whether or not an individual's food consumption patterns contribute to climate change, environmental degradation or animal suffering. If there were no causal link, then it would be impossible to ascribe responsibility. However, there are individual contributions, maybe very small ones. Yet, responsibility does not disappear just because an individual played only a minor part. In this case, we speak of shared responsibility. Institutions help individuals to take their responsibility and are therefore a manifestation of collective responsibility. So far, two institutions have been mentioned: market and state. The idea of ethical consumerism is commonly understood as building on virtuous consumers with purchasing power to force more sustainable market solutions. This assumption has been rightly challenged. The last chapter therefore looks at the institutional context of GFC.

Institutional context of green food consumption

Institutions link individual action to a collective effort. They facilitate taking responsibility by providing a Can to a moral Ought. From such a broad perspective, GFC has an individual as well as a social ethical side. While the former reflects an individual's options, it presupposes the latter, which sets possibility conditions (Mieth, 2011).

Discussions on green consumerism reproduce the age-old social science discussion whether or not the state or the market is the supreme institution to solve collective action problems (Ostrom, 2003).

However, neither one is the only saving institution. Moreover, both are just two of many institutions allowing to deal with collective action problems (Ostrom, 2010). Institutions operate under certain conditions and are built on normative assumptions, which have to be made explicit (Immergut, 2011). They can be appropriate for some problems and fail with regard to others. Some of the market's shortcomings have been mentioned in this paper. While it is true that democratically legitimised politics is the best way to bring about SD and facilitate GFC, the state itself does not always produce the best policy solutions: Politics might, for instance, be locked in joint decision traps that lead to decisions taken at the lowest common denominator, as with the Common Agricultural Policy. Even democratically legitimised governments might come up with morally questionable policies as in the case of intensive animal husbandry (Misereor, 2011). State solutions also tend to one-size-fits-all-solutions that often fail to solve problems in practice (Ostrom, 2010).

The paper argues for a wider discussion of social institutions for GFC. The focus on purchasing power of virtuous consumers needlessly narrows institutional options to state and market. However, there are more institutions such as deliberative policy making (Immergut, 2011) or Commons (Ostrom, 2010). They can be found in the shape of community-shared agriculture, intercultural and urban gardening, collaborative cooking classes, etc. None of them alone might produce SD but then no single institution really can. Institutional diversity seems to be the best way to achieve SD (Ostrom, 2010). This implies that individuals innovate an institutional solution that fits a specific context and that needs to be accepted by the state and possibly protected from markets.

Above, the paper adopted a more comprehensive understanding of GFC that included shopping, cooking, eating and possibly growing one's own food and that in the end leads to different, more sustainable lifestyles. Based on the concept of SD, GFC attributes responsibility to individuals for their food consumption patterns. Institutions create the possibility conditions for moral individual actions. Institutional diversity pays tribute to the fact that humans play different roles (parents, teachers, etc.) in everyday life. As such, they can contribute in these roles in different contexts to SD and GFC. Social ethics together with different social sciences can reason on institutions that facilitate taking over responsibility and that regard GFC as part of a good life.

References

Bilharz, M., Fricke, V. and Schrader, U. (2011). Wider die Bagatellisierung der Konsumentenverantwortung. Gaia 20: 9-13.

Eser, U. (2012). Bewahrung und Entwicklung: Nachhaltige Entwicklung als dialektische Figur. In: Studierendeninitiative Greening the University e.V., Tübingen (ed.). Wissenschaft für Nachhaltige Entwicklung! Multiperspektivische Beiträge zu einer verantwortungsbewussten Wissenschaft. Marburg, Germany, pp. 27-44.

Ethical Consumer (n.y.). Ethical Consumer. A Beginner's Guide. Available at: www.ethicalconsumer.org/linkclick.aspx?fileticket=9SvJhnu632s%3d&tabid=155.

Grunwald, A. (2012). Ende einer Illusion. Warum ökologisch korrekter Konsum die Umwelt nicht retten kann. München, Germany.

Immergut, E. (2011). Democratic theory and policy analysis: four models of 'policy, politics and choice'. Der moderne Staat 4: 69-86.

Meisch, S. (2012). Verantwortung für den Frieden: Welche Fragen stellen sich bei der Umsetzung von Zivilklauseln? In: Nielebock, T., Meisch, S. and Harms, V. (eds.) Zivilklauseln in Forschung, Lehre und Studium. Hochschulen zum Frieden verpflichtet. Baden-Baden, Germany, p. 23-52.

Mieth, D. (2011). Sozialethik. In: Düwell, M., Hübenthal, C. and Werner, M.H. (eds.) Handbuch Ethik. Stuttgart; Weimar, Germany, p. 517-520.

Misereor (2011). Instrumente im Sinne einer nachhaltigen, klimafreundlichen Fleischproduktion. Aachen, Germany.

Ostrom, E. (2003). Rethinking institutional analysis and development: the bloomington school. Rethinking governance systems and challenging disciplinary boundaries. Interview. Available at: http://mercatus.org/publication/rethinking-institutional-analysis-interviews-vincent-and-elinor-ostrom.

Ostrom, E. (2010). A multi-scale approach to coping with climate change and other collective action problems. Solutions 1: 27-36.

Ott, K. (1997). Ipso Facto. Zur ethischen Begründung normativer Implikate wissenschaftlicher Praxis. Frankfurt/Main, Germany.

Ott, K. and Döring, R. (2008). Theorie und Praxis starker Nachhaltigkeit. Marburg, Germany.

Popp, A., Lotze-Campen, H. and Bodirsky, B. (2010). Food consumption, diet shifts and associated non-CO_2 greenhouse gases from agricultural production. Global Environmental Change 20: 451-462.

Ropohl, G. (2009). Verantwortung in der Ingenieursarbeit. In: Maring, M. (eds.) Verantwortung in Technik und Ökonomie. Karlsruhe, Germany, p. 37-54.

Sandberg, J. (2011). My emissions make no difference: climate change and the argument from inconsequentialism. Environmental Ethics 33: 229-248.

Sinnott-Armstrong, W. (2005). It's not *my* fault: global warming and individual moral obligations. Perspectives on Climate Change 5: 293-315.

Stoecker, R. (2007). Das Pilatus-Problem und die Vorzüge eines dynamischen Verantwortungsbegriffs. In: Berendes, J. (eds.). Autonomie durch Verantwortung. Impulse für eine Ethik in den Wissenschaften. Paderborn, Germany, p. 147-160.

Tugendhat, E. (2010). Willensfreiheit und Determinismus. In: Tugendhat, E. (ed.) Anthropologie statt Metaphysik. München, Germany, p. 57-73.

Voget-Kleschin, L. (2012). Employing a normative conception of sustainabilty to reason and specify green consumerism. In: Potthast, T. and Meisch, S. (eds.) Climate change and sustainable development. Ethical perspectives on land use and food production. Wageningen Academic Publishers, Wageningen, the Netherlands, p. 361-366.

Voget-Kleschin, L. (2013). Employing the capability approach in assigning individual responsibility for collective sustainable development. In: Röcklinsberg, H. and Sandin, P. (eds.) The ethics of consumption: the citizen, the market and the law. Wageningen Academic Publishers, Wageningen, the Netherlands. p. 83-88.

Werner, M.H. (2011). Verantwortung. In: Düwell, M., Hübenthal, C., Werner, M.H. (eds.) Handbuch Ethik. Stuttgart; Weimar, Germany, p. 541-548.

Shaping the context and content of food choices

K. Grill
Umeå University, Sweden; kalle.grill@umu.se

Abstract

Food choices have great impact on our health and on our planet. However, while comprehensive regulation is in place to avoid direct poisoning of either ourselves or our environment, we have done little to combat more indirect and long-term harms. I propose that one good reason for our passivity in this regard is that reasonable people differ in their preferences for both environmental preservation and good health, as weighed against the taste and the symbolic value of food. From a liberal perspective, respect for reasonable preference for unhealthy and eco-destructive food provides a strong reason against frustrating those preferences by prohibition or other heavy-handed regulation. However, if preferences are our concern we must note how choices depart from preferences and we must distinguish between preferences for different sorts of things. Because of cognitive limitations and biases, our preferences are not always satisfied by our choices. Furthermore, preferences are typically in internal conflict. Preferences over foods should be distinguished from preferences over choice contexts and over choice contents in the form of available alternatives. Design of choice contexts, as well as the context of preference formation, does not necessarily frustrate our food preferences, but may rather influence what food preferences we have. People may reasonably prefer that both choice contexts and choice contents be conducive to choices that satisfy their long-term preferences for health and environmental preservation.

Keywords: preference, health, long-term, short-term, environment

Introduction

The impact of food production and consumption on our health and on our planet makes food an urgent political issue. However, while we have few qualms about regulating food to avoid direct poisoning of either ourselves or our environment, regulation to avoid more indirect and long-term harms is highly controversial. In the following section, I will briefly describe the practical problem. In the section after, I will argue that respect for people's reasonable preferences warrants caution in addressing the problem. In the remaining sections, I will argue that there are several preference-based reasons for nonetheless shaping the context and content of food choices and so address the problem. Throughout, I presume a liberal perspective that ascribes great importance to people's preferences for how their own lives turn out.

Background – what is the problem?

What we eat affects our health. For example, a large recent global comparative risk assessment study attributes 254 million DALYs annually, including 12,5 million deaths, to dietary risk factors and physical inactivity, with diets the more important factor (A DALY is a Disease-Adjusted Life Year, which is way of measuring loss of quality of life; the same DALY or death can be attributable to more than one risk factor). These numbers are for 25+ year olds and so do not include childhood undernutrition. The study is one of the original outcomes of the Global Burden of Disease Study of 2010 (Lim *et al.*, 2013; for an overview over the study and some controversy surrounding it, see Cohen, 2012).

The two largest individual dietary risk factors cited in the study are a diet low in fruits and a diet high in sodium. These individual risk factors, as well as a diet low in nuts and seeds, a diet low in whole grains, a diet low in vegetables, and a diet low in seafood omega-3 fatty acids, all rank higher than for example drug use, occupational injury and sanitation. The trend, moreover, is towards greater importance of diets relative to other risk factors (Lim *et al.*, 2013).

While foodborne illness is an immense health problem, especially for children and especially in developing countries, the global negative health effects of poor diets are arguably greater than those of contaminated food. This relationship is naturally much stronger in developed countries. Moreover, foodborne illness is mainly transmitted by meat and poultry. In fact, 75% of all new infectious diseases start in animals and animal products, mostly during food production (World Health Organization). This makes a non-vegetarian diet a major risk factor for both dietary risks and food safety risks. This background factor is not considered in the risk assessment study just cited. The study does list as a more direct risk a diet high in processed meat, ranking it above for example sanitation and intimate partner violence.

Regarding the environmental effects of food, as of 2000 the livestock sector alone had contributed 18% of anthropogenic greenhouse gases and 63% of reactive nitrogen mobilization, and consumed 58% of directly used human-appropriated biomass. These numbers are very likely to grow (Pelletier and Tyedmers, 2010). While non-vegetarianism is the singularly greatest environmental problem, transportation, over-fertilization and waste are immense problems in their own right.

These massive health and environmental problems are largely the direct or more indirect result of consumer choice. Most people most of the time have ample opportunities to make food choices that are healthy and eco-friendly, rather than unhealthy and eco-destructive. To the extent that substantial opportunities for choice are lacking, they could in most cases be rather easily provided by improved city planning (such as to bring supermarkets to poor neighbourhoods where they are lacking).

That individual choices in the aggregate have very bad consequences seems on the face of it to call for political or other coordination. Why, then, have we not done more to address these problems? One reason, of course, is that large corporations in the food industry actively seek to prevent change towards more healthy and eco-friendly choices, because such choices are less profitable for them (Moodie *et al.*, 2013). Profits are threatened also by regulation to prevent direct harms such as from food poisoning and concentrated and localized pollution. However, corporations have for the most part lost this battle and adjusted, at least in developed counties. The battle over indirect and long-term harms may have only begun. Corporations may also have had weaker reason to oppose regulation of direct and immediate harms, since people tend to respond more negatively to these obvious harms than to indirect and long-term ones, and so the former are a greater source of badwill, even when legal.

These and other historical contingencies partly explain why we regulate direct and short-term harm and not indirect and long-term. However, I propose that we also have some moral reason for not doing more about the massive harms of poor food choices. I will explain this claim in the next section. I then go on to argue why, despite this reason, we ought to do more to shape and thereby change food choices.

Why not regulate? – respecting reasonable preferences

One obvious and important difference between food that causes food poisoning and food that causes diabetes or heart failure is that there is little demand for contaminate food, but massive demand for what I will call unhealthy food (food that tends to contribute to disease and poor health). Banning products is perhaps more acceptable when few people demand them in the first place. However, I propose that this cannot be the whole story. As liberals, we have reason to be concerned with individual liberty even when it is not much in demand. That only very few people prefer to walk backwards through the park or to honour some particular god is no reason to forbid these practices.

I propose that the reason for why we should hesitate to regulate unhealthy food more than contaminate food is that it is more reasonable to prefer unhealthy food. It is reasonable because consuming unhealthy food has valuable positive consequences. The most obvious such consequence is simply that it tastes good – that it is enjoyable or pleasant to eat. Moreover, that it tastes good is not just due to successful advertising but our bodies are hard-wired to appreciate fat and sweetness. Consumption of unhealthy food is also valuable because it has deep symbolic meaning, either in itself or as part of social and cultural practices.

In contrast, contaminate food does not taste good and only very rarely has symbolic meaning. This contrast between directly and indirectly harmful food may admittedly to some extent depend on the former not being predictably available. If contaminate food was sold in stores there would no doubt arise some new subculture somewhere where its consumption had symbolic meaning, perhaps even deep symbolic meaning. However, this does not change the fact that contaminate food now has little symbolic meaning. The status quo is morally relevant because it determines people's values and people's values are important.

Arguments for regulating the consumption of unhealthy food often compare the harms of poor diets to other harms, such as from transmittable disease. Such comparison is incomplete, however, if the benefits of unhealthy food are not considered. It is in the light of these benefits that it is reasonable to prefer the taste or the ease or the symbolic importance of consuming certain non-poisonous foods over the small gain in health or in environmental preservation of more healthy alternatives.

As for environmental destruction, even obvious and direct effects are typically diffused so do not directly affect choosers. This makes it somewhat more reasonable to prefer the taste, ease and symbolic value of eco-destructive foods over more eco-friendly alternatives. To a large extent, this is a collective action problem. In many cases, only lack of coordination makes these preferences reasonable. It is arguably not reasonable for prawn eaters as a group to prefer king prawns, relative to for example sea shrimp, over the immediate preservation of large areas of mangrove forests. It may be reasonable, however, for a single prawn eater to prefer the relative goodness of prawns to shrimp over her own miniscule contribution to the destruction of such forests. That this is a collective action problem does not change the fact that preferences are reasonable in the current situation. It does mean that political action should be considered to solve the problem, but the situation is more complex than with products that are not even reasonably preferred.

Not all eco-destructive choices result from collective action problems. Indirect and long-term environmental effects may not obviously harm any particular person (partly because of the so called non-identity problem – people who come to exist only because of our destruction of the environment can hardly be said to be harmed by this destruction). Even when these effects do harm particular people, these people will not be identifiable at the time. It is reasonable both to hold that only effects on particular persons (and animals) are morally relevant, and that harms to identifiable people are much more important than harms to non-identifiable people (this is why we tend to spend more to save a single trapped miner by immediate action than to save one statistical future miner by improving safety generally). Because prawns are tasty and because exotic animal products are symbolic of refined taste or an adventurous spirit, these may reasonably be preferred by food consumers to the long-term preservation of biotopes or species.

This talk of reasonable and unreasonable preferences will strike some as illiberal. If actual markets interactions are the uniquely appropriate way to treat each other with respect, it would make little sense to call unreasonable some preferences revealed on markets (White, 2010). Putting such reverence for markets to one side, we may consider whether distinguishing reasonable from unreasonable preferences is inappropriately perfectionist and so in violation of liberal neutrality. I propose that it is not, since the unreasonable preferences are unreasonable because they fail to properly appreciate *neutral* goods like health and environmental preservation (which is instrumental to economic development and physical safety).

Possible lingering controversy over the nature and status of alleged unreasonable preferences should not detract from the point that some food choices, though unhealthy and eco-destructive, are reasonable and should therefore be respected. That these choices are reasonable does not mean that they are morally right or rationally required, it merely means that there is reasonable disagreement in a liberal society over the relative importance of the values advanced by these choices.

Food choices and food preferences – lessons from behavioural science

Respect for people's reasonable preference for unhealthy and eco-destructive food provides a strong reason against frustrating those preferences by prohibition or other heavy-handed regulation. However, if preferences are our concern, we must note how choices depart from preferences and we must distinguish between preferences for different sorts of things.

Behavioural science shows ever more clearly how the choices we make are heavily influenced by aspects of the choice context that are obviously irrelevant to the content of the choice (for an overview, see Kahneman, 2011). Ego depletion, just to take one example, means that we are more prone to make unhealthy food choices if we have recently used self-control in some other area (Baumaister *et al.*, 1998). Shopping after straining our will-power on work or exercise will typically make us shop more on impulse. Cognitive load too, to take another example, makes us more prone to choose on impulse, and so, as the authors of one study put it, when trying to remember something completely unrelated to shopping, 'the consumer is more likely to choose the alternative that is superior on the affective dimension but inferior on the cognitive dimension (e.g. chocolate cake)' (Shiv and Fedorikhin, 1999). It can hardly be maintained that we prefer chocolate to fruit when we try to remember a six digit number, but prefer fruit to chocolate when the number has only two digits (unless one confuses choice and preference, on which see Hausman, 2011, Chapter 3). Food choices do not necessarily reveal food preferences.

Behavioural science also shows that our preferences are heavily influenced by various biases that we would not identify with on reflection. One example is the optimistic bias that makes us misjudge our relative health status:

> Optimistic bias appears to be a fairly pervasive phenomenon when individuals consider their comparative chances of experiencing nutrition-related health problems and also when they assess their standing on nutrition-related risk factors (Miles and Scaife, 2003: 5)

Because of this and other biases, our long-term preferences for avoiding health problems do not transfer into food preferences that can satisfy these long-term preferences.

Even without biases, our preference structures are quite complex, with long-term preferences for health conflicting with short-term preferences for pleasure or gratification. Oftentimes, we have both of these preferences, and none of our alternatives can satisfy both. Moreover, the long-terms preference does not automatically organize the short-term preferences with which it conflicts. I may have a short-term preference to eat much chocolate every day, and a long-term preference to eat much chocolate only some days. It is not clear how these two levels of preferences should be squared. Should I eat chocolate today or not? In practice, our bias towards current pleasure and future (or past!) sacrifice makes the situation even more difficult.

Preferences for choice context and choice content

Given the discrepancy between preference and choice, we may reasonably have preferences over the choice contexts that, together with our preferences (and our beliefs), determine what we choose. Given the discrepancy between long-term and short-term preferences, we may reasonably have preferences over the contexts in which our short-termed preference are formed. Choice-determining and preference-forming contexts include public advertising, product displays in stores, and portrayal of products in mass-media (heavily influenced by public relations efforts by food producers). These are aspects of our living environment that we may prefer to have one way or another, and not for any other reason than that they affect what we prefer and what we choose.

The mentioned contexts are of course already heavily regulated and could of course be more heavily regulated, or just regulated differently. As always, good regulation must consider a host of complicated empirical factors, including compliance and unwanted side-effects. However, it is important to note that regulation need not conflict with the preferences or choices it is intended to affect. This is because these preferences are shaped in the regulated contexts and so are not prior to these contexts. A person may today prefer bottled water over tab water because of advertising, but if the advertising stops he may tomorrow prefer (inexpensive and eco-friendly) tab water over bottled.

Phenomena we most typically associate with choice context can also influence choice content. For example, seeing an ad prior to consuming food can affect the taste of that food, which is of course a central property of that alternative (Elder and Krishna, 2011). If an ad causes unhealthy food to taste better, and this improvement in taste causes us to prefer the unhealthy food, we may reasonably, prior to seeing the ad, prefer not to see it. Generally, to the extent that we can affect it, we typically prefer the content of our choices to be such that we form immediate and short-term preferences that are in harmony with our long-term preferences.

More obviously, choice content is affected by the addition or subtraction of alternatives from the alternative set. Even such more drastic shaping of our choices may not conflict with any of our preferences. Sometimes we prefer an unhealthy or eco-destructive option if it is available, but do not prefer it and do not miss it, if it is not available. Since our preferences are susceptible to changes in the alternative set in this way, we can by conscious restriction of the choice content change our choices without frustrating our preferences. Even if removing unhealthy options does frustrate our preferences at the time of choice, we may of course still prefer to have them removed, in order to satisfy our long-term preference for health and environmental preservation.

Conclusion – shaping the context and content of food choice

Preferences over foods should be distinguished from preferences over choice contexts and over choice contents in the form of available alternatives. People typically have long-term preferences for such things as good health, preserved biodiversity and limited climate change. Since these things are affected by food choices, people may prefer that both choice contexts and choice contents be conducive to choices that satisfy their long-term preferences.

A proactive approach to the design of the context and content of our food choices may help satisfy our long-term and most important preferences. Such design does not necessarily frustrate our food preferences, but may rather influence what food preferences we have and what choices we make. We can affect this design both as consumers and as citizens.

References

Baumeister, R.F., Bratslavsky, E., Muraven, M. and Tice, D.M. (1998). Ego depletion: is the active self a limited resource? Journal of Personality and Social Psychology 74: 1252-1265.

Cohen, J. (2012). A controversial close-up of humanity's health. Science 338: 1414-1416.

Elder, R.S. and Krishna, A. 2010. The effects of advertising copy on sensory thoughts and perceived taste. Journal of Consumer Research 36: 748-756.

Hausman, D.M. (2011). Preference, value, choice, and welfare. Cambridge University Press, Cambridge, UK, 168 pp.

Kahneman, D. (2011). Thinking, fast and slow. Penguin Group, London, UK, 499 pp.

Lim, S.S., Vos, T., Flaxman, A.D., Danaei, G., Shibuya, K., Adair-Rohani, H., AlMazroa, M.A., Amann, M., Anderson, H.R., Andrews, K.G., Aryee, M., Atkinson, C., Bacchus, L.J., Bahalim, A.N., Balakrishnan, K., Balmes, J., Barker-Collo, S., Baxter, A., Bell, M.L., Blore, J.D., Blyth, F., Bonner, C., Borges, G., Bourne, R., Boussinesq, M., Brauer, M., Brooks, P., Bruce, N.G., Brunekreef, B., Bryan-Hancock, C., Bucello, C., Buchbinder, R., Bull, F., Burnett, R.T., Byers, T.E., Calabria, B., Carapetis, J., Carnahan, E., Chafe, Z., Charlson, F., Chen, H., Chen, J.S., Cheng, A.T.-A., Child, J.C., Cohen, A., Colson, K.E., Cowie, B.C., Darby, S., Darling, S., Davis, A., Degenhardt, L., Dentener, F., Des Jarlais, D.C., Devries, K., Dherani, M., Ding, E.L., Dorsey, E.R., Driscoll, T., Edmond, K., Ali, S.E., Engell, R.E., Erwin, P.J., Fahimi, S., Falder, G., Farzadfar, F., Ferrari, A., Finucane, M.M., Flaxman, S., Fowkes, F.G.R., Freedman, G., Freeman, M.K., Gakidou, E., Ghosh, S., Giovannucci, E., Gmel, G., Graham, K., Grainger, R., Grant, B., Gunnell, D., Gutierrez, H.R., Hall, W., Hoek, H.W., Hogan, A., Hosgood, H.D., Hoy, D., Hu, H., Hubbell, B.J., Hutchings, S.J., Ibeanusi, S.E., Jacklyn, G.L., Jasrasaria, R., Jonas, J.B., Kan, H., Kanis, J.A., Kassebaum, N., Kawakami, N., Khang, Y.-H., Khatibzadeh, S., Khoo, J.-P., Kok, C., Laden, F., Lalloo, R., Lan, Q., Lathlean, T., Leasher, J.L., Leigh, J., Li, Y., Lin, J.K., Lipshultz, S.E., London, S., Lozano, R., Lu, Y., Mak, J., Malekzadeh, R., Mallinger, L., Marcenes, W., March, L., Marks, R., Martin, R., McGale, P., McGrath, J., Mehta, S., Memish, Z.A., Mensah, G.A., Merriman, T.R., Micha, R., Michaud, C., Mishra, V., Hanafiah, K.M., Mokdad, A.A., Morawska, L., Mozaffarian, D., Murphy, T., Naghavi, M., Neal, B., Nelson, P.K., Nolla, J.M., Norman, R., Olives, C., Omer, S.B., Orchard, J., Osborne, R., Ostro, B., Page, A., Pandey, K.D., Parry, C.D.H., Passmore, E., Patra, J., Pearce, N., Pelizzari, P.M., Petzold, M., Phillips, M.R., Pope, D., Pope, C.A., Powles, J., Rao, M., Razavi, H., Rehfuess, E.A., Rehm, J.T., Ritz, B., Rivara, F.P., Roberts, T., Robinson, C., Rodriguez-Portales, J.A., Romieu, I., Room, R., Rosenfeld, L.C., Roy, A., Rushton, L., Salomon, J.A., Sampson, U., Sanchez-Riera, L., Sanman, E., Sapkota, A., Seedat, S., Shi, P., Shield, K., Shivakoti, R., Singh, G.M., Sleet, D.A., Smith, E., Smith, K.R., Stapelberg, N.J.C., Steenland, K., Stöckl, H., Stovner, L.J., Straif, K., Straney, L., Thurston, G.D., Tran, J.H., Van Dingenen, R., van Donkelaar, A., Veerman, J.L., Vijayakumar, L., Weintraub, R., Weissman, M.M., White, R.A., Whiteford, H., Wiersma, S.T., Wilkinson, J.D., Williams, H.C., Williams, W., Wilson, N., Woolf, A.D., Yip, P., Zielinski, J.M., Lopez, A.D., Murray, C.J.L. and Ezzati, M., 2012. A comparative risk assessment of burden of disease and injury attributable to 67 risk factors and risk factor clusters in 21 regions, 1990-2010: a systematic analysis for the Global Burden of Disease Study 2010. The Lancet 380: 2224-2260.

Miles, S. and Scaife, V. (2003). Optimistic bias and food. Nutrition Research Reviews 16: 3-20.

Moodie, R., Stuckler, D., Monteiro, C., Sheron, N., Neal, B., Thamarangsi, T., Lincoln, P. and Casswell, S. (2013). Profits and pandemics: prevention of harmful effects of tobacco, alcohol, and ultra-processed food and drink industries. The Lancet 381: 670-679.

Pelletier, N. and Tyedmers, P. (2010). Forecasting potential global environmental costs of livestock production 2000-2050. Proceedings of the National Academy of Sciences 107: 18371-18374.

Shiv, B. and Fedorikhin, A. (1999). Heart and mind in conflict: the interplay of affect and cognition in consumer decision making. Journal of Consumer Research 26: 278-292.

White, M.D. (2010). Markets and dignity: the essential link (with an application to health care). In: White, M.D. (ed.) Accepting the invisible hand: market-based approaches to social-economic problems. Palgrave Macmillan, New York, NY, USA, 216 pp.

World Health Organization. (2009). 10 Facts on food safety. Available at: www.who.int/features/factfiles/food_safety/en.

The consumer does not exist: overcoming the citizen/consumer paradox by shifting focus

S. Aerts
KAHO Sint-Lieven, Hospitaalstraat 23, 9100 Sint-Niklaas, Belgium & Centre for Science, Technology and Ethics, KU Leuven, Kasteelpark Arenberg 1, P.O. Box 2456, 3001 Leuven, Belgium; stef.aerts@kahosl.be

Abstract

A common framework to describe the difficult situation of today's agricultural markets is the different (perceived) attitudes of the citizen (as found via surveys) and that of the consumer (assessed via buying patterns). It is said that the citizen is demanding for ever increasing animal welfare and environmental efforts (usually at an increased production price), but at the same time the consumer in the shops is heading for the cheapest product, regardless of the production system. This dilemma (or paradox) has been confirmed many times by scientific and market research. This way of framing does however not align with the complexity of contemporary food production. It starts from the – flawed – assumption that food producers (farmers) and consumers interact. But only on rare occasions they still do. Except in short-chain commercialisation systems, consumers do not buy food as it is produced by farmers, but instead they buy – even fresh produce – from retailers that have acquired it from auction markets. For other food products many more intermediate steps exist. In this paper we will show that in most instances in which conventional food production systems made progress towards more welfare friendly or more sustainable production this has been done by circumventing the citizen-consumer paradox instead of addressing it. Examples are the shift to non-battery eggs in the Low Countries in 2005-2006 and the current steps towards non-castration pig production. We conclude that focusing on the consumer in order to restore balance is – if not futile – overly optimistic. It assumes a consumer position that is not in line with reality. Furthermore, it obscures the important leverage of other actors in the food chain, some of whom may not be unsympathetic to this 'invisible' position.

Keywords: consumption, market power

Introduction

An often-heard argument in the discussion about animal welfare in agriculture or about health effects of food is that there is a great dichotomy between the way people voice their concerns about certain practices in surveys and their buying decisions. As de Bakker and Dagevos (2012) put it: 'Is there not a yawning gap between our responsible intentions as citizens and our hedonic desires as consumers?'

This situation results in a difficult situation. Based on the public opinion the government pushes for more welfare friendly production systems, or increases the requirements for food products, often with an increased production cost as a result. At the same time in the shops the main reason to buy a specific product remains its price. Research indeed shows that most consumers are most sensitive to price when quality remains the same, regardless of information given (Verbeke and Vackier, 2004; Verbeke and Ward, 2006). Moreover, although consumers state a high interest in animal welfare, they have little interest in information about it (Verbeke, 2009).

It is clear that this situation proves a serious hurdle for (especially) European producers that are confronted with increased production prices, constant (if not decreasing) product prices, and import from countries with different (if not lower) production standards. This situation is commonly known as the 'citizen/consumer paradox', the 'citizen/consumer dilemma' or the 'double moral standards of the citizen/consumer'.

In this paper we will argue that the citizen/consumer paradox is a flawed concept, not only because the difference between the (perceived) attitudes of the citizen (as found via surveys) and that of the consumer (assessed via buying patterns) is not as big as commonly thought, but more fundamentally because it starts from a flawed assumption that food producers (farmers) and consumers interact. Therefore, progress in the food chain has to made by circumventing the citizen-consumer paradox instead of by addressing it.

Exploring the citizen/consumer paradox

An important distinction has to be made between what we call the 'citizen/consumer' in this paper, and the 'citizen-consumer' or the 'consumer citizen' as it appears in other works (e.g. Lockie, 2009; Davies, 2010; MacRae, 2012; Thirkell, 2012). The former stresses the division between the citizen and the consumer; it thus refers to the general situation that we described earlier where there is a gap between surveyed intentions and observed buying behaviour. The latter emphasises the close relation between the roles as citizen and as consumer, and therefore refers to the (relatively small?) number of people that go 'shopping for change' (Johnston and Szabo, 2011) and practice the idea of 'voting with your dollar' (Johnston, 2008) or euro, in our case.

The citizen/consumer paradox is present in many different areas of the modern society (a quick literature search will reveal that), but it is certainly present in food and agriculture. In the Low Countries a decade ago there has been a vivid debate about the citizen/consumer paradox (or dichotomy) (see e.g. Dagevos and Sterrenberg, 2003; Lips, 2004; Visak *et al.*, 2004; Aerts, 2005; Brom *et al.*, 2005). They have identified several reasons why consumers are not expressing the behaviour they indicate to find important. Availability, priorities, and non-rational arguments all seem to account for some of the apparent dichotomy. Revealing in that respect are the findings of the annual 'Consumer Behavior Monitor' reported by the Belgian OIVO (see e.g. OIVO, 2012). In this studies, typical 'ethical' purchasing criteria (such as animal welfare, fair trade, GMOs, working conditions, environment) are all reported to be important, but only when specifically asked. None of them were given spontaneously.

There are some distinct views on whether the consumer will be able to overcome this dichotomy. De Bakker and Dagevos (2012) state: 'Assuming that there is no clear distinction between citizens and consumers, still leaves us with the objection that 'citizen inspired' consumption behavior is substantially blocked by institutional conditions that shape our shopping environment. Are sustainable food choices not always hindered by the economic interests of big food companies, marketing strategies of large retailers and dominant cultural norms? From our point of view this would be an underestimation of the critical capacities and selfhood of modern people.' On the other hand Visak *et al.* (2004) point to the importance of the (physical, time, and mental) distance between the act of purchasing a food product and its consequences for animals and the environment. They refer to Milgram (1974), who suggests that the loss of individual responsibility is the most important reason for 'socially organised evil' in today's society. This is exactly what the FAO (2004) suggested: '[...] if something is everyone's responsibility in general, it tends to be seen as no one's responsibility in particular.'

A general conclusion by the abovementioned authors seems be that there are many reasons to question the existence of a real dichotomy between the citizen and the consumer in a single individual, but that the difficult shift to a more sustainable consumption is nevertheless a reality that begs addressing. As Brom *et al.* (2005) put it: 'The policy impact of this ideal, therefore, cannot be the reproach of 'double standards' towards consumers who do not live up to the ideal; the impact should be finding ways to stimulate consumers in recognizing this ideal and stimulate that it has a practical impact.' Nevertheless, earlier the same authors insisted responsible consumption is not a mere marketing problem (Visak *et al.*, 2004).

We could conclude that there is not a citizen/consumer paradox *on the individual level*, but there definitely is a dichotomy *on the societal level*. Therefore, the problem remains.

Some examples of 'solved issues' in the low countries

Following the discussion in the Low Countries about the citizen/consumer paradox, some interesting shifts have occurred in the consumer food market. Since 2005-2006 – years before EU legislation banned cage egg production – nearly all table eggs sold in The Netherlands and Belgium have been produced in non-battery systems. A similar move is seen in the pig production sector were major retail firms have pledged to refrain from selling meat from castrated pigs.

By and large, this shift to more 'animal friendly' products has not been a result from pressure group action towards consumers or government, but targeted and below-the-radar negotiation with those retail firms, supported of course by broader media actions. It seems that by this new approach the animal welfare advocates have shown that it is easier to convince five (or fifteen) buying directors than 5 (or twenty five) million consumers. A similar move is continuing in the industrial egg market, judging by the regularly appearing press releases by food processors that they will 'say no to battery eggs' (e.g. Flanders today, 2012).

Knowing the actual 'hourglass' shape of the agriculture and food market (see e.g. Grievink, 2006) that becomes increasingly narrow (compare Grievink, 2003), one wonders why so much time has been spent convincing the end consumer? If this new approach is taken with due attention to all parties involved, it even seems a good solution to the problem (Aerts and Lips, 2010).

Indeed, it does deliver:
- higher coherence between citizen concerns and available products;
- higher coherence between citizen concerns and purchased products;
- better market access for 'better' production systems.

Lessons to be learned

The hourglass model of the food (retail) industry illustrates quite directly that in most cases there is no direct contact between producers and consumers. The notion that 'they ask us to produce according to A, and then buy B' is not what happens in the majority of the market. The important leverage in the market is not with the consumer, but a step upstream.

Looking at the combined market share of the top-5 retail firms we see that in most countries this exceeds 2/3 to 4/5 of the food market (see e.g. Hendrickson and Heffernan, 2005). We would argue that the adagio 'We sell what they buy' should in most cases be replaced by 'They buy what we sell'. Lips (2004) already argued that buying a product when it is available is not the same is wanting it to be available: 'One can only be sure whether all these people really want these products to be offered when they would explicitly ask for them when they would not be available (anymore)' [translation my own]. He went on to state that this type of research was not yet available. Meanwhile, the changes in the egg market have provided a real-life experiment of precisely this type, and we have not seen any decline in the table egg consumption, nor any demonstrations for the continued availability of battery eggs. Anecdotal evidence suggests most consumers didn't even notice the change, and some retailers even communicated the change only months after it had been introduced.

Focusing on the consumer to drive the change is like talking to the person in the passenger seat. This may work when that person is called Hyacinth Bucket, but not in the majority of cases.

Conclusion

From the analysis above, we conclude that the food chain is not consumer-driven in the real sense of the word, nor is it producer driven. It is retail driven. The consumer is in essence a decision-taker, more than a decision-maker.

Therefore, the citizen/consumer paradox could in fact be said to be irrelevant. One should not focus on solving the paradox, but on solving the problem. Somewhat simplified, one could say 'if it's not there, the consumer won't buy it'. At least in some cases, the consumer will not even ask for it.

We suspect that in many cases the problem of 'high' societal demands and 'low' consumer support can be solved by circumventing the citizen/consumer paradox. Ethicists, welfarists, and agriculture lobbyists should focus on the decision makers. Retail companies hold the key to real change, change that can shift the market instead of changing it. A shift that can be made with due attention to the needs of all parties involved.

References

Aerts, S. (2005). Dierenwelzijn en consumptie. In: De Tavernier, J., Aerts, S. and Lips, D. (eds.) Dier en welzijn. LannooCampus, Leuven, Belgium.

Aerts, S. and Lips, D. (2010). Dierenwelzijn en consumptie. In: De Tavernier, J., Aerts, S. and Lips, D. (eds.) Dier en welzijn, 2nd edition. LannooCampus, Leuven, Belgium.

Brom, F.W. A., Visak, T. and Meijboom, F. (2005). Food, citizens and market. The quest for responsible consuming. In: Pothast, T., Baumgartner, C. and Engels, E-M (eds.) Die richtigen Masse für die Nahrung. Biotechnologie, Landwirtschaft und Lebensmittel in ethischer Perspective. Francke Verlag, Tübingen/Basel, Germany/Switzerland, p. 115-132.

Dagevos, H. and Sterrenberg, L. (2003). Burgers en consumenten – tussen tweedeling en twee-eenheid. Wageningen Academic Publishers, Wageningen, the Netherlands, p. 160 pp.

Davies, J.R. (2010). The European consumer citizen: a coherent, tangible and relevant notion of citizenship? PhD, University of Leicester, Leicester, UK.

De Bakker, E. and Dagevos, H. (2012). Reducing meat consumption in today's consumer society: questioning the citizen-consumer gap. Journal of Agricultural and Environmental Ethics 25(6): 877-894.

Flanders Today (2012). Food industry says no to battery eggs (May 23, 2012). Available at: http://www.flanderstoday.eu/content/food-industry-says-no-battery-eggs.

Food and Agriculture Organisation. (2004). The ethics of sustainable agricultural intensification. FAO Ethics Series 3. Food and Agriculture Organisation of the United Nations, Rome, Italy.

Grievink, J.W. (2006). The changing face of the global food industry. Brands, products and concepts under retail dominance? Available at: www.cooperatie.nl/download/17-grievink.pdf.

Grievink, J.W. (2003). The changing face of the global food industry. Presentation at the OECD Conference 6 February 2003, The Hague, the netherlands. Available via: www.oecd.org/tad/agricultural-trade/oecdconferencetoexplorechangesinthefoodeconomythehague6-7february2003.htm.

Hendrickson, M. and Heffernan, W. (2005). Concentration of agricultural markets. University of Missouri, Columbia, USA.

Johnston, J. (2008). The citizen-consumer hybrid: ideological tensions and the case of Whole Foods Market. Theory and Society 37(3): 229-270.

Johnston, J. and Szabo, M. (2011). Reflexivity and the Whole Foods Market consumer: the lived experience of shopping for change. Agriculture and Human Values 28(3): 303-319.

Lips, D. (2004). Op zoek naar een meer diervriendelijke veehouderij in de 21ste eeuw. Aanzet tot het ontwikkelen van win-winsituaties voor dier en veehouder. Dissertationes de Agricultura, 609. PhD, KU Leuven, Leuven, Belgium.

Lockie, S. (2009). Responsibility and agency within alternative food networks: assembling the 'citizen consumer'. Agriculture and Human Values 26(3): 193-201.

MacRae, R., Szabo, M., Anderson, K., Louden, F. and Trillo, S. (2012). Empowering the citizen-consumer: re-regulating consumer information to support the transition to sustainable and health promoting food systems in Canada. Sustainability 4: 2146-2175.

Milgram, S. (1974). Obedience to authority, Tavistock, London, UK.

Onderzoeks- en informatiecentrum van de verbruikersorganisaties (2012). Consumer behavior monitor. OIVO, Brussels, Belgium.

Thirkell, P.C. (2012). The Consumer as critical player in moving towards a sustainable marketing system: refining the agenda. In: Haase, M. and Kleinaltenkamp, M. (eds.) Proceedings of the 37th Macromarketing Conference. June 13 – 16, 2012. Berlin, Germany. pp. 392-398.

Verbeke, W. (2009). Stakeholder, citizen and consumer interests in farm animal welfare. Animal Welfare 18(4): 325-333.

Verbeke, W. and Vackier, I. (2004). Profile and effects of consumer involve-ment in fresh meat. Meat Science 67(2): 159-168.

Verbeke, W. and Ward, R. (2006). Consumer interest in information cues denoting quality, traceability and origin: an application of ordered probit models to beef labels. Quality and Preference 17(6): 453-467.

Visak, T., Meijboom, F.L.B, Kalis, A. and Brom, F.W.A. (2004). Verantwoord consumeren: De dubbele moraal als vertrekpunt voor een discussie over normatieve vooronderstellingen en verantwoordelijkheid Verkennend Essay, NWO Programma Ethiek, onderzoek & Bestuur. Ethiek Instituut, Universiteit Utrecht, Utrecht, the Netherlands.

Consumer citizenship: a self-contradictory concept?

A. Kallhoff
University of Vienna, Department of Philosophy, NIG, A 0306, Universitätsstraße 7, 1010 Vienna, Austria; angela.kallhoff@univie.ac.at

Abstract

The concept of consumer citizenship relates to two discourses in contemporary political philosophy. In particular, the concept of citizenship is currently in three central traditions in political philosophy: the traditions of political liberalism, of civic republicanism and of deliberative democracy. Yet, against each of these backgrounds, consumer citizenship looks like a degraded form of citizenship. Because of this result of the *first section*, I shall take a different stance on consumer citizenship. This concept is in one line with recently explored 'qualified concepts of citizenship', i.e. 'ecological citizenship'. *Section two* discusses the implications of this approach to consumer citizenship. A *third section* first states that as a result of the first two approaches, consumer citizenship looks like a self-contradictory concept. Yet, a more thorough look at the tensions in that concept contributes to another interpretation: Consumer citizenship highlights tensions of the concept of citizenship more generally. In particular, consumer citizenship provides an antipode to the concept of consumer sovereignty.

Keywords: citizenship, consumer citizenship, qualified citizenship, public goods

Introduction

Citizenship has, of course, been a key concept in political philosophy since Aristotle's *Politics*. According to Aristotle, a citizen is a person who participates in political life in a distinct way: 'But the citizen whom we are seeking to define is a citizen in the strictest sense (...) and his special characteristic is that he shares in the administration of justice, and in offices' (Pol. III.1, 1275a 22-24). In particular, in Aristotle's *Politics* the definition of citizenship is tied both to the constitution, which hammers out the distinct capacities of the state, and to the competences of persons who participate in political life. Aristotle thinks that only those persons deserve the title of a 'citizen', who gets actively involved in executive power and legislation; moreover, a citizen should possess virtues, foremost the virtue of justice (Pol. III.4, 1276 b 20-34).

Since Aristotle's *Politics*, the concept of a citizen has been scrutinized from various perspectives. Until today, there is not one single approach to citizenship. Instead, there is a debate about how 'citizenship' shall best be categorized. This debate is related to the 'big programs' in current political philosophy: political liberalism, civic republicanism, and deliberative democracy.[12] In the first section, I shall give a short sketch of each of these approaches to citizenship. Yet, more recently, the debate on citizenship has also been stimulated by – what might be called – qualified conceptions of citizenship. 'Cyber citizenship', 'green citizenship', and 'economic citizenship' are now discussed as concepts which express citizenship duties and citizenship rights in specified areas of social life. The second section is dedicated to introducing that debate. Both sections suffice to set the stage for the discussion of consumer citizenship. The third section discusses the implications for 'consumer citizenship'.

[12] Communitarians would be a further group, including for example the positions of Michael Walzer, Charles Taylor, Will Kymlicka. Yet, from the beginning, the identification of that group of thinkers as 'communitarians' suffers from several shortfalls – some even question whether or not they belong into one camp. What holds this group together is the assumption, that political communities and values which constitute those communities have priority over individual conceptions of the good – at least so regarding political society.

Citizenship in political philosophy

In the western tradition of philosophy, citizenship has been tied to the notion of rights. Since Marshall's (1964) contribution to the development of rights, three groups of rights have been distinguished. These are rights of the first generation: civil rights and political rights; and rights of the second generation: socio-economic rights, comprising i.e. a right to education, to welfare and to culture. *Political liberals* are particularly interested in defending the rights of persons as citizens. In the 20th century, this approach to citizenship was re-established by John Rawls. In *Political liberalism* (1996), Rawls interprets citizens as free and equal persons. In particular, citizens regard each other as free in that they regard themselves as 'self-authenticating sources of valid claims. That is, they regard themselves as being entitled to make claims on their institutions so as to advance their conceptions of the good [...]' (Rawls, 1996: 32). In particular, each citizen should be guaranteed an adequate system of basic rights which is compatible with the same set of rights of each citizen.[13] In short, in political liberalism citizenship is characterized by normative claims of persons against the state and against each other. These normative claims are rights which guard persons from unfair restrictions.

Different from that approach, civic republicans underscores the 'common good' of a citizenry (Pettit, 1997; Sunstein, 1988; Viroli, 1995). Authors in that camp interpret citizens as free persons, too. Yet, freedom has a different meaning. A citizen is as free as his nation state, the republic. Moreover, the freedom of the republic needs to be defended continuously. In order to achieve this goal, two characteristics of the republic citizen are pivotal: The republic citizen is virtuous, and the republic citizen puts the common good at the center of her concern. As an approach to political philosophy, republicanism has a long tradition. Recent authors in the field of civic republicanism try to re-establish republicanism in a specified way. They wish to portray a republicanism which stresses the common good in processes of legitimation of political power; and they wish to see a republicanism which includes civic virtues.

A third important group in discussing citizenship is deliberative democracy. In that perspective, citizens are distinguished as persons who unfold political power in the public sphere. Following Jürgen Habermas (1998), the public sphere is a realm which is constituted by exchange of arguments on items of common interest. In particular, it is a 'metatopical realm' (Taylor, 2004), which has some specific characteristics: It is an open realm to which each citizen is invited; it is a realm of deliberation, that means that through various media, persons exchange their rational arguments on issues of common interest; and it is important for generating political power which controls the governments and their institutions.[14] Before commenting on these alternatives, I shall proceed with qualified concepts of citizenship.

Qualified concepts of citizenship

Different from the 'big programs', recent authors have contributed to specified concepts of citizenship. The underlying hypothesis is that in our days, societies are highly differentiated. In particular, there are many specified spheres of society which follow their own rules. In order to defend basic rights in those spheres, and in order to defend institutions which pay tribute to 'equity' and to 'justice', political philosophy needs to discuss those qualified conceptions of citizenship too. Yet, due to the diversity of the sub-spheres of society, it is difficult to recall a common theme in those approaches. Instead, scholars who work on qualified concepts of citizenship appear to share some common concerns rather than a singular methodological approach to citizenship. In particular, two points have been emphasized. In introducing

[13] For the precise outline of the first principle of justice, see Rawls (1996: 5).

[14] This summary of current debates on citizenship gives only a short sketch of key positions. There are further interesting approaches to citizenship, e.g. by Faulks (2000). For a discussion of citizenship competences, see also Kallhoff (2013).

qualified concepts of citizenship, scholars wish to articulate a critique of recent developments in society. Moreover, they wish to get away from the big programs and instead favor a dialogue about concrete drawbacks and nuisances of existing institutions. I shall give two examples for underpinning these claims.

In discussing 'green citizenship' or 'ecological citizenship', authors wish to say that it is now the time to address the duties and freedoms of citizens anew. Some authors argue that political institutions need to react to global ecological problems. Therefore, we need a post-global concept of citizenship which takes ecological duties seriously (Dobson, 2003; Hailwood, 2005). Kwame Anthony Appiah has recently (2007) defended a concept of 'cosmopolitan citizenship'. The claim that it is reasonable to interpret citizens as 'citizens of the world' is – of course – not new. Yet, authors now defend the more radical claim that 'global citizenship' is no longer a choice, but rather a fact. Political philosophy needs to react to this. Another group of authors works on rights and duties of citizens as related to the economic sphere.[15] Citizenship here means that the rights of persons in the economic sphere, i.e. the right to work, the right to participate in the gains in the economic sector, should be discussed anew.[16]

Consumer citizenship

I am now in a situation to introduce consumer citizenship into current debates on citizenship. In particular, I shall proceed in *two steps*. I shall *first* give an interpretation of 'consumer citizenship' against the background that has been elaborated so far. I shall discuss if 'consumer citizenship' can be conceived of as either a qualified concept of citizenship or as related to the big programs in political philosophy. *Secondly*, I shall discuss the consequences of that interpretation. Against the background of political philosophy, one might argue that consumer citizenship is a self-contradictory concept. Yet, I shall demonstrate that it appears to have a function different from that: it contributes to highlighting tensions between spheres of citizenship more generally.

Consumer citizenship against the background of political philosophy

In order to introduce consumer citizenship into the debates of political philosophy, I shall start with the qualified concepts of citizenship. A qualified concept of citizenship has three characteristics (section two): It relates the concept of a political person to a specified sphere of society; it articulates a critique of recent developments in society; and it focuses on concrete drawbacks and nuisances of existing political institutions.

As a consequence, in order to discuss consumer citizenship, we need to focus on the sphere of consumption. Persons are participants in the economic sphere. In particular, as consumers, they choose items according to their needs and desires; and they buy items for money. In times of service economies, it is not only shoes and cars which persons buy. Instead, we buy services of various types; we pay for education, for health-care, and for insurances. The sphere of society, which consumer citizenship addresses, is one side of the market sphere.

For articulating critique and for defending related normative claims, it is helpful to start with 'injustice'. Despite the fact that this is a broad notion and moreover a key notion in political philosophy, some rather obvious things come to my mind. It is unfair that some items are particularly cheap, others are costly. It is not fair that consumption is tied to economic means, means which are distributed very arbitrarily in this world. Too many things are now traded in the market sphere; some items simply should not be

[15] There is not a good concept available here, perhaps economic citizenship. Corporate citizenship has received another meaning: it denominates the commitment of the private sector to support civil society in communal projects.

[16] This claim has been defended in Ulrich (2001).

marketized, and so forth.[17] On the other hand, scholars also criticize that the market sphere is not free at all. Instead, governments impose ever new regulation on that sphere. As a consequence, consumers are no longer free to choose.[18]

In order to give an adequate interpretation of consumer citizenship, a third aspect needs to be underscored. This aspect draws on the traditions of political philosophy (section one). Citizens are persons whose rights and duties are defined by political institutions, not by individual wishes or desires. Moreover, citizenship has always *two sides*. On the one hand, citizens are endowed with *rights*. Even though the interpretation of that side varies, political philosophers underscore that citizens have a right to be protected from unfair practices – practices of governments as well as of fellow citizens. This aspect is particularly clear in the writings of John Stuart Mill. Following his no-harm principle, there is no right whatsoever to infringe on actions of persons in an arbitrary and harmful way. Put the other way round, persons are free as long as their actions do not harm another fellow person (Mill, 2003: 94).

On the other hand, citizens have *duties*. Citizens are responsible for actions which effect fellow citizens. In civic republicanism, they are particularly responsible for contributing to the common good. Yet, in political liberalism the duties of citizens are also highlighted. In Rawls's contribution to political liberalism, citizens are held responsible for supporting the common good as well. He says: 'The third respect in which citizens are viewed as free is that they are viewed as capable of taking responsibility for their ends and this affects how their various claims are assessed' (Rawls, 1996: 33).

I shall now pull the strands of thought together. In my view, it is not particularly helpful to give a definition of consumer citizenship at this point. Instead, I wish to highlight three characteristics of that notion from the perspective of political philosophy:

1. Consumer citizenship is a notion which highlights a 'fair share' of persons in the sphere of consumption.
2. Consumer citizenship stresses the (limited) freedoms of persons in participating in the market sphere as consumers.
3. Consumer citizenship emphasizes the duties of persons as responsible actors in the sphere of consumption.

So far, I have elaborated some normative implications of the concept of citizenship against the background of current debates in political philosophy. I shall now turn to a discussion of these aspects. In particular, I shall ask: Is consumer citizenship a self-contradictory concept?

Consumer citizenship as a self-contradictory concept?

At first glance, all three normative claims about consumer citizenship are plausible. Different from a person, a citizen is a member of the political community. She has rights and duties – both regarding fellow-citizens and regarding the government. Yet, a more thorough discussion demonstrates that the claims pull into different directions. The first claim (1) is in one line with arguments for social justice. Even though there is a broad debate on how social justice should be defined, authors in that field are all interested in defining a 'fair share' of persons. Yet, it also needs to be noted, that contemporary contributions conceive of social justice as a corrective to the market sphere. The first normative claim, instead, sounds like claiming a comprehensive regulation of the market sphere – at least regarding a minimum share.

[17] This echoes one critique which has been elaborated in Kallhoff (2011).

[18] This perspective has frequently been articulated by scholars who underwrite what might be called the Chicago school of economics. See Friedman (2002).

The second claim (2) introduces freedoms of consumers. Actually, this is what most persons will regard as critical to interpreting consumption in normative contexts. No person should be limited in choosing goods according to her preferences. This is the market dogma which even those persons, who underwrite social justice, do not wish to give up. This claim is encapsulated in the notion of consumer sovereignty. It says that persons should be free to choose consumption goods according to their preferences. Yet, this freedom is not an unlimited freedom. One perspective for highlighting the limitation of that view results from a theory of public goods. Markets are not suitable for providing public goods. Instead, public goods need to be supported by means of public finance. This says that there are limitations of the consumers' freedoms. A normative theory, instead, needs to discuss justifications for the adequate set of limitations.

In this respect, (1) could serve as a reasonable constraint. Even though persons should be free to choose their consumption goods, social justice is a normative constraint. This may result in a variety of policies. It could mean to defend a sphere of non-marketized goods. Public goods are suited much better for guaranteeing an equal share to each citizen. Therefore, it would be right not to privatize public goods.[19] Another approach might defend a limitation of the market sphere in terms of keeping it separate from, i.e. spheres of life which should not follow market laws.[20] In short, claims of justice could serve as a constraint of consumer freedoms.

But what about the third claim (3)? It states that consumer citizenship emphasizes the duties of persons as responsible actors in the sphere of consumption. Is this notion coherent with the interpretation of consumer citizenship as highlighting the limited rights of citizens to profit from the market sphere as consumers? Actually, the debate on consumer responsibility is far more radical than the debate about a fair share of goods. Many now claim that it is not the market, but the consumer who could achieve a turn-around in the economic sphere. Since the whole system of the market depends on consumer choices, a shift of preferences and responsible behavior would be a very powerful mechanism in order to change market structures. In particular, consumers could demand information about the economic value creation chain; the consumer could claim that firms reveal the ecological footprints of their products; consumers could claim that a firm pays respect to human rights, etc. Yet, these powers do not have to be claimed; instead, the consumers already have these powers. All that a concept of consumer citizenship could add is a perspective which highlights another point: Persons are not only individuals; but as citizens they are part of a political community which nowadays also includes a cosmopolitan community. In this context, responsibility is not perfectly voluntary. Instead, the responsibility of citizens is part of its membership in the community of citizens worldwide. True, this community is a community of fate – not a chosen one. But, it is up to each citizen to contribute to transforming that community of fate into a political community.[21]

To summarize my argument, I do not think that consumer citizenship is a self-contradictory concept. True, it mirrors a characteristic trait of each concept of citizenship: Citizenship is not a monolithic notion; instead, it comprises elements which pull into various directions. It is part of the challenge to discuss citizenship: one has to explain the right balance between elements such as freedom, duties, and

[19] For a thorough discussion of the underlying distinctions, see Kallhoff (2011).

[20] This is one of the claims which Habermas encapsulates in his critique of the 'colonization of life-spheres' (see Habermas, 1987: chap. VI.2). In a way, it also relates to Michael Walzer's arguments for 'complex equality' in his *Spheres of justice*; see Walzer (1983: 19-20). In Walzer's view, equality is not an isolated notion. Instead, spheres of justice can be distinguished regarding both the goods on which they focus and principles of fair distribution respectively. Complex equality says that an undue overlap of principles between the spheres should be avoided.

[21] This is a central claim of the approach to citizenship by Van Gunsteren (1998).

rights.[22] More importantly, as a normative concept, consumer citizenship is a systematic antipode to consumer sovereignty.

References

Appiah, K.A. (2007). Cosmopolitanism. ethics in a world of strangers. Penguin Books, London, UK.

Aristotle (1921). Politics. In: Ross, W.D. (ed.) Transl. by Benjamin Jowett. Clarendon Press, Oxford, UK.

Dobson, A. (2003). Citizenship and the environment. Oxford University Press, Oxford, UK.

Faulks, K. (2000). Citizenship. Routledge, London and New York, UK and USA.

Friedman, M. (2002) [1962]. Capitalism and Freedom. 40th anniversary ed. with a new pref., University of Chicago Press, Chicago and London, USA and UK.

Habermas, J. (1987). The theory of communicative action. Vol. 2. Life-world and system. A critique of functionalist reason. Transl. by Thomas McCarthy. Beacon Press, Boston, MA, USA.

Habermas, J. (1998). Three normative models of democracy. In: Habermas, J. (ed.) The inclusion of the other. Studies in Political Theory. The MIT Press, Cambridge, MA, USA, p. 239-252.

Hailwood, S. (2005). Environmental citizenship as reasonable citizenship. In: Dobson, A. and Valencia Sáiz, A. (eds.) Citizenship, environment, economy. Routledge, London, UK, p. 39-54.

Held, D. (1996). Models of democracy. 2nd ed. Stanford University Press, Stanford, CA, USA.

Kallhoff, A. (2011). Why democracy needs public goods. Lexington, A division of Rowman and Littlefield publ., Lanham, MA, USA.

Kallhoff, A. (2013). Politische Philosophie des Bürgers. Verlag Böhlau, Wien, Köln, Weimar, Germany, 216 pp.

Marshall, T.H. (1964). Class, citizenship, and social development. Garden City, Doubleday & Company, New York, NY, USA.

Mill, J.S. (2003). On liberty. In: Warnock, M. (ed.) Utilitarianism and on liberty. Blackwell Publishing, Malden, MA, USA.

Pettit, P. (1997). Republicanism. A theory of freedom and government. Oxford University Press, Oxford, UK.

Ralws, J. (1996). Political liberalism. With a new introduction and the 'reply to Habermas'. Columbia University Press, New York, NY, USA.

Sunstein, C.R. (1988). Beyond the republican revival. The Yale Law Journal 97: 1539-1590.

Taylor, Ch. (2004). Modern social imagineries. Duke University Press, London, UK.

Ulrich, P. (2001). Integrative Wirtschaftsethik. Grundlagen einer lebensdienlichen Ökonomie. 3. rev. Aufl. Haupt, Bern, Switzerland.

Van Gunsteren, H. R. (1998). A theory of citizenship. Organizing plurality in contemporary democracies. Westview Press, Boulder, CO, USA.

Viroli, M. (1995). For love of country. An essay on patriotism and nationalism. Clarendon Press, Oxford, UK.

Walzer, M. (1983). Spheres of justice. A defense of pluralism and equality. Basis Books, New York, NY, USA.

[22] This is one key insight in the political philosophy of citizenship of Kallhoff (2013).

The impossibility of an ethical consumer

J. Karlsson
Lund University, CTR, Trollebergsvagen 99, 227 31 Lund, Sweden; jerker@allotetraploid.se

Abstract

The thesis of this article is that the notion of an ethical food consumer is untenable unless it is coupled with a conception of food citizenship. The main arguments delivered against the notion of ethical food consumption are that consumption does not take the operations of moral psychology into account, nor afford means to tackle structural problems inherent in the relation between consumer and producer. The notion of an ethically aware food citizen is on the other hand capable of handling these problems head on. Hence, a realistic food ethics has to operate with an integrative view of consumption and citizenship.

Keywords: food citizen, moral psychology, social cognitive theory, virtues

The division between public and private life has a long-standing history in western thought and culture. It can be traced with historic certainty back to the division made in ancient Greece between *polis* and *oikos* and the distinction has played some role or other in political theorizing ever since. A contemporary theory that relies heavily on holding these two selves apart is Rawls' (2001) *Justice as fairness*. While Rawls is not alone in letting the distinction between public and private carry a major theoretic weight, we can still use his account to get a better grip of what notions the terms 'public' and 'private' are intended to capture in a modern liberal political theory.

According to Rawls (2001: 10) the 'public' is instantiated by the duties and rights connected to a politically construed citizenry, while the 'private' is instantiated by individual aspirations and commitments; i.e. the things that individuates us as persons. The latter is built around the scaffolding of the former and is contained and thus controlled by it. Yet, the private retains a high degree of freedom by only being legitimately restrained by what Rawls (2001: 28, 182-183) termed *constitutional essentials*, i.e. the fundamental principles of the political process and the basic rights and liberties of citizenship. We are allowed, so to speak, to play whatever game we please as long as we keep the ball in the park and, importantly, do not force others to play by our self-imposed rules.

Given the philosophic, historic, and legal role the public and private dichotomy has played in the west, it is no surprise that it is also reflected in how we have shaped our ethical models, as for example the ethics of food consumption. It is consequently not a coincidence that the distinction between food producer and food consumer has a sibling. In corporate governance, we separate food producer and food industry where the former is the object of CSR while the latter is the object of legislation. It thus seems right and proper that the producer engage the consumer while the industry and citizenry remains apart from their relation.

We end up with two separate marriages. The consumer bounds with the producer in a marriage of life-style and choice while the citizen bounds with the industry in a marriage of convention and propriety. These antenuptial plans are a strange form of bigamy, because the bride and the groom are the same in both marriages. The only thing that differs is that the two marriages have separate prenuptial agreements. The consumer and producer are to live by private discourse in unsanctioned institutions while the citizen and the industry are to live by public discourse in official institutions.

Our present conception is shaped by an overarching ideology of contractarianism that restrains the legitimate use of state power. The state, so it is maintained, has no legitimate right to stir in the pots and pans we have placed on our stoves. What pertains to our home (*oikos*) is simply not a matter for the public (*polis*). In present market liberal discourse, we, for better and worse, conceive of the marketplace (*agora*) as an extension of our private sphere (*oikos*). In the market, I exercise my individual freedom as a private citizen, unencumbered by my role as a public citizen. Here I create a lifestyle and enact what Rawls termed a *comprehensive doctrine*, i.e. the idiosyncratic moral and aesthetic commitments that make my life my own and worth living. As a public citizen on the other hand, I am duty bound to uphold the basic social and legal structures that make our freedom as private citizens possible. What these structures demand in detail is a contested issue among political philosophers and politicians alike. There are minimalist libertarians and comprehensive social liberals that wage a perennial war over the extent of private freedom and the weight of public duty. Nevertheless, they seem to agree that the distinction between public and private is essentially meaningful and worth upholding. I would like to suggest that it is not, at least not in a realistic food ethics.

We need first to call to mind that the mode of interaction between consumer and producer is established through the legal system that defines the responsibilities and rights for each party. We do not live in a primitive barter economy but in a highly specialized and technically advanced market economy that exists because of a material and legal infrastructure, that is the direct result of public policy making. Availability and price of goods are as much, if not more, due to investments in infrastructure, subsidies, and transnational trade agreements as they are due to unadulterated consumer demand. Also, accountability and rights change at each step some goods change hands and thus legal responsibility and legal right does not necessarily overlap with moral responsibility and moral right.

As consumers, we find ourselves placed in a market with a preset availability and price structure we had no role in creating. Yet, unless we intend to starve ourselves to death in the shopping mall, we had better start buying. Something similar holds for the prospective producer who is restrained by the present market structure and this is not restricted to the formal threshold for entering the market. Beyond the paperwork, licenses and subsidies the producer must acquire channels for processing, distribution, and marketing. In short, a prospective producer must first navigate the market structure before her product can satiate or kindle a consumer demand.

Given the interwoven and complex market structure we, like Wilkins (2004) has suggested, should move from being passive food consumers to become active food citizens; i.e. be policy makers and community builders rather than inactive recipients of pre-made policies and loyal subjects to the *ancien régime*. Behind this admonition lies the realization that there simply is not a free and open market ready to supply us with the kind of goods we desire and neither is there a straight path between prospective producers and potential customers that can be easily traversed to remedy present imperfections in the market. We have played the old game far too long, so if we want to change the setup we cannot confine ourselves to move our own piece. We must modify the rules by which the game is played by all players. This contention is of cause adverse to the ideology and rhetoric of ethical food consumption.

Johnston (2008) has shown that ethical consumer discourse is, at least in some major instances, based on an idea that dollar voting can accomplish major progressive change. It construes the consumer as a citizen that has replaced her voting bill with a dollar bill and by so doing strikes an uneven balance between individual freedom and collective responsibility. As we have previously argued, the private is in principle construed as free and unrestrained by other than private concerns and we are not supposed to be held accountable for our private preferences. Since the state is not allowed to meddle in our private affairs, we are left as sole individuals to face the market with our voting dollar bill ready at hand. The absurdity of this scenario is somewhat mitigated by being coupled with an idea of public action such as consumer committees and other NGOs informing the consumer and lobbying the industry. Still, this is a system with limited perspicuity and no direct accountability.

Neither NGOs nor consumers have any formal right to inspect production facilities or hold companies accountable if they fail to keep their ethical commitments. Essentially, implementation of ethical principles equals the good will of the industry and organizing a boycott is the only legal means to punish perceived wrongdoers. Yet, the economic impact of consumer boycotts on targeted firms remains unclear (Lindenmeier and Tscheulin, 2008) and boycotting, as a societal institution, has no court of appeal for the accused party and provides no reparations for those innocently condemned. It should also be noted that consumer activism through NGOs and action groups are collective undertakings that in many ways mimic those of state intervention. If consumer action should prove effective, they have no less restraining effect on the freedom of the producer and other consumers than would state action sanctioned by law. The line between private and public that the ethical consumer theory was supposed to safeguard is thus blurred.

There are also damaging questions to be raised concerning the practical feasibility of ethical consumption. An ethical consumer operates with a high cognitive load and economic stressors that is bound to make formation of virtues in consumption pattern well-nigh impossible. As social psychologists (e.g. Bandura, 2001: 8-9; Zimbardo, 2009: 446-456) amply attest, ethically sensitive behavior hinges on patterns of reactivity to external cues that is easily displaced by situational forces. Morally relevant properties must be made salient for us and we need a sense of personal responsibility and self-efficacy for our moral senses to operate functionally. Deindividuation, lack of personal accountability, and remoteness between action and morally relevant effect, all contribute to numb our moral awareness. In situations where our moral vision is blurred, we instead begin to attend to cues that are presently salient, such as immediate needs and wants or peer behavior. These conditions are further aggravated by the ease with which consumer behavior can be manipulated in the shopping mall, engineered to the standards of psychology of marketing that aims to make us buy and not to make us aware of ethically damning qualities of products or feel personally accountable for our choices.

Virtues, as any set of dispositions, take practice and patience to form. Acquiring virtues is essentially a social learning activity where we by contemplation of our own behavior in response to the behavioral cues of others consciously reprogram our automated responses. The market is inhospitable to this kind of process. At the grocery store, my behavior is not directed towards other living beings but packages and bundles that do not respond to my actions. As I walk down the ledges, I move in a carefully controlled and planed environment where messages like 'sale', 'outlet', 'special offer', and 'two for the price of one' are made conspicuous while 'animal welfare' and 'deforestation' are nowhere to be seen. All the while my only social guidance is the mute behavior of my fellow consumers, that silently respond to the environmental cues of their surrounding in their own (statistically predictable) way. The feedback of my shopping experience is a brief transaction with a clerk and being handed a receipt. A modest amount on the receipt and a full shopping bag tell me I did well and the clerk will smile at me as to confirm this.

To be an ethical consumer is cognitively taxing. Ethical consumption demands that I actively evaluate different alternatives based on previously acquired information while I am at the shopping mall. The information required to make an informed choice can be hard to come by and difficult to understand. How many, without some prior exposure to high-school biology and chemistry, knows about chlorofluorocarbons, phosphates, sodium dichloroisocyanurate and Triclosan? And even if they do happen to know that these substances are to be avoided, they have no means to know if these have been used during the production of the foodstuff in front of them unless these are declared as direct ingredients.

Of course, I can rely on marketing messages such as 'environmentally friendly' and 'green alternative' but such messages are as non-committal as they are likely to be instances of green washing. Granted, there exist green-labels but they are a bewildering lot that according to Horne's (2009) review offer no clear standard of reference to the real impact a product has on the environment nor how well one product performs in relation to another. In the end, the clerk will give me a warm smile no matter what product I buy and if my purchase is motivated by a label, it is most likely its salience rather than its factual meaning, of which I am unaware, that motivates my choice.

The situation is quite different when I talk with friends and imagine deforestation and *a silent spring*. Here consequences become cognitively tangible and through the exchange of ideas and images, I learn to respond with moral indignation and might even sign a petition against animal cruelty or join a rally against deforestation. My commitment to the environmental cause is met with praise while indifference is derided as irresponsible. In short, I acquire a competence to display a virtuous pattern of behavior pertaining to environmental issues in a social setting. However, there is no immediate connection between the pattern of behavior displayed in a social setting and the pattern of behavior displayed in the shopping mall. The reason for this is not that my views change the moment I enter the mall but the shopping mall is a different setting that prompts some behavioral cues while suppressing others.

The contradictory behavioral patterns I display most likely lack malignant premeditation. While remaining hypocritical in light of my environmentalist commitments, inconsistent behavioral patterns can simply be cases of moral incontinence and as such, argued previously by Brom *et al.*, (2005), may not reflect an alternate moral standard but instead attest to the frailty of goodness. As St. Paul lamented, '[f]or the good that I would I do not: but the evil which I would not, that I do' (Romans 7:19). We, however, need not revert, like St. Paul, to blame our moral shortcomings on sin. It is sufficient to know that moral sensibility, like other evolved functions, have weaknesses and can be overridden, particularly when confronted with situations different from those they once evolved to handle. Indeed, the spirit is willing but the flesh is weak.

There are also reasons to believe that the moral sensibility of the producer is equally ill suited to function in an ethics of food consumption. The relationship between the producer and the consumer is deindividuated, which makes it difficult to form the type of bond that is necessary for the proper operation of our moral senses. The distance between producer and end consumer diffuses the sense of personal responsibility towards the end consumer while responsibility to the board of directors, shareholders and business partners are closer to home and thus more likely to have a direct impact on behavior. Likewise, if the consumer can be bewildered by the fauna of chemicals and production methods, so can the producer. There is no reason to believe that small-scale food producers possess knowledge of the environmental impact of their business other than what is common knowledge. Deeper insight into the environmental impact of a business demands advanced knowledge of biology and chemistry as well as tools and resources to perform an impact analysis and keeping up to date with new research. There is good reason to believe that only major actors wield the human and financial capital needed to actively engage these issues on their own accord, provided that they have the will to do so.

If social psychology has its research right then it poses a major threat to the distinction between public and private self. The private may namely be less under our volitional control than we have traditionally assumed. The market where we were supposed to exercise our individual freedom as consumers is primarily ruled by situational forces outside our direct control rather than by our conscientious and self-conscious deliberate choices. Going shopping is an activity that is quite different from going to the ballot box. The shopping mall is not a place that encourages ethical sensitivity and deliberate choices, nor is it a place where consumers can educate and inform themselves or where their choices are met with morally relevant feedback. The mall is rather a place where our self-efficacy and locus of control are pushed aside and we are led by perceptually salient cues to interact with objects in a pre-arranged manner. What we hear, see, and sense are the result of environmental engineering that is fine-tuned to affect our psychological machinery. Our evolved moral sensibility has little to offer in terms of resistance when it is faced with situations it is not adapted to handle. Yet, contrary to the aforesaid, the rhetoric and ideology of ethical food consumption demand that the shopping mall should be the place where we exercise our virtues and make informed choices.

In light of what has been said, some of the ideas about human moral psychology that underlie the idea of ethical consumption appear to be blatantly false. Chief among these false ideas is that free and rational choice characterize us as consumers. Consumer choices are not premeditated and calculated in a higher degree than is to be expected from a newly evolved primate species. Behavior is most often on autopilot and relies on previously learned patterns and reactivity to salient stimulus. While humans can learn and unlearn an ethically desirable reactive pattern, such changes are not accomplished by a simple act of volition but as all instances of learning, demand time, attention, and appropriate feedback. While ethical thoughts and considerations can awake in us in some contexts, where the damaging effects on the environment are made salient through arguments and images, there is no automatic carryover to a new and adaptive pattern of consumer behavior. As we have previously argued, several obstacles prevent such a carryover effect from occurring. The basically social nature of our moral sentiments are one such factor, but we also have to wrestle with institutional obstacles such as the inherent inertia in the industry as well as public policy measures that skew competition and affect the overall structure of the market. The market in itself lacks perspicuity and formal accountability in an ethics of consumption model. The same model also puts unrealistic burdens on the individual in terms of acquiring information about products and competence to evaluate this information in relation to those environmental concerns the consumer holds.

However, none of the problems that have been raised in this article is impossible to surmount if the idea of an ethical consumer is abandoned in favor of a notion of an ethically sensitive food citizen. Ethical consumption is problematic chiefly because it relies on a naïve appraisal of consumer power and capabilities and fails to observe the psychological and sociological factors that govern human behavior. As long as we confine ourselves within the consumer discourse, any move to address these problems will be seen as an illegitimate attempt to interfere in the private lives of citizens. Yet is seems most natural to construe environmental concerns and popular call for government action as a supplication for help, a cry amounting to 'help me, for I am beside myself!' Ethically sensitive food citizens may be well aware of their own shortcomings as consumers and their bedazzlement over the complexity of the issue at hand. A call for policy change might therefore be a call for the means to do the good that I desire and help to avoid the evil that I shun. In short, confronted with my own impotence as a consumer, I, in the capacity of citizen, plead for arrangements that empower me to act virtuously.

The marriage between the ethical consumer and corporate sociable responsibility has thus proved to be an unhappy one. It is not just that the marriage is on unequal terms that put one party at a constant disadvantage. The main fault with the marriage is that neither party is allowed to engage the other in an ethically meaningful and constructive way, because neither consumer nor producer is allowed to meet face to face under circumstances that promote moral cognition. The public sphere that could have functioned as a bridechamber for such a meeting has been deemed anathema and they are both cast out on the town square (*agora*) to spend their nuptial night.

Ethical issues can arise on each side of the demarcation line between public and private, but they do not fall neatly on either side for the simple reason that human beings are continuous entities across their public and private selves. Few questions of ethical import can be said to belong to the public-self or the private-self exclusively. By their very nature, ethical questions extend outwards and affect third party while originating in the idiosyncrasies of self-perception and character formation. For this reason, if none else, it is doubtful if a distinction between food consumer and a food citizen can carry any weight in food ethics. Consequently, a theory that construes the relationship between consumer and producer thus, risks making a mockery of the whole affair.

References

Bandura, A. (2001). Social cognitive theory: an agentic perspective. Annual Review of Psychology LII: 1-26.

Brom, F.M., Meijboom, F. and Visak, T. (2005). Food, citizens, and market – the quest for responsible consuming. In: Potthast, T. Baumgartner, C. and Engels, E.M. (eds.) Die richtigen Maße für die Nahrung – Biotechnologie, Landwirtschaft und Lebensmittel in ethischer Perspektive. Francke Verlag, Tübingen, Germany, p. 115-132.

Horne, R.E. (2008). Limits to lables: the role of eco-lables in the assessment of product sustainability and routes to sustainable consumption. International Journal of Consumer Studies 33: 175-182.

Johnston, J. (2008). The citizen-consumer hybrid: ideological tensions and the case of Whole Foods Market. Theory and Society XXXVII(3): 229-270.

Lindenmeier, J. and Tscheulin, D.K. (2008). Konsumentenboykott: State-of-the-Art und Forschungsdirektiven. Zeitschrift für Betriebswirtschaft V(78): 553-579.

Rawls, J. (2001). Justice as fairness. Harvard University Press, Cambridge, MA, USA.

Wilkins, J.L. (2005). Eating right here: moving from consumer to food citizen. Agriculture and Human Values 22: 269-273.

Zimbardo, P. (2009). The lucifer effect: how good people turn evil. Rider, London, UK.

Getting the message across: the importance of information in Fair Trade marketing

R. Schleenbecker and U. Hamm
Organic Agricultural Sciences, Department of Agricultural and Food Marketing, University of Kassel, Steinstr.19, 37213 Witzenhausen, Germany; r.schleenbecker@uni-kassel.de

Abstract

Consumers are confronted with information asymmetry in products from Fair Trade since they are unable to check whether the promised product quality is adhered. Therefore, information on the product is crucial, but on the other hand, extent and content of consumers' expectation towards that information is unknown. Therefore, the focus of this survey is on consumers' information search when purchasing a Fair Trade product. A mixed methods approach was applied: Focus group discussions and Information Display Matrix experiments were performed in order to gain insights into consumers' information preferences and decision making in the case of Fair Trade products. A complementary questionnaire was conducted. Results show that participants search for information quite extensively, with an emphasis on the Fair Trade label, but also information on price, production method and further ethical criteria are important.

Keywords: ethical consumption, information search behaviour, Information Display Matrix, IDM

Information search and ethical consumption

When discussing how to bring about a change in consumption patterns towards ethical consumption, research has shown that consumer information is playing a central role. However, Fair Trade (FT) products as a form of ethical products are characterised by the fact that they are so called 'credence goods'. Consequently, consumers are not able to verify if the process related FT criteria are adhered. Therefore, consumers need to trust the advertised credence characteristics. Information is decisive here. However, information needs to be adapted to consumers' needs so as to avoid information overload. Hence, the goal of the research project was to identify the information needs as regards extent and content of information on FT product features, including the role of FT labels. Beyond that, the role of price and organic quality were examined.

Methodology

A multi-method approach was chosen to analyse the questions: First, an exploratory study based on three focus group discussions (FGD) was conducted in order to gain insights into FT consumers' knowledge on the products and the underlying standards. Next, Information Display Matrix (IDM) experiments were conducted with 389 FT consumers. The IDM is a process tracing method which is suitable for identifying information search and decision behaviour. The IDM experiments were supported by a questionnaire, both were carried out on laptop computers. In the following, the FGD method will be highlighted, followed by a description of the empirical survey. Next, an introduction into the IDM method will be given. The research design of the quantitative study is described subsequently.

Focus group discussions

In market research, FGDs are mostly used to explore opinions, to discover behaviour patterns and attitudes as well as to detect underlying consumer motives (Lamnek, 2005; Mayerhofer, 2007). FGDs aim at generating insights into relevant topics on which information is scarce (Blank, 2007; Bryman, 2008; Mayerhofer, 2007). In this study, three discussion rounds were performed. They were carried out in three German cities in October 2010. The participants were approached randomly on central public places. Screening questions were applied to ensure participants were FT consumers. The groups majorly consisted of occasional buyers of FT products. The majority of the totally 27 participants (20; 71%) stated that they buy at least three different FT products from time to time or more often. About half of the participants (15; 55%) were female, and two thirds (18) of them were between 18 and 44 years old. Each discussion round lasted about 1.5 hours. An expense allowance of 20 € was distributed to each participant.

Qualitative content analysis based on the approaches of Mayring (2003) and Gläser and Laudel (2006) was used to analyse the transliterated material. Text categories and subcategories were created, based on the discussion guideline and reassessed by theoretical substantiated research questions. Additionally, category creation was revised after a first perusal of the transcripts. Further categories were derived from the material if expedient. After text extraction, paraphrasing of the text took part. According to Gläser and Laudel (2006), the categories were adapted again during the attribution to the text units, if expedient. In the last step, the evaluation of the categorised and paraphrased text material was performed.

Information display matrix experiments

The IDM is a quantitative research method which is used to analyse consumers' information search and decision behaviour. Participants are offered various products including defined product information. The goal is that participants decide for one product they purchase subsequently. The IDM is designed in a matrix form. The matrix consists of different products and general product attributes in columns and rows respectively. The fields of the matrix are results of the combination of the products and the attributes. The fields contain more specific information on the attributes. The specific information is accessible via clicking on fields with the computer mouse. Participants in this study were able to access as much information as needed in order to make their purchase decision. The whole information search process was recorded by the computer, so the complete decision making process including tracking of information search, evaluation of alternatives and purchase decision at the end was stored. Consequently, amount, content and sequence of information search as well as the final purchase decision can be analysed.

The most important advantage of the computer based application of the IDM – compared to direct inquiries – is that the bias between actual and indicated behaviour (Berekoven *et al.*, 2006; Jacoby *et al.*, 1978) is minimised since participants interact with a computer instead of a person (cp. Bailey, 1994). Further, preferences are not stated, but are surveyed in an experiment which is designed similar to real buying procedures, e.g. in online shops. The social desirability bias which is often reported in direct inquiries and especially occurs in ethical consumption is minimised. In this study, the IDM was conducted in a real shopping environment. Therefore, the negative implications of artificial experimental situations were minimised while still having the advantage of controlled circumstances. Referring to prior research, the IDM is a suitable method to identify consumers' search and decision behaviour especially since nowadays consumers are used to information presentation in a diagrammed way as in online shops but also in product test magazines (cp. Aschemann-Witzel and Hamm, 2011; Zander and Hamm, 2010). The IDM is suitable to the analysis of products where a conscious and targeted information uptake is necessary (Kuß, 1987; Schopphoven, 1996). Ethical products, and thus FT products are characterized by a rather intensive information search due to a higher involvement.

Research design of the quantitative study

As an exemplary product, FT coffee was chosen due to its popularity. The survey was conducted in five retail stores in Germany where FT coffee was available: two conventional supermarkets, two world stores and one organic food shop. Participants were approached randomly and screened according to the requirement that they purchase FT products at least occasionally.

Participants could virtually choose from nine different FT coffees which were equipped with different ethical attributes. Those ethical attributes were chosen based on the results of the FGD, literature and information as given by FT marketers. The attributes were: protection of children (such as tackling the problem of child labour), producer income (such as a price which secures producers standard of living), trade relations (such as a guaranteed purchase of harvest) and social projects (such as the establishment of a health centre). Further, products were equipped with a FT label, plus information on prices[23] and production method.

The specific information was hidden behind the fields as can be seen in Figure 1. The information on ethical attributes varied in three manners: information was either presented in precise or imprecise manner, or the field contained no information. Further, prices varied according to the quality of information given. Four products were equipped with a faked label and five with the well-known Transfair label. With regard to the production method, products were either labelled as organic or no information was given (=conventional). The combination of the different attribute specifications resulted in 18 different products which were designed based on considerations as regards content. The 18 different products were spread across two different 7×9-matrices (two times nine products and seven arguments). Thus, two different groups of participants existed. Each set of products contained a product

Figure 1. IDM with opened field.

[23] Prices ranged from 4.99 € till 6.49 € per 500g in supermarkets and organic food stores and per 250g in worldshops due to an approximately doubled price level of coffee in worldshops.

which was equipped only with information on the FT label and the price. Attributes and products were arranged randomly in order to avoid biases due to common reading habits.

Prior to the experiment, participants were introduced into the task and informed that they were supposed to choose and buy one of the coffees displayed. They were told that their purchase decision was binding. The task was supported by a monetary incentive of 5 €. Participants were allowed to access as much information as they wanted and in the order they preferred, by clicking on a field which opened subsequently and contained the specified attribute information (Figure 1). Fields could be opened repeatedly, but only one field at a time was accessible. Participants had the possibility to mark interesting attribute specifications. At the end of the process, participants put their product of choice into a virtual shopping basket as in online-shops.

The additional questionnaire surveyed consumers' motivations to purchase FT products, additional information on participants' information search habits and shopping behaviour and on socio-demographics.

Results and conclusions

The results display the importance of information in FT marketing. Information search was performed quite extensively as the results of the IDM experiments show: 51.45 fields were accessed averagely. The average utilisation level was 88% of the total matrix, and 70.2% of participants opened all fields of the matrix at least once. On the other hand, there are a number of people who searched for information to a far smaller extent, as the SD of 26.70 implies. Therefore, a regression analysis was conducted so as to explain the factors leading to differences in information search extent. It was hypothesised that consumers being more concerned with ethical products search for information more extensively (cp. Beatty and Smith, 1987). Further, it was hypothesised that age, gender and education influence information search intensity. However, this regression model could not explain the variations in information search extent. Presumably, involvement, which was not surveyed here, plays a major role in explaining information search extent, as explained by Kardes *et al.* (2010: 195).

First and repeated accessions as well as product purchase in the IDM were interpreted to gain insights into FT consumers' information preferences. It could be shown that it is not only ethical attributes which are important to consumers, but foremost information on the FT label, the production method and price. Results of the FGDs emphasise the importance of a label as well. When it comes to the recognition of FT quality, a label seems to be most important. A FT label is further important to assure that the promised quality is adhered. A few participants, however, attached smaller importance to the label: Few participants stated that for them, a FT label was not crucial since not all products from FT carry a label but they can make sure that the product is from Fair Trade nevertheless.

Additional information on the FT aspects of the products appeared to be less relevant than label information due to fewer and later accession of the information. Further, a clear ranking of attributes could hardly be elaborated, except for the dominance of the attribute 'protection of children'. Since ethical information was accessed nevertheless, it can be concluded that displaying additional ethical information is less important, but still crucial even though the concrete design of the information is not decisive.

Furthermore, most consumers have chosen a product containing imprecise information on the ethical attributes. It appears that consumers are overburdened with precise information due to their rather low state of knowledge on FT. They therefore prefer content with general, imprecise information. Two logistic regression analyses were applied in order to explore variables affecting purchase decision. It could be shown that the probability to choose a product which is equipped with imprecise ethical attributes falls with an increasing number of accessed information fields. Furthermore, the probability of choosing a product with imprecise attributes rises the more credible consumers perceive the faked label. This implies that a product with imprecise attributes is preferred if consumers' knowledge on FT labelling is low and is less preferred by consumers who search for information quite intensively. However, both variables only explain a small part of the total variance (Nagelkerkes R^2: 0.030 and 0.056 respectively).

Information search strategies (cp. Ball, 1997; Beatty and Smith, 1987; Payne *et al.*, 1978; Sauermann, 2004) give insights into consumers' information processing behaviour. Attribute-wise information search was applied most often. Totally, search strategies with reduced cognitive effort were applied predominantly.

In order to match product information with consumers' needs, marketers should offer information which is reduced to the central aspects of FT and does not go too much into detail regarding the background of the products since most consumers seem to be overburdened with precise, detailed information. Furthermore, marketers should consider that consumers are directly comparing single product attributes with competing products rather than considering the product as a whole. Marketers should also be aware that there is a considerable amount of FT consumers whose knowledge on FT labels and ability to recognise a FT label is not profound. Therefore, further efforts should be invested in the creation of label awareness and knowledge as well as consumers' ability to differentiate between labels. As organic turned out to be a major criterion, ongoing attempts to equip FT products with an organic label as well should be expanded.

References

Aschemann-Witzel, J. and Hamm, U. (2011). Measuring consumers' information acquisition and decision behavior with the computer-based information-display-matrix. Methodology 7: 1-10.

Bailey, K. (1994). Methods of social research. Free Press, New York, NY, USA, 533 pp.

Ball, C. (1997). A comparison of single-step and multiple-step transition analyses of multiattribute decision strategies. Organizational Behavior and Human Decision Processes 69: 195-204.

Beatty, S.E. and Smith, S.M. (1987). External search effort: an investigation across several product categories. Journal of Consumer Research 14: 83-95.

Berekoven, L., Eckert, W. and Ellenrieder, P. (2006). Marktforschung. Methodische Grundlagen und praktische Anwendung. Gabler, Wiesbaden, Germany, 448 pp.

Blank, R. (2007). Gruppendiskussionsverfahren. In: Naderer, G. und Balzer, E. (eds.) Qualitative Marktforschung in Theorie und Praxis – Grundlagen, Methoden und Anwendungen. Gabler, Wiesbaden, Germany, pp. 281-301.

Bryman, A. (2008). Social research methods. Oxford University Press, Oxford, UK, 748 pp.

Gläser, J. and Laudel, G. (2006). Experteninterviews und qualitative Inhaltsanalyse. Verlag für Sozialwissenschaften, Wiesbaden, Germany, 336 pp.

Jacoby, J., Chestnut, R., Weigl, K. und Fisher, W. (1976). Description of a process methodology research: paradigm, and pilot investigation. Advances in Consumer Research 3: 306-314.

Kardes, F., Cronley, M., Cline, T. (2010). Consumer behaviour and managerial decision making. South Wester Cengage Learning, Mason, OH, USA, 412 pp.

Kuß, A. (1987). Information und Kaufentscheidung. Methoden und Ergebnisse empirischer Konsumentenforschung. de Gruyter, Berlin, Germany, pp. 205.

Lamnek, S. (2005). Gruppendiskussion – Theorie und Praxis. Beltz, Weinheim, Germany, 312 pp.

Mayerhofer, W. (2007). Das Fokusgruppeninterview. In: Buber, R. and Holzmüller, H. (eds.) Qualitative Marktforschung. Konzepte, Methoden, Analysen. Gabler, Wiesbaden, Germany, pp. 477-490.

Mayring, P. (2008). Qualitative Inhaltsanalyse: Grundlagen und Techniken. Beltz, Weinheim, Germany, 135 pp.

Payne, J.W., Bettman, J.R. and Johnson, E.J. (1993). The adaptive decision maker. Cambridge University Press, Cambridge, UK, 330 pp.

Payne, J.W., Braunstein, M.L. and Carroll, J.S. (1978). Exploring Predecisional Behavior: An Alternative Approach to Decision Research. Organizational Behavior and Human Performance 22: 17-44.

Sauermann, H. (2004). Vocational choice: a decision making perspective. Journal of Vocational Behaviour 66: 273-303.

Schopphoven, I. (1996). Messen von Entscheidungsqualität. Peter Lang, Frankfurt/Main, Germany, 364 pp.

Zander, K. and Hamm, U. (2010). Consumer preferences for additional ethical attributes of organic food. Food Quality and Preference 21: 495-503.

Zander, K. and Hamm, U. (2012). Information search behaviour and its determinants: the case of ethical attributes of organic food. International Journal of Consumer Studies 36: 307-316.

Closer to nature: the ethics of 'green' representations in animal product marketing

S. Borkfelt[1], S. Kondrup[2] and M. Gjerris[2]
[1]Department of Aesthetics and Communication, Aarhus University, Jens Chr. Skous Vej 4, 8000 Aarhus C, Denmark; [2]Institute of Food and Resource Economics, Faculty of Science, University of Copenhagen, Rolighedsvej 25, 1958 Frederiksberg C, Denmark; engsb@hum.au.dk

Abstract

Empirical cases from the Danish food market are examined in order to critically discuss the respective modes of communication in light of the premises of socially responsible consumer marketing. This analysis suggests that specific marketing instruments are used to sell animal products by blurring the difference between the paradigms of animal welfare used by producers, and the paradigms of animal welfare implicit in the public understanding of the concept. These instruments rely on the ethical, political and sustainable consumption discourses in order to sell one image of animal welfare in intensive animal production while the production at the same time presupposes a quite different paradigm of animal welfare. Two cases are used to illustrate this: (1) the Danish dairy company Arla Foods' campaign with the tagline 'Closer to nature'; and (2) selected 'quality brands' that present themselves as welfare-oriented alternatives to conventionally produced animal products, but with only marginal improvements. The rhetoric of both cases specifically manifests a deep coherence between nature, farm, animal and end product, and thereby creates associations of production tied to lives living *in* nature – thus attempting to display a green, eco-, climate-, and animal friendly production. The tension between marketing and the idea of ethical consumerism is apparent as the need for independent information to make value-based choices is challenged by the liberal rules of the market and more specifically by the lack of a restrictive food labelling policy. The relationship between the ways in which animal welfare is communicated and emphasized through food marketing, and commonly held perceptions of acceptable standards for animal welfare, is discussed and the need for transparency in the area of animal welfare stressed.

Keywords: advertising, consumers, transparency, naturalness, animal welfare

Introduction

As animal welfare has come to play an increasing role in public debate and in the awareness of consumers, it has also become possible to profit from using the concept in marketing animal products. Thus, products are being marketed with distinct reference to ideas of 'natural' lives and to the well-being of animals, even as definitions of such concepts remain debated and reference to them is therefore often highly ambiguous.

Due to the very nature of such 'green' concepts there is a distinct risk of confusing consumers, when similar rhetoric is used in the marketing of a wide spectrum of animal products, from organic products at one end to products with only marginal welfare improvements at the other. Use of these concepts thus creates a tension between consumer interpretations and the reality behind such 'green' marketing communication, where communicative focus is on consumer attitudes rather than on the reality of production standards, which potentially misleads consumers.

Paradigms of animal welfare

The welfare of animals in intensive animal production systems is debatable. Different notions of what parameters should be used to assess animal welfare often form the background of these debates. The public discussion is between producers who typically see animal welfare as closely linked to parameters relevant for production rates such as growth, litter size, disease rates and mortality (Te Velde *et al.*, 2002) and animal- and/or environmentalism-oriented NGO's and citizens that focus on parameters such as suffering and naturalness; often understood as the ability to perform species-specific behaviour (Miele, 2010).

This is reflected in debates on animal welfare where three basic paradigms can be found. The first paradigm (A) focuses on the health and functioning of the animal's biological systems. Poor welfare is indicated by 'reduced life expectancy, impaired growth, impaired reproduction, body damage, disease, immunosuppression, adrenal activity, behaviour anomalies, and self-narcotization' (Broom, 1991). The second paradigm (B) focuses on the subjective mental states, both positive and negative, in the animals (Duncan, 1996). The last paradigm (C) focuses on the connection between the animal and its environment, emphasizing the possibility for the animal to perform species-specific behaviour in an environment to which it is biologically adapted (Rollin, 1993). Different ways of modifying and combining these can be found (e.g. Yeates *et al.,* 2011).

Worth noticing in this context is the discussion of what the relationship is between animal welfare and animal ethics. Typically, animal ethics is seen as the broader concept embracing concerns that are traditionally not seen as part of the animal welfare discussion, such as animal rights, integrity, etc. This relationship can be translated into the abovementioned paradigms as the place where paradigms B and C differ the most. Where B is solely looking at mental experiences, C is also concerned with the perceived naturalness of the behaviour. To the extent that this is done even if the behaviour in question is of no relevance to the mental experience of the animal, one could say that paradigm B is a position where debates on animal welfare and animal ethics meet.

In the following we will discuss the value conflicts that apply to the marketing of two different food suppliers producing and selling dairy and meat products respectively. Our aim is to show how the industry strategically employs the concepts of animal welfare held by the public and exemplified in paradigms B and C when telling the story of their products, despite the fact that concerns with animal welfare in intensive animal production systems are almost exclusively based on paradigm A.

Dairy as 'closer to nature'

In 2008, the Danish dairy company Arla Foods launched the slogan 'Closer to nature' as a platform for the global Arla brand. More than just a campaign, the launch of the 'Closer to nature' platform was meant to start a reconceptualization and rebranding of the company as 'green'. As such, Arla Foods has chosen to apply the idea of the company as closer to nature in several areas, including an emphasis on natural ingredients, lessening the company's environmental footprint, and an overall wish for a production that happens in 'harmony' with nature (Arla Foods, 2011).

According to Arla's sustainability webpages, the nature-friendly strategy affects every part of the company's supply chain and also includes activities for both Arla employees and consumers (Arla Foods, 2013a). In other words, the company aims to make the naturalness imbedded in the campaign into a perceived intrinsic quality of the entire company, or, as one of the employees implementing the campaign, comments: 'When people see the Swedish car brand Volvo, most automatically think of safety.

In much the same way, it is our wish that our customers will automatically associate Arla with "Closer to nature"' (Arla Foods, 2011, our translation).

From a communicative as well as an ethical perspective, this aspiration necessarily represents a challenge. Where the safety connected to a car brand is arguably a statistically quantifiable or testable quality whose meaning is relatively fixed for both producers and consumers, the tagline of 'Closer to nature' must be seen as open to different interpretations. Indeed, the mere word and concept of 'nature' has been recognized as one of the most complex in our culture, with multiple possible meanings and nuances, not least in advertising (Salvador, 2011; Soper, 1995). Thus, consumers and producers are likely to have different ideas of what the concept of 'nature' means as it is used in the campaign, and to adhere to different paradigms of animal welfare, as consumers think of naturalness as connected to animal welfare paradigm C. This problem is not lessened by the fact that dairy cows do not live in the wild and thus do not have a 'natural' habitat one can refer to in trying to determine what 'natural' surroundings are for the species. Any idea of 'naturalness' for the cows must therefore be confined to the (agri)cultural setting of the dairy farm itself, which may not fit consumers' ideas of what nature is.

Moreover, while the 'nature' that Arla (and by extension the consumer of its products) is supposedly closer to is far from clearly defined, the very idea of being 'closer' is itself ambiguous and begs the question of what it is closer than. Just as the definition of nature itself, this missing referent is left to the consumer to decide and could be anything from the products of competitors, or previous standards at Arla itself, to contemporary urban lifestyles.

Once deconstructed, therefore, the association of naturalness, which the campaign tries to connect with consumption of Arla's products, is based on referents that are far from clear, thus allowing consumers to draw their own conclusions, which through the contextual associations of the word 'nature' to welfare paradigm C are likely to be distinctly different from the standards used in production. This can also be concluded from the fact that the section on animal welfare in the company's own guidelines for dairy farmers clearly connect this concept mostly to questions of hygiene, medicine and 'basic physiological and behavioural needs', and is thus more closely connected to paradigm A, with reference to elements of paradigm B (Arla Foods, 2012).

The website and commercials connected with the campaign leave little doubt that the kind of animal welfare paradigm communicated in the campaign is one in which animal welfare equals naturalness in the production and is clearly connected to ideas of animals kept on pasture, in (natural) surroundings that please them. While the text on the website only mentions animal welfare directly in headlines and in passing, and tends to emphasize environmental issues and the use of 'natural' ingredients in the products, the visual rhetoric employed in the site's images generally draws on outdoor scenery, often depicting cows on pasture. In other words, the campaign utilizes the abovementioned welfare paradigm C, and is thus in its communication largely consistent with consumer expectations about naturalness and animal welfare, but inconsistent with the paradigm actually employed in Arla's dairy production: three out of four cows in Denmark never get outside (Videncentret for Landbrug, 2013), more than 20% have *digitalis dermatitis*, a painful hoof infection (Krogshede, 2013), and the calf is taken from the mother within 24 hours of birth.

This impression is further strengthened by the television commercials attached to the campaign in many countries, which generally show images of the great outdoors in the form of green fields, often with grazing cows. The commercials also emphasize the inherent naturalness of milk as a product coming straight from nature, which – while on the surface level about the naturalness of product ingredients – makes the implicit assumption that cows on Arla dairy farms live in natural surroundings, whether they get to experience outdoor environments or not. Thus, for Arla Foods, the cows and dairy farms themselves constitute 'nature', which gives their products an inherent naturalness, as is also apparent when the company invites consumers to 'get Closer to nature' by visiting selected Danish dairy farms (Arla Foods, 2011, 2013b). As such, this simplification of both 'nature' and animal welfare in the company's marketing communication stands in contrast to the inherent complexity of 'nature' and connected concepts in our culture.

Welfare quality meat brands

The notion of 'welfare quality brands' is used as a general term for the range of different animal products that are marketed as expressing higher animal welfare and meat quality standards than other products without being linked to specific welfare certification labels. The welfare quality brands cover various types of meat and most Danish supermarkets carry at least one type – usually placed separately in the cold counter accompanied by distinct brand markers. 'Den go'e gris', 'Antonius' and 'Bornholmergrisen' are three so called welfare-pigs produced for the Danish market by Europe's largest meat processing company, Danish Crown (Danish Crown, 2013a). These products are all branded as specific forms of differentiated quality meat clearly distinctive from conventional production. They are marketed with taglines emphasizing both the end product, i.e. taste and quality, and the production process, i.e. housing conditions, and their marketing generally derives from the premise that the welfare of the animal is connected to the taste of the meat. Directly translated the three taglines read: 'Happy pigs taste better' (Danish Crown, 2011, our translation), 'The taste of well-being and quality' (Danish Crown, 2012, our translation), and 'More taste and welfare' (Lenskjold, 2009, our translation).

Although different problematic issues are at stake here, we will narrow our focus to the production related questions regarding the animal, excluding the more product-oriented emphasis on meat quality. As previously stated, there are competing definitions of animal welfare and it is therefore a controversial concept inherently connected to ethical challenges in terms of continually changing societal values. The welfare quality brands apply words such as happiness, well-being and welfare allegedly to denote the quality of the pig's life whereas taste is specifically consumer-directed. With this in mind it is not evident that well-being and welfare unequivocally denote the experience of the animal and not of the consumer. Thus, the taglines 'The taste of well-being and quality' and 'More taste and welfare' clearly contain expressions that are ambiguous and value-laden in ways that enable them to be perceived as related to the health, welfare and quality of life for both humans and pigs.

According to the information available at Danish Crown's website the 'Antonius-pigs' are produced with a 'wide-reaching respect for the pigs' natural needs and behaviour' (Danish Crown, 2011, our translation) while the 'Bornholmer-pig' 'has a better life because it has extra space for moving around' and 'More space meets ... the natural needs of the pig and increases the welfare of the individual animal' (Danish Crown, 2013b, our translation). This information specifically constructs the link between animal welfare and the unfolding of an inherent nature applied in welfare paradigm C, and is thus in clear compliance with the ways in which consumers perceive animal welfare.

As such, the main problem is not a discrepancy between the stories told by Danish Crown, describing what a pig-life must consist of to be a good pig-life, and consumer notions of animal welfare. The marketing instruments used by Danish Crown subscribe to the public opinion of nature and animal welfare as mutually constitutive partners in the conceptual discussions of farm animal welfare: animal welfare arises when an animal life is *closer to nature*, borrowing from Arla's slogan, although as previously discussed this pairing of nature and animal welfare reveals no clear ideas of what is meant by either concept. The main problem is that the welfare quality brands essentially profit from the present 'green' or 'ethical' consumption discourse with its positive development in consumer attitudes to animal welfare by using almost the same rhetorical expression as highlighted by organic and free range animal products. The welfare quality brands target consumer segments specifically selecting animal products with information that tells a story about the good and natural animal life – stories that correspond to consumers' ideas of animal welfare.

Nevertheless, there is a tendency to brand meat products as *welfare meat* solely on the basis of a few marginal improvements in the production systems, compared to the minimum criteria for accepted animal welfare standards. In other words, the conventional production of pigs generally marks an acceptable minimum of welfare standards, which means that all systems positively differing even minutely from this can be branded as welfare meat. For instance, in the specific housing conditions behind the welfare quality brands used in this paper, the individual pig has only 0.2 m^2 more space than conventional finishing pigs, while there is no demand for outdoor access and fixation is still allowed for farrowing and lactating sows (Anonymous, 2011). The welfare quality brands do contain minor improvements compared to conventional production systems, but it is difficult to see how these improvements exceed the basis of welfare paradigm A. In order to satisfy the criteria of natural behaviour from the consumers' perspective the pigs should have the freedom to move around, access to pasture, the freedom to root, to play, to develop own sleep patterns, to lie in the mud and to give parental care (Boogaard *et al.*, 2011). As discussed in the case of Arla Foods, consumers and producers rely on different interpretations (and effectuations) of animal welfare. The criteria of natural behaviour as a prerequisite for animal welfare is found only in consumers' expectations and are not met by the welfare quality brands. In sum, Danish Crown is in risk of misleading consumers, as indicated by a study from the Danish Consumer Council (2011) in which almost one third of respondents thought pigs labelled Antonius to be free-range.

Conclusion

As both Arla Foods' 'Closer to nature' campaign and the marketing of welfare quality meat brands testify to, the current liberal rules of the market enable marketing of products through highly ambiguous uses of the concepts of naturalness and animal welfare, even when animal welfare standards tied to the production are only marginally improved compared to the minimum standards found in the industry. Moreover, such marketing relies heavily on consumers' belief in notions of animal welfare based on naturalness that differ markedly from the welfare paradigm employed by farmers and producers of the products.

The marketing of animal products such as those discussed above thus exploit a clear discrepancy between industry standards and consumer ideas of animal welfare, and therefore risks misleading consumers in the process as long as no restrictive policy regarding the labeling of products by reference to animal welfare exists.

References

Anonymous (2011). Sådan lever grisen bag mærket. Tænk, July 2011.

Arla Foods (2011). Tættere på Naturen – fra ide til virkelighed. Available at: http://lfw.dk/ad5g.

Arla Foods (2012). Quality assurance programme Arlagården. Version 3.3. Available at: http://lfw.dk/ess6.

Arla Foods (2013a). Closer to nature. Available at: http://lfw.dk/w4ao.

Arla Foods (2013b). Activities. Available at: http://lfw.dk/7lt8.

Boogaard, B.K., Boekhorst, L.J.S., Oosting, S.J. and Sørensen, J.T. (2011). Socio-cultural sustainability of pig production: citizen perceptions in the Netherlands and Denmark. Livestock Science 140: 189-200.

Broom, D. (1991). Animal welfare: concepts and measurement. Journal of Animal Science 69: 4167-4175.

Danish Consumer Council (2011). Danish Crown på kanten af vildledning. Available at: http://lfw.dk/hv89.

Danish Crown (2011). Antonius. Available at: http://lfw.dk/o2qx.

Danish Crown (2012). Den go'e gris. Available at: http://lfw.dk/73xw.

Danish Crown (2013a). Svinekød. Available at: http://lfw.dk/4tgj.

Danish Crown (2013b). Bornholmergrisen. Available at: http://lfw.dk/s5lj.

Duncan I.J.H. (1996). Animal welfare defined in terms of feelings. Acta Agriculturae Scandinavica. Section A. Animal Science Supplementum 27: 29-35.

Krogshede, M.B. (2013). Hver femte ko har digitalis dermatitis. Maskinbladet February 22, 2013.

Lenskjold, M. (2009). Bornholmergrisen er populær. Landbrugsavisen. Available at: http://lfw.dk/e8os.

Miele, M. (2010). Report concerning consumer perceptions and attitudes towards farm animal welfare. European Animal Welfare Platform, Belgium, 16 pp.

Rollin, B. (1993). Animal welfare, science and value. Journal of Agricultural and Environmental Ethics 6 (Suppl. 2): 44-50.

Salvador, P. (2011). The myth of the natural in advertising. Catalan Journal of Communication & Cultural Studies 3: 79-93.

Soper, K. (1995). What is nature? Blackwell, Oxford, UK, 289 pp.

Te Velde, H., Aarts, N. and Van Woerkum, C. (2002). Dealing with ambivalence: farmers' and consumers' perceptions of animal welfare in livestock breeding. Journal of Agricultural and Environmental Ethics 15: 203-19.

Videncentret for Landbrug (2013). Hvor mange køer er på græs i Danmark?. Available at: http://lfw.dk/emuq.

Yeates, J., Röcklinsberg, H. and Gjerris, M. (2011). Is welfare all that matters? A discussion of what should be included in policymaking regarding animals. Animal Welfare 20: 423-32.

Towards a broader understanding of citizenship in policy debate on food advertising to children

C.L. Mah[1], B. Cook[2], S. Hoang[3] and E. Taylor[3]
[1]University of Toronto/Centre for Addiction and Mental Health; 33 Russell Street, Toronto, ON, M5S 2S1, Canada; [2]Toronto Public Health; 277 Victoria Street, Toronto, ON, M5B 1W2, Canada; [3]Centre for Addiction and Mental Health, 33 Russell Street, Toronto, ON, M5S 2S1, Canada; catherine.mah@utoronto.ca

Abstract

Contemporary food policy often focuses on 'downstream' elements of the food system, particularly the roles and responsibilities of individual members of a consuming public. For example, to address the effects of food and beverage advertising on child health, in the absence of widespread agreement on the most appropriate form of collective action, policy debate has tended to revolve around moral reasoning about how children should behave and interact with the world around them. In this paper, we attempt to broaden this debate by sharing results from in-depth interviews (n=35) carried out as part of our Food Advertising to Children: Ethics for Policy study, funded by the Canadian Institutes of Health Research. We will discuss how 'food citizenship' can be viewed not only in terms of consumption roles (e.g. children's behaviour), but the expectations for participation embedded within policy actor roles (e.g. health professionals, 'government' broadly defined). Such framing is important for how diverse policy actors understand and incorporate citizenship into their practices. Whether health professionals are construed as only 'program delivery agents', for example, or active citizens, can affect their power to influence policy processes; it also conditions the range of policy options deemed suitable for public debate.

Keywords: moral reasoning, food citizenship, health professionals, policy participation

Citizens, food and citizenship

Public health literature has increasingly highlighted the relationships between individuals' diets and broader food systems change (Drewnowski and Popkin, 1997; Lang and Heasman, 2004; Rayner and Lang, 2012). For example, nutrition behaviours are now readily acknowledged to be a function of food environments, over and above individuals' knowledge, capacity, and propensity to act (e.g. Story *et al.*, 2008).

In the policy arena, however, public health interventions continue to focus heavily upon directing changes to individual purchase and consumption of food. This deficit in addressing 'upstream' or structural conditions influencing public health nutrition has been recognized in food policy for some time (Caraher and Coveney, 2004). It is also a reflection of what continues to constitute valid 'end-points' for public health policy interventions. Decisions about whether an intervention is worth adopting are often mired in a search for 'enough' evidence about 'effectiveness', where predicted changes in risk-factor behaviours as a result of the intervention are predominant outcomes of interest rather than a start to the policy conversation (Tannahill and Kelly, 2013). As a result, individuals are constructed as self-regulating members of a consuming public, rather than agents in public health policy change.

Concurrently, a body of work on 'food citizenship' has emerged, as an important characteristic to cultivate in healthier and more sustainable food systems. Situated within 'food democracy' as 'demand for greater access and collective benefit from the food system' (Lang, 1999: 217), food citizenship is a reinterpretation of what it means to consume food in a complex, interconnected, and dynamic system. It acknowledges eating as inherently political: imbued with duties, responsibilities, and rights (Wilkins, 2005). Consumers are more than eaters to be influenced – they are participating agents in food systems change (Baker, 2004; Welsh and MacRae, 1998).

By extension, as we discuss in this paper, 'food citizenship' is a concept that is relevant not only for consumption roles or even actions of 'citizens' per se, but in the different expectations for participation embedded within different policy actor roles. What is the role of health professionals or school administrators, for example, in negotiating the interface between citizens and systems? What types of 'citizen' participation are expected from people who take on 'expert' roles in their professional work? In this paper, we will address these questions through examples from our FACE (Food Advertising to Children: Ethics for Policy) study, funded by the Canadian Institutes of Health Research. The purpose of our study was to develop an understanding of the factors underlying policy inaction on food advertising, amidst strong evidence of its effects on health, a growing consensus in terms of high-level policy recommendations from influential organizations, and reasonable policy options forwarded. Without widespread agreement on the most appropriate form of collective action, policy debate has tended to revolve around moral reasoning about how children should behave and interact with the world around them (and how parents should govern such behaviour). In our investigation into this moral reasoning, we sought to understand how diverse policy actors situate roles and responsibilities for different actors in policy change. Here, we will share a selection of results from our findings demonstrating how diverse policy actors understand and incorporate citizenship into their social and moral constructions.

The issue: food and beverage advertising directed toward children

Concerns about the impact of food and beverage advertising directed towards children have been raised for decades. Recently, a critical mass of state, civil society, and some industry organizations have called for action to mitigate its effects on individual and population health. Major systematic reviews commissioned by the US Institute of Medicine (McGinnis *et al.*, 2006) and the UK Food Standards Agency (Hastings *et al.*, 2003), updated most recently for the WHO (2009), have concluded that food and beverage marketing has a clear effect on child health and wellbeing. These syntheses have found that foods advertised to young people are predominantly those that are energy dense and nutrient poor, and that advertising influences knowledge, preferences, purchase requests, and dietary habits of even younger children, adversely affecting health status. Further, it has been observed that existing research 'almost certainly underestimates the influence of food promotion' (WHO, 2009: 36). The World Health Organization (WHO) has gone on to release a list of recommended actions (WHO, 2010) and a framework for implementing these actions (WHO, 2012).

In Canada, these developments have not catalyzed any shift in policy. The province of Quebec (1980) has had a consumer protection statute in place for decades and at the national level, an industry self-regulatory process has been in place guiding advertising since the 1990s. As noted above, in the absence of policy change, ongoing debate has tended to centre on moral reasoning. For example, the argument that children should be granted access to all types of information as consuming persons, including advertising, is often situated against arguments about social capital, where children are vulnerable and must be protected to ensure their future potential. Ostensibly these are different imperatives, but we contend that they both direct attention to a narrow range of collective solutions to manage liberty and personal choices. As such, we were interested to learn whether the policy community's conception of the issue is actually as narrow as it appears to be, and to understand how roles and responsibilities for different actors are constructed through moral discourse.

Methods

We conducted a series of semi-structured interviews with policy actors active in the issue of food advertising to children from across Canada in May-June 2012. We recruited individuals purposively from five key groups of policy actors: academics, civil society organizations, health care practitioners, public health decision makers, and industry (including private sector food firms, and communications/ marketing). We did not seek a representative national sample but aimed for a good distribution of participants across professional roles and from different regions across Canada (Table 1). Recruitment was carried out in waves, with interviews ongoing, and completed upon conceptual saturation (n=35).

We created a series of developmental vignettes that served as the interview guide. Each vignette identified a central adult character faced with a dilemma where he or she needed to act upon advertising in the life of a child/children under their responsibility. Vignette central characters included, for example, Mandy and her 6-year-old daughter Olivia; Mohammed, a primary school principal; and Mary, a public health nurse leading an after-school program for youth. Interviewees were asked to approach the vignettes with a 'whole life' perspective, that is, to incorporate both their professional and personal (parental) roles when considering the moral dilemmas at hand. Using vignettes allowed interviewees to move fluidly between their ideal scenarios (what 'should' happen) to what they perceived to be the 'real-world' likelihood of such action (what 'would' happen). The vignettes were piloted with three individuals, also policy actors in related fields, and refined accordingly. All interviews were conducted by telephone, lasting 40-90 minutes, with most interviews taking about 60 minutes, were digitally recorded and transcribed by a third party. Transcriptions were then dual coded by the research team, with regular team debriefings to check and refine interpretations. We also consulted throughout the process with an advisory committee consisting of four individuals selected for their particular expertise in, respectively, public health ethics, marketing including food advertising, work in federal government, and consumer advocacy.

Table 1. Characteristics of the interview sample (n=35).

Sociodemographic characteristics	n
Professional roles	
Academics	2
Civil society organizations	11
Health care practitioners	7
Public health decision makers	10
Industry	5
Geographic representation	
West	7
Ontario	23[a]
Quebec	3
Atlantic	2
Gender	
Female	26
Male	9

[a] The Ontario figure is high, but was reasonable for this policy community; many individuals working at the national level reside in Ontario, e.g. federal government and civil society organizations with national offices.

Citizens as responsible, informed consumers

Ideal citizens

We found that while citizens are understood as more than eaters and shoppers, the range of potential actions they can take to shape policy are relatively narrow. Even 'ideal' citizens are conceived of as policy actors principally through forms of consumer demand.

We need to keep on asking for governments to step forward and at the same time we need to ask companies to cut down voluntarily. We need to make our positions known. We can do that by writing letters to the editor, by boycotting certain TV stations and favouring the ones without advertising. [P 30]

Awareness as action and resistance

The concept of citizen 'awareness' was correspondingly prominent. It was instructive what did not emerge: how awareness could lead to actions constitutive of policy change – behaviour changes, relationship building, communicating and negotiating, formal organizing or political participation. This could be interpreted in two ways. First, it reflects a dominant model for action in public health, where awareness-raising is viewed as a key (and sometimes only) step in engaging citizenry towards shaping social norms, and thus a mediating factor for policy change. Second, we posited whether awareness was conceived of as active resistance.

> M'hmm well again regulations ... the public always needs to kind of know what's going on because if they don't know what's going on no one is going to get upset about it and I think that's often a problem ... [P 35]

> Yes, exactly so it's a first step, you got to do it. I mean it's just things don't happen without some public awareness right ... generally speaking especially if an issue is controversial you need to have the public onboard. [P 2]

> Well I think it's strictly related to advocacy for civil society and providing evidence and mobilizing, raising public awareness too for civil society. For industry it's about being corporate, demonstrating some corporate social responsibility however as we know the profit margin usually the profits trump CSR most of the time ... [P 2]

A disconnect between citizens and governments

Whereas the notion of the ideal citizen as informed consumer defines citizens' relationship to markets, we found that citizens' capacity to engage politically is seen to be mediated through their representation by civil society organizations, rather than in relationship to the state.

> Well I think frankly first of all citizens have to become more demanding, more demanding for accountability. ... Now I do think this is where civil society can give leadership and I think if civil society organizations can bring in partners from the private sector and from government and start to make change on a neighborhood community basis then I think that can create momentum. [P 33]

> I think it needs to be a public issue. I think it needs to be brought to the public and made a movement and I think that organizations that work in the public interest writ large should be involved in this. I think politicians need to be educated and sensitized to this issue and add their voices to it and I'm not saying that it should be government led by any means. [P 26]

Health professionals as service delivery agents

We found that health professionals are constructed principally in their role as service providers. Within this role, they are at risk of infringing upon the parent-child relationship when it comes to individual behaviour change. To the extent that health professionals should participate in policy, it is through supporting consumer advocacy on a population level.

> I think Public Health in Canada I mean we have core competencies and advocacy is one of those core competencies so I think you know that falls upon Public Health to look at what is our role in advocacy around different topic areas ... [P 3]

> Well Mary has limited control right. I mean she's an educator, a Public Health nurse right. Patrick has his home environment, people who shop for him, I'm sure he buys stuff at the store so it's not as though she can have direct control over him and therefore force him to eat healthily despite what he would want so important to acknowledge that right off the bat therefore what are her options and I think she's really left with the options of educating and supporting. [P 29]

> Mary has a role as a Public Health nurse to not only help individuals but to change the social environment so part of her role is to work to advocate for health promoting social change and if that includes advocating for restrictions on advertising of those things that are inconsistent with promoting health than you know that should be considered part of her role. I think the problem is, is that Mary's bosses may or may not support that and you know so yeah. [P 5]

Towards a broader understanding of food citizenship

The abbreviated selection of themes we have introduced here can be understood as representative of two social phenomena. First, it was evident among this policy community that the broader concepts embedded in 'food citizenship' – itself we see as informed both by classical notions of participation in the public sphere and revitalized in literature on deliberative democracy (Fischer, 2009) – have not suffused public health discourse. Rather, citizens are viewed even in their ideal state as responsible consumers, defined in their relationship to markets. To some extent, this finding could be reflective of the specific policy issue we have addressed in our study. Second, we observe that our findings can be interpreted within the shift from 'government to governance', where complex policy problems are seen as requiring interconnected solutions across sectors and professional roles (Miller and Rose, 2002). The shift has also been framed, however, as redressing 'government failure', or problems 'too complex' for government (Salamon, 2002:8). The changing role of public sector professionals in relationship to this context is evident in our study, where policy participation among health professionals was narrowly conceived. Yet an important understanding is emerging in policy studies regarding how actors meaningfully negotiate policy through practice in 'networked governance' contexts (Wagenaar and Cook, 2003). Like others, then, we suggest that to advance public health policy in the domain of food, we need a better understanding of what constitutes and enables active, reflexive citizenship among both consumers and professionals (Tonkens and Newman, 2011). Whether health professionals are construed as only 'program delivery agents', for example, or active citizens, can affect their power to influence policy processes, as well as the types of policy levers they have at their disposal. Such inquiry will be valuable to inform how diverse policy actors can incorporate citizenship into their practices. In framing terms, it will also condition the range of policy options that are deemed suitable for public debate.

References

Caraher, M. and Coveney, M. (2004). Public health nutrition and food policy. Public Health Nutrition 7(5): 591-598.

Drewnowski, A. and Popkin, B.M. (1997). The nutrition transition: new trends in the global diet. Nutrition Reviews 55(2): 31-43.

Fischer, F. (2009). Democracy and expertise: reorienting policy inquiry. Oxford University Press, Oxford, UK, 339pp.

Hastings, G., Stead, M., McDermott, L., Forsyth, A., MacKintosh, A.M., Rayner, M., Godfrey, C., Caraher, M. and Angus, K. (2003). Review of research on the effects of food promotion to children. Centre for Social Marketing, Glasgow, UK, 208 pp. Available at: www.foodstandards.gov.uk/multimedia/pdfs/foodpromotiontochildren1.pdf.

Lang, T. (1999). Food policy for the 21st century: can it be both radical and reasonable? In: Koc, M., MacRae, R., Mougeot, L.J.A. and Welsh, J. (eds.) For hunger proof cities: sustainable urban food systems. Centre for Studies in Food Security, Ryerson University, Ottawa, Canada, pp. 216-224.

Lang, T. and Heasman, M. (2004). Food wars: the global battle for minds, mouths, and markets. Earthscan, London, UK, 36 5pp.

Miller, P. and Rose, N. (2008). Governing the present: administering economic, social, and personal life. Polity, Cambridge, UK, 246 pp.

Rayner, G. and Lang, T. (2012). Ecological public health: reshaping the conditions for good health. Routledge, London, UK, 409 pp.

Salamon, L.M. (ed.) 2002. The tools of government: a guide to the new governance. Oxford University Press, Oxford, UK, 669 pp.

Story, M., Kaphingst, K.M., Robinson-O'Brien, R. and Glanz, K. (2008). Creating healthy food and eating environments: policy and environmental approaches. Annual Review of Public Health 29(1): 253-272.

Tannahill, A. and Kelly, M.P. (2013). Layers of complexity in interpreting evidence on effectiveness. Public Health 127: 164-170.

Tonkens, E. and Newman, J. (2011). Active citizens, activist professionals. In: Newman, J. and Tonkens, E. (eds.) Participation, responsibility and choice: summoning the active citizen in Western European welfare states. Amsterdam University Press, Amsterdam, the Netherlands, pp. 201-216.

Wagenaar, H. and Cook, D.M. (2003). Understanding policy practices: action, dialectic and deliberation in policy analysis. In: Hajer, M.A. and Wagenaar, J. (eds.) Deliberative policy analysis: understanding governance in the network society. Cambridge University Press, Cambridge, UK, pp. 139-171.

Welsh, J. and MacRae, R. (1998). Food citizenship and community food security: lessons from Toronto, Canada. Canadian Journal of Development Studies 29: 237-255.

WHO (World Health Organization) (2009). The extent, nature, and effects of food promotion to children: a review of the evidence to December 2008. Prepared by Cairns, G., Angus, K., and Hastings, G. WHO, Geneva, 117 pp. Available at: http://whqlibdoc.who.int/publications/2009/9789241598835_eng.pdf.

WHO (World Health Organization) (2010). Set of recommendations on the marketing of foods and non-alcoholic beverages to children. WHO, Geneva, 14 pp. Available at: http://whqlibdoc.who.int/publications/2010/9789241500210_eng.pdf.

WHO (World Health Organization) (2012). A framework for implementing the set of recommendations on the marketing of foods and non-alcoholic beverages to children. WHO, Geneva, 61 pp. Available at: www.who.int/entity/dietphysicalactivity/MarketingFramework2012.pdf.

Wilkins, J.L. (2005). Eating right here: moving from consumer to food citizen. Agriculture and Human Values 22: 269-273.

Section 5. Animal husbandry and innovative meat production

Comfort, health and production: Portuguese dairy farmers talk about animal welfare

S. Silva[1], M. Magalhães-Sant'Ana[2], J. Borlido Santos[1] and I.A.S. Olsson[2]
[1]Office for Science Communication, IBMC – Instituto de Biologia Molecular e Celular, Porto, Portugal; [2]Laboratory Animal Science, IBMC – Instituto de Biologia Molecular e Celular, Porto, Portugal; olsson@ibmc.up.pt

Abstract

Farm animal welfare has increasingly become a concern and a subject of public debate within Europe, with different stakeholders emphasizing different aspects of animal welfare. In Portugal, the discussion of animal welfare issues is relatively recent and the available data about how different stakeholder groups conceptualize animal welfare is very scarce. In this study we explored Portuguese dairy farmers' representations of animal welfare, its significance and how farmers perceive other stakeholders' roles in the welfare of dairy cattle. The sample population was chosen considering that farmers, in their role as caregivers, are central actors in the promotion of animal welfare, and considering the importance of dairy farming within agricultural activity in the region. During the period October 2012-January 2013 we carried out semi-structured interviews with dairy farmers (n=22) from the north-west region of Portugal, randomly selected from the public records made available by local cooperatives. The interview guide was outlined to prompt an open discussion about animal welfare. It started by exploring the general requirements of dairy farming and the farmer's personal history. Targeted questions included what is considered important for success in dairy farming, the current welfare status of their cows, how animal welfare is defined and how to assess it, its importance within daily routine and how they perceive their own role in promoting animal welfare. The interviewees were further presented with three visual stimuli: the Five Freedoms; a list of other actors for dairy cow welfare and a picture of the Compassion in World Farming poll 'Where do you want our milk to come from?'. Results presented here are preliminary as the full content analysis is under process at the time of writing (April 2013). All farmers considered animal welfare centrally important for their activity and the reasons that were pointed out have mainly to do with improving production. The farmers' overall perspective on welfare was, however, broader and more complex. It encompasses concerns with subjective experiences, biological functioning and to some extent also natural behaviour, within both resource-based and animal-based approaches. Several tensions became evident in the discussion of the CiWF poll, i.e. between: (1) own current conditions and practices versus representation of an optimal production system; (2) consumers' opinion versus willingness to pay; and (3) different aspects of animal welfare.

Keywords: dairy cow, semi-structured interview, qualitative study, Portugal

Introduction

Farm animal welfare has increasingly become a concern and a subject of wide public debate for the European citizens. Different stakeholders emphasize different aspects of animal welfare, and these different representations contribute to the current societal discussion and to the definition of future initiatives intended to improve farm animal welfare (Kauppinen *et al.*, 2010; Vanhonacker *et al.*, 2008).

However, and despite that farmers are one of the most important stakeholder groups (as well as major actors in the promotion of animal welfare in their role as caregivers), research has only recently started to direct attention to farmers' representations of animal welfare (European Commission, 2007; Kauppinen *et al.*, 2010). The WelfareQuality project represents a major step toward providing systematically collected information on how farmers think about animal welfare. That major interdisciplinary research project included assessments of attitudes among a range of stakeholders, farmers being one of them. Farmers in Hungary, France, United Kingdom, Italy, The Netherlands, Norway and Sweden were interviewed about their participation in animal welfare schemes, perception and definition of animal welfare and perception of market and society and belief in animal friendly products (Kjærnes *et al.*, 2007). Farmers in Norway, The Netherlands and Italy participated in 'farmers' juries' consisting of 2-day events of focus group discussions on animal welfare and welfare assessment tools (Bock *et al.*, 2010).

The discussion of animal welfare issues is relatively new in Portugal and there is very limited data on the view of different stakeholder groups. The aim with the present study was to provide a first mapping of how Portuguese farmers think and act in regard to animal welfare. The sample population – dairy farmers in the north-west of the country – was chosen for a combination of geographical and farming practice reasons. Dairy farming is the most important agricultural activity in the region north of Porto where the research team is located and around 40% of the total number of dairy cows in Portugal is found here (INE, 2011). We further considered dairy farmers a particularly interesting group to interview about animal welfare for several reasons. Firstly, on dairy farms animals are kept for long periods of time; dairy cows are usually born and reared on the farm where they will enter into production in their second year of life and remaining productive for a number of subsequent years. Secondly, during the milking procedure attention is given to individual animals at least twice a day, and with the exception of highly modernized farms with automatic milking systems, milking also involves direct human-animal interaction. Thirdly, unlike for most other farm animal species in Portugal, dairy cows are not subject of specific animal protection legislation. All these factors combined lead us to expect that dairy farmers' perception of animal welfare would be based primarily on their own experience with animals and less influenced by enforced regulations.

Sample and methods

Data collection for this study took place through semi-structured interviews with dairy farmers (n=22) in the north-west of Portugal between October 30th 2012 and January 4th 2013.

The farmers were randomly selected from the public records made available by local cooperatives. The north-west region of Porto has three geographical hubs: Póvoa do Varzim, Vila Nova de Famalicão and Vila do Conde. The number of farms selected from each hub was proportional to the total number of dairy farms existing at each region. The advising veterinarians served as intermediaries in establishing contact with the farmers. All interviews, conducted face-to-face, took place at the farms and were carried out by the same person (SS). The interviewer had been trained in performing semi-structured interviews and three exploratory interviews were performed with farmers on the same subject and in a neighbour region. These exploratory interviews were also used to develop and refine the interview protocol. Each interview took 1-1.5 h and was recorded on a digital voice recorder for later transcription. All interviews were carried out in Portuguese.

At the start of the interview, the farmers were introduced to the nature (procedures) and purposes of the study, in order to obtain their informed consent. The interview protocol, however, was outlined to enable a gradual introduction of the topic of animal welfare, starting with general questions about the requirements of dairy farming and the farmer's personal history, in order to coax interviewees into speaking freely about animal welfare.

Regarding animal welfare, the interview protocol intended to address farmers' perceptions of what is important to be successful in dairy farming, the current welfare of their cows, the main problems and difficulties that they face in terms of animal welfare, and which sources of information they make use of to solve welfare issues. The protocol also addressed farmer representations about animal welfare, its importance within their professional routine and how they perceive the evolution of this concept, and the farmer's role in the promotion of animal welfare.

The farmers were further presented with two visual stimuli as part of the interview. They were asked to express their view of the Five Freedoms as these are presented on a fact sheet published by the national farmers' association. They were also asked to comment on the poll 'Where do you want our milk to come from?' that the Compassion in World Farming (CiFW) was promoting on their website. For this, the interviewer presented the two images – one of cows on pasture and one of cows in an indoor loose housing system – used in the poll.

The interview protocol also included questions related to the farmer's perception about production diseases, the rules and legislation on animal welfare, and the role of other stakeholders in this field, ending with questions about the farmers' own welfare.

Results

Of the 22 farmers interviewed, 19 were men and 3 women. Their age ranged from 25 to 61, with a mean of 42 years of age. Regarding the level of education, half of the farmers (11) had completed middle school/lower secondary education; 9 farmers had completed high school/upper secondary education and 2 had only completed primary/elementary/basic education. The herd size ranged from 55 to 326 cows (25 to 144 cows in production). Nearly all interviewees were the main responsible for the dairy herd management and the owners of the farm.

The farmers all expressed that animal welfare was important because of its crucial role for production. This was often expressed in terms such as animals which are not faring well are not producing well. The importance of good animal health and welfare was also mentioned in the context of the farmer's own situation, with reference to that problems with animal health and welfare make the farmer's life difficult.

The farmers' overall perspective on welfare was, however, broader and more complex. We constructed a matrix combining the three concepts of animal welfare as outlined in David Fraser's and Ian Duncan's classical paper (Duncan and Fraser, 1997) with the two approaches to measure welfare, in terms of what resources are provided ('resource-based') and in terms of the animals themselves ('animal-based'). Viewing the interviews against this matrix showed that indeed all possible combinations are represented in the way that farmers talk about dairy cow welfare. That is, they express issues belonging to subjective experiences (comfort, feeling well, not being unhappy), biological functioning (health, milk yield, food consumption) and to some extent also natural behaviour, within both resource-based and animal-based approaches and often cutting across categories, as illustrated in Figure 1.

In response to the question of what is required to be a successful dairy farmer, among the frequently mentioned aspects were to feed the cows well and to provide comfortable conditions for them, in particular as regards bedding. Many farmers referred to the importance of engaging with the cows in terms of being dedicated to them, paying attention to them, respecting them and also liking them.

Several tensions became evident in the discussion of the CiWF poll (Figure 2), i.e. between: (1) own current conditions and practices versus representation of an optimal production system; (2) consumers' opinion versus willingness to pay; and (3) different aspects of animal welfare.

When questioned about the current welfare of their cows or when confronted with the CiWF poll most farmers expressed in general terms that it was good for cows to be free on pasture. However, when

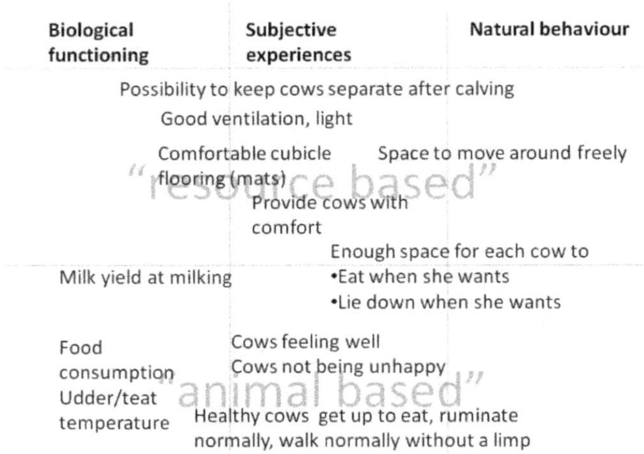

Figure 1. This matrix illustrates how different references to animal welfare made by dairy farmers in the interviews can be seen against a matrix combining three theoretical dimensions of the animal welfare concept with two approaches to measuring welfare.

Figure 2. Poll that the Compassion in World Farming (CiFW) was promoting on their website (adapted from http://www.ciwf.org.uk) and that was presented to the farmers as visual stimuli during the interview.

asked if they would be willing to adopt an extensive management system, the farmers presented motives related with the level of production and financial and contextual constraints to prefer an indoor loose housing system.

For example, one farmer said that 'I'd say that I would like that the milk came from here'(pasture), but when questioned if this would be possible replied 'No, not [here] (...) or rather, it would be possible (...) if I could live with 70 cows (half of the herd), then I would have space on my dairy farm'. Other interviewees were unwilling to adopt an extensive management system despite recognizing that 'the animals fare much better [in the extensive system]' or 'I prefer this picture (pasture), but I would adopt that one (...) this is what I think is beautiful, to have the animals grazing (...) but to manage in a profitable way, I believe that it has to come from here' (the picture of the intensive management system).

As mentioned, tensions related with consumers' opinion versus willingness to pay also became evident in the discussion of the CiWF poll. In the opinion of one farmer, 'Everyone says that this one is better (cows on pasture) (...), now, it ought to say below how much a litre of milk from one and from another would cost (...) if they show the difference, they would go for this' (intensive). Another farmer expressed a similar opinion: 'There's a fallacy here (...) because they don't say that you would have to pay at least three times more (...) this isn't about telling people 'what is it that you want?', this is about 'how much are you willing to pay?''.

The position of other stakeholders towards animal welfare was not an issue that the interviewees had thought about before. Most farmers considered the cooperatives to be knowledgeable and engaged, but were generally dismissive as regards to the retailers, for being merely interested in the price of milk. The view of consumers was more mixed. The general impression was that the consumers were poorly informed, but several farmers thought they were nevertheless concerned with animal welfare.

Discussion

The preliminary results of this interview study indicate that Portuguese dairy farmers pay attention to and think about animal welfare. They primarily justify their attention to the welfare of their cows in terms of its relevance for production, but it is clear that they have a broader notion of animal welfare. A number of farmers also make direct or indirect reference to the subjective experience of the cows. These observations are very much in agreement with what the WelfareQuality project showed for Dutch, Norwegian and Italian farmers (Bock *et al.*, 2010).

As regards how they see consumers as animal welfare stakeholders, the interviewees in the present study tended to see consumers as un- or misinformed and generally not willing to pay the extra cost that for example a pasture-based system would require. This is much like the view of the Italian farmers in the WelfareQuality study (Bock *et al.*, 2010), indicating that this may reflect a generalized Southern European perspective. Interestingly, the Eurobarometer on animal welfare indicates that consumers in this region of Europe indeed would like to be better informed about animal welfare (EC, 2007).

Acknowledgements

'Biosense – Science engaging society: Life sciences, social sciences and publics' (PTDC/CS-ECS/108011/2008 – FCOMP-01-0124 -FEDER-009237) is a project financed by FEDER Funds through 'COMPETE – Programa Operacional de Fatores de Competitividade' and by national funds through FCT (Fundação para a Ciência e Tecnologia). The authors wish to thank the veterinarians Andreia Santos, José Pedro Azevedo, António Manuel Ventura, Eduardo Torre e Manuel Paulo for their help in establishing contact with farmers and extend their thanks to the farmers participating in the study.

References

Bock, B.B., Swagemakers, P., Jacobsen, E. and Ferrari, P. (2010). Dialogue between farmers and experts regarding farm animal welfare. Farmers' Juries in Norway, the Netherlands and Italy. Welfare Quality Reports no. 17. Cardiff University, Cardiff, UK.

Duncan, I.J.H. and Fraser, D. (1997). Understanding animal welfare. In: Appleby, M.C. and Hughes, B.O. (eds.) Animal Welfare. Cambridge University Press, Cambridge, UK, pp. 19-31.

European Commission (2007). Attitudes of EU citizens towards Animal Welfare, Special Eurobarometer 270.

INE (2011). Recenseamento Agrícola 2009 Análise dos principiais resultados. Instituto Nacional de Estatística, I.P. Lisboa, Portugal.

Kauppinen, T., Vainio, A., Valros, A., Rita, H., and Vesala, K.M. (2010). Improving animal welfare: qualitative and quantitative methodology in the study of farmers' attitudes. Animal Welfare 19: 523-536.

Kjærnes, U., Miele, M. and Roex, J. (2007). Attitudes of consumers, retailers and producers to farm animal welfare. Welfare Quality Reports no. 2. Cardiff University, Cardiff, UK.

Vanhonacker, F., Verbeke, W., Poucke, E.V. and Tuyttens, F.A.M. (2008). Do citizens and farmers interpret the concept of animal welfare differently? Livestock Science 116: 126-136.

Farmers' views on the impact of breeding traits on profitability, animal welfare and environment

A. Wallenbeck[1], T. Ahlman[1], L. Rydhmer[1] and H. Röcklinsberg[2]
[1]Swedish University of Agricultural Sciences, Department of Animal Breeding and Genetics, P.O. Box 7023, 750 07 Uppsala, Sweden; [2]Swedish University of Agricultural Sciences, Department of Animal Environment and Health, P.O. Box 7068, 750 07 Uppsala, Sweden; anna.wallenbeck@slu.se

Abstract

Taking the growing consumer and citizen interest in animal welfare and the environment into account, a closer look at the production animals used and their genetic potential is necessary. Development of long term sustainable animal production systems need to involve breeding goals. The farmers' views on different traits are fundamental for the understanding of how to balance competing breeding goals. The aim of this web questionnaire study was to investigate pig and dairy farmers' views on a range of breeding traits in relation to farm profitability, animal welfare and environmental impact of production. The results show similar patterns for pig and dairy farmers. Most farmers (>80%) considered all given traits to be related to profitability. Fewer traits were regarded important for animal welfare, expressed as longevity, health and roughage consumption. Even fewer traits were considered to have an impact on the environment; all of them were related to production, feed conversion, roughage consumption or survival/longevity. Furthermore, no differences of importance were seen between farmers with conventional and organic production. Longevity, disease resistance and roughage consumption seem to be considered as key traits with impact on both animal welfare and environment by both dairy and pig farmers. Consumers' and citizens' interest and concerns about animal welfare and environment seem to be shared by farmers. However results show that key traits considered to influence both animal welfare and environment were parallel to profitability.

Keywords: pig, dairy, genetic selection, ethical values, organic production

Introduction/background

Consumer and citizen interest in animal welfare and environment is high in Europe (Eurobarometer, 2007, 2011), and animal welfare is essential for sustainable development of animal production. However, sustainable development of animal production needs to include economic, environmental and social aspects (e.g. Dockes *et al.*, 2011; Gamborg and Sandoe, 2005) and in the development of animal production systems, improvement of animal welfare is not sustainable without also considering the economics and the environment. The increased consumer interest in sustainable food production is also reflected in the increased interest in organic production (e.g. Eurobarometer, 2010). Organic production aims for a cyclical use of resources, balance between plant and animal production on farm level, and a higher animal welfare than in conventional farming as well as social responsibility (IFOAM, 2012). Improvement of organic and other high animal welfare production systems to date has mainly focused on changes in production environment and management (e.g. input of resources), while less focus has been directed towards the foundation of production, i.e. the genetic of the animal breeds placed in these production environments.

In most cases, the breeding goal of the animal material presently used in organic and/or animal welfare oriented production systems are not related to the production environment, indicating a need for development of breeding strategies for such production systems. Development of long term sustainable breeding goals needs to consider the requirements of different stakeholders in the food chain, e.g.

farmers, retailers and consumers (Gamborg *et al.*, 2005). However, there is in general a lack of knowledge about stakeholder preferences, in i.e. interests and values hold among farmers and breeding organisations. There are some reports of differences in underlying values between organic and conventional farmers (Lund *et al.*, 2002) as well as between consumers appreciating organic products or not (Liljenstolpe, 2008), However, as consumer interests in e.g. high animal welfare and decreased environmental impact is outspoken, knowledge about farmers' preferences are essential. How do farmers value animal welfare in relation to other issues? Looking at their preferences will give an insight into what characteristics they regard important for improving animal welfare and meet environmental concern.

The aim of this study was to investigate conventional and organic pig and dairy farmers' views on relationships between different breeding traits and farm profitability, animal welfare and environmental impact of production. The choice of investigating both pig and dairy farmers views are based on the facts that the breeding structure and the production form is very different in pig and dairy production. Thus similarities and discrepancies in views between these groups of farmers can provide important knowledge for the understanding of underlying values.

Material and methods

In a novel attempt to assess farmers views on breeding traits, two advanced web questionnaires about traits important for animal production were developed, one for pig and one for dairy farmers. Invitations were sent to 1,481 Swedish dairy farmers, i.e. one-fourth of all dairy farmers in Sweden and 522 pig farmers, i.e. half of the pig producers in Sweden. The aim of the questionnaire, expressed in the invitation, was to assess traits of importance for sustainable production. The invitation included a farm-specific link to the questionnaire that anyone with access to the e-mail account could use. The respondents could enter the questionnaire several times as long as they had not submitted it. Once the respondent had finished and submitted his/her answer, the farm-specific link to the questionnaire was closed. The questionnaire was finished and submitted by 120 pig farmers and 468 dairy farmers.

The questionnaire consisted of five steps: (1) the farmer stated what traits they intuitively consider important in their herd; (2) the farmer ranked 15 given species specific traits against each other; (3) the farmer weighed traits against each other given the estimated genetic gain; (4) the farmer marked whether they considered each of the 15 given traits to have impact on farm profitability, animal welfare and the environment; and (5) the farmer answered general questions about him/herself and the herd he/she worked in (e.g. organically certified farm and gender of the respondent). The 15 species specific traits were chosen in order to represent both production and functional traits and to include both traditional and potential future breeding traits. This study focuses on information from step 4 and 5. In the questionnaire it was stated that in order to be sustainable in the long term a breeding strategy should include several different aspects, e.g. farm profitability, animal welfare and environmental impact. Farm profitability, animal welfare and environmental impact were, however, not defined to the respondents; thus the respondents' reply in step 4 is based on individual perceptions of these expressions.

Differences in views on the impact of traits on animal welfare, farm profitability and the environment between organic and conventional farmers were statistically analysed as binomial models (having or not having impact) using SAS Proc GLIMMIX (SAS, 2008) and the statistical model included the effects of production system (conventional or organic), gender and age of the respondent. Descriptive statistics were analysed using Proc FREQ.

Results

Most farmers (>80%) considered almost all given traits to have impact on farm profitability. However, only 63.6% and 32.0% of the dairy farmers considered temperament and methane production, respectively, to have impact on profitability. Among pig farmers only 62.7% considered roughage consumption to have impact on profitability. The animal welfare related traits most unanimous regarded important by both pig and dairy farmers were related to health, offspring survival, longevity and roughage consumption. Traits considered to have an impact on the environment were related to production, feed conversion, roughage consumption and longevity (Figure 1 and 2).

The proportion of farmers considering a trait to have impact on farm profitability, animal welfare, and the environment differed ($P<0.05$) in a few cases between organic and conventional dairy farmers, but not between organic and conventional pig farmers. The differences seen between organic and conventional dairy farmers were differences in scale, but without significant effect of the ranking of traits, i.e. even though there was a difference in the proportion of farmers considering the trait to have impact, the average order of traits were the same for organic and conventional farmers.

Discussion

The results show similar patterns in views between pig and dairy farmers in whether traits impact profitability, animal welfare and environment. This indicates that these groups of farmers share the same general views about farming practices and needs. We also found similar patterns in views between organic and conventional farmers, for both dairy and pig farmers. In Sweden organic dairy farms have on average a larger number of animals than conventional farms, hence both organic and conventional farmers live under conditions demanding profitability, often due to large investments, whereby animal

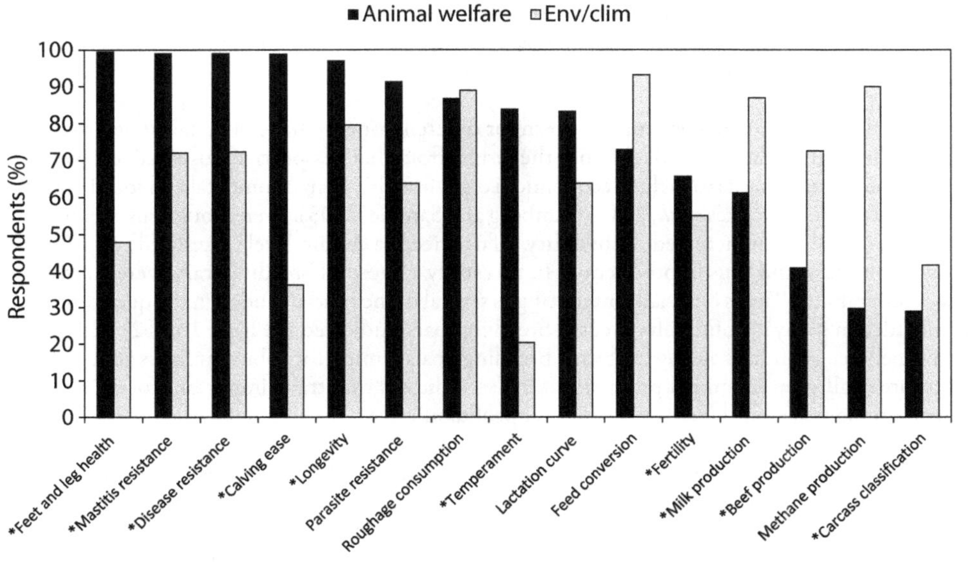

*Figure 1. Proportion of dairy farmers considering the 15 given traits to be related to animal welfare and environment/climate. *indicates traits included in the breeding goals of the breeds currently used in Swedish dairy production.*

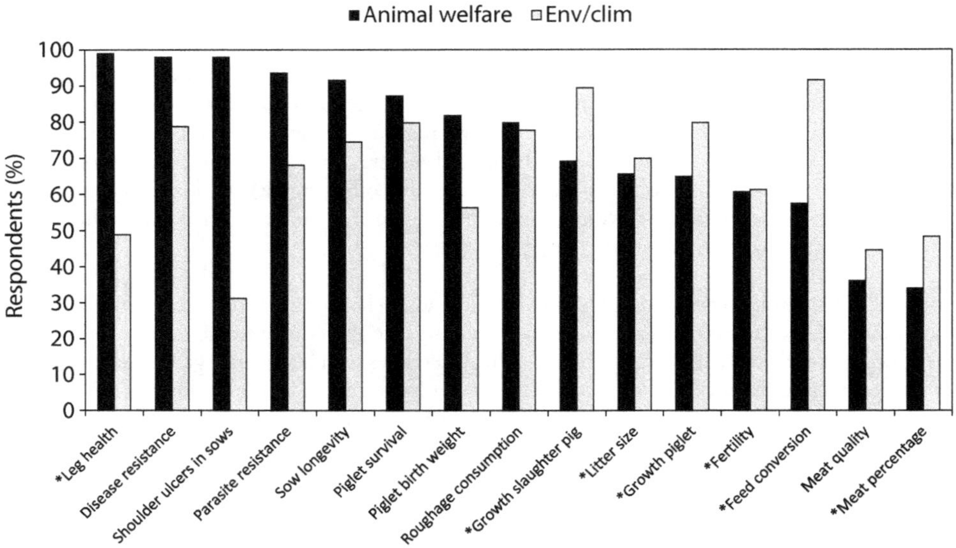

*Figure 2. Proportion of pig farmers considering the 15 given traits to be related to animal welfare and environment/climate. *indicates traits included in the breeding goals of the breeds currently used in Swedish pig production.*

welfare and environmental concerns generally are more difficult to take into practical concern. Hence similarities between the systems could be a sign on either small factual differences between the systems when it comes to profitability, animal welfare and environment, or, and more interesting if would be the case, since the same breeds are used, genetic differences are small and the same effects of the traits are seen in both organic and conventional production systems.

Given the above mentioned consumer concern for different aspects that can be subsumed in the terms sustainability and welfare, we will look into the results from that perspective. Sustainable development of animal production needs to include economic (i.e. profitability), environmental and social aspects (i.e. animal welfare) (e.g. Dockes *et al.*, 2011; Gamborg and Sandoe, 2005). Even though sustainability does not increase linear with increased profitability, a farm needs a certain level (threshold) of profitability to stay in business, and thus in production. In this study, almost all breeding traits were considered to impact profitability. This is especially interesting as several of the traits included in the questionnaires are not included in today's profitability focused breeding goals (indicated in Figure 1 and 2). This indicates that farmers see economic needs for broad breeding goals aiming for robust animals to meet current and future challenges in animal production. It also indicates that traits important for animal welfare and environment are considered important for profitability.

The trait clusters farmers considered to impact animal welfare (health, offspring survival, longevity and roughage consumption) and environment (production, feed conversion, roughage consumption and longevity) are logic in relation previous reports (e.g. Hörning, 2006; Pryce *et al.*, 2004; Rodenburg and Turner, 2012). Our results show a trend for interesting agreement in that both longevity, disease resistance and roughage consumption was considered to have impact on both environment and animal welfare by a large proportion of both dairy and pig farmers. To start with roughage, its impact on environment/climate as an efficient local feed resource is relatively clear for dairy production, but the corresponding view among pig farmers is a novel finding. The fact that pig farmers consider pig roughage consumption to be a trait of importance for environment is interesting and further studies on the potential of roughage as a local feed source for pigs is needed. These findings could reflect farmers' positive experiences of fulfilling the species specific needs of ruminants (cows) and omnivores (pigs) to graze and forage. As to the point of longevity and disease resistance, its high score on environmental impact can be interpreted in terms of animals staying in production for many years also being healthy and efficient. Poor animal health, reproduction and feed efficiency causes increased emissions of greenhouse gases and nutrient leakage (Lundström *et al.*, 2009). It is also possible however, that there has been confusion between animal welfare concerns and animal ethics. From a certain animal ethics point of view, ensuring a long life for animals is relevant, but hard to see from an animal welfare perspective. Given that animal welfare is related to the state of the animal (interpreted as its function/health, its feelings or its possibility to perform species specific behaviours), as usually done in animal science, longevity is no issue, but rather the welfare of an animal during life (Yates *et al.*, 2011).However, as animal welfare is essential in animal ethical theories focussing on preferences, respondents might have mixed welfare as a scientific measure and welfare as an ethical goal. Do the farmers value traits enhancing a long animal life as it is regarded important in itself, or are they interested in improving the animal's capacity to live a long life at a certain welfare level? Both views are possible, and both interesting as pig farming aims at producing meat, and Swedish dairy cow have an average age of 4 years at slaughter. Further information about farmer's view is thus needed, and will be gained through focus groups.

Taken together, the results of this study indicates a general agreement in farmers' views on breeding traits' impact on farm profitability, animal welfare and environment, both between dairy and pig farmers and between farmers with organic and conventional production. Consumers' and citizens' interest and concern about animal welfare and environment seem to be, at least partly, shared by farmers as general trait clusters influencing animal welfare and environment were evident and key traits considered to impact both profitability, animal welfare and environment were identified (longevity, disease resistance and roughage consumption). More research is needed to understand the similarities between farmers with different production systems with regard to scoring of breeding traits.

References

Dockes, A.C., Magdelaine, P., Daridan, D., Guillaumin, A., Remondet, M., Selmi, A., Gilbert, H., Mignon-Grasteau, S. and Phocas, F. (2011). Farming and breeding goals for sustainable animal breeding – views and expectations of stakeholders from the production chains and the general public. Inra Prod. Anim. 24: 285-296.

Eurobarometer (2007). Attitudes of EU citizens towards animal welfare. In Special Eurobarometer, p. 49.

Eurobarometer (2010). Europeans, Agriculture and the common agricultural policy. Special Eurobarometer 365.

Eurobarometer (2011). Attitudes of European citizens towards the environment. Special Eurobarometer 365.

Gamborg, C. and Sandoe, P. (2005). Sustainability in farm animal breeding: a review. Livestock Production Science 92: 221-231.

Gamborg, C., Olsson, A. and Sandoe, P. (2005). Farm animal breeding related ethical concerns and tools for implementation CODE_EFABAR ethical report.

Hörning, B. (2006). Organic livestock husbandry and breeding, In: Kristiansen, P., Taji, A. and Reganold, J. (ed.) Organic agriculture – a global perspective. Comstock Publishing Associates, Ithaca, New York, NY, USA, p. 151-163.

IFOAM (2012). The IFOAM norms for organic production and processing. Available from: www.ifoam.org/about_ifoam/standards/norms/IFOAMNormsVersionAugust2012withcover.pdf.

Liljenstolpe, C. (2008). Consumer valuation studies and structural modelling of the pig industry: a focus on animal welfare. Acta Universitatis agriculturae Sueciae.

Lund, V., Hemlin, S. and Lockeretz, W. (2002). Organic livestock production as viewed by Swedish farmers and organic initiators. Agriculture and Human Values 19: 255-268.

Lundström, J., Albihn, A., Gustafson, G., Bertilsson, J., Rydhmer, L. and Magnusson, U. (2009). Lantbrukets djur i en föränderlig miljö – utmaningar och kunskapsbehov. Rapport. Fakulteten för veterinärmedicin och husdjursvetenskap, SLU, Uppsala, Sweden.

Pryce, J.E., Conington, J., Sørensen, P., Kelly, H.R.C. and Rydhmer, L. (2004). Breeding strategies for organic livestock, In: Vaarst, M., Roderick, S., Lund, V. and Lockeretz, W. (ed.) Animal health and welfare in organic agriculture. CABI Publiching, Oxon, UK, pp. 357-388.

Rodenburg, T.B. and Turner, S.P. (2012). The role of breeding and genetics in the welfare of farm animals. Animal Frontiers 2: 16-21.

SAS (2008). Sas OnlineDoc, Version 8.

Yates, J., Röcklinsberg, H. and Gjerris, M. (2011). Is welfare all that matters? A discussion of what should be included in policymaking regarding animals. Animal Welfare 20: 423-432.

The welfare of dairy cattle: perspectives of industry stakeholders

B.A. Ventura, M.A.G. von Keyserlingk and D.M. Weary
Animal Welfare Program, Faculty of Land and Food Systems, University of British Columbia, 2357 Main Mall, Vancouver, BC, V6T 1Z4, Canada; bethventura2@gmail.com

Abstract

The aim of the current study was to describe the perspectives of stakeholders within the dairy industry on key issues affecting the welfare of dairy cattle. A secondary aim was to examine if these stakeholders believed that people outside of the industry should also have a voice in formulating solutions to these issues. Five heterogeneous focus groups were conducted during a dairy cattle industry meeting in Guelph, Canada in October 2012. Each group contained between 7-10 participants and consisted of a mix of dairy producers, veterinarians, researchers, students, and industry specialists. The 1-h facilitator-led discussions focused on participants' perceptions of the key welfare issues and the role of different stakeholder groups in addressing these concerns. Discussions were audio-recorded and transcribed verbatim, and the resulting transcripts coded and the themes identified. Lameness was uniformly recognized as the most important welfare issue facing dairy cattle; cow comfort, painful procedures (like dehorning) and production diseases (like mastitis) were also commonly discussed. Participants had mixed views on the roles of different stakeholders in formulating solutions. Most felt that producers and others working within the dairy industry (particularly veterinarians) should be primarily responsible, but many participants acknowledged that the general public, as consumers and as citizens, also play an important role. Participants seemed to focus on a two-fold knowledge deficit – first between researchers and producers, and second between dairy industry groups and the public – and agreed that improved knowledge translation was required to develop solutions to welfare concerns. These results indicate that many people within the dairy industry see value in more inclusive engagement with non-industry stakeholders about dairy cattle welfare. Future work will assess perspectives of people outside of the dairy industry to identify areas of shared concern and provide a basis for policy solutions that better incorporate societal values.

Keywords: stakeholder attitudes, dairy cattle, animal welfare, engagement

Introduction

The dairy industry faces rising pressure to address societal concerns related to the care and handling of cows and calves. In this paper we present preliminary results from a focus group study with primarily North American dairy industry stakeholders. We describe how these individuals conceptualize problems related to dairy welfare and how they interpret their own role and that of other stakeholders in achieving socially sustainable solutions to these problems.

Much of the interaction to date between industry and non-industry groups has been polarized and simplistic (Croney, 2010), with non-industry calling for more 'humane' treatment and industry responding that this is already a priority. Such discourse often generates frustration on both sides and may lead to erosion of public trust in dairying (Brom, 2000).

Some of the disagreement between different stakeholders is likely rooted in a divergence in values and expectations around what constitutes good animal welfare. Lay citizens often conceptualize a good life for animals in terms of ability to achieve a 'natural' life experience, while producers and others working within agriculture tend to consider animal health and functioning the priority (Lassen *et al.*, 2006; Te Velde *et al.*, 2002; Vanhonacker *et al.*, 2008). However, little work to date has focused on dairy cattle, and to our knowledge no work has been done on the views of North American industry stakeholders. This gap in science has resulted in a lack of data on the specific issues of most concern to these stakeholders, and their perspectives on how these issues should be resolved.

Aims

Our aim was to facilitate and describe the discussions on the welfare of dairy cattle by stakeholders within the dairy industry, including producers, veterinarians, researchers, industry specialists, and students. As is typical of such stakeholder consultations, we sought to create a sense of priority issues to the participants (Rowe and Frewer, 2005). A secondary aim was to explore how these participants conceptualized the roles of different groups, including people external to the dairy industry (e.g. consumers, citizens, animal rights advocates, etc.), in addressing these issues.

Methods

Five heterogeneous focus group sessions took place at the Dairy Cattle Welfare Symposium in Guelph, Canada in October 2012. Each group consisted of 7 to 10 conference attendees and included dairy producers, veterinarians, researchers, service providers, dairy and marketing board members, and students. Discussions lasted approximately 1 hour and were facilitated by trained moderators. Moderators followed a script of questions focused on: (1) dairy welfare challenges, barriers to improvement, and solutions; and (2) the role of different stakeholder groups in addressing these challenges. All discussions were audio-recorded and transcribed verbatim, and the resulting transcripts coded and prominent themes identified.

Findings

Analysis is ongoing, but below we summarize some of the emerging results.

Issues of concern

Lameness was uniformly recognized as the most important welfare issue facing the dairy industry, both because it was considered painful but also because it was a primary factor in decisions to cull cattle. Cow comfort, painful procedures like dehorning, and production diseases like mastitis were also frequently discussed.

Roles of stakeholders

Participants had mixed views on the roles of the different stakeholder groups in formulating solutions. There was general agreement that producers and others with an on-farm presence, particularly veterinarians, were the primary actors responsible for ensuring the welfare of cattle. Perceptions of the role of researchers was varied, with some participants expressing frustration that academics were too distant from the realities of farming, and others acknowledging that research had an important role to play in informing best practices. Many participants also acknowledged that the general public, primarily as consumers, also have a say in the way in which dairy cattle are raised and handled. Depending on the way in which participants interpreted this effect, attitudes toward perceived consumer influence ranged from frustration to resignation or acceptance.

Perceived knowledge deficits

Participants often expressed frustration related to knowledge deficits, including those between researchers and producers. Some participants argued that researchers did not understand what issues were relevant for farmers, and many participants felt that farmers were often unaware of key findings generated by researchers. Participants agreed that increased extension efforts were necessary to improve knowledge transfer to the farm.

Participants were almost unanimous in considering the public to be largely ignorant of the realities of farming. Many considered this ignorance to be problematic in that consumer demands – particularly for more naturalistic environments – were deemed to be unrealistic. However, some participants argued that the dairy industry helped to perpetuate certain 'myths', for example, through pastoral depictions of cows in advertising.

Implications

Addressing perceived public ignorance of dairy practices may be problematic for the industry, in part because improved knowledge of animal care practices may actually lead to strengthened opposition. Moreover, the 'deficit model' of public understanding has increasingly been called into question (Schiele, 2008), not least because farm animal production is a public good that cannot reasonably be expected to operate in isolation of societal expectations. It should also be noted that the complex relationship between public knowledge and concerns around animal farming is not well understood, but one-way public educative efforts are unlikely to fully resolve societal concerns around dairy farming. Rather, we suggest that the dairy industry engage in real dialogue with the public, in order to better understand mainstream societal values regarding animal care and ultimately modify practice to better accommodate these values (Callon *et al.*, 2009; Castle and Culver, 2006).

Most participants viewed improved engagement between and among industry stakeholders in a positive light. Indeed participants commented that their participation in the focus groups was an example of the type of engagement needed. We suggest that the methodology described in the present study may thus serve as a good model of one way in which stakeholder engagement might function at the industry level, as a way to improve collaboration between stakeholders at each stage of dairy production so that the industry can work more effectively to address welfare challenges.

Reason to be optimistic

We suggest that this type of stakeholder engagement can help generate new policy that helps to address welfare issues and bridge divergent concerns. Our participants' prioritization of lameness was unsurprising, but what is of note is that their comments around lameness, along with dehorning, were often expressed in relation to the pain experienced by the animals. We anticipate that concerns about pain will be shared by people external to the dairy industry even if they do not express concerns about lameness per se (perhaps due to their ignorance of the problem, though this warrants future research). Thus, reducing the pain experienced by farm animals may be a common goal that diverse groups can agree is a priority issue for future policy recommendations.

The current focus group study is the first piece of a larger research effort to describe diverse perspectives on dairy cattle welfare. Future work will focus on participants outside of the dairy industry; we hope this will identify areas of shared concern and provide a basis for the dairy industry to develop policy solutions that better incorporate societal values.

Acknowledgements

We thank Ken and Barbara Leslie and Amy Stanton, without whose support these focus groups would not have occurred. Thanks also to our moderators who generously donated their time in order to train and execute the focus groups – we could not have carried out this project without them. Finally, many thanks especially to our wonderful participants who shared their time and their views with us.

References

Boogaard, B.K., Oosting, S.J. and Bock, B.B. (2008). Defining sustainability as a socio-cultural concept: citizen panels visiting dairy farms in the Netherlands. Livestock Science 117: 24-33.

Brom, F.W.A. 2000. Food, consumer concerns, and trust: food ethics for a globalizing market. Journal of Agricultural and Environmental Ethics 12: 127-139.

Callon, M., Lascoumes, P. and Barthe, Y. (2009). Acting in an uncertain world: an essay on technical democracy. MIT Press, London, UK.

Castle, D. and Culver, K. (2006). Public engagement, public consultation, innovation and the market. Integrated Environmental Assessment and Management 6: 137-152.

Croney, C.C. (2010). Words matter: implications of semantics and imagery in framing animal-welfare issues. Journal of Veterinary Medical Education 37: 101-106.

Croney, C.C., Apley, M., Capper, J.L., Mench, J.A. and Priest, S. (2012). The ethical food movement: what does it mean for the role of science and scientists in current debates about animal agriculture? Journal of Animal Science 90: 1570-1582.

Ellis, K.A., Billington, K., McNeil, B. and McKeegan, D.E.F. (2009). Public opinion on UK milk marketing and dairy cow welfare. Animal Welfare 18: 267-282.

Lassen, J., Sandoe, P. and Forkman, B. (2006). Happy pigs are dirty! Conflicting perspectives on animal welfare. Livestock Sciemce 103: 221-230.

Rowe, G. and Frewer, L.J. (2005). A typology of public engagement mechanisms. Science Technology and Human Values 30: 251-290.

Schiele, B. (2008). On and about the deficit model in an age of free flow. In: Cheng, D., Claessens, M., Gascoigne, T., Metcalfe, J. Schiele, B. and Shi, S. (ed.) Communicating science in social contexts. Springer, Dortrecht, the Netherlands, p. 93-117.

Te Velde, H.T., Aarts, N. and Van Woerkum, C. (2002). Dealing with ambivalence: farmers' and consumers' perceptions of animal welfare in livestock breeding. Journal of Agricultural and Environmental Ethics 15: 203-219.

Vanhonacker, F., Verbeke, W., Van Poucke, E. and Tuyttens, F.A.M. (2008). Do citizens and farmers interpret the concept of farm animal welfare differently? Livestock Science 116: 126-136.

Gnawing doubt: eating animals and the promise of cultured meat

L. Ursin
Department of Philosophy, Norwegian University of Technology and Science, 7491 Trondheim, Norway; lars.ursin@ntnu.no

Abstract

Envisioned technologies can make and be motivated by ethical spaces. In this paper I will explore the ethical spaces that motivates and are made by the emerging technology of cultured meat. Proponents of cultured – or *in vitro* – meat argue that it has the potential to present a solution to several of perceived problems of factory farming: Animal welfare issues, environmental concerns, and global access to meat products. Critics of in vitro meat technology argue that the solution it introduces just will aggravate an alienation from farm animals caused by factory farming. In the eyes of the critic, cultured meat would be a technological quick fix that just eases the consequences of a flawed way relating to both food and animals. In this paper I will use the technology of cultured meat as a tool to analyze the moral relation between animals as fellow creatures versus animals as food.

Keywords: *in vitro* meat, eating animals, human nature, factory farming, food ethics

Introduction

We eat meat at an unprecedented rate today. The Food and Agriculture Organization of the United Nations (FAO) has statistics on global meat consumption per capita from 1961until today. Over these last 50 years most countries in the world have had a considerable growth in meat consumption per citizen (FAO, 2012). According to the 2010 forecasts of the FAO, global meat consumption amounted in 2012 to 42.5 kilograms of meat per capita. In developing countries, the average meat consumption was 32.4 kg., while the industrialized countries averaged 79.2 kg (FAOSTAT, 2010).

The situation today, even the situation 50 years ago, is in one way quite different from the situation in the 18th and 19th century in Europe. 200 years back, meat consumption was significantly lower than today. According to Vincent Knapp (1997: 544), the average annual per capita consumption of meat in Germany was 12.7 kg in 1800. The access to meat was poor for the average citizen, and even poorer than this number suggests, since the distribution of meat was quite uneven among the social classes.

The social inequality of meat eating points to a similarity with the global situation today: The rich ate and eat more meat than poor people. A graphic description of the social status of meat consumption is given by Knapp: 'Prior to 1800, meat consumption in Europe was mainly a matter of privilege. (...) In a single sitting on 27 June 1749, Frederick the Great of Prussia and his guests consumed chicken, beef, lamb, sausage, veal, mutton, goose, and turkey, a greater variety of meat than most eighteenth-century peasants ate in a lifetime' (Knapp, 1997: 542).

The historical increase in meat consumption in the 19th and 20th century was made possible by agricultural, industrial and logistical improvements. The increase was motivated by the social prestige of eating meat, but also by the socially and medically motivated political movements of providing every citizen with a healthy diet. The importance of meat as a vital source of nitrogen/protein was acknowledged in the 19th century, and made the provision of meat to the population a basic constituent of public health policy for governments in Europe (Franklin, 1999: 4).

Meat and modernity

The relationship between humans and animals is described in the Bible in Genesis 1:28: 'Rule over the fish in the sea and the birds in the sky and over every living creature that moves on the ground.' In Genesis 9:3, the legitimacy of killing and eating animals is given: 'Everything that lives and moves about will be food for you. Just as I gave you the green plants, I now give you everything' (Bible, 2011).

In ancient times, the ideal of food ethics was temperance. Food intake should be according to bodily needs, in harmony with the natural order. In medieval times the ethics of food was part of the purification of the soul, with opposing means of purification through either ascetic fasting or gluttonous eating to numb the bodily desires (Zwart, 2000). The relationship to animals as food was rather uncomplicated. As described in the Bible, humans and animals share a moral universe created by God. Animals are created to be ruled by and eaten by humans.

In medieval times, animals were slaughtered by the butcher in the middle of town. But from the 19th century onwards, the *abattoirs* (from fr. *abattre* – to fell trees) were introduced in order to 'have slaughtering performed in municipal establishments built far from urban centers, in other words the *dissociation of slaughtering and butchery*. This was the measure that "cleared" the butcher and made him "innocent"' (Vialles, 1994: 17). Noilie Vialles describes the way euphemistic metaphors of harvesting ('felling trees') are used in the slaughtering process. Moreover, the slaughtering process is increasingly divided into steps of stunning and bleeding that obscures the process of killing an animal. The result of the dividing process is that no person clearly has killed the animal to be eaten – it is harvested.

This complication of the relationship between humans and animals as food appears already in the 18th century. The change is visible in the Norbert Elias' description of the norms of carving meat on the table: 'From a standard of feeling by which the sight and carving of a dead animal on the table are actually experienced as pleasurable, or at least not at all unpleasant, the development leads to another standard by which reminders that the meat dish has something to do with the killing of an animal are avoided to the utmost' (Elias, 2000: 102). The relation between the animal as a fellow being and a source of food should be obscured. To highlight the fact that eating meat presupposes the killing and cutting of a dead animal is no longer a part of the pleasure of meat eating.

The evolving moral obligation of dissociation of meat from animal, as exemplified by Elias and Vialles, can be seen as part of a first few steps of the modern project of separating the natural from the societal. An illustrative description of this process is given by Syse and Bjørkdahl in a comparison of two different editions of a standard Norwegian cookbook: *Gyldendals store kokebok* from 1955 and 2002. While the 1955 edition starts the chapter on preparation of meat with illustrations and descriptions of animals, the 2002 edition starts with illustrations and descriptions of pieces of meat (Syse, 2012).

Rather than being part of a common divine moral universe, animals are gradually from the 18th century on objects of a realm of nature, while humans are part of a qualitatively different realm of society. Humans alone are sensitive to the moral dimension of the social realm. In order to distinguish humans from animals, humans should treat animals in a humane and civilized way, not in the brutal way animal predators relate to their prey. The rupture between human and animal nature should be made clear in keeping the animals in separate places, slaughtering animals in separate places, and making animals edible in separate places.

The introduction of a modern scientific perspective on food further transforms the eating of animals into consumption of an object of nature. In the 19th century, the question of eating meat is understood in the novel language of *nutrition*: Calculating the optimal intake of calories and amino acids for human bodies.

As Hub Zwart observes, the modern scientific approach to food represents a return to the ancient ideal of temperance, but in a novel form: 'Rather than on subjective experience, contemporary dietetics relies on exact measurement (weight watching) as well as on labeling practices, informing food consumers about the ingredients and components of food products' (Zwart, 2000: 123). One might view the modern approach as putting the ancient end of harmony with the natural order to work at pursuing the further medieval end of setting the mind free from bodily failure. In such a perspective, food consumption is basically a question of preventive medicine.

Gnawing doubt

The double separation of animals from humans and meat from animals has in the 20th century been accompanied by a growing concern with animal welfare. Adrian Franklin and Robert White emphasizes four aspects of the growing concern for animal welfare in the 20th century (Franklin and White, 2001: 223-224): *Firstly*, hunted animals are at risk of extinction and domestic animals are threatened by untamed market structures leading to illegitimate living conditions at factory farms hard pressed for profit. *Secondly*, a growing misanthropy is based on human treatment of animals and a growing concern for a lack of sustainability in our ways of dealing with the environment. *Thirdly*, humans establish more close and strong relationships with animals in the role of pets – blurring the line between humans and animals as companions. *Lastly*, Franklin and White observe a totemization of and emotional engagement with wild animals.

These new views and valuations on animals and meat has taken place alongside an exponential growth in the world's population and a steady growth in the meat consumption per capita, that has led to an enormous growth in the world's population of livestock. The parallel processes of separating meat from animals on the one hand, and a growing concern for animal welfare on the other, has led to a possible tension in our relation to eating meat.

This tension is described by Carolien Hoogland and colleagues in this way:

> When combined, these two trends may interact to allow people to consume in ways that actually conflict with their personal values. As a result of the dissociation of the animal from the food that is consumed, people's buying behavior may unintentionally provide incentives for activities which they actually dislike – such as factory farming – because their concern for animal welfare does not translate into corresponding food choices (Hoogland *et al.*, 2005: 16).

Such a psychological ambivalence can result in an unresolved ethical dilemma for modern carnivores: Meat is nutritionally beneficial and tasteful, but comes with an unpleasant ethical after-taste. The promise of cultured meat is to remove such a gnawing moral doubt by means of technological innovation.

Cultured meat

In regenerative medicine, an ability to cultivate any replacement tissue and organs *in vitro* would create a new medical and ethical situation for organ replacement in humans. Shortage of genetically and ethically available replacement organs would no longer be the main issue for helping patients with organ failure.

In analogy with this, the introduction of cultivated muscle tissue from non-human species in the form of *in vitro* meat promises a new situation in meat production. Cultured meat promises a solution to the main ethical, existential and political problems of factory farming: Animal welfare issues, environmental concerns, and global access to meat products (Hopkins and Dacey, 2008). The current technology of in vitro meat attempts to make stem cells grow into muscle cells in a bioreactor. Using scaffolding or

3D printing techniques, the aim of in vitro tissue engineering is to be able to reproduce the structure of muscle tissue quite indistinguishable from muscle tissue from a slaughtered animal.

If cultured meat technology is successful in accomplishing this aim, is scalable, and makes it possible to produce globally affordable meat, then the potential befits are huge. The environmental advantages of cultured meat compared to livestock is in a life cycle assessment by Tuomisto and Teixeira de Mattos estimated to be:

> In comparison to conventionally produced European meat, cultured meat involves approximately 7-45% lower energy use (only poultry has lower energy use), 78-96% lower GHG emissions, 99% lower land use, and 82-96% lower water use depending on the product compared. Despite high uncertainty, it is concluded that the overall environmental impacts of cultured meat production are substantially lower than those of conventionally produced meat (Tuomisto and Teixeira de Mattos, 2011).

As cultured meat just involves the use of stem cells from animals, no animals needs to be kept and slaughtered for their meat. The grounds for concern for animal welfare for livestock for the modern carnivore are thus dissolved.

In the grocery store, fruits and vegetables are often presented very much like they did at the field. Herbs and lettuce are sold while still growing in pots. Farm animals, on the other hand, should be transformed into meat products in a way that appropriately mask their origin. Chickens are not sold alive in supermarkets, and the link between the bacon and the pig is weak or absent. The ultimate masking of the animal origin of meat would be cultured meat. The introduction of cultured meat would liberate us from the unfortunate need to slaughter animals in order to survive (Welin and Van der Weele, 2012). Meat could be produced like bread in the in shop bakery, or beer at the microbrewery – or even like art (Driessen and Korthals, 2012). In line with this perspective, one of the leading organizations promoting research on the development of in vitro meat is characteristically called New Harvest (www.new-harvest.org). Harvesting meat has finally become like felling trees.

Technological quick fix

Critics of cultured meat argue that the promotion of cultured meat rather would be restricted to a mitigation of some unpleasant symptoms of the flawed modernist way of relating to nature, animals and food. The rupture between animals as fellow creatures versus animals as food should be seen as contingent and unnecessary. In this view, we should strive to clarify the metamorphosis of an animal into food. We should make the travel from farm to fork traceable and visible. It is deeply problematic to aim for an escape from our past and traditions in hunting and farming animals for their meat.(Fairlie, 2010: 231) We should rather be concerned with and proud of our place in nature as bodily beings in complex interaction with other species, and reflect on the proper – limited – scope of justice in nature. We should make the travel from farm to fork traceable and visible, and aim to give the consumer of meat products a richer experience of the origin of the products.

One could argue with Jacob Metcalf that cultured meat is the moral and nutritional dream of modernism, utilitarianism, sentienism and nutritionalism: 'Like many biotechnologies, cultured meat thrives in a discourse of inevitability and requires the speculative horizon of biotech ethics as much as it requires nutrient solutions.'(Metcalf, 2013: 78) Cultured meat would primarily act to further untie the crucial cultural bond between man and nature that separates food from 'foodlike substances', reducing food to nutrition:

> What cultured meat disrespects is not 'nature' or the 'inherent dignity' of animals, but the thick, intra-active, historical, material and discursive relating that sustain ecologies of food. As problematic as

> those relating are in their current instantiations, cultured meat and other promising biotechnologies seek to solve those problems by making them ever more invisible. It is the de-worlding of the relations that sustain human and non-human life alike that is disconcerting, not the technology itself (Metcalf, 2013: 83).

In a qualitative study of the views on Norwegian consumers on the status of animals, Maria Guzman, the informants reject a strict separation of humans and animals (Jacobsen *et al.*, 2003). Humans and animals are seen to belong to separate spheres – the social/moral and the natural – but animals are nevertheless viewed as subjects, and moreover as subjects with a similar physical and psychological structure as humans. Thus the ontological status of animals is neither that of a natural object or a social subject, but of a *natural subject*. Guzman's informants thus steer clear of anthromorphizing animals on the one hand, and an objectification of animals on the other.

To uphold the special ontological status of animals could be a way to resolve the gnawing doubt of the meat eaters. In order to be a satisfactory solution it would, however, entail quite high standards on the living conditions of animals. Champions of cultured meat argues that such standards are unattainable, given the current global demand for meat (Welin and Gold, 2012). The technology of cultured meat, if successful, could make the division between animals as companions and as a source of food a practical, not just a psychological, reality.

References

Bible (2011). New international version. International Bible Society 775, 42.

Driessen, C. and Korthals, M. (2012). Pig towers and in vitro meat: disclosing moral worlds by design. Social Studies of Science 42(6): 797-820.

Elias, N. (2000). The civilizing process: sociogenetic and psychogenetic investigations: Wiley-Blackwell, West Sussex, UK.

Fairlie, S. (2010). Meat: a benign extravagance: Chelsea Green Publishing Company, White River Jct., VT, USA.

FAO (2012). Food outlook November 2012. Available at: http://www.fao.org/docrep/016/al993e/al993e00.pdf.

FAOSTAT (2010). Available at: http://faostat3.fao.org/home/index.html.

Franklin, A. (1999). Animals and modern cultures: a sociology of human-animal relations in modernity: SAGE Publications, Thousand Oaks, CA, USA.

Franklin, A. and White, R. (2001). Animals and modernity: changing human–animal relations, 1949-98. Journal of sociology 37(3): 219-238.

Hoogland, C.T., De Boer, J. and Boersema, J.J. (2005). Transparency of the meat chain in the light of food culture and history. Appetite 45(1): 15-23.

Hopkins, P.D. and Dacey, A. (2008). Vegetarian meat: could technology save animals and satisfy meat eaters? Journal of Agricultural and Environmental Ethics 21(6): 579-596.

Jacobsen, E., Almås, R. and Johnsen, J.P. (2003). Den politiserte maten: Abstrakt forlag.

Knapp, V.J. (1997). The democratization of meat and protein in late eighteenth and nineteeth century Europe. Historian 59(3): 541-551.

Metcalf, J. (2013). Meet shmeat: food system ethics, biotechnology and re-worlding technoscience. Parallax 19(1): 74-87.

Syse, K. and Bjørdahl, K. (2012). Dyret i kogebogen. In: Gjerris, M. (ed.) Kød: en antologi.

Tuomisto, H.L. and Teixeira de Mattos, M.J. (2011). Environmental impacts of cultured meat production. Environmental science & technology 45(14): 6117-6123.

Vialles, N. (1994). Animal to edible: Cambridge University Press, Cambridge, UK.

Welin, S. and Gold, J. (2012). Kød uden dyr? In Kød: En antologi.

Welin, S. and Van der Weele, C. (2012). Cultured meat: will it separate us from nature? Climate Change and Sustainable Development 348.

Zwart, H. (2000). A short history of food ethics. Journal of Agricultural and Environmental Ethics 12(2): 113-126.

Innovation and recognition of food and farming styles

M. Korthals
Applied Philosophy, Wageningen University, Hollandseweg 1, 6706 KN Wageningen, the Netherlands; michiel.korthals@wur.nl

Abstract

The 'productionist paradigm' in agriculture emphasizes high tech inputs and high material output. Its assumed main ethical value is the concept of justice as fair distribution of outputs, with the main aim to reduce hunger and to optimalize sustainability. Although this paradigm can be proud of its enormous success to feed the world, it also produces huge problems, which make it unsustainable and not fair. The approach decreases biodiversity, increases erosion, is animal unfriendly and depletes scarce resources. It has an huge impact on climate change and its economic and patenting system increases the gap between rich and poor, thereby reducing chances for rural livelihood, producing unemployment, inherently instable social relations and food insecurity. Probably, it can produce food for even more than eight billion in 2050 with enormous cost. The agrosystem approach, its alternative, starts with local farming practices and innovation activities. It implicitly acknowledges the ethical value of how to lead a good life. However, variants of this approach either give farmers full priority in what and how to produce (as in agrarianism) or give consumers the lead (as in urban gardening). The current utilitarian and deontological ethical approaches are quite blind for how to deal with these strategies that cut across justice, good life and other values. Due to the dynamics of social, economic and cultural systems, agricultural priorities cannot anymore be established one sided by either farmers (and producers) or consumers. Sustainability and climate change require changing consumer food styles (in expanding urban areas) and new farming styles. To improve the approach of agrosystem into one of plural and democratic agrosystems, I propose to make room conceptually for the negotiation and cooperation of farmers and consumers about their interests, preferences and values. In this revised agrosystem approach a concept of justice is assumed not as fair distribution but as fair recognition of farming and food styles. This ethical notion enables to take into account the pluralism of farming and food styles and various interactive ways of intensification and innovation with consumers.

Keywords: productionism, agrarianism, types of intensification, food democracy

Introduction

Not only do we have to prepare to feed 9 billion people in 2050, in 2075 we still have to produce enough nutritious and adequate food for them. And what will be the world population in 2100? So which (land based) agricultural practices might succeed in meeting the future societal demands? How to make informed and sensible guesses about best practices in agriculture? With 'best practices' are meant integrative approaches, that take into account humans, nature and environment, productivity, economics, ethics, culture and politics. 'Best practices' of two different forms of agriculture will be here discussed and evaluated on a number of decisive indicators like their effect on soil quality, on biodiversity, ecosystem services and (long term) sustainability, on employment and rural liveability, on consumer friendliness and global estimations of productivity. I have chosen for these indicators, because more abstract ethical theories like utilitarian and deontological theories are not really helpful in making choices for one or another system of production and consumption. They start with abstract principles and not with practices of people be it farming or cooking and consuming (food styles), in which ideas of a good life and of justice are incorporated.

In this paper I will first develop ethical indicators to assess different agricultural systems. Next I will analyze and assess the current 'productionist paradigm' of agriculture with its high tech inputs and high outputs and its alternative, often called Conservation Agriculture or Agrosystem approach. Although the first paradigm can be proud of its enormous success to feed the world, it also produces huge problems, which make it unsustainable and socially not fair. The second has a positive sustainability record, but is very much labour (farmer) oriented. Both approaches fail to take into account the wider social-cultural context in which agrifood sectors are playing a role. I will finally develop some ethical ideas which make it necessary to relativize the claims of these two approaches and provide some thoughts how they can co-exist and develop fruitfully together. My main thrust is that in order to produce sustainable good food for all, we need pluralist approaches that take seriously the best practices of small and medium farmers and innovators.

Ethics and agriculture

It is a long standing criticism of people involved in agrofood that environmental philosophy is not the most fruitful discipline to tackle problems of agriculture and food (Korthals, 2004). With its focus on wild nature, and its plea for a non-interfering role of humans, environmental philosophy is not an appropriate starting point to understand what agriculture means and can mean. The ethical theories of famous ethicists like Singer, Rawls or Pogge mostly refrain from a detailed analyses of the place of agriculture and food in the good life of humans. Their attention to agro food is directed only by the question of food security and fair distribution of food in the light of global justice. Hunger and malnutrition, indeed the two most severe problems of agrofood are tackled by the utilitarian Peter Singer with the construction of a global government; the Rawlsian-Kantian solution of Pogge is concentrating on global institutions. With respect to Western food consumers they mostly appeal to change wasteful food styles towards health and soberness. The main problems of these top down global ethics proposals is their lack of attention for the dynamics of farming, agrofood technologies and the wider social cultural context (urbanisation) that is necessary to take into account when agricultural practices and food styles are confronted by the challenge to construct a more sustainable and just food system. Moreover, the ethical dilemmas of environment, poverty trap mechanisms and agency and farming styles of poor people in the food producing developing world is not taken into account. I will therefore use not only justice and good life criteria but also criteria directly distracted from agricultural practices, like the effect on soil quality, on biodiversity, ecosystem services, and (long term) sustainability, on employment and rural liveability, and global assessments of productivity; which means, I will give prima facie recognition to food and farming styles.

High tech intensive agriculture and its problems

The current agricultural regime with high (artificial and natural) inputs and high outputs are seen by many as unsustainable; it decreases (agro-)biodiversity, it increases erosion, it is animal unfriendly and it depletes scarce resources. It has a huge impact on climate change and its economic and patenting system increases the gap between rich and poor, thereby producing inherently instable social relations and enhancing food insecurity. Probably, it can produce food for even more than eight billion in 2050 but the cost will be enormously. The Netherlands is a good example of intensive farming, less than 2% of the labour force is working in agriculture and its yearly use of chemicals is one of the biggest in the world per hectare.

The 'productionist paradigm' in agriculture emphasizes high technological inputs and high outputs. The ethical value it assumes is the concept of justice as fair distribution of outputs (yields), with the aim to reduce hunger and to optimalize sustainability (Borlaug, 2000; Paarlberg, 2009). Probably, it can produce food for even more than eight billion in 2050 however, the problems it faces seem to be enormous (Lang, 2009; Roberts, 2008). On the indicator global food security this approach therefore scores well. However with respect to the other indicators, this approach scores dramatically low. The approach decreases biodiversity, increases erosion, is animal unfriendly and depletes scarce resources. It has an huge impact on climate change and its economic and patenting system increases the gap between rich and poor (Glenna, 2009), thereby reducing chances for rural livelihood, producing unemployment, inherently instable social relations and food insecurity.

Regarding the first and second indicators, the use of chemicals implies decreasing of (agro)-biodiversity due to current agricultural system firstly implies radical decline of insects like bees, and of birds, distortions of food chains and water pollution (Stevens, 2004; Swaay, 2008). Biotechnologically driven breeding results in reduction of alleles and the establishment of monocultures. Wheat now covers not more than five variants from the 5,000 original ones; the result is increasing vulnerability to pests and other negative effects on yields, which in turn requires more chemicals to destroy these pests. The use of mineral oil in producing chemicals for fertilizers and pesticides is large (in US app. 37% of the energy used in agriculture), which contributes to global warming. The use of other minerals (phosphate) is probably already on peak level. The pollution of surface and groundwater due the use of these chemicals is still increasing (OECD, 2008, Environmental Performance of Agriculture in OECD Countries since 1990, Paris.)

Regarding the other indicators, the need to constantly increase investments in newer chemicals, equipment and (patented) seeds puts farmers in a treadmill of up scaling (larger farms, larger monoculture plots), which implies more dependency on loans and the world market, which is since a few years quite a volatile structure. Intellectual property rights that make seeds dependent on licensing policies of a few large seed companies increase the burden of debts of farmers. This economic treadmill (Cochrane, 1993) drives smaller farmers out, and makes the larger ones more vulnerable to economic up and downs; quite an uncertain factor in food security. The tendency to extradite farmers from land increases the gap between rich and poor (Glenna, 2009), thereby reducing chances for rural livelihood, producing unemployment (World Bank, 2013), inherently instable social relations and food insecurity (Clover, 2010). Extradited farmers are moving towards to the slums of cities and cannot find jobs, caught in the urbanization trap.

One of the main assets of this paradigm is the claim that intensification of farming sets free more 'wild nature'. Originally, Borlaug (2000) invented this claim of land sparing instead of land sharing and it is repeated for many years by Rabbinge and others. However, it is a claim that totally refrains from ecosystem dynamics and from social structures. Flora and fauna are affected by the intensification with chemicals and genetic modified seeds on intensified plots. Monarch Butterfly, one of the most important butterflies in the US is according to NYT (March 13, 2013) rapidly declining because its feed is targeted by GM-maize. The consequence is that birds predating on this butterfly also are reduced in number. Water and air pollution doesn't stop with the fences of agricultural areas. Besides the neglect of ecosystem interaction, the claim also neglects social dynamics. Social science research makes it clear that this type of intensive farming, due to the treadmill of increasing investments and decreasing profits, is expanding in wild nature. Studies analyzed land expanding trends in various countries and couldn't identify decrease of cropland or stablization in countries where fast rising yields take places. In the study of Ewers (2009) the same conclusion is drawn: 'land sparing is a weak process that only occurs under a limited set of circumstances.' On most indicators, current intensive farming scores very low or even negative.

Agrosystem and farmers

The approach of agrosystem, its alternative, starts with local farming practices (Critchtley, 2010). Rotation, intercropping and agrodiversity play a big role to reduce risks of the failure of one crop in hard times of drought or heat. In most developing countries this type of farming is done by small and medium farmers. They don't have the money for chemicals or patented seed. It is a form of integrated management of biological resources, water, some external outputs and labour. The surplus of their harvest is sold on local markets, and the family life and the community they live in supplies them with assistance if they need this. In many cases these small farmers strive for increase of production, but when there is no access to a larger market, it is useless, or when the surplus is directly destined for relatives and friends, be it in a friendly way or in a hostile way. Mostly also, all types of plants, bushes, trees, and animals are integrated in farming activities; animals are kept not only for their milk or meat, but also for their manure and energy; the role of trees in adding nutrients in the soil is recognized, water, mostly very scarce is, is used very careful. Intensive human labour is treated as essential, in transport, sowing, weeding, harvesting and storing. Agrosystem farming is everywhere different: so there are many variants, depending on local circumstances; this is a big difference with intensive farming.

Agrosystem as agroecology is intensifying these interconnections, by taking into account all these circuits (ecosystem services), selected technologies and aims at revitalization of small farms (Altieri, 2011; Davis, 2012). In the theoretical agronomic literature and broader, sociological and philosophical reflections on this type of farming, this approach is called agrarianism. In this literature all emphasis is put on farmers; they are the labourers of the land, and determine what crops and animals are to be selected, which means what type of food is offered to farmers' family and, when there is a local market, to the community. As a matter of fact, exchange of products can take place, and so farmers can enrich their diet. But the farmer orientation is crucial. Agrarianism has a got a political expression in food sovereignty movements like La Via Campesina (Wittman, 2009). But in its more philosophical version, the virtue of agrarian life is stressed, with its direct contact with natural processes, its community ties, its time horizon and its low ecological foot print. Berry (2010) argues that these farmers really care for the earth; he distrust a farmer when he uses a computer, so modern technology should only be used scarcely (or least, technology that is visible as modern technology; 'traditional' technology is often not any more visible as technology). Paul Thompson has written a large book, *The agrarian vision* (2010) in which he makes a plea for this approach. A different variant is that of urban community gardening that give consumers the lead (as in urban gardening), however, it remains to be seen in how far farming in urban cultures is a sustainable (in the social and the environmental sense) solution.

Assessing this type of farming on the indicators of sustainability, biodiversity, employment and ecological footprint, gives only positive marks (De Schutter, 2009). However, what about the emphasis on the often back bone breaking physical labour, the steering of the farmer, the disregard of new urban life styles (consumers), social contexts needed for innovation (Hounnoukou, 2012), and the sometimes low productivity? The ideal of the good life that is envisioned here is quite narrowly restricted to a community, or even a family farm, and doesn't consider broader social contexts, for example that more than 50% of the global population lives in cities and that even small farmers want at least some modern technologies that can make their daily toil decrease and improve their products. I am not arguing for existing, often very unsustainable, life styles of consumers, but for a prima facie recognition of their daily concerns with having a job, housekeeping and dealing with relationships. All in all, the social and cultural blindness that rightly can be identified in the current high tech system, is also partly present in this type of approach.

Towards a pluralism of food and farming styles

Both approaches fail to take into account the divergent and changing preferences and cultural concerns of urban populations and the wider social-cultural context in which the agrifood sectors of our globalizing world are playing a role. Current intensive agriculture in general is not viable anymore, but the alternative is not well adapted to new circumstance, e.g. that the majority of people now live in cities, and that food choices are rapidly changing, partly due to globalization and changing households.

In my view these two approaches are in need of severe and structural improvements, and they can learn a lot of each other, but both lack a consideration of opportunities for new relationships between farmers and consumers. The second approach has as a big advance due to its recognition of small and medium famers, but lacks the recognition of food styles of consumers, and the first is unsustainable. It is therefore necessary to develop a framework to relativize the claims of these two approaches and provide some ideas how they can co-exist fruitfully together and simultaneously be made responsive to consumers food styles and these changing contexts (including science and technology dynamics). My main thrust is here that in order to produce good food for enough people, we need pluralist approaches that take seriously the best practices of small and medium farmers and innovators, that comply to the above mentioned indicators.

Due to the dynamics of social, economic and cultural systems, agricultural priorities cannot anymore be established one sided by either farmers and producers or consumers. Sustainability and climate change require changing consumer food styles (in expanding urban areas) and new farming styles, which however cannot be dictated by the management of food industries, research labs or farmer organizations. To improve the approach of agrosystem into one of plural and democratic agrosystems, one must make room conceptually for the negotiation and cooperation of farmers and consumers about their interests, preferences and values. In this revised agrosystem approach a concept of justice is assumed not as fair distribution but as fair *recognition of farming and food styles*. This ethical notion enables to take into account the pluralism of farming and food styles and their various interactive ways of intensification and innovation with consumers.

Conclusion

I have firstly developed some relevant indicators to asses two different agricultural sustems, Sustainability but also liveability, employment and considerations of urban population played here a role. Social Justice not the only principle, and distribution of food and food security is not the only principle for agriculture and food; respecting the values people have concerning farming and food styles is of equal importance. Next I analyzed the current 'productionist paradigm' of agriculture with its high inputs and high outputs and its alternative, often called Conservation Agriculture or Agrosystem approach. Although the first paradigm can be proud of its enormous success to feed the world, it also produces huge problems, which make it unsustainable and socially not fair. The second has a positive sustainability record, but is very much hard labour intensive and farmer oriented. Both approaches fail to take into account the wider social-cultural context in which agrifood sectors are playing a role. I finally developed a framework to relativize the claims of these two approaches and provide some ideas hoe they can co-exist fruitfully together and can be improved in the direction of a more democratic system, that recognizes both farming styles and food styles. My main thrust is here that in order to produce good food for enough people, we need pluralist approaches that take seriously the best practices of small and medium farmers and innovators. Let's take into account diversity, complexity, differences, dynamics and provisionality. The global agrofood ethics from below I envision, considers networks and studies cases to learn from comparing narratives about what can work and what not in a certain area. This narrative benchmarking of good practices of agrosystems is however only fruitful when deficiencies are targeted, like urban and social blindness and the neglect of the hard toil many small farmers nowadays experience. It needs an innovative platform for *continuously* adapting ethics and technologies and correspondingly narrative benchmarkings of practices for an agrofood sector that recognizes both the work of farmers and the food styles of consumers.

References

Berry, W. (2010). What Matters? Counterpoint, Berkeley, CA, USA.

Borlaug, N.E. (2000). Ending world hunger: the promise of biotechnology and the threat of antiscience zealotry. Plant Physiology 124: 487-490.

Clover, D. (2010). Exploring the resilience of Bt cotton's 'pro-poor success story. Development and Change 41(6): 955-981.

Cochrane, W. (1993). The development of American agriculture. University of Minnesota Press, Minneapolis, MN, USA.

Critchley, W. (2010). More people, more trees. Environmental recovery in Africa. Practical Action Publishing Ltd. Warwickshire, UK.

Davis, A.S., Hill, J.D., Chase, C.A., Johanns, A.M. and Liebman, M. (2012). Increasing cropping system diversity balances productivity, profitability and environmental health. PLoS ONE 7(10): e47149. doi:10.1371/journal.pone.0047149.

De Schutter, O. (2009). Seed policies and the right to food: enhancing agrobiodiversity and encouraging innovation. Report presented to the UN General Assembly (64th session) (UN doc. A/64/170).

Ewers, R.M., Scharlemann, J.P.W., Balmford, A. and Green, R.E. (2009). Do increases in agricultural yield spare land for nature? Global Change Biology 15: 1716-1726.

Glenna, L.L. and Cahoy, D.R. (2009). Agribusiness concentration, intellectual property, and the prospects for rural economic benefits from the emerging biofuel economy. Southern Rural Sociology 24(2): 111-129.

Hounkonnou, D., Kossou, D., Kuyper, T.W., Leeuwis, C., Nederlof, E.S., Röling, N., Sakyi-Dawson, O., Traoré, M. and Van Huis, A. (2012). An innovation systems approach to institutional change: smallholder development in West Africa. Agricultural Systems 108: 74-83.

International Assessment of Agricultural Knowledge, Science and Technology for Development (IAASTD) (2008). Global summary for decision makers. Available at www.unep.org/dewa/agassessment/docs/IAASTD_GLOBAL_SDM_JAN_2008.pdf.

Korthals, M. (2010). Global justice and genomics: toward global agro-genomics agency. Genomics, Society and Policy 6: 13-25.

Korthals, M., (2004). Before dinner. Springer, Dordrecht, the Netherlands.

Lang, T., Barling D. and Caraher, M. (2009). Food policy. Oxford University Press, Oxford, UK.

Paarlberg, R. (2009). The ethics of modern agriculture. Society 46: 4-8.

Roberts, P. (2008). The end of food. Bloomsbury, London, UK.

Stevens, C.J., Dise, N.B., Mountford, J.O. and Gowing, D.J. (2004). Impact of nitrogen disposition on the species richness of grasslands. Science 303: 1879-1875.

Thompson, P. (2010). The agrarian vision, University Press of Kentucky, Lexington KY, USA.

Van Swaay, C.A.M., Nowicki, P., Settele, J. and Van Strien, A.J. (2008). Butterfly monitoring in Europe: methods, applications and perspectives. Biodiversity and Conservation 17: 3455-3469.

Wittman, H. (2009). Reworking the metabolic rift: Via Campesina, agrarian citizenship, and food sovereignty. Journal of Peasant Studies 36(4): 805-826.

World Bank (2013). World development report 2013. World Bank Washington, DC, USA.

Large scale insect rearing and animal welfare

D.M. De Goede[1], J. Erens[1], E. Kapsomenou[1] and M. Peters[2]
[1]Wageningen University, Sub-department Communication, Philosophy and Technology, Hollandseweg 1, 6706 KN Wageningen, the Netherlands; [2]Vereniging Nederlandse Insecten Kwekers (VENIK), Helvoirtsestraat 13, 5268 BA Helvoirt, the Netherlands; douwe.degoede@wur.nl

Abstract

In Europe, the intentional cultivation of insects for feed or food has long been uncommon. Recently however, a trend of insect consumption as an environmentally friendly alternative to the consumption of traditional animal livestock emerges, most notably in the Netherlands. Possible benefits include recycling of organic waste into biomass of high nutritional value, lower ecological footprint, and increased food security. This paper discusses the potential of large scale insect rearing in the Netherlands in relation to the new Dutch Wet Dieren (Animal Act). Current legislation on welfare of production animals typically refers to Brambell's five freedoms, and explicitly excludes invertebrates. In the absence of specific welfare regulations for insects, and through trial and error, pioneering insect farmers have individually developed best production practices for their niche markets and insect farmers are hesitant to share their expertise when independent authorities and welfare regulations are not available yet. The Animal Act not only includes several insects as production animals, but also asks farmers to respect the intrinsic value of the insects in terms of Brambell's five freedoms. In our paper we address the question if and how the existing framework of Brambell's five freedoms can be applied to measure insect welfare. We observe that scientific information about insect welfare and used rearing methods is limited. We argue that an explicit declaration of applicability of Brambell's framework to insects is perceived as a scientific problem, but that it also suggests a deliberate moral extension, leading to the inclusion of insects as a category of moral objects in Dutch animal welfare legislation.

Introduction

The enormous biodiversity within the taxonomical class of Insecta provides a huge potential for human utilizations, such as honey and raw silk production, as natural enemies in pest control, medical applications and as a source of animal protein (Ramos-Elorduy, 1997; Ramos-Elorduy *et al.*, 2002; Wang *et al.*, 2006). To increase sustainability of livestock production systems, it is argued that the latter has high potential, because high quality protein sources for animal feed are becoming scarcer and prices of raw animal protein continue to increase. The so called 'Commissie-Van Doorn', a think tank commissioned by the Dutch Provincial Excecutive (Provinciale Staten) to develop a long term strategy for Dutch animal husbandry, recommended to aim for at least 50% European produced high-protein animal feed by 2020. Utilizing new protein sources such as insects may be necessary to achieve this target (Van Doorn *et al.*, 2011). FAO (2012) as well has insisted on developing large scale insect rearing to produce sustainable, alternative protein sources. Concurrently, the development of large scale insect rearing for animal feed may create chances to work out a sustainable alternative for livestock breeding: the intentional cultivation of insects for human consumption. Ramos-Elorduy (2009) made mention of a vast range of around two thousand recorded edible insect species. Indeed, as a source of proteins, vitamins and minerals, insects have constituted, and in many countries still constitute part of the human diet. Nonetheless, in western societies, insects are nowadays generally not conceived as a conventional source of food, and insect consumption is rather regarded as primitive and distasteful. As a new technology, insect consumption also leads to fear and resistance, e.g. because insect derived proteins are considered foreign proteins, or because insects are strongly associated with inconveniences. The western stigma surrounding entomophagy – human consumption of insects – is however increasingly being challenged, remarkably in the western world itself, as awareness of the potential ecological, nutritional and economic benefits that insects have to offer grows (DeFoliart, 1999; Oonincx and de Boer, 2012; Oonincx *et al.*, 2010). Some of the most relevant arguments are:

1. Bioconversion. Insect rearing can be realized on residual waste streams, including manure of poultry, swine and cow as well as fish waste and other low-grade biomass not suitable for human consumption (e.g. Ocio *et al.*, 1979; Ramos-Elorduy *et al.*, 2002, Rumpold and Schlüter, 2013). By recycling organic waste into biomass of high nutritional value insects can play a role in 'closing the loop' in the food production chain.
2. High feed conversion efficiency. Due to their poikilothermic nature, insects can allocate a large portion of their food into biomass increase, with a relatively low energy input (Nakagaki and Defoliart, 1991; Van Huis, 2013).
3. Low space requirement. Land use in insect rearing is far less than that in the livestock industry (Oonincx and de Boer, 2012) and can easily be realized on multiple floors piled on top of each other (Veldkamp *et al.*, 2012).
4. Low emissions. Both greenhouse gasses and ammonia emissions of commercially reared insects are relatively low compared to conventional livestock (Oonincx *et al.*, 2010; Rumpold and Schlüter, 2013). It is also assumed that insect rearing has a relatively low water-footprint, but this has not yet been confirmed scientifically.
5. Minimal food safety risk. Due to the large phylogenetic distance between insects and humans and mammals it is argued that the risks of transmission of diseases from insects to humans are very low (Van Huis, 2013).
6. Biodiversity protection. Insects can be used as a substitute for fishmeal and it has been argued that this could reduce pressure on other species (Rumpold and Schlüter, 2013).

The potential of large scale insect rearing in the Netherlands in relation to the recently enforced 'Wet Dieren' (hereafter called Animal Act) is unclear. In this paper we aim to list some challenges for developing large-scale insect rearing, and concentrate on the applicability of Brambell's five freedoms that are at the basis of the Animal Act to measure insect welfare.

Challenges for developing large-scale insect rearing

In the Netherlands, insect rearing probably started as a leisure activity to produce feed for reptiles, birds and other insectivores that are kept as household pet. Insect farmers supply private customers, pet shops and zoos. To facilitate this niche market, 29 insects were included on a 'positive list' of animals to be kept for production. Article 34 of the Dutch Animal Health and Welfare Act namely states that it is not allowed to keep animals for production purposes without explicit permission (Ministerie van Volksgezondheid, 1998). It is now argued that some of these insects show potential for large-scale rearing in the Netherlands, most notably yellow mealworm (*Tenebrio molitor*), lesser mealworm (*Alphitobius diaperinus*), superworm (*Zophobas morio*), migratory locust (*Locusta migratoria*), house cricket (*Acheta domesticus*), black soldier fly (*Hermetia illucens*), and house fly (*Musca domestica*). However, some obstacles would need to be removed in order to establish large-scale insect rearing.

In the first place, there is remarkably little scientific information available on how these insects should be reared. The techniques that the farmers currently employ have instead been developed through a process of 'trial and error' at farm level. This is mainly a scientific problem, although insect farmers are particularly hesitant to share their expertise because independent authorities are not available yet.

Secondly, the market for insect derived products as part of the food chain is restricted as a consequence of the European feed ban on processed animal protein (EC No 999/2001) that was enforced after the Transmissible Spongiform Encephalopathies (TSE) crisis in 1994. Recently Veldkamp *et al.* (2012) concluded that according to existing EU regulations EC No 1069/2009 and EU No 142/2011 it is prohibited to use insects in the food chain. Although reconsideration about the authorization of insect protein as ingredient in fish, pig and poultry diets by the European Commission (EC) is expected (Veldkamp *et al.*, 2012), potential investors exercise restraint. Likewise, the admission of insects as novel foods to the market is opaque (Van Wagenberg *et al.*, 2012).

Thirdly, insect farmers state that current animal welfare legislation is unclear with respect to insects. While European laws regulate multiple health and welfare aspects of animals in husbandry systems, most of these regulations explicitly exclude invertebrates, and specific legislation about insect rearing is missing. As a consequence, insect farmers operate in legislative grey zones. At the same time, public interest – although marginal – in the welfare of insects seems to appear because the public starts to associate large-scale insect breeding with factory farming and its associated animal welfare and public health issues. Dutch politician Marianne Thieme for instance questioned the public support of multiplying the number of animals being killed for the benefit of factory farming (Tweede Kamer, 2012).

The Dutch Animal Act – insect welfare related to Brambell's five freedoms

The Dutch Animal Act is a regulatory framework under development. It will in phases replace existing laws and regulations concerning animal health and animal welfare. At its basis are the so-called Brambell's Five Freedoms, formulated in 1965 by the UK Farm Animal Welfare Advisory Committee with respect to intensively farmed veal calves, pigs and chicken and gradually adopted in European legislation on animal welfare (Veissier *et al.*, 2008). Article 1.3.3 of the Dutch Animal Act (enforced on 01-01-2013) admits the Five Freedoms as a specification of the legal obligation to recognize and respect the intrinsic value of animals (Wet Dieren, 2013) Specifically, this implies that all production animals should be held free from:

1. thirst, hunger and inappropriate feed;
2. physic and physiological inconvenience;
3. pain injury and disease;
4. fear and chronic stress;
5. limitations of their natural behaviour.

Scientific information about farming of conventional livestock in relation to these points is abundant, and there seems to be general consensus about their relevance for sentient animals.

This is very different for insects. Scientific information about how insects should be reared is hardly available, let alone in relation to their welfare. For the sector this creates fear to harm the intrinsic value of insects, i.e. a fear of doing something morally wrong. The sector aims to develop welfare standards and meet welfare requirements, also in relation to increased consumer suspicion about the origin and production methods of their food. We will shortly discuss the five freedoms below, and formulate some suggestions for debate.

The freedom from thirst, hunger and inappropriate feed

The natural diet of insects can consist of multiple plant and animal sources, and many insects are largely polyphagous in their feeding behaviour. Insects can be reared on multiple plant and animal sources, such as grasses, grains, poultry waste, eggs and manure and also human food waste (Diener *et al.*, 2009; Ocio *et al.*, 1979). Studies of optimal diets for insects in breeding facilities are limited, but seem to suggest mixed diets (e.g. Cooper *et al.*, 2004). Although their feed provides already sufficient water to many species (most notably herbivores), insect farmers indicate that careful water provisioning is important to prevent dehydration in some species (e.g. *M. domestica*). It has also been suggested that low humidity negatively influences food conversion efficiency (Fraenkel and Blewett, 1944). The appropriateness of feed is likely to become an issue of scientific and moral debate, especially when waste disposal is directly connected to food production.

The freedom from physic and physiological inconvenience

It is expected that the provision of an appropriate rearing environment includes at least sufficient sheltering possibilities. These should be established to provide darkness and cover enough area to potentially shelter all individuals in the colony. This freedom closely relates to the freedom from limitations in natural behaviour as insects may be more inclined to hide during the day. Some commercially reared insect species are primarily nocturnal, and largely active at dusk and during the night, suggesting that a day/night rhythm should be provided. In addition, also the provisioning of an appropriate temperature range for the poikilothermic insects will for instance be of high influence for this freedom.

The freedom from pain, injury and disease

Although there is general consensus that insects are nociceptive (Tracey Jr. *et al.*, 2003), i.e. insects demonstrate a sensory perception of potentially damaging noxious stimuli, this does not necessarily imply the sensation of pain, which, according to the International Association for the Study of Pain, is always an emotional state of the brain. Indeed, the observation that insects do not exhibit protective behaviour towards damaged body parts has been regarded as evidence that insects do not feel pain. Illustrative examples include a locust which went on feeding while it was being eaten by a praying mantis and a tsetse fly that kept foraging after a large part of it has been removed by dissection (Eisemann *et al.*, 1984). There is however no consensus about pain perception in insects, and generalizations are most probably inaccurate.

As insects are primarily seen as pests, insect killing techniques mainly revolve around insecticides and methods of controlling harmful insects such as mosquitoes. In the context of insect rearing and insect welfare, little is known about humane killing methods. Because insects are poikilothermic, hypothermia or dry-freezing, i.e. a dehydration process which allows water to be extracted from the frozen insect when the surrounding pressure is reduced, is generally considered a natural and humane way to kill them. It is the main killing technique currently employed by insect rearing companies. Alternative methods include boiling, spraying with hot water after stunning with CO_2, mechanical crashing through centrifugation and shredding. Nonetheless, humane methods of killing insects have been under-researched and general guidelines are needed, especially in relation to insect welfare.

Some pathogens that frequently infest European commercial rearing facilities have been reported such as the *M. domestica* salivary gland hypertrophy virus (MdSGHV, see Lietze *et al.*, 2009)) and the *A. domesticus* densovirus (AdDNV, see Szelei *et al.*, 2011). However, underlying disease transmission mechanisms are not yet fully understood, and for other species such as *Z. morio* where infections have led to massive mortality, empirical information on the diseases, let alone potential treatment, is not even available. Farmers endorse that when facilities are infected, the virus spreads rapidly and causes high mortality rates. All in all, the low state of scientific knowledge suggests that it is practically infeasible to provide insects in rearing facilities the freedom from diseases.

The freedom from fear and chronic stress

Although it has been argued that many insect species possess a kind of believe-desire psychology that could make them appropriate objects of sympathy and moral concern (Carruthers, 2007), this does not necessarily mean that they can suffer mentally. It could however be investigated whether particular procedures used in insect rearing facilities are a potential source of stress for insects, e.g. sifting insects to separate them from their feeding substrate, and whether there are alternatives to these.

The freedom from limitations in natural behaviour

Approaching the natural conditions in breeding facilities might be the most reliable way to ensure that the reared insect species will reside under optimal conditions, thereby minimizing mortality and increasing productivity. Without trying to generalize, we observe that natural behaviour and life-history of some commercially reared insect species deviates from other production animals in several ways:

1. High density clustering. Insect species are often observed to aggregate in high population densities. For instance, larvae of T. militor are naturally attracted to live in close proximity with conspecifics (Weaver and McFarlane, 1989). M. domestica larvae are even thought to facilitate one another in digesting their feeding substrate (W. Jansen, personal communication). Insect farmers however consider the inclination to cluster at high density as one of the main limiting factors in keeping high mealworm numbers, because it easily leads to overheating. Studies on temperature effects on different insect species are therefore needed in mediating within this trade-off and infer optimal rearing densities for larvae (e.g. Salin *et al.*, 1999). In this respect, it could also be debated whether this freedom should relate to individuals, or populations.
2. Poikilothermic animals. An insects' body temperature is determined by the surrounding environmental temperature. Hence, a temperature gradient in insect enclosures could be provided to allow active thermoregulation of the species. This has also been shown to be beneficial in mitigating diseases (Ouedraogo *et al.*, 2004).
3. Metamorphosic life-stages. Different stages within the lifecycle of insect species (larval, nymph and adult) may require completely different rearing conditions. Unclear is whether Brambell's freedoms are equally valid to the different stages.
4. Migratory behaviour and gregarious state. Locust (L. migratoria) occurs in two distinct forms: a solitary phase and a gregarious phase, with completely different behavioural characteristics. They are permanently being reared in the gregarious phase as this phase is elicited by an increase in population density. Evidently, the naturally occurring extensive migratory and flying behaviour associated with this phase is impossible to allow fully whilst in captivity. It is unclear what the possible welfare effects are of artificially keeping of locusts in a permanent gregarious phase.
5. Short lifecycle. Most insect species have lifecycles that are significantly shorter than conventional livestock species. The housefly *M. domestica* for instance has a lifecycle of 14 to 18 days.

Insect biodiversity is too large to generalize upon welfare standards. Therefore, it seems that regulations implementing insects and their welfare are facing a comprehensive task in meeting the specific requirements of a multitude of species. It might therefore be advisable to specify a range of living conditions that meet insect welfare whilst also leaving room for the exploration of optimal rearing conditions. Especially when taking in mind the currently unstandardized methods among breeders. This needs, however, transparency from the breeders themselves to shed light on their current breeding methods in order to assess welfare and determine future rearing requirements.

Are the five freedoms a relevant concept for insect welfare?

In this paper we have until now mainly focussed on the issue of insect welfare as a scientific problem. It is understandable that insect farmers see the challenge to up-scale insect rearing in accordance to animal welfare regulations as a scientific problem. It is true that legislature aims to base its animal welfare regulations on scientific evidence, and that scientific knowledge about insect welfare is hardly available. Especially because large scale insect rearing is positioned as a sustainable alternative supply channel for animal protein, scientific evidence, as well as transparency, i.e. the degree of shared understanding of and access to product related information as requested by a supply chain's stakeholder without loss, noise, delay or distortion' (Hofstede *et al.*, 2004), is absolutely needed.

But the basis of this scientific problem is of a different kind. That is, Brambell's freedoms reflect an ethical, rather than a science based approach (Korte *et al.*, 2007). Brambell's freedoms were formulated on behalf of 'higher animals' (veal calves, pigs and chickens), towards which moral concern is by general consensus required. The explicit declaration of applicability of this ethical approach to all production animals suggests a deliberate moral extension, leading to the inclusion of insects as a category of moral objects in Dutch Animal welfare legislation. If such a moral extension is deliberate, the upscaling of insect rearing in accordance to animal welfare is indeed a scientific challenge. The Dutch Ministry of Economic Affairs has meanwhile consulted entomologists from the Wageningen University to develop a protocol for insect farmers to demonstrate that their insects are reared, transported and killed with due respect for insect welfare. However, the idea that we are required to give insects moral concern by analogy with 'higher animals' may be hard to accept. We therefore argue that not only scientific evidence, but also consensus on the moral status of insects is needed. Until then, the Five Freedoms are a suggestive concept for insect welfare.

References

Carruthers, P. (2007). Invertebrate minds: a challenge for ethical theory. The Journal of Ethics 11(3): 275-297.

Cooper, T.M., Mockett, R.J., Sohal, B.H., Sohal, R.S. and Orr, W.C. (2004). Effect of caloric restriction on life span of the housefly, *Musca domestica*. The FASEB Journal 18(13): 1591-1593.

DeFoliart, G.R. (1999). Insects as food: why the western attitude is important. Annual Review of Entomology 44: 21-50.

Diener, S., Zurbrügg, C. and Tockner, K. (2009a). Conversion of organic material by black soldier fly larvae: establishing optimal feeding rates. Waste Management & Research 27(6): 603-610.

EC No 999/2001. Regulation of the European Parliament and of the Council of 22 May 2001 laying down rules for the prevention, control and eradication of certain transmissible spongiform encephalopathies. Offical journal of the European Communities.

EC No 1069/2009. Regulation of the European Parliament and of the Council of 21 October 2009 laying down health rules as regards animal by-products and derived products not intended for human consumption and repealing regulation (EC) No 1774/2002. Official Journal of the European Union.

Eisemann, C.H., Jorgensen, W.K., Merritt, D.J., Rice, M.J., Cribb, B.W., Webb, P.D. and Zalucki, M.P. (1984). Do insects feel pain? – a biological view. Experientia 40(2): 164-167.

EU No 142/2011. Commission Regulation of 25 February 2011 implementing Regulation EC) No 1069/2009 of the European Parliament and of the Council laying down health rules as regards animal by-products and derived products not intended for human consumption and implementing Council Directive 97/78/EC as regards certain samples and items exempt from veterinary checks at the border under that Directive. Official Journal of the European Union.

FAO (2012). Assessing the potential of insects as food and feed in assuring food security, summary report. FAO Technical Consulatation Meeting, Rome, Italy.

Fraenkel, G. and Blewett, M. (1944). The utilisation of metabolic water in insects. Bulletin of Entomological Research 35: 127-139.

Hofstede, G.J., Schepers, H., Spaans-Dijkstra, L., Trienekens, J. and Beulens, A. (2004). Hide or confide: the dilemma of transparency. Reed Business Information, 's-Gravenhage, the Netherlands.

Korte, S.M., Olivier, B. and Koolhaas, J.M. (2007). A new animal welfare concept based on allostasis. Physiology & Behavior 92(3): 422-428.

Lietze V.U., Sims K.R., Salem T.Z., Geden C.J., Boucias D.G. (2009). Transmission of MdSGHV among adult house flies, *Musca domestica* (Diptera: Muscidae), occurs via oral secretions and excreta. Journal of Invertebrate Pathology 101: 49-55.

Ministerie van Volksgezondheid, W. e. S. (1998). Gezondheids- en welzijnswet voor dieren, Hoofdstuk III: De zorg voor het welzijn van dieren. STB. 1998, 169. Ministerie van Volksgezondheid, Welzijn en Sport, Den Haag, the Netherlands.

Nakagaki, B.J. and Defoliart, G.R. (1991). Comparison of diets for mass-rearing *Acheta domesticus* (Orthoptera: Gryllidae) as a novelty food, and comparison of food conversion efficiency with values reported for livestock. Journal of Economic Entomology 84(3): 891-896.

Ocio E., Viñaras R. and Rey J.M. (1979). Housefly larvae meal frown on municipal organic waste as a source of protein in poultry diets. Animal Feed Science and Technology 4(3): 227-231.

Oonincx, D.G.A.B. and De Boer, I.J.M. (2012). Environmental impact of the production of mealworms as a protein source for humans – a life cycle assessment. PLoS ONE 7(12): e51145.

Oonincx, D.G.A.B., Van Itterbeeck, J., Heetkamp, M.J.W., Van den Brand, H., Van Loon, J.J.A. and Van Huis, A. (2010). An exploration on greenhouse gas and ammonia production by insect species suitable for animal or human consumption. PLoS ONE 5(12): e14445.

Ouedraogo R.M., Goettel M.S. and Brodeur J. (2004). Behavioral thermoregulation in the migratory locust: a therapy to overcome fungal infection. Oecologia 138: 312-319.

Ramos-Elorduy, J. (1997). Insects: a sustainable source of food? Ecology of food nutrition 36(2-4): 247-276.

Ramos-Elorduy, J. (2009). Anthropo-entomophagy: cultures, evolution and sustainability. Entomological Research 39(5): 271-288.

Ramos-Elorduy, J., González, E.A., Hernández, A.R. and Pino, J.M. (2002). Use of *Tenebrio molitor* (Coleoptera: Tenebrionidae) to recycle organic wastes and as feed for broiler chickens. Journal of Economic Entomology 95(1): 214-220.

Rumpold, B.A. and Schlüter, O.K. (2013). Potential and challenges of insects as an innovative source for food and feed production. Innovative Food Science & Emerging Technologies 17(0): 1-11.

Salin C., Vernon P. and Vannier G. (1999). Effects of temperature and humidity on transpiration in adults of the lesser mealworm, *Alphitobius diaperinus* (Coleoptera: Tenebrionidae). Journal of Insect Physiology 45: 907-914.

Szelei, J., Woodring, J., Goettel, M.S., Duke, G., Jousset, F., Liu, K.Y., Zadori, Z., Li, Y., Styer, E., Boucias, D.G., Kleespies, R.G., Bergoin, M. and Tijssen, P. (2011). Susceptibility of North-American and European crickets to *Acheta domesticus* densovirus (AdDNV) and associated epizootics. Journal of Invertebrate Pathology 106: 394-399.

Tracey Jr., W.D., Wilson, R.I., Laurent, G. and Benzer, S. (2003). Painless, a *Drosophila* gene for nociception. Cell 113(261-273): 262-217.

Tweede Kamer, 2012. Vragen van het lid Thieme (PvdD) aan de staatssecretaris van Economische Zaken, Landbouw en Innovatie en aan de minister van Volksgezondheid, Welzijn en Sport over insecten als eiwitbron voor varkens- en kippenvoer. Tweede Kamer der Staten-Generaal. Den Haag, the Netherlands.

Van Doorn, D., Klink, A., Hectors, A., Rottenberg, G., Naaijkens, J., Vos, M., Scholten, M., Winsemius, P. and Artz, R. (2011). Al het vlees duurzaam. De doorbraak naar een gezonde, veilige en gewaardeerde veehouderij in 2020. Report Commissie Van-Doorn Den Bosch, the Netherlands.

Van Huis, A. (2013). Potential of insects as food and feed in assuring food security. Annual Review of Entomology 58(1): 563-583.

Van Wagenberg, C.P.A., Eppick, M.M., Janssens, S.R.M., Van der Roest, J., Van der Sluis, A.A. and Van der Spiegel, M. (2012). Ontwikkeling en vermarkting van nieuwe eiwitten – ervaren belemmeringen en oplossingen. Report. LEI, part of Wageningen UR, The Hague, the Netherlands.

Veissier, I., Butterworth, A., Bock, B. and Roe, E. (2008). European approaches to ensure good animal welfare. Applied Animal Behaviour Science 113(4): 279-297.

Veldkamp, T., Van Duinkerken, G., Van Huis, A., Lakemond, C.M.M., Ottevanger, E., Bosch, G. and Van Boekel, M.A.J.S. (2012). Insects as a sustainable feed ingredient in pig and poultry diets – a feasability study report. Wageningen UR Livestock Research, Lelystad, the Netherlands.

Wang, Y., Kim, H.-J., Vunjak-Novakovic, G. and Kaplan, D. L. (2006). Stem cell-based tissue engineering with silk biomaterials. Biomaterials 27(36): 6064-6082.

Weaver, D. and McFarlane, J.E. (1989). Aggregation in yellow mealworms, *Tenebrio molitor* L. (Coleoptera: Tenebrionidae) larvae. 15(5): 1617-1627.

Wet Dieren, 2013. Wet van 19 mei 2011, houdende een integraal kader voor regels over gehouden dieren en daaraan gerelateerde onderwerpen (Wet dieren). Stb., Den Haag, the Netherlands 2012: 659.

Section 6. Animal philosophy

Beneath the surface: killing of fish as a moral problem

B. Bovenkerk and F.L.B. Meijboom
Ethics Institute, Utrecht University, Janskerkhof 13a, 3512 BL Utrecht, the Netherlands; b.bovenkerk@uu.nl

Abstract

Are we morally justified in killing fish and if so, for what purposes? We will not focus on the suffering that is done during the killing, but on the question whether death itself is harmful for fish. We need to distinguish two questions: (1) can death be considered a harm for fish; and (2) if so, how much of a harm is it? We will explore three lines of reasoning: (1) fish desire to stay alive; (2) death deprives fish of future happiness or goods; (3) something valuable is lost when fish are killed. Some argue that a being can form a desire to stay alive only when it has the capacity to be aware of itself as a distinct entity existing over time. We will cast doubt on this view: Do we value continued life because it is desirable or do we desire continued life because it is valuable? It seems more plausible that it is not the desire to live that matters, but being able to enjoy goods, and death thwarts future opportunities for enjoyment. This would entail that a being can have an interest in continued life, without actively being interested in it. Next, we will discuss the question *how* harmful death is for fish. A widely shared intuition is that it is worse to kill a human being or mammal than a fish, because human or mammal life is more valuable. But can we really account for this intuition?

Keywords: harm of death, right to life, desires

Introduction

Fish are routinely killed by humans for consumption, as by-catch, for the aquarium industry, in animal testing, and during sports fishing. The amount of fish killed by us annually far outnumber the amount of terrestrial mammals or birds killed. While the question whether we are allowed to kill farm animals such as cows and chickens is hotly debated, it appears to be taken for granted that killing fish is morally less problematic or even unproblematic. Some self-proclaimed vegetarians even make an allowance in their diets for fish. Also, it has been argued that from an environmental perspective eating fish is best, as this would contribute less to global warming than other sources of protein (Kiessling, 2009, but see Röcklinsberg, 2012, for a critical discussion of this viewpoint). In this paper we want to raise the question whether this notion of killing fish as an unproblematic practice is morally justified. Are fish indeed so different from other animals, including ourselves, that this warrants differential treatment? Our central question is, therefore, are we morally justified to kill fish and if so, for what purposes? This question is distinct from the question whether we are allowed to make fish suffer in the process of killing. We will not focus on the suffering that is caused during the killing, but on the question whether death itself is harmful for fish. In order to answer this question, we need to distinguish two questions: (1) can death be considered a harm for fish; and (2) if so, how much of a harm is it? In other words, would it be sufficiently harmful to plea for an end to fish consumption, sports fishing, the aquarium industry or animal experiments?

First, we should mention here that when we speak about 'the harm of death', we are talking about harm in a moral sense. Of course death harms fish in the sense that their body is damaged – in the same sense as a plant can be harmed when it is cut – but is this a harm that matters morally? We will discuss three lines of reasoning that can be put forward to argue that death is a harm for animals. After that, we will discuss the question *how* harmful it is: Most of us seem to have a moral intuition that it is worse to kill a horse than it is to kill a fish, but can this intuition be justified?

The desire account

When we consider why it would be morally wrong to kill a person, one of the first reasons that comes to mind is that this person does not want to be killed. The person has a desire or preference for staying alive. Is this a convincing reason and if so, does it apply to fish as well? Whether this is a convincing reason, amongst other things, depends on what moral theory one adheres to. From a classical utilitarian point of view, where all that matters is the total amount of happiness or pleasure in the world, a desire for staying alive is irrelevant. After all, if only experiences count, death doesn't harm one, for when one is dead one ceases to have experiences. As Singer (1980) points out, killing as such is not wrong in classical utilitarianism; it could only be wrong indirectly, when those who stay behind worry about being killed next and this worrying decreases the total amount of happiness in the world. This, however, entails that if someone were killed in secret, death would not be a harm to this person. To avoid this counterintuitive implication, Singer adopts preference utilitarianism, arguing that when a being has a preference for staying alive, this is an important factor to take into consideration even if this preference can ultimately be overruled by other preferences in a utilitarian calculus. The question for a preference utilitarian is, then: What beings have a preference to stay alive? When confronted with avoidance behaviour of animals in danger, such as fish struggling on a hook, at first sight we might interpret this as a fear of death. Singer (1980), however, warns us against taking this to mean a preference for continued existence. Rather, we should interpret this as a desire to stop the pain or the threatening situation, and of course this desire can also come about by killing the animal. In his view, a being can form a preference to stay alive only when it has the capacity to be aware of itself as a distinct entity existing over time.

If we look at the 'harm of death question' from a rights perspective, we see that the whole range of positions in this debate can be taken, but that all depend on the interpretation of the criterion of self-consciousness. Regan (2004) argues that an animal has a right not to be harmed – and he regards killing as such a harm – when an animal is a subject of its own life. This requires consciousness, but only limited self-consciousness. Tooley (1972) on the other hand, argues that a being can only have a right to life if it has a desire to live and that only self-conscious beings have a desire to live. This is because you need to conceive of yourself as an entity that exists over time to have a desire to live. Other than Singer he thinks this requires self-consciousness. Cigman (1981) also argues that self-consciousness is required when she argues that death is only a harm for beings with the capacity for categorical desires. Her argument goes as follows: It is not necessary for a creature to *actually* have a desire not to die in order to have a right to live, but the creature must at least have the *capacity* for the desire not to die. When does one have this capacity? Cigman draws on Williams' concept of a categorical desire to make the distinction between having the desire to go on living, or values that are simply instrumental in living. Life as a categorical desire answers the question whether or not 'one wants to remain alive' (Cigman, 1981: 58). This refers to reasons one may have to go on living, or those things that make a life worth living or give it meaning. Examples are raising children or writing a book.

What do we know about fish' capacities for: (1) awareness of themselves as distinct entities existing over time; (2) self-consciousness; and (3) categorical desires? We will assume that fish can experience pain and distress. This has been argued for a number of species that are commonly used by humans (Braithwaite, 2010; Sneddon, 2003) and moreover, there is a high degree of consensus about this among marine biologists. To be able to experience pain and distress some form of consciousness is necessary. However, this does not yet tell us whether fish have awareness of themselves over time, let alone whether they are self-conscious or have categorical desires. In fact, this is not a question that can be answered yet. Not enough research on fish capacities has been carried out. Such research is made difficult by the fact that over 35,000 different species of fish exist, all with their own evolutionary adaptations. Also, it is difficult to design the tests that could tell us what goes on in fish' brains. Tests have been carried out in some species that show they have a memory span much larger than the proverbial three seconds (Nilsson *et al.*,

2008). Some tests show an ability for complex mapping in certain fish (Braithwaite and De Perera, 2006), and there is evidence that certain fish of different species cooperate to reach a common goal (Bshary *et al.,* 2006). However, this research does not tell us much about self-consciousness or categorical desires.

The foregone opportunities account

Even though it seems to be a stretch of the imagination to conceive of self-conscious fish, one can raise the question whether these capacities are really necessary to be harmed by death. One can raise doubts about the argument that death harms us because we desire continued life. Do we value continued life because it is desirable or do we desire continued life because it is valuable? If we take the latter point of view – which we think we should – we can ask what is valuable about continued life. This question could be answered either in an objective or in a subjective sense. According to an objective value theory, a certain good is in a being's interest even if this being does not value this good. This would imply that these non-moral goods are valuable in themselves even if nobody experiences them as valuable (Kaldewaij, 2011). Even if fish would not value anything as good in their lives, their death would still be harmful; perhaps death would not be harmful *to* these fish, but their death would still be a bad state of affairs. What criterion would we use to decide whether this state of affairs is bad? Examples of possible criteria could be beauty or biodiversity. So, in the case of beauty it would be bad to kill fish if this would diminish the beauty in the world. However, if we were to create extra fish and then kill them, this would not diminish the beauty in the world compared to the prior situation. Also, who is to say what fish are beautiful? On colourful tropical fish we might agree, but what about unappealing creatures, such as catfish? Similar objections could be raised to other proposed candidates of an objective list of the good.

Let's therefore turn to a subjective value theory. From the point of view of the animal in question we could say that the animal derives pleasure from certain goods in its life and this makes that the animal has an interest in the continuation of these goods. According to DeGrazia (2002: 61), 'death forecloses the valuable opportunities that continued life would afford'. This so-called 'foregone opportunities account' of the harm of death follows Nagel (1991) in that it takes life as instrumentally valuable for all beings that can have experiential wellbeing. According to Kaldewaij (2011: 61) a benefit of Nagel's view is 'that it can explain the magnitude of the harm of death: death takes away the possibility of ever experiencing, doing or accomplishing anything you value again'. One could object that animals are not aware of these foregone opportunities. However, this view on the harm of death does not require that individuals are aware of their lost opportunities. A being, it is argued, can have an interest in continued life, without actively being interested in it. All a creature needs in order to be harmed by death in such an account, is the ability to have experiences that matter to it and that it would be deprived of when dead. Life is then instrumentally valuable for animals to the extent that they can have valuable experiences that make their lives worth living. As animal welfare scientists have shown, animals do not just have simple desires such as eating when they are hungry, but they actually derive pleasure from eating as well and this makes their life worth living (Duncan, 2006).

A question that the foregone opportunities account raises is whether we can really be deprived of something if we do not exist anymore. After all, when we are dead, we don't know what we are missing. This problem has spurned a philosophical debate too complex to discuss within the scope of this paper, but it still casts doubt on our preliminary conclusion that death is a harm for fish. Suffice it to say that according to Nagel (1991) it is possible to be harmed by something even though we do not experience it. He uses the example of betrayal: if you are betrayed behind your back you are still harmed, even though you don't realise you are betrayed. It has been objected that in this case you still could find out that you have been betrayed and suffer harm as a result of that, whereas when you are dead it is impossible to ever experience the betrayal (see Fischer, 1997). Yet, Nagel's claim is that one has unpleasant experiences when betrayed because betrayal is bad and not that betrayal is bad because it generates unpleasant experiences, and therefore a betrayed person that will remain ignorant of the betrayal is still harmed (Nagel, 1991: 5).

How harmful is death for fish?

Suppose we cast aside our doubts and assume that death is harmful for fish, this still does not settle the question of whether we are allowed to kill fish for consumption or other purposes, such as experimentation or recreation. A widely shared intuition is that it is worse to kill a human being than a fish. But how can we account for this intuition?

DeGrazia (2002) starts from the equal consideration of equal interests-principle, but argue that some animals have a higher interest in continued life than others. If we consider life as instrumentally valuable for the goods that are valuable for a being, it appears that different species can have different interests in life if they differ – either qualitatively or quantitatively – in the goods that are valuable for them. Those beings that have richer opportunities to function or to experience enjoyment stand to lose more from death. Philosophers like Mill (1861) and Singer (2011) have argued that a human life has qualitatively more value for a human than an animal's life has for that animal, because if one could take the vantage point of both the human and the animal one would choose to be the human. Kaldewaij (2011) argues, however, that this type of 'intersubjective' argument fails, as nobody can really be acquainted with both types of lives. How can we judge whether an animal enjoys the goods in its life less than a human does? There is no intersubjective vantage point from which to compare the lives of humans and animals. Simmons (2010) is not convinced by this reasoning, however, because we do share many types of pleasures with animals, in particular sensory pleasures (associated with food, sex, warmth, etc.) and we can pose the question whether a life with only these sensory pleasures or a life with both sensory pleasures and pleasures derived from our intellect and creativity is more valuable.

One can wonder whether this view of the greater value of life due to creative and intellectual pleasures is not overly anthropocentric. Since these traits are a great source of value for us we assume that they would be a great source of value for other beings as well. But how would we know we are not biased? How can we really judge that, say, reading a book or listening to a beautiful piece of music would be more valuable to, say, a bird than the pleasure it derives from gliding through the air (if it could do both)? The quality of life of animals may simply be made up of different elements than the quality of life of humans. Our point, then, is not epistemological, but rather moral. It is not that we cannot know how animals experience certain sensations, but rather that we cannot judge whether one experience of one being is qualitatively better than a similar experience of another being, just as we cannot judge whether going to the opera is more valuable to me than going to a football match is for my neighbour. It appears that if we want to argue that most human lives are more valuable than most animal lives we would need to resort to an objective value theory that could show that regardless of the value of life as experienced by a particular being, certain goods and enjoyments are objectively superior to other ones. However, what could be the basis of such an objective claim? Perhaps consciousness could provide a common yardstick to measure the quality of life of different species (and within species) by. We could conceive of consciousness as a gradual notion; some animals have more and some less consciousness. Could those who have more consciousness experience the goods in their lives to a higher degree and therefore their life could have more value?[24]

Whether or not we could say that life has more value for humans than for other animals, from an unequal *value* of life we cannot automatically conclude an unequal *interest* in continued life, let alone draw

[24] Of course, here we may run into the same problem that it is difficult to assess levels of consciousness, but in theory this yardstick does seem to have the advantage of providing an objective yardstick.

specific action-guides.[25] If we were to consider that levels of consciousness or of creative and intellectual pleasures would make certain lives more valuable and that those with more valuable lives have a greater interest in continued life, would we then not have to conclude that some humans are harmed more by death than other humans? And from the fact that some are harmed more by death, would we not have to conclude that some have a greater interest in continued life? Here again we can wonder whether there is a neutral vantage point to compare individuals' interests by. But even if there were, most people would not want to draw the conclusion that in a case where we would have to choose between saving a more intelligent or a less intelligent person, for example, we have to save the more intelligent one. This argument does not only refer to marginal humans, but also to normal, healthy adults, who also differ in their levels of consciousness and creative and intellectual powers. Assuming that fish have a lower quality of life than other animals, then, the question whether or not we are allowed to kill them is not settled yet. Still, our intuition remains that if we have to choose between killing a fish or killing a human or other mammal, we should kill the fish.

How can we explain this intuition? Either it could be understood as a predictable bias on the part of humans and our intuition is simply wrong. The intuition could simply be influenced by our culture or tradition. Or perhaps this intuition is based on the acknowledgement that self-conscious beings suffer psychologically from the idea that they will die. This, however, is not the type of harm that we are dealing with in this paper; we are addressing the question whether death *as such* is harmful for fish. After all, we think we should not kill humans even if they do not fear death or even when we would kill them in their sleep. Another option is to grant the option outlined above that those with a higher quality of life have a greater interest in continued life and that this should lead us to treat them differently. In that case we would also have to acknowledge that it is worse to kill normal adult humans than marginal humans, and moreover, that it is worse to kill humans who can experience greater intellectual pleasures than humans with 'lower' intellectual pleasures. Yet, even if we would take this last, dubious avenue, this would still leave open the question whether killing fish for consumption, recreation or animal experimentation is justified. Even if sentient animals and marginal humans would have a *weaker* interest in continued life, this does not mean they have *no* interest in it. In order to determine the justifiability of the use of fish, we would have to weigh different interests against each other. Following Van de Veer (1979), we could distinguish basic from serious and peripheral interests and argue that basic interests should trump peripheral ones. We consider recreation to be a peripheral human interest, which does not trump fish' basic interest in life. In the case of animal experimentation according to this line of reasoning it may be allowed to kill fish if we can reasonably predict that this would save human lives. In the case of consumption, our judgment would have to rely on the specific situation at hand; in each case we would need to weigh how serious the human interest in fish consumption (and the revenue from fish production) is.

Conclusion

In this paper we have examined several arguments for the claim that death is a harm in a morally relevant sense and we have applied these to the case of fish. It is unlikely that fish are self-conscious or have categorical desires. However, these capacities are relevant only for a desire-account of the harm of death and in our view this account is flawed. A more promising take on the harm of death is offered by the foregone opportunities account, which argues that death is a harm because life is instrumentally valuable for all beings that can have experiential well-being. This means that sentience is a sufficient capacity to be able to be harmed by death and fish are (most likely) sentient. Next, we have examined how one could

[25] This point is analogous to the point made by Simmons (2009), who argues that from unequal quality of life we cannot conclude unequal *right* to life. Here, however, we focus not so much on a right to life, but only on the harm of death and the concurrent view that animals would have an interest in continued life.

account for the commonly shared intuition that it is worse to kill a human being than to kill a fish. Some argue that even if we apply the equality of interest-principle, it is still possible to differentiate between different species, because some have a higher interest in continued life than others. However, even if we could argue that some lives contain greater value than others, this does not necessarily mean that some have a greater interest in continued life than others, nor does it settle the question whether fish may be killed. We have suggested that the latter depends on a weighing of basic, serious, and peripheral interests of humans in killing fish for consumption, recreation, or experimentation against the basic interest of fish in survival.

References

Braithwaite, V. (2010). Do fish feel pain? Oxford University Press, Oxford, UK.

Braithwaite, V.A. and De Perera, T.B. (2006). Short-range orientation in fish: how fish map space. Marine and Freshwater Behaviour and Physiology 39(1): 37-47.

Bshary, R., Hohner, A., Ait-el-Djoudi, K. and Fricke, H. (2006). Interspecific communicative and coordinated hunting between groupers and giant Moray eels in the Red Sea. PLoS Biology 4(12): e431.

Cigman, R. (1981). Death, misfortune and species inequality. Philosophy & Public Affairs 10(1): 47-64.

DeGrazia, D. (2002). Animal rights: a very short introduction. Oxford University Press, Oxford, UK.

Duncan, I.J.H. (2006). The changing concept of animal sentience. Applied Animal Behaviour Science 100: 11-19.

Fischer, J.M, (1997). Death, badness, and the impossibility of experience. The Journal of Ethics 1(4): 341-353.

Kaldewaij, F. (2011). Animals and the harm of death. In: Armstrong, S.J. and Botzler, R.G. (eds.) The animal ethics reader. Routledge, Oxford, UK.

Kiessling, A. (2009). Feed – the key to sustainable fish farming. Fisheries, Sustainability and Development. Royal Swedish Academy of Agriculture and Forestry: 303-322.

Mill, J. S. (1861). Utilitarianism. Bobbs-Merrill, Indianapolis, IN, USA.

Nagel, T. (1991). Mortal questions. Cambridge University Press, Cambridge, UK.

Nilsson, J., Kristiansen, T.S., Fosseidengen, J.E. Fernö, A. and Van den Bos, R. (2008). Learning in cod (*Gadus Morhua*): long trace interval retention. Animal Cognition 11 (2): 215-222.

Regan, T. (2004). The case for animal rights. University of California Press, Berkely, CA, USA.

Röcklinsberg, H. (2012). Fish for food in a challenged climate – ethical reflections. In: Potthast, T. and Meisch, S. (eds.) Climate change and sustainable development. Ethical perspectives on land use and food production, Wageningen Academic Publishers, Wageningen, the Netherlands, pp. 326-334.

Simmons, A. (2009). Animals, predators, the right to life, and the duty to save lives. Ethics & the Environment 14(1): 15-27.

Simmons, A. (2010). Do all subjects of a life have an equal right to life: the challenge of the comparative value of life. Paper Presented at Tom Regan: A Celebration. Faculty Scholarly Dissemination Grants. Paper 103. Available at: http://scholarworks.gvsu.edu/fsdg/103.

Singer, P. (1980). Animals and the value of life,[w:] matters of life and death, Rd. T. Regan. Random House, New York, NY, USA.

Simmons, A. (2011). Practical ethics. Cambridge University Press, Cambridge, UK.

Sneddon, L.U. (2003). The evidence for pain in fish: the use of morphine as an analgesic. Applied Animal Behaviour Science 83(2): 153-162.

Tooley, M. (1972). Abortion and infanticide. Philosophy & Public Affairs 2: 37-65.

VandeVeer, D. (1979). Interpecific justice. Inquiry 22(1-2): 55-70.

In awe of fish? Exploring animal ethics for non-cuddly species

C.P.G. Driessen
Department of Philosophy, Utrecht University, Janskerkhof 13a, 3512 BL Utrecht, the Netherlands; Applied Philosophy Group, Wageningen University, Hollandseweg 1, the Netherlands; clemens.driessen@wur.nl

Abstract

Though rarely counted as individuals, fish are by far the most consumed animals in the world, outnumbering all other food animals combined. These widely eaten animals are generally conceived of as dull and mindless creatures, 'swimming protein to be plucked from rivers and seas'. Over the past several years, however, evidence has been amassed which indicates that fish are more sentient and intelligent than their alleged 'three second memory' would imply. Thus far these findings have had only limited success in arousing public interest in fish welfare. Today still fish are rarely discussed or treated as sentient beings that feel pain and suffer. A series of exploratory workshops with professionals working in a variety of fish-related fields suggested that the reason that people have a hard time appreciating the perspective of fish is not so much a lack of knowledge about fish sentience. Instead it may have more to do with the perceived inability of fish to strike up meaningful relations with humans: they are quintessentially non-cuddly animals, cold, slimy, and with their unblinking and sideways directed eyes they don't have a 'face' to us. Many people however know that some fish species display amazing abilities, such as being able to swim to the Sargasso Sea and back to the river they came from, as the eel does. Such animals seem to generate awe and perhaps respect not for to some extent resembling humans, but by their being different and quintessentially 'other' – precisely for us having trouble to imagine their life form. The type of affect generated by this sense of otherness differs markedly from the common modes of caring about animals based on nearness, empathy and direct interaction, and is more akin to aesthetic experiences evoked around environmental concerns. In discussion with recent work in relational animal ethics, environmental aesthetics, and more-than-human geography, this paper explores the potential of the experience of awe as motivating and guiding an ethics for (farmed) fish. By attending to discourses and material practices concerned with fish, fisheries and fish farming, the challenges of particular understandings of an 'ethics of awe' are considered, including the question of whether and in what ways awe could be expressed as and translated into consumer preferences.

Keywords: fish welfare, animal aesthetics, geographies of affect, aquaculture

Introduction: the neglect of fish welfare

Even though they are rarely counted as individuals instead of in tons, fish are by far the most consumed animal in the world, outnumbering all other food producing animals combined. Nevertheless, or perhaps somehow because of this, many people still conceive of fish as dull and mindless creatures, 'swimming protein to be plucked from rivers and seas' (Hanlon, 2010). Correcting this view, over the past years evidence has been amassed which indicates fish are more sentient and intelligent than their alleged 'three second memory' would imply (Braithwaite, 2010). Though the nature of their cognitive abilities is still debated (Allen, 2013), it has been argued that these claims to sentience, including the ability to feel pain and to suffer, should be enough to give fish 'the benefit of the doubt' (Lund *et al.*, 2007) and bring them into our moral community to concern ourselves with their welfare (Bovenkerk and Meijboom, 2012). Thus far however these findings have had only a limited success in arousing public interest in fish welfare.

Debates about fish production and consumption tend to focus on stock depletion from overfishing or the environmental effects of aquaculture. Alarming reports warn of the ecological destruction caused by industrial fisheries, sketching disconcerting effects of decades of 'fishing down the food web' (Pauly,

1998). A few more years of insufficiently regulated fishing may leave us with oceans filled only with jellyfish and plastic soup. In these reports, the rapid depletion of fish communities is mainly described as loss of biomass with ecosystemic effects (Myers and Worm, 2003). Within this ecological perspective, the suffering of fish subjects in the process of being decimated remains absent. Meat, the flesh of land based animals, in some societies has come under increasing moral scrutiny and is now for many laden with ambivalence, leading to people to self identify as e.g. 'part time vegetarians' for a variety of reasons. As part of this change however, fish is widely considered to be an attractive alternative (Dagevos *et al.*, 2012). Consumer concerns over the welfare of marine life have so far mainly centred on our mammalian kin. It is now common to find cans labelled 'dolphin friendly tuna', whereas there seems to be no market for 'tuna friendly tuna'. Even today fish are rarely discussed as 'experiencing subjects' that are able to e.g. feel pain and suffer. It may very well be that emerging information about the capacity of fish to suffer and the effects on them of fishing methods would contribute to improving their lot, but so far the term 'fish welfare' has not yet caught on in the public imagination.

Why is the welfare of fish commonly neglected?

Informing this extended abstract are three exploratory workshops that were organized as part of a series of interdisciplinary one day courses on 'fish welfare', each involving 25 participants with various professional interests in fish (e.g. fisheries, aquaculture, laboratory animal research, retail, NGOs, governmental agencies). From our discussions it appeared that the main reason for the public neglect of the welfare of fish may not (only) be lack of awareness of the recent findings on fish sentience, but the perceived inability of fish to communicate and strike up meaningful relations with humans. It was estimated that for many people fish are quintessentially non cuddly animals, cold, slimy, and with their unblinking and sideways directed eyes they don't have a 'face' to us. They are genetically and physiologically remote from us. They do not communicate in ways that we are responsive to, they don't show emotions in ways that we understand, they don't scream when hurt. They live in a radically different world that is difficult to fathom. On top of this it is hard for us to see (let alone recognize) them as individuals, especially when they live in shoals.

What was striking in these responses was the large number of reasons that participants in the workshops came up with to explain the lack of human interest in fish welfare. Most of these reasons did not centre on the general lack of knowledge on fishes' ability to suffer, but emphasized barriers conceived as part of more relational understandings of animal ethics. In animal ethics, relational approaches have recently become prominent that not only describe human moral concerns over animals as in practice often based on their closeness, dependence, reciprocity or some other relational factor, but also argue for such considerations to be ethically relevant to determine our obligations to animals around us. These types of relational understandings of ethics tend not to favour fish. It is probably no coincidence that the three recent major books developing relational and contextual approaches to living with animals all feature a dog on their cover (Donaldson and Kymlicka, 2011; Haraway, 2008; Palmer, 2011). If at all, these works only marginally discuss fish. Fish seem physiologically challenged in generating rapport with humans that could have moved us into an empathic relationship, or to give rise to the kind of personal embodied interspecies encounter which allows us to become attuned to their behaviours, needs and desires. Very few people attempt to imagine what it is like to be a fish, or think of fish as having an 'inner life' at all, as being experiencing individuals. In our culture there are not many spaces which facilitate the experience or conceptualisation of fish subjectivity. From popular culture to supermarket advertising it seems we live in affective worlds that are geared towards ignoring the sentience of non-mammalian marine life.

While fish sentience is mostly overlooked, there are some societal practices that actively engage with fish as agents. Perhaps angling is the most widely engaged-in form of interacting with wild animals. Bear & Eden have drawn attention to the fact that many recreational anglers (in the UK) do not view fish merely as 'alien bodies living in another world, behind an "impassable screen"' (Bear and Eden, 2011: 337). They describe in detail how affective and transformative relations can arise between humans and fish; and not just with species or groups of fish, but even with particular individuals. However, even though in angling circles fish welfare is increasingly a matter of debate and policy making, the acknowledgement of fish as conscious beings is contested, probably as such recognition has the potential to jeopardize the very practice.

Frying Nemo

Several participants in the fish welfare workshops brought up the popular animated film *Finding Nemo* as a potential tool to raise awareness regarding the plight of fish. The film narrates the story of an endearing clownfish on an epic journey to liberate his son Nemo, who has been abducted from their home in the Great Barrier Reef to live in a fish tank at a Sydney dental practice (Stanton, 2013). The narrative structure invites the audience to empathize with the lead characters, further enabled by a slight adjustment of the placement of the eyes to create more expressive faces, facilitating anthropomorphic projection. The movie won an academy award and is the best selling DVD of all times, with a sequel currently underway. Reflecting the hope this tale might raise interest in the welfare of fish, US animal rights organization PETA produced a poster titled 'Frying Nemo', showing a distressed cartoon clownfish on his side in a pan. The animated film however does not seem to have done for the popular understanding of fisheries what Bambi did for the appreciation of hunting. A cursory Google search brought up at least eight fish-and-chips shops the world over actually named 'Nemo'. Two more are even called 'Frying Nemo', retroactively highlighting how 'roasting Babe' or 'grilling Bambi' never really caught on to entice consumption of pigs or deer. Moreover, even though a Moorish Idol in the film states 'Fish aren't meant to be kept in a box, kid. It does things to you', Nemo actually caused a frenzy in the aquarium trade of clown fish (Prosek, 2010). It appears empathy with fish is just beyond human capacity.

In awe of fish?

People may have a hard time appreciating fish perspectives and acknowledging their ability to suffer, but some do know that various fish species have amazing abilities, such as being able to swim to the Sargasso Sea and back to the river they came from, as the eel does. These animals seem to generate awe and perhaps respect not by resembling humans, but by being different and other; precisely for us having trouble to imagine their life form. The remainder of this extended abstract will offer a brief and preliminary foray into the potential of an 'ethics of awe' for (farmed) fish. When thinking of charismatic species, first spring to mind the kind of spectacular megafauna that are impressive due to their size and behaviours: whales, elephants, tigers – the favourite 'flagship species' of environmental protection organisations. But also more modest and everyday species such as insects and birds can be (in different ways) charismatic and evoke a variety of affective experiences, for example, as being mysterious, enchanting, or repulsive (Lorimer, 2007).

One way of appreciating the awesomeness of creatures beyond those that look impressive or pretty is by imagining them as the (current) endpoint of millions of years of Darwinian struggle (Gessert, 2010). A related understanding of 'animal aesthetics' is one that emphasizes the aesthetic quality of 'looking fit for function' (Parson, 2008). A competing view of animal aesthetics focuses instead on the expressive beauty of animals (Brady, 2009), highlighting the ways in which we interpret the looks and behaviours of animals as culturally meaningful in both their and our world. In the light of this, the eel and its mysterious migration can be considered not merely as a magnificent adaptation to ocean gulf streams and a series of different ecological niches, but to render these sites more meaningful also for humans, by connecting in unexpected ways geographically remote areas.

An ethics of awe would be a way to move beyond mere *human extensionism*, whereby moral status is distributed based on some hierarchy in which semblance to humans is the sole criterion. These animals do us the favour of revealing how humans are not the pinnacle of evolution. An ethics of awe then does not call for a new hierarchy of moral status based on the complexity and degree of otherworldliness of features of particular fish species, as an alternative to a moral status based on qualities such as sentience or being a 'subject of a life'. Instead, the sense of awe may invite us to question the very idea of moral status ascription, implying a different kind of understanding of ethics. Awe could thus highlight the subjective and relational-ecological dimension of moral status ascription, emphasizing how 'we ascribe moral status as beings who are already part of the world, who are already "environmental"' (Coeckelbergh, 2012: 6). So awe as a mode of ethics is not to be understood as supervening on properties of the objects of our concern, but as the outcome of us being struck by something. Here awe can be considered to *disclose* a moral world, whereby not just a particular moral decision, but also what it means to be a moral subject is at stake (Driessen and Korthals, 2012). And this means awe, or other affective modes of relating to others, is not just to be taken as a way of motivating people to do the right thing; a position that would reserve determining what the right thing to do actually is to more rational and non-subjectivist modes of ethical reasoning.

Some of charismatic fish species that are killed in especially gruesome ways do generate protests that involve trying to comprehend fish perspectives. Notably the cutting of shark fins, leaving them to bleed to death or drown (Spiegel, 2000) and the slow killing of eels by desliming their skin in salt baths (Lambooij *et al.*, 2002), have been the subject of protest and, at least in some countries, regulation. What is interesting here is that reports on these practices involve the evocative description of these processes of killing which inform readers about particular features of the animals, and at the same time implicitly encourage them to imagine what it is like to die in the circumstances described. Thus, the fact that eels take 13 minutes to stop displaying vital signs after their head is severed from their body not just raises problems for designing humane slaughter methods (Van De Vis *et al.*, 2003), but also could bring a sense that we are dealing with a completely otherworldly life form of which we understand very little.

Fisheries are often marketed and generally still thought of as a romantic pursuit in which rugged men go out to sea and after a battle with the elements haul in nets full of fish. The realities of efficient industrial processes of non-stop trawler operations with automated fish processing and offshore canning may instead give rise to a profound sense of unease perhaps best summarized by the term 'factory fishing'. Likewise in aquaculture, the scale of operations and high stocking densities can be perceived as disenchanting and alienating. Some aquaculture companies try to address both the concern over welfare and a sense of alienation or lack of meaning by modelling the role of their 'site managers' on that of knowledgeable and caring farmers who know what is best for their stock. They 'do not take decisions on questions of husbandry from behind a desk and computer screen' (Scottish Seafarms). Here the welfare of farmed fish is configured as the outcome of human management that is attuned to animal and environmental signals and that integrates a variety of considerations in practice (cf. Driessen, 2012).

Conclusion: awe and consuming fish?

This abstract has briefly discussed three sources of moral consideration of fish: sentience, empathy, and awe. The food product that perhaps alienates us the most from all these three sources is *fish fingers* (or *fish sticks* as they are called in the US). Commercialised in the 1950s and originally marketed with the slogan 'no bones, no waste, no smell, no fuss', the rectangular bite size encrusted product is presented without even a hint of the kind of fish inside. This can be considered the ultimate disenchantment of fish. Fish fingers completely undo any 'sensuous immediacy' towards nature, as advocated by Theodor Adorno to help us stop seeing nature as merely an object of consumption, in his aesthetization of ethics (cf. Bennett, 2001). At the same time, fish are one of very few food animals that are on sale and display in Western European supermarkets with their head still attached. Their welfare still not being a commonly appreciated issue could be taken as a sign that fish make for very peculiar experiences of 'things becoming food' (Roe, 2006).

In the confines of this extended abstract some tentative ideas on alternative ways to appreciate the welfare of fish have been discussed. These start from the argument that raising public interest over fish may require more than merely arguing for consistency in how we treat sentient beings and more than attempts at making people empathize with the suffering of fish. Thus, in our affective geography in which so far there has been little space for acknowledging the subjectivity of fish, the implication of animal aesthetics for animal ethics would be that moral status is to be seen as something to be 'evoked' and 'experienced' besides just being 'claimed' or 'derived' from more theoretically founded moral arguments (cf Haraway, 2008). This involves the deploying of different genres besides arguments in animal ethics, for which the early work of Rachel Carson might be an inspiration (Carson, 2010; cf. Lockwood, 2012). In her 'Undersea', she pioneered in the 1930s the affective power of literary language to describe the intricate food webs of hidden and mysterious marine ecologies. This may also involve thinking of ways of designing and staging human-fish encounters to generate the type of experiences that promote moral imagination and awe in a way that subverts common species hierarchies that draw exclusively on terrestrial notions of suffering, meaning and sociability. Perhaps a mixed approach of 'shock and awe', of displaying the gruesome reality of immense amounts of suffering in combination with evocative accounts of enchanting creatures, would make for an effective route for those who are interested in raising consumer concerns over fish welfare.

References

Allen, C. (2013). Fish cognition and consciousness. Journal of Agricultural and Environmental Ethics 26(1): 25-39.

Bear, C., and Eden, S. (2011). Thinking like a fish? Engaging with nonhuman difference through recreational angling. Environment and Planning D: Society and Space 29(2): 336-352.

Bennett, J. (2001). The enchantment of modern life: attachments, crossings, and ethics. Princeton University Press, Princeton, New Jersey, USA.

Bovenkerk, B. and Meijboom, F.L.B. (2012). The moral status of fish. The importance and limitations of a fundamental discussion for practical ethical questions in fish farming. Journal of Agricultural and Environmental Ethics 25(6): 843-860.

Brady, E. (2)009. Aesthetic appreciation of expressive qualities in animals. Postgraduate Journal of Aesthetics 6(3): 1-14.

Braithwaite, V. (2010). Do fish feel pain? Oxford University Press, Oxford, UK.

Carson, R.. (2010). Undersea. Nawakum Press, Santa Rosa, CA, USA.

Coeckelbergh, M. (2012). Growing moral relations: critique of moral status ascription. Palgrave Macmillan, Basingstoke, UK.

Dagevos, H., Voordouw, J., Van Hoeven, L., Van der Weele, C. and De Bakker, E. (2012). Vlees vooral(snog) vanzelfsprekend; consumenten over vlees eten en vleesminderen. Report 2012-029. LEI, Wageningen UR, the Hague, the Netherlands.

Donaldson, S. and Kymlicka, W. (2011). Zoopolis: a political theory of animal rights. Oxford University Press, Oxford, UK.

Driessen, C. (2012). Farmers Engaged in Deliberative Practices; an Ethnographic Exploration of the Mosaic of Concerns in Livestock Agriculture. Journal of Agricultural and Environmental Ethics 25(2): 163-179.

Driessen, C. and Korthals, M. (2012). Pig towers and in vitro meat: disclosing moral worlds by design. Social Studies of Science 42(6): 797-820.

Gessert, G. (2010). Green light: toward an art of evolution. MIT Press, Boston, MA, USA.

Hanlon, M. (2010). Do fish have feelings too? It's a slippery question for science'. Daily Mail, 8 March edition.

Haraway, D.J. (2008). When species meet. University of Minnesota Press, Minneapolis, MN, USA.

Lambooij, E., Van de Vis, J.W., Kloosterboer, R.J. and Pieterse, C. (2002). Evaluation of captive needle stunning of farmed eel (*Anguilla anguilla*): suitability for humane slaughter. Aquaculture 212(1): 141-148.

Lockwood, A. (2012). The affective legacy of silent spring. Environmental Humanities 1: 123-140.
Lorimer, J. (2007). Nonhuman charisma. Environment and Planning D 25(5): 911-932.
Lund, V., Mejdell, C.M., Rocklinsberg, H., Anthony, R. and Hastein, T. (2007). Expanding the moral circle: farmed fish as objects of moral concern. Diseases of Aquatic Organisms 75(2): 109-118.
Myers, R.A. and Worm, B. (2003). Rapid worldwide depletion of predatory fish communities. Nature 423(6937): 280-283.
Palmer, C. (2011). Animal ethics in context. Columbia University Press, New York, NY, USA.
Parson, G. (2008). The aesthetic value of animals. Environmental Ethics 29(2): 151-169.
Pauly, D. (1998). Fishing down marine food webs. Science 279(5352): 860-863.
Prosek, J. (2010). Beautiful friendship'. National Geographic Magazine.
Roe, E.J. (2006). Things becoming food and the embodied, material practices of an organic food consumer. Sociologia Ruralis 46(2): 104-121.
Spiegel, J. (2000). Even Jaws deserves to keep his fins: outlawing shark finning throughout global waters. Boston College International and Comparative Law Review 24: 409-438.
Stanton, A. (2013). Finding Nemo (2003). Pixar.
Van De Vis, H., Kestin, S., Robb, D., Oehlenschläger, J., Lambooij, B., Münker, W., Kuhlmann, H., Kloosterboer, K., Tejada, M., Huidobro, A., Ottera, H., Roth, B., Sørensen, N.K., Akse, L., Byrne, H. and Nesvadba, P. (2003). Is humane slaughter of fish possible for industry. Aquaculture Research 34(3): 211-220.

Fish welfare, environment and food security: a pragmatist virtue ethics approach

R. Anthony[1]*, M. Gjerris*[2] *and H. Röcklinsberg*[3]
[1]*Department of Philosophy, 3211 Providence Drive, University of Alaska Anchorage, AK 99508, USA;* [2]*Institute of Food and Resource Economics, Faculty of Science, University of Copenhagen, Rolighedsvej 25, 1958 Frederiksberg C, Denmark;* [3]*Department of Animal Environment and Health, Box 7068, Swedish University of Agricultural Sciences, 750 07 Uppsala, Sweden; rxanthony@uaa.alaska.edu*

Abstract

With continued population growth, potentially negative impacts of climate change, and potential impacts of food access and security among poorer and more vulnerable communities fish as food resource is becoming more and more important. The growing recognition of fish as sentient beings must be considered in tandem with other concerns such as the relative weighting of their welfare, human equity, environmental protection, food security and food safety. Sensible environmental practices are needed and must be tied closely to effective policies around food security as well as regulation that take into account the issues of fish welfare. We suggest a pragmatic virtue ethics approach would be relevant and fundamental for such policies – both regarding process and substance – and show how it can contribute to the discussion on how to relate fish welfare to environmental concern and the issue of equity in a more secure global food system.

Keywords: fish farming, fish capture, fish welfare, pragmatism, virtue ethics

Introduction

If fish are capable of conscious experience (as much of the recent literature suggests), how then should their welfare be reflected in the regulation of capture fisheries and aquaculture? How should we balance or prioritize fish welfare concerns against other interests, such as the livelihoods of fish farmers and fishermen/women, long distance trade of fish products, increasing demands for animal proteins, etc? A related factor in prioritizing such issues in legislation or trade regulations is environmental aspects, raising, for example, the question of how the increase in demand for fish will impact protection efforts of our oceans and coastal areas?

We begin by briefly discussing fish welfare as presented in the published literature (Section 2) followed by a short description of environmental and food security issues (Section 3). We then outline a pragmatic ethics framework with a virtue ethics turn. While utilitarian and deontological approaches have their place in elevating fish into the moral community, other ethical approaches like pragmatism and virtue ethics offer new perspectives on resolving fish-human-environment dilemmas (Section 4). In the last section we explore a pragmatist cum virtue approach by applying it to some central areas associated with global consumption of fish as food (Section 5).

There are at least five central areas that we discuss here. They include: (1) consideration of fish welfare; (2) equity for those whose livelihoods are contingent on fish farming/aquaculture and the capture fisheries; (3) Environmentally sensible practices; e.g. avoid overfishing, protect coastal areas, rivers, lakes and oceans as sustainable sources of food and biodiversity; (4) consumption of fish and aquatic products as part of global food security, e.g. waste reduction through holistic systems; and (5) food safety, i.e. health aspects of human consumption. Although all the areas are important, due to limitations of space, the focus of this paper will be on fish welfare, environmental concerns and equity aspects of food security (see also Röcklinsberg, 2012 for a discussion of the intersection of fish welfare, climate concern and food security).

Do fish matter morally?

There are approximately 55,000 species of aquatic animals and 32,000 species of fish. Approximately 1,500 species are caught globally, in inland, coastal, and marine waters. There is compelling evidence that many of the fish we consume (e.g. tilapia, trout and salmon) are sentient and/or possess higher order cognitive abilities such as pain experience and consciousness (Braithwaite, 2010). As these capabilities are typically found to warrant that mammals should have their welfare taken into consideration it seems arbitrary not to include the relevant fish species in the ethical community as well. They have a welfare that matters to them and thus, we should factor this seriously into our moral deliberations (Lund *et al.*, 2007).

The welfare of fish is connected to mental experiences of fish such as sentience and cognition, capacities underscoring their interests and moral status. Scientists and philosophers rely on different arguments and evidence including research on brain and physiological structures and functions, analogy, interpretations of behavioral capacities, and evolutionary accounts where emotion and cognition have adaptive value and contribute to their fitness. More specifically, arguments regarding capacities in fish tend to turn on demonstrating that: (1) fish possess neuro-anatomical structures that give rise to affective states like pain, pleasure and other emotions like fear, for example, a lateral pallium; (2) fish possess physiological indicators beyond nocioceptors; and (3) sophisticated mentation in some species of fish is exemplified in behaviors such as future planning, learning (inferential reasoning and possession of beliefs and desires), aversion, frustration, self awareness and image and cooperative hunting (see for example Allen, 2013; Chandroo *et al.*, 2004; Sneddon *et al.*, 2003).

Notable dissenting views (include Rose, 2002,2007; Cabanac *et al.*, 2009) hold that fish and mammalian brains differ significantly and thus fish do not experience conscious pain, emotions and other affective states because they lack important structures like a neocortex or cerebral cortex. Furthermore, Rose (2002) argues that nociception (that is, sensory receptiveness to harmful stimuli) in fish is not accompanied by consciousness. Others like Cabanac *et al.* (2009) point out that consciousness emerged only in early *Amniota* (the first vertebrates fully to be independent of bodies of water), e.g. species that include mammals, birds, and reptiles but not fish.

The necessity for similar structures in different species begs the question of course. Absence of evidence (of these structures) is not evidence of absence (of consciousness or other capacities otherwise used as criterion for moral status) (Algers *et al.*, 2009; Lund *et al.*, 2007). Research focusing on capacities and behavior of animals themselves instead of relying on interspecific analogies show that other relevant structures, such as the lateral pallium that serve the same purpose in fish that the mammalian brain structures, do for humans (Braithwaite *et al.*, 2013). Where there is empirical uncertainty, vulnerable beings should be given the benefit of the doubt as a matter of morality (Braithwaite, 2010; Lund *et al.*, 2007). If mammals are granted moral standing based on being sentient, there is thus no reason to deny fish the same moral consideration.

Environmental and food security concerns

Fisheries and the environment

Growing consumption of food from animal sources and increases in the global population are straining already waning resources, some of which are non-renewable. Soil erosion, unscrupulous water extraction and waste discharge practices that plague inland fish farms, and ocean acidification, pollution and degradation of ocean floors from exploitative commercial endeavors challenge us to show more restraint in how we use or exploit resource (www.weforum.org/reports/global-risks-report-2011). The demand by aquaculture for small pelagic species as food is greater than global imports for human food uses (Tacon and Metian, 2009). Approximately 80% of 523 world fish stocks are nearly fully or over-exploited and concerns about recovery in order to maintain current output from capture fisheries is either pessimistic or uncertain (Worm *et al.*, 2009). In some cases, illegal and unregulated fishing is rampant (Agnew *et al.*, 2009). Unreported fishing, environmental, governance and management challenges also pose threats to food security (O'Leary *et al.*, 2011). Furthermore, increases in both herbivorous and omnivorous finfish output in aquaculture settings will likely require additional fish stocks as feed for larger fish (e.g. Atlantic salmon) and also boost the demand for greater fish processing waste and production of suitable replacement ingredients (Bostock *et al.*, 2010).

Global and local food security in relation to equity

Food security is defined by the FAO (2011) as, 'a situation that exists when all people, at all times, have physical, social and economic access to sufficient, safe and nutritious food that meets their dietary needs and food preferences for an active and healthy life'. Availability, accessibility, adequacy, acceptability and agency are dimensions brought into focus as fish or aquatic foods gain attention as important contributors to the global food supply and food security for a global population approximating 9 billion by 2050. Food security is also influenced by rising prices and increases in demand. But, can those without means, either because they have inadequate or marginal incomes, gain access to or afford these products? The vulnerable stakeholders here include fish farmers in developing economies and those who already have limited access to good sources of nutrition (especially if prices increase) (FAO, 2012a,b). Geopolitically, some regions will become more favored for fisheries and aquaculture than currently is the case. A delicate balance of regulation and inability to fish due to depleted wild stocks may not only lead to nutritional losses for many engaged in fisheries or aquaculture, but may also have devastating negative social and economic implications for small to medium commercial producers. Food security will involve generating adequate political will and appropriate strategies to not only protect and rebuild depleting fishery stocks, but more importantly, and where possible to improve resilience of important oceanic regions (Pikitch *et al.*, 2012).

A pragmatic and virtue based approach to issues of fish ethics

In the following we will employ an ethics framework inspired by pragmatists like John Dewey (2008: 1932) on the 'fish dilemma'. Such a methodology begins with an examination of 'focal practices' (Borgmann, 1984) or 'central habits' (McKenna, 2004) (and their related social institutions) seen as the basis of human flourishing and socio-economic-political-environmental-technological challenges. Further, pragmatists understand that a moral dilemma or dilemmas (or existing 'indeterminacy,') entail unresolved issues with moving parts, namely, they are 'wicked problems' (Rittle and Webber, 1973).

How can a pragmatic approach contribute to *framing* wicked problems associated with human consumption of fish so that *processes of inquiry* lead to viable strategies and solutions? At least three important ethical concerns need to be addressed when we consider the moral status of fish relative to the other concerns raised above. They include: (1) what are our habits or patterns of conduct and governance?; (2) where are these current habits and trajectories leading us?; (3) how are we to orient ourselves in relation to the five areas of fish welfare, human equity, environmental protection, food security and food safety? Given an intention to increase respect for fish as sentient animals, we argue these questions will encourage *active engagement* within a *community of interested interlocutors/stakeholders* from multiple backgrounds and expertise. When the relevant stakeholders deliberate about the 'fish dilemma' they must consider both substantive and procedural issues – that is the content of the issues need to be mapped and discussed, and the structure of the discussion should be open and promote *refinement* of the moral dilemma. Consideration of these issues will pave the way for the interlocutors to identify the conflict of duties and interests, and the epistemological gaps, associated risks and uncertainties, and strategies for resolution in latter stages of *meaning clarification* and *the context of ethical assessment*. It should be noted that the pragmatic method of addressing wicked problem is not static but rather open to revision as new dilemmas present themselves.

The framework we propose is both inclusive and integrative in nature. It is inclusive in that it considers the interests of relevant stakeholders such as the fish themselves, those who rely on capture fisheries and farm fishing as part of their livelihoods, the environment and consumers of fish products. Further it reflects *an ethics of engagement* since not only is it sensitive to the ways in which we are connected within the food, economic, political, moral and ecological chain that involves fisheries but it also seeks to integrate these facets in our moral deliberations, and to other dimensions of morality such as character, judgment and motivation.

Finally, it should be noticed that such a framework rests on the assumption that the deliberators are interested in promoting the good of all stakeholders involved (including the fish themselves). Thus, a virtues oriented perspective focusing on habits and trajectories that contribute to a life worth living or flourishing is a suitable point of departure. The categories of virtue we have in mind here include measured consumption habits, efficacy and social relations (Liszka, 2002). Our contention is that these virtues can motivate pragmatic discourses around good governance of capture fisheries and aquaculture. While we do not tease out the specific virtues here but lay the groundwork for more specific discussions, arguably, they include private and civic virtues. The approach provides a more holistic gaze on both the nature and impacts of our dominant production and consumption habits or focal practices, the question of the morality of animal-based diets, and the challenges of linking future fish supplies, from both capture fisheries and aquaculture with resource systems, and human and institutional capacities. The set of virtues that underpins the framework we suggest must be sensitive to: (1) impacts of farm and capture procedures on fish; (2) stakeholder preferences; (3) quality of life issues; and (4) contextual features (these questions are inspired by Jonsen *et al.*, 1998).

Applying the framework to the five areas

In this section we will show one possible way to apply the pragmatist virtue approach to three of the abovementioned five areas; Fish welfare, environmental concerns, and food security. Due to space limitations, this will serve as a sketch of possibilities rather than an in-depth elaboration. Also, whether or not fish use (farmed or captured) should be considered a resource in solving food insecurity or be abandoned as inherently wrong practice in the light of our knowledge of fish as sentient beings, is an issue worth considering in more detail. For the sake of making the starting point and application of the ethical approach as clear as possible, our point of departure here will be the line of argumentation by FAO (2012a,b), i.e. fish is and will be an important contribution to minimizing global starvation and nutritional deficits. The ethical task set forth here is thus to discuss how this can happen in a more sustainable and ethically sound way than is occurring today.

These issues can be elaborated by the following procedural steps (in italics):

1. *Framing* could be taken up this way: If fish supplies are to be increased to meet future global demands, namely, expected population growth, consumer preferences for a variety of animal protein sources and capacity to consume due to rise in incomes, in a way that marries well profit with animal care and welfare, adaptive changes in policies, practices and fisheries governance need to occur beyond private consumer decision-making. What changes should be taken up?

2. In undertaking the *process of inquiry*, we should consider asking what is our vision for the future of fisheries? Here, the *community of interlocutors* must understand and plan for social, economic, political and cultural transformations that will be impacted by innovations in capture fisheries and aquaculture need to be well organized, transparent, scientifically validated, morally just and participatory. An integrated regulatory, policy and program scheme that seeks to prevent contentious partisanship will improve cooperation between all the stakeholders and affected communities and industries.
3. During *meaning clarification*, core values, aims and central background conditions to promoting both private and civic virtues should be raised and scrutinized. Perceived and actual drivers (e.g. epistemological and ethical) that influence good governance and which may corrupt the industries should be identified as well as the standard for balancing economic opportunities, fish welfare and related environmental and food security risks with industry practices.
4. In the *context of ethical assessment*, interventions or strategies (e.g. policies, regulations, procedures and best practices) should be tested to see if they produce satisfactory solutions. The set of values motivating each stakeholder will likely be revealed as well as the nature of the interplay between the drivers. The approach suggested here will challenge the interlocutors to identify and potentially attempt to reconcile conflicting values and priorities. In cultivating a culture of responsibility, we argue that the processes of framing and inquiry in the current situation, should center on actions by industry to ensure better fish welfare standards, and on governmental agencies and the research community to address the interconnections between welfare, environment and food security. Above this, their communicating with each other and the public will improve overall acceptance of proposed interventions.

By nudging the pragmatic approach to decision making towards virtue ethics, we can take a broader gaze on what is necessary to encounter and engage the fish issue. A virtues approach recognizes the importance of ensuring direct respect for fish and their welfare, and the goals of an ethically viable and practically sustainable, safe and secure global food system (with, without or with less dependence on animals in our diets) and the necessary conditions for its cultivation. Our policies, principles and practices should flow from our view of what contributes to human flourishing in order that we produce and reproduce good practices that better relate to the non-human world. A pragmatic framework that is steeped in virtue theory can promote a kind of moral ecology that challenges us to participate in a continuing interconnectedness (reciprocal engagement) between humans and the natural world. A virtue ethics account applied to food engenders modes of conduct tied to food that can help promote a more intimate knowledge of the practices that cultivate and sustain the food upon which a well-functioning and equitable society may rely. There is a life that is better than focusing on consumption, and which guards against exaltation of the human, while promoting understanding and adoption of habits that are durable, responsible and authentic (Borgmann, 1984). Every member of the fish-as-food moral community plays an integral role in contributing to the health and balance of our lived spaces and should be respected accordingly.

References

Agnew, D.J., Pearce, J., Pramod, G., Peatman, T. and Watson, R. (2009). Estimating the worldwide extent of illegal fishing. PLOS ONE 4(2): E4570.

Algers, B., Blokhuis, H.J., Bøtner, A., Broom, D.M., Costa, P., Domingo, M., Greiner, M., Hartung, J., Koenen, F., Müller-Graf, C., Morton, D.B., Osterhaus, A, Pfeiffer, D.U., Raj, M., Roberts, R., Sanaa, M., Salman, M., Sharp, J.M., Vannier, P. and Wierup, M. (2009). Scientific Opinion of the Panel on Animal Health and Welfare on a request from European Commission on General approach to fish welfare and to the concept of sentience in fish. The European Food Safety Authority Journal 954: 1-26.

Allen, C. (2013). Fish cognition and consciousness. Journal of Agricultural and Environmental Ethics 26(1): 25-29.

Borgmann, A. (1984). Technology and the Character of Contemporary Life: A Philosophical Inquiry. Chicago: University of Chicago Press, 1984.

Bostock, J., McAndrew, B., Richards, R., Jauncey, K., Telfer, T., Lorenzen, K., Little, D., Ross, L., Handisyde, N., Gatward, I. and Corner, R. (2010). Aquaculture: global status and trends. Philosophical. Transactions of the Royal Society B Biological Sciences 365(1554): 2897-2912.

Braithwaite, V. (2010). Do fish feel pain? Oxford University Press, Oxford, UK.

Braithwaite, V.A., Huntingford, F. and Van den Bos, R. (2013). Variation in emotion and cognition among fishes. Journal of Agricultural and Environmental Ethics 26: 7-23.

Cabanac, M., Cabanac, A.J. and Parent, A. (2009). The emergence of consciousness in phylogeny. Behavioral Brain Research, 198: 267-272.

Chandroo, K.P., Duncan, I.J.H. and Moccia, R.D. (2004). Can fish suffer?: perspectives on sentience, pain, fear and stress. Applied Animal Behaviour Science 86: 225-250.

Dewey, J. (1985/1932). Ethics (vol. 7). Southern Illinois University Press, Carbondale, IL, USA.

Dewey, J. (2008). The middle works of John Dewey, Vol. 14, 1899-1924 Human Nature and Conduct, 1922 1st Edition Edited by Jo Ann Boydston. Southern Illinois University Press, Carbondale, IL, USA.

FAO (2001). The state of food insecurity in the world. Food and Agriculture Organization of the United Nations, Rome, Italy.

FAO (2012a). The state of world fisheries and aquaculture. Food and Agriculture Organization of the United Nations, Rome, Italy.

FAO (2012b). The state of food insecurity in the world. Food and Agriculture Organization of the United Nations, Rome, Italy.

Jonsen, A.R., Siegler, M. and Winslade, W.J. (1998). Clinical ethics: a practical approach to ethical decisions in clinical medicine. 4th edition. McGraw-Hill, Inc., New York, NY, USA.

Liszka, J. (2002). Moral competence: an integrated approach to the study of ethics. 2nd edition. Prentice Hall, Upper Saddle River, NY, USA.

Lund, V., Mejdell, C., Röcklinsberg, H., Anthony, R. and Håstein, T. (2007). Expanding the moral circle: farmed fish as objects of moral concern. Disease of Aquatic Organisms: Disentangling socio-political, moral, ethical and scientific aspects 75: 109-118.

McKenna, E. (2004). Pragmatism and the production of livestock. In: McKenna, E. and Light, A. (eds.) Animal pragmatism: rethinking human-nonhuman relationships. Indiana University Press, Bloomington, IN, USA, p. 160-178.

O'Leary, B.C., Smart, J.C., Neale, F., Hawkins, J.P., Newman, S., Milman, A.C. and Roberts, C.M. (2011). Fisheries mismanagement. Marine Pollution Bulletin 62(12): 2642-2648.

Pikitch, E., Boersma, P.D., Boyd, I.L., Conover, D.O., Cury, P., Essington, T., Heppell, S.S., Houde, E.D., Mangel, M., Pauly, D., Plagányi, É., Sainsbury, K. and Steneck, R.S. (2012). Little fish, big impact: managing a crucial link in ocean food webs. Lenfest Ocean Program. Washington, DC, USA, p. 108.

Rittel, H. and Webber, M. (1973). Dilemmas in a general theory of planning. Policy Sciences 4: 155-169.

Röcklinsberg, H. (2012). Fish for food in a challenged climate – ethical reflections. In: Potthast, T. and Meisch, S. (eds.) Climate change and sustainable development. Ethical perspectives on land use and food production, Wageningen Academic Publishers, Wageningen, the Netherlands, p. 326-334.

Rose, J. (2002). The neurobehavioral nature of fishes and the question of awareness and pain. Reviews in Fisheries Science 10: 1-38.

Rose, J.D. (2007). Anthropomorphism and 'mental welfare' of fishes. Disease of Aquatic Organisms 75: 139-154.

Sneddon, L.U., Braithwaite, V.A. and Gentle, M.J. (2003). Do fish have nociceptors: evidence for the evolution of a vertebrate sensory system. Proceedings of the Royal Society 270: 1115-1121.

Tacon, A.G.J. and Metian, M. (2009). Fishing for feed or fishing for food: increasing global competition for small pelagic forage fish. Ambio: A Journal for the Human Environment 38(6): 294-302.

Worm, B., Hilborn, R., Baum, J.K., Branch, T.A., Collie, J.S., Costello, C., Fogarty, M.J., Fulton, E.A., Hutchings, J.A., Jennings, S., Jensen, O.P., Lotze, H.K., Mace, P.M., McClanahan, T.R., Minto, C., Palumbi, S.R., Parma, A.M., Ricard, D., Rosenberg, A.A., Watson, R. and Zeller, D. (2009). Rebuilding global fisheries. Science 325: 578-585.

Animal welfare, consumer behaviour, and public policy

J.L. Harfeld
Centre for Applied Philosophy, Aalborg University, Sohngaardsholmsvej 2, 9000 Aalborg, Denmark; jlh@learning.aau.dk

Abstract

This paper investigates the normative aspects of animal food production and retailing and takes as a presumption that food consumers deliberately choosing animal-welfare products act normatively (on a moral basis) and thereby include values that either supervene or supplement the values of product price, taste, nutrition, etc. The animal welfare aspect of a given product is a 'credence attribute' and can as such not be directly experienced via the product. Thus, information external to the product becomes the only conditions from which to evaluate the product and motivate possible behaviours. There is, however, a well-documented attitudinal ambivalence in food product shopping where we experience inconsistencies between what could be defined as general citizen attitudes or values on the one hand and consumer behaviour patterns on the other. In other words, in anonymous studies people will generally claim a higher level of concern for animal welfare than is subsequently obvious from the sales records of grocery stores. There is, however, also a measurable connection between costumers' willingness to pay (extra) and moral imperatives derived from values about animal welfare. The paper will focus on analysing the relation between the animal welfare related ethical values of consumers and some of the extents and limits of consumers' ethical actions. It will, furthermore, normatively evaluate the concrete social systems of urban life – economic, political and cultural – which constitute the environmental extents and the limits of individual consumers' abilities for ethical deliberation. Thus the main question becomes: What is the connection between the consumers' ability for ethical deliberation (what we with Aristotle could call phronesis) in food choices and the cultural, economic and political framing of consumer deliberation? This question leads to a normative and constructive sub-question which asks what kind of systems of public policy a society needs in order to positively enable the individual consumer to deliberate phronetically. In other words: how do we create a society with genuine possibilities for ethical consumerism?

Keywords: animal production, ethical consumerism, public policy

Introduction

The industrialisation and on-going intensification of agriculture has not only changed the structure of agriculture itself, it has also been problematic in relation to some aspects of animal welfare. It has, furthermore, changed the very demographics of Danish society. The countryside still gives the impression of an agricultural nation and farming might still be considered part of the national identity, but the individual involvement and relations with agriculture have been gradually disappearing. In 1950, roughly a quarter of all adult Danes were employed in agriculture. Today it is about 3% and declining. A paradigmatic occupational shift of such magnitude must necessarily lead to a gradual estrangement between the general public and the agricultural industry. First and second hand knowledge of animal based food production is simply no longer common place among the public. At the same time, however, the production and retail of animals, eggs, and milk has increased dramatically and the Danes have become the number one per capita meat consumers in the world. In other words, we consume more animal products and know increasingly less about the animals.

On the other hand – or maybe also as a reaction to this – the latter part of the 20th Century has seen the birth and development of a new type of political food consumer focusing on ethical aspects of the origin (primary production) of what we eat. When Ruth Harrison, indignantly inspired by the plight of animal life in modern agriculture, wrote her famous book, Animal Machines (Harrison, 1964), she was but one voice in a choir of a slowly but surely increasing awareness of, and interest in, animal rights and welfare in relations to what we eat. Today the state certified organic (which includes significant welfare parameters) production of meat in Denmark is a 40 million EURO business and almost 20% of eggs produced are organically raised – both numbers are slowly rising.

When food consumers deliberately choose animal-welfare products on the basis of animal welfare they include values in their deliberation which either supervene or supplement the values gained through product price, taste, nutrition, etc. These 'credence attributes' (Verbeke, 2009) are not something the consumer can possibly acquire any knowledge about by buying, handling, tasting and digesting the food. They are in all ways external to the physical product and any evaluation of them – and any accompanied consumer behaviour – relies on the knowledge and understanding achieved 'behind the project'. As an ethical credence attribute, animal welfare is a motivational factor in consumer behaviour to such an extent that consumers are willing to pay significantly extra for food from production systems labelled as animal welfare friendly (Bennett *et al.*, 2002; Makdisi and Marggraf, 2011).

However, credence attributes about animal welfare, although influential factors of consumer behaviour towards, for example, organic products, do not necessarily entail a movement from attitudes to action. In other words, studies have shown that citizens generally claim a higher concern for animal welfare in food production than is actually discernible in the analysis of grocery store receipt records (Verbeke, 2009).

Systems as ethically relevant

If the above description of the consumer-production situation is correct then we would have to look at the possibilities for true ethical/political consumerism in a sceptical light. I suggest that we need what could be called ethical facilitation. In other words, ethical and animal welfare inspired consumer behaviour in the purchase of food can be facilitated through at least two distinct avenues of policy and politics. The first avenue is a heightened focus on adequate, easily accessible and correct information about the products – both as general easily accessible knowledge and as personal experience. The second avenue is less concrete but possibly much more powerful in its effects and focuses on establishing a political society that supports and encourages ethical considerations for food production animals. In the following I will address mainly this second avenue.

To take an animal ethical view when delving deeper into the complicated issue of consumer knowledge, perception, understanding and deliberation is to recognise the animal welfare problems as both individual and systemic. On the scale on individuals acting in certain ways, animal welfare problems can be seen as deriving from clearly identifiable subjects – they are matters of what Slavoj Zizek would call 'subjective violence' (Zizek, 2009). In such instances the perpetrator of harm is easily identifiable because he or she is an individual – a subject which we can point to and say 'you did it'. In such instances, possible punishment, deterrent or other measures to avoid future harm can be adopted and applied to the subject/individual. On the other hand, many animal welfare problems derive from an 'objective violence' which is the consequence of sometimes inscrutable systems of human organisation, policy and culture. Indeed, objective violence is harm originating from a less questioned backdrop of our society. It is, as Zizek points out: 'violence inherent to the 'normal' state of things' (Zizek, 2009: 2). As individuals (as both citizens and consumers) in different societies we are under the influence of a number of 'normal states of things' in the shape of common background societal ideologies. We are, so to speak, not only embedded in interrelational contexts but also in spheres of thought deriving from powerful and power asserting social, economic and political systems. We are, as Aristotle phrased it, *zoon politicon* and thus for better and for worse embedded in the social fabric of the society in which we live. We cannot be understood merely as individual consumers – hermetically closed from the systems in which we consume, act, and buy.

As consumers – or consuming citizens – we are embedded in the system of markets and market thinking. National and global markets go beyond and across borders and remain unquestioned in their essence and often unseen in their extent. All markets have different levels of national and international freedom, restriction and regulation; but no matter how they function, their function is mechanistic, i.e. the market does not *want* anything. The invisible hand (Smith, 2007: 351) is mind- and heartless. When the markets work 'well' it is a matter of efficiency in providing trade between individuals. There are, however, restrictions on the market which we take for granted. We are, for example, not allowed to sell certain things freely in the market. Cocaine, guns and children are entities which we do not accept as proper market entities. Some other value or set of values than the freedom and efficiency of the market is at stake here. The danger of markets is, in Zizek's terminology, the objective violence of an unquestioned (or little questioned) market-type evaluation which can only appraise the animals in production systems within an instrumental and functionally based category. This makes it impossible – or at least hard – to conceptualise animals in agricultural market systems with an ethical approach of intrinsic animal value. We inevitably make an evaluative mistake when approaching the animal in the system of the market only. Mark Sagoff calls it a category mistake (Sagoff, 1989: 10) and it is a mistake about the kinds of value we are talking and asking about. If the question of the value of a pig can only be answered in terms of market value in money then we are unable to talk meaningfully about animal ethics.

Changing systems, changing views, changing behaviour

Given this kind of systematic problem of society, what kind of systems of public policy would a society need to positively enable the individual to ethical consumer behaviour? How do we create a society (how would a society have to be) in which the capability for phronesis and ethical conduct is enabled for the citizens – in such a way that they can choose properly when grocery shopping? A number of answers could be given to these questions. We could increase public awareness through government information campaigns, we could expand the range of labelling to include both negative and positive areas, and we could prohibit certain kinds of food or food practices. The most interesting – and perhaps the most radical – is, however, to focus on changing the society construct by enabling an inclusion of animals in the system of the citizen state. In other words, change the backdrop and not things against the backdrop. To achieve what Rosalind Hursthouse calls 'a shift of moral vision' (Hursthouse, 2000: 165) towards animals we need to change the background against which we are presently seeing them.

I suggest the Nordic welfare state model has potential to include animals in a way which would constitute such a change in background. The Nordic welfare state's emphasis on the relational properties of rights and citizenship is mirrored in its concept of solidarity. Solidarity is commonly defined as having both a descriptive characteristic of a certain common-ground structure in a social group, and as a normative characteristic concerning members of a social group having 'mutual obligations to aid each other, as and when should be necessary' (Bayertz, 1999: 3).

Concerning the former, the problem quickly becomes one of community cohesion. Solidarity, as well, I believe, as Aristotelian communitarian based theories, relies on a given group's participants having some kind of common linkage by which each member can be ascertained as 'one of us.' Whatever this linkage is anchored in, it be rationality, shared values, common fate or common history, it implies that: (1) there is a limit to the community and thus someone or something must be external to it; and (2) the characteristics of those who are external to the community provide, to a certain extent, a definition of the community members through their otherness. In relation to our present question about farm animals this raises the question of inter-species descriptive solidarity – i.e. is it meaningful to understand the animals as integrated into human social groups and, indeed, can people and animals understand themselves as belonging to the same social groups? In other words: does descriptive solidarity necessarily entail bonds that could not exist between man and animal? I do not believe this is the case. We simply have too much evidence of group-oriented human-animal bonding to reject the entire notion of inter-species communities. Companion animals such as dogs are commonly referred to and understood as parts of families and, conversely, dogs understand themselves as part of the family (pack) (Wells, 2009). It seems to me, that there are no logical species boundaries in descriptive solidarity relating to humans' understanding of the relationship. We are, as humans, obviously capable of '*feelings* of belonging together or sympathy' (Wildt, 1999: 217) towards animals of many different species.[26] This, however, is most likely not a reciprocal understanding and besides a few examples – most noticeably dogs – it would be difficult to attribute a general cross-species feeling of belonging from animals in general to humans in general. On the other hand, it is hardly a necessity for community membership (descriptive solidarity) that all participating members are rational beings and, because of this rationality, understand their connectedness to the whole that is society. The community and levels of solidarity are defined from the stand-point of those rational beings who understand the concept of the communal connectedness. There is no logical reason why such a community should not include beings which do not understand their part in the community, nor partake knowingly in its structuring.

We encounter comparable issues when addressing the normative characteristics of solidarity – and, to a certain extent, comparable answers. Understanding solidarity as a moral relation between individuals is, as its communitarian roots suggest, a very old perception. Although Theophrastus (approx. 371-287 BC) to a large extent continues the philosophy of his teacher Aristotle, he rejects Aristotle's dismissal of animals as belonging in the moral community and maintains 'that animals enjoy kinship (*oikeiotês*) with us, so that it is unjust to kill them' (Sorabji, 1993: 177). With this moral notion of kinship, Theophrastus introduces a non-speciesist normative solidarity (as well as non-speciesist communitarian ethics). It is the matter of belonging which is stressed instead of rational (contract-) abilities and as such it supports an asymmetrical solidarity that could be extended to include the non-human parts of the community as rights beneficiaries – albeit not as obligation-havers. One could possibly make an argument supporting the claim that animal rights and animals' inclusion in the moral community is impossible due to the limits of moral psychology. Bernard E. Rollin states that we have a deep-seated '(perhaps biologically based) intuition that we favour those made close to us by bonds of blood, friendship or love' (Rollin, 2005: 110) and this is the foundation of what he calls the 'reasonable partiality' of moral psychology. We are limited by our psychological make-up to such a degree that it renders us incapable of being universally impartial. Rollin does, however, claim that partiality, based on Aristotle's concept of *philia*,[27] has already been extended to many companion animals (Rollin, 2005: 120) and that there is no reason why, with time, such extensions of moral psychology could not encompass other animals. And by this 'encompasion' we change our perception as consumers and citizens and this correlates directly to our positive or negative attitudes and behaviour towards them. Attitudes and their corresponding behaviour are not fixed entities and can change, reflecting new beliefs and opinions (Knight *et al.*, 2010). Such changes, seen not individually but on a societal level, are the development of what Rollin calls 'Ethics' – i.e. 'the set of beliefs that society, individuals, or subgroups of society hold about good and bad, right

[26] Our feelings of connectedness will, however, decline with every step away from the characteristic ways of life that we share with the animals. As easy as it is to understand and feel connected to a number of the aspects of being a dog or a pig, it is equally difficult to understand and feel connected to being a spider. Arguing along similar lines it would most likely be impossible to conceive of a community in which rational beings came to understand themselves as vitally connected to inanimate objects. The otherness of the rock is, as Martin Heidegger described it, that it has no world (Heidegger, 1995: 19), and without a world to compare to our own the rock eludes the grasp of community.

[27] The concept of *philia* is most often translated to 'friendship' or 'brotherly love' and has in Aristotle a wide ranging number of qualities including being the mark of a just society (Schwarzenbach, 1996: 97-128).

and wrong, justice and injustice, fairness and unfairness' (Rollin, 2006: 31). Changing these beliefs is the key to changing the lives of animals and increasing animal welfare among the animals in our society.

References

Bayertz, K. (1999). Solidarity. Kluwer Academic Publishers, Dordrecht, the Netherlands.

Harrison, R. (1964). Animal machines: the new factory farming industry. Stuart, London, UK.

Heidegger, M. (1995). Phänomenologische Interpretation von Kants Kritik der reinen Vernunft. Vittorio Klostermann, Frankfurt am Main, Germany.

Hursthouse, R. (2000). Ethics, humans and other animals: an introduction with readings. Routledge, London, UK.

Knight, S., Bard, K., Vrij, A. and Brandon, D. (2010). Human rights, animal wrongs? Exploring attitudes toward animal use and possibilities for change. Society & Animals 18(3): 251-272.

Makdisi, F. and Marggraf, R. (2011). Consumer willingness-to-pay for farm animal welfare in Germany – the case of broiler. In: Proceedings of the 51st Annual Conference, Halle, Germany, September 28-30, German Association of Agricultural Economists (GEWISOLA).

Rollin, B.E. (2005). Reasonable partiality and animal ethics. Ethical Theory and Moral Practice 8(1): 105-121.

Rollin, B.E. (2006). Science and ethics. Cambridge University Press, Cambridge, UK.

Sagoff, M. (1989). The economy of the earth: philosophy, law, and the environment. Cambridge University Press, Cambridge, UK.

Schwarzenbach, S. (1996). On civic friendship. Ethics 107(1): 97-128.

Smith, A. (2007) (orig. 1776). Wealth of Nations. Cosimo Classics, New York, NY, USA.

Sorabji, R. (1993). Animal minds and human morals: the origins of the western debate. Duckworth, London, UK.

Verbeke, W. (2009). Stakeholder, citizen and consumer interests in farm animal welfare. Animal Welfare 18: 325-333

Wells, D.L. (2009). Behaviour of dogs. In: Jensen, P. (ed.) The ethology of domestic animals: an introductory text. CABI Publishing, Wallingford, UK.

Wildt, A. (1999). Solidarity: its history and contemporary definition. In: Bayertz, K. (ed.) Solidarity. Kluwer Academic Publishers, Dordrecht, the Netherlands.

Zizek, S. (2009). Violence. Macmillan, New York, NY, USA.

The 'secret' of killing animals

M. Huth
Messerli Research-Institute, Vienna, Austria; martin.huth@vetmeduni.ac.at

The basic question of this talk would be the one if phenomenology can contribute to awareness of 'meat production' regarding especially citizen's responsibility in dealing with and possibly changing current conditions. Therefore I will describe the notion of a problematic but unsolved inconsistency in the treatment of different (but similar) animals in current animal ethics and in our society. But we know that for a long time – why didn't we change our practices rationally yet? I will recur on phenomenology as method to reveal structures of normatively relevant significances qua 'matters of concern' (Latour, 2004: 231). These structures constitute our views on animals and their differences for us – there is no animal per se, thus, radical consistency and equality is something utopian. Although ethical considerations have to proceed from a *diversity* of practices and thus different treatments, there is also *one fundamental structure* behind all our experiences of animals: their vulnerability (Negative Integrity, to be explained in the abstract) which constitutes a moral demand that *cannot not* be answered. But this vulnerability is hidden in meat production (and lab experimentation); we perceive – under current conditions – rarely what happened before we grab the packed meat in the shelf. The citizen's responsibility is thus a critical distance to our practices deriving from an *enabled experience* of animals reified and instrumentalized (although vulnerable) in meat production. The consequence would not be an equal, but at least a more humane treatment by altering our practices. Our practices are always tied to a practical tradition (and tradition of experience). So they cannot be altered radically instantly (as Regan, Singer and others would demand from us) – but they can be changed proceeding from the normalities we live in.

I want to exemplify phenomenology's specific competence in concentrating on the apparent inconsistency between our practices in meat production and our treatment of pets qua companion animals. While we want to treat the latter humanely until the very last phase of their lives (e.g. euthanasia tied to certain rituals), meat production is usually first of all a matter of economic efficiency; society in general cares only *more or less* about the well-being of these animals and do moreover not really think about prolonging (and sometimes even improving) their lives like in pets. We keep them in order to kill them. Economy and nutrition seem to justify the killing and the keeping conditions as well as the 'collateral damages' of pollution, etc. In this case the citizen's possibilities to change common and traditionally grown practices seem moreover very limited. Although there are a lot of opposing voices, they regularly do not have much influence on a practical level. An example would be the prohibition of farrowing crates in Austria; due to apparent 'economic and structural constraints' the period until they are probably banned lasts 20 years. It is to be suspected that a similar proceeding in welfare of companion animals would be publicly condemned – while the treatment of pigs seems not or only to a certain degree a matter of public interest.

This appears as a veritable inconsistency since there are similar, even equal animals which are treated in part extremely differently. Dogs and pigs share a lot of socio-cognitive abilities (see Benz-Schwarzburg, 2012), but dogs are considered as family members, pigs are mostly considered as future Schnitzel. However, these abilities are often acknowledged as an anchor of moral considerations. There are a lot of current approaches which ask for an elimination of such inconsistencies (e.g. O'Sullivan, 2011), but in fact this was also a claim of P. Singer (Singer, 1979, 2011) and T. Regan (Regan, 2004) more than three decades ago. Why is there no significant turn in our practices when we know well that there are obvious logical/moral inconsistencies? Is there a problem with our ethical consideration, is there a problem with our ability to change the circumstances including our own behaviour? Is there a problem with changing our habits rationally? Or is there a lack of interest in being rational since our current practice is too comforting?

Cora Diamond and others indicate that there might be a fundamental problem with 'classical' approaches to animal ethics. They concentrate on the animals 'per se' but not on the relation we have to them and the significance they have for us, and this leads in part to counterintuitive consequences (Diamond, 2012: 39). Singer wants to find 'neutral grounds' from which we can see that animals are to be treated equally on conditions of self-awareness and the ability to suffer (Singer, 2011: 92f). As rational beings we could take that into consideration and have criteria for a right treatment of all living beings. Any significance by our usual practices is bracketed out. But according to Singer's approach it would be only a matter of being too squeamish not to eat humans that have died in car accidents and by lightning because they are not harmed and so the criterion of suffering is not applicable any more (Diamond, 2012: 86). Singer cannot explain why for instance a dead human body is simply 'nothing to eat' in our and nearly every other societal context – and he claims that the different treatment of similar animals is irrational and leads in part to speciesism. But this is not the way we deal with ethical problems within our practices and habits. We do not overthrow automatically our *orthopraxies* (Husserl) if there is a purely rational argument which opposes them, our habits are not to be altered by blunt arguments, no matter if they are logically sound. This is also shown by the fact, that there were even demonstrations against Singer when he was speaking publicly in Germany, not directly because of his animal ethics approach but because of his position concerning severely disabled new born. But these two issues have a fundamental connection since Singer uses the criteria of self-awareness and ability to suffer undermining common understandings of matters of concern and stating indirectly that being a human is per se morally totally irrelevant – a matter of fact, but not of concern. Demonstrations were not only a matter of ideological eagerness but an opposition against an approach that claims obligatory what is against fundamental convictions.

Phenomenology is a philosophical approach that provides the possibility of revealing the fundamental traits of our experiences of (killing) animals by bracketing out what does not stem from experience itself. Against the background of a confusing, overcomplicated situation we can draw our attention to these traits instead of theoretical constructions like self-awareness that has apparently to be proved by experiments and methodological observations. Phenomenology is a philosophical approach that aims to describe (and by that interpret, there is no merely objective description) our fundamental experiences, but also practices and habits well (e.g. Merleau-Ponty, 1966: 18).

Here we can say that there are two branches, here as (a) and (b), of a proper description of the situation, one concerning the multidimensionality or diversity in our practices, the other one focusing a fundamental structure behind all experiences of animals.

a. There are very diverse experiences of animals. Since we cannot perceive any animal per se, the consideration of them is tied to certain practices which constitute certain points of view. There are cute animals, dairy cows, dangerous animals, guide dogs, companion animals as family members, animal pests, and many more images of animals. We treat animals according to the significance they have in our point of view, and we cannot leave our perspective behind since we are corporeal beings that are not purely rational; instead we have or are a kind of embodied rationality (Merleau-Ponty, 1966: e.g. 401). In acting and thinking we can only proceed from the practice we are already in, and these practices also constitute normative claims. We can bring ourselves in critical distance to these normatively meaningful practices, but we stay always in a perspective basing on these kinds of habits.
b. But we can also describe a fundamental structure of all experiences of animals. Proceeding from Merleau-Ponty's concept of intercorporeality and interanimality (Merleau-Ponty, 2000: 227) we can state that nonhuman living beings affect and address me in their 'mere' appearance. Our and their environments (see Uexküll's concept of *Umwelt,* see also Merleau-Ponty, 2000: 232) have an intersection, we cannot really chose if we want to be affected by them. This leads to another phenomenological approach: Emmanuel Levinas described the ethical situation of the Other that appears as vulnerable being and therefore as a being with a demand to which *I cannot not response* (Levinas, 1992, passim). There is something categorical in the demand of the Other constituted

by her or his vulnerability and mortality. A general positive definition of vulnerability is however impossible because every species and every individual might have a certain constitution and a certain way of being-in-the-world that can be harmed intentionally or eventually. The impossibility of defining vulnerability as basis for what is usually (especially in the Dutch debate) call integrity leads me to the concept of *Negative Integrity* (Huth, 2013). This refers to the corporeal being as a whole including 'bodiliy' as well as intellectual, behavioural, species specific and individual traits. Integrity as expression of being unhurt intact (described in Rutgers and Heeger, 1999) *is derived from* what I call Negative Integrity. It is our 'ability' (actually a kind of passivity, not of spontaneity) to be harmed (Derrida, 2010: 55). Another problem in defining (negative) integrity explicitly is named in the following quote: 'You cannot point at this integrity directly. (...) You begin to see integrity when you see how he whole informs every part' (Craig Holdrege in Schmidt, 2008: 161f). Nevertheless, vulnerability is something that you cannot neglect, something you cannot not be responsive to.

Our ethical considerations proceeding from our habits and common practices do have a fundamental problem when dealing with meat production. I would argue that we are not rigidly stuck in our routines, as Singer and Regan would probably state. The problem here is that we have something like a hole in the net of our everyday experience regarding this topic (Ricoeur, 1995: 223). Horkheimer uses the metaphor of a skyscraper to describe our society, and in the cellar there is the slaughterhouse (Horkheimer, 1987; 329f). I think that this metaphor has a double meaning. Firstly, some animals are excluded from a lot of political and also moral considerations, 'pushed to the margins and down to the cellar'. They are just animals (while pets are family members). Secondly, they are (made) invisible. Missing experiences lead to unreflect-able practices.

Walter Burkert shows in his book *Homo Necans* that in ancient rites there were topoi included claiming that animals were nodding in approval to their being sacrificed (Burkert, 1997: 11). This is not 'true' but shows that there has been a certain awareness of the vulnerability I described above – and a way of dealing with the brutality of ending an Other's life and possibly causing pain. Today, the killing of animals is out of our experience, it happens outside the cities, outside of usual spiritual practices, too. There is no more prosaic institution than a slaughterhouse. Here we have no *dealing with* vulnerability, we face a veritable hiddenness (especially for the public view, but also for the workers there). Of course there are numerous shock documentaries, but our everyday lives do not consist of a being confronted with such movies or articles but of grabbing packed meat in the shelf.

Meat production is not the only issue where vulnerability is hidden by a reification; in the lab we find a very similar matter – as the following quote should show heuristically: 'They did not like having rats in clear cages "because the animals could look at you"' (Linda Birke in Acampora, 2006: 100). We see that our practices (in)tend to happen offside the experience of vulnerability.

Hence, the situation of the citizen is in a way a precarious one. We are prevented and we prevent ourselves from recognising Negative Integrity in the context of meat production. As Iris Marion Young (Young, 2011, passim) and others found out, morally problematic issues do not only stem from intentional actions of autonomous subjects which want to do 'something bad'. Structures and events are constituted not only intentionally, but at least the maintaining needs not to be merely eventually. Citizens (as consumers) can and should gain awareness that the hidden demand of 'meat animals' is only hidden but not inexistent, and furthermore, they can try to change their habits; but only by turning their gaze intentionally to the production conditions.

Thus citizen taking increased responsibility will start with the task to get informed about procedures of 'meat production' and to develop awareness about the diverse formations in which animals and their treatment (including their death) occur. We should try to fill the hole in our experience. Here we meet the approach of the Critical Theory stating that any reification of living beings might be grounded in oblivion (Adorno and Horkheimer, 1989: 270-289) or, as I would say, hiddenness.

It is not our task to compensate global or just certain cultural structures from which we cannot fully emancipate ourselves; we are socialized in a certain context of normality and normativity which we can reflect only from a perspective, and therefore never totally. There is no pure reason determining our actions. We are not capable of being totally autonomous (in a Kantian sense) since we are not in the position to be independent of the world we live in and of what is going on in common practices. *Ultra posse nemo obligatur*. Paraphrasing Derrida (2010), a good conscience builds a mere fiction and is nevertheless something we should strive for. Personal responsibility seems limited but neither arbitrary nor weak. To take responsibility would not mean that there is an obligation to end all inconsistencies but to change and deal with them in a way that we can call a viable compromise. This must be established in a discourse, which in times of so-called post-democracy cannot rely on fixed institutions; but we see e.g. NGOs and interest groups working for such a compromise.

Changing means changing and refining our common practices, our orthopraxies and the normativity tied to them, but proceeding from these practices and significances, not from pure (practical) reason. This would not end all 'inconsistencies' in our common practices, but would change habits at least in part and could lead to an ethically reflected dealing with these inconsistencies. Animals would be treated according to their respective 'matter of concern' – that would not lead to a total equality (animals are not to be treated offside our practice they are part of), but a matter of concern is locked against radical reification and instrumentalization. Phenomenology can therefore contribute to meat consumption a kind of responsiveness or attentiveness regarding the Negative Integrity that opposes practices and structures that conceal the vulnerability behind technical expirations and apparently forced properties. This would be a more practical issue than concentrating on (human) traits in animals that we can only assume but not perceive and that build more a matter of fact than of concern. The citizen's task to fill the hole in experience is not comforting at all but not as supererogative (and utopian) as Singer's, Regan's or e.g. Francione's claim to save the animal per se (see also Donaldson and Kymlicka, 2011: 5-16).

References

Acampora, R. (2006). Corporeal compassion. Animal ethics and philosophy of body. University of Pittsburgh Press, Pittsburgh, PA, USA.

Adorno, T.W. and Horkheimer, M. (1989). Dialektik der Aufklärung. Reclam, Leipzig, Germany.

Benz-Schwarzburg, J. (2012). Verwandte im Geiste – Fremde im Recht. Sozio-kognitive Fähigkeiten bei Tieren und ihre Relevanz für Tierethik und Tierschutz. Harald Fischer Verlag, Erlangen, Germany.

Bruno Latour (2004). Why has critique run out of steam? From matters of fact to matters of concern. Critical Inquiry 30: 225-248.

Burkert, W. (1997). Homo necans. De Gruyter, Berlin, Germany.

Derrida, J. (2010). Das Tier, das ich also bin. Passagen, Wien, Austria.

Diamond, C. (2012). Menschen, Tiere und Begriffe. Suhrkamp, Frankfurt am Main, Germany.

Donaldson, S. and Kymlicka, W. (2011). Zoopolis. A political theory of animal rights. Oxford University Press, New York, NY, USA.

Horkheimer, M. (1987). Dämmerung. Notizen in Deutschland. Gesammelte Schriften Band 2. Suhrkamp, Frankfurt am Main, Germany.

Husserl, E. (1950). Husserliana XV. Zur Phänomenologie der Intersubjektivität. Dritter Teil. Texte aus dem Nachlass. Springer, Dordrecht, the Netherlands.

Huth, M. (2013). Negative Integrität. Das Konzept der Leiblichkeit in der Ethik der Mensch-Tier-Beziehung. Tierethik 1: 108-128.

Levinas, E. (1992). Jenseits des Seins oder anders als Sein geschieht. Alber, Freiburg im Breisgau, Germany.

Merleau-Ponty, M. (1966). Phänomenologie der Wahrnehmung. De Gruyter, Berlin, Germany.

Merleau-Ponty, M. (2000). Die Natur. Aufzeichnungen von Vorlesungen am Collège de France 1956-1960. Fink, München, Germany.

Regan, T. (2004). The case for animal rights. University of California Press, Berkeley, CA, USA.

Ricoeur, P. (1995). Le juste. Galimard, Paris, France.

Rutgers, B. and Heeger, R. (1999). Inherent worth and respect for animal integrity. In: Dol, M., Fentener van Vlissingen, M., Kasanmoentalib, S., Visser, T. and Zwart, H. (eds.) Recognizing the intrinsic value of animals. Van Gorcum, Assen, the Netherlands, p. 41-51.

Schmidt, K. (2008). Tierethische Probleme der Gentechnik. Zur moralischen Bewertung der Reduktion wesentlicher tierlicher Eigenschaften. Mentis, Paderborn, Germany.

Singer, P. (2011). Practical ethics. Cambridge University Press, Cambridge, UK.

Young, I.M. (2011). Justice and the politics of difference. Princeton University Press, Princeton, NJ, USA.

More than harm: a critical analysis of the harm principle in Regan's thinking

K. Keerus[1], M. Gjerris[2] and H. Röcklinsberg[3]
[1]University of Tartu, Jakobi 2, 51003, Tartu, Estonia; [2]University of Copenhagen, Rolighedsvej 25, 1958 Frederiksberg C, Denmark; [3]Swedish University of Agricultural Sciences, Box 7068, 750 07 Uppsala, Sweden; kulli.keerus@ut.ee

Abstract

Tom Regan encapsulated his principle of harm as a prima facie direct duty not to harm those individuals who have an experiential welfare. However, his consideration of deprivational harm, the examples of which is loss of freedom or death, can easily be interpreted as harm which is not experienced by its subject. This creates a gap between Regan's criterion for moral status and his account of what our duties are. We discuss how Regan's understanding of harm relates to his claim that animals should be taken into consideration because it matters to them what happens to them and compare his account of harm with three basic paradigms of welfare known in animal welfare science: the feelings-based, functioning-based and natural-living-based paradigms. We argue that Regan's account coheres with feelings-based paradigm and his account of deprivational harm does not exclude the possibility to interpret this as experiential, too. We will show a potential source for the confusion: in convincing his readers that certain cases of deprivation constitute harm to individuals even if they never experience it as such, Regan avails of the evaluative use of the term 'harm': something is called harm only if it is (implicitly) decided to be wrong. Due to his ambition to argue for direct duties towards animals, Regan too easily casts off the alternative and – from our perspective – better reasons for considering the alleged acts of unexperienced and unknown deprivation wrong.

Keywords: Tom Regan, animal welfare, harm as deprivation, moral status

Introduction

Historically, growing knowledge of the mental capabilities of animals has led to an increased focus on animal welfare. What we do to animals and how we act towards them is often held morally important because of what their capacities and capabilities are. Animals are beings, individuals, to whom it matters what happens to them (Regan, 2004: xvi). Basically, it matters to animals if they suffer or feel pain. Thus, the obligations we are said to have towards animals are frequently related to the 'harms' that we might cause them. Tom Regan formulates it as the principle of harm:

> We have (...) a prima facie direct duty not to harm those individuals who have an experiential welfare (Regan, 1983: 262).

This raises three questions: (1) what counts as 'a harm'; (2) what is the relation of the terms 'harm' and 'wrong'; and (3) is it true that each and every act of harm, committed by a moral agent, is a moral wrong against its subject, and *vice versa*?

As Tom Regan defines harm as diminished welfare, we analyse his account of harm in light of three basic paradigms of welfare used within animal welfare science today, and then analyse Regan's understanding of 'harm' as 'deprivation'. We argue that Regan's account of why we should avoid harm that is not felt by its subject is flawed. The reason is from our point of view that Regan does not disconnect the meanings of terms 'wrong' and 'harm'.

Descriptive and evaluative interpretations of 'harm'

Our point of departure is that the notion of harm can be used both descriptively and for evaluation. Compare two cases. In the first one, a leg of a dog will be amputated out of curiosity to see how long does it take for a dog to get on its feet again. In another case, the amputation will be undertaken for the sake of the dog: to avoid sepsis, for example. In both cases, the previous functional ability of the dog to move and run will be (almost) restored by prosthesis and the animal will be trained to use it. Thus, the life of the dog may become as good as before both in functional and experiential sense. However, many of us tend to think that the amputation in the first case would still be wrong, because it is unnecessary *harm* to the bodily integrity of the animal. In the second case, the amputation to save the life of the dog may be considered morally right. The rightness of it, however, makes some to hesitate whether it is proper to call the procedure harmful at all in the second case.

Wouldn't we rather speak about saving the dog than harming it in this second case? If one agrees that the notion of 'harm' is not appropriate here, one ascribes negative evaluative meaning to the term. 'Harm' and 'wrong' are used as synonyms. If an act is right or neutral, it cannot be called harmful. This usage of the term might involve a reversible explanatory relationship between 'harm' and 'wrong': it is not clear whether something is called wrong because it is harmful or something is considered harmful because it is thought to be wrong, and from this perspective it doesn't matter. If one disagrees however that 'harm' and 'wrong' might be synonyms, one leaves the possibility open for the descriptive use. In that case, amputation might be described as harm to bodily integrity, for example, but at the same time it is denied that the procedure was morally wrong. It might, of course, be possible to think about the amputation as a lesser harm compared to loss of life, in which case the evaluative sense is still there. However, we regard it important to leave open the possibility that 'harm' can also be used neutrally, without any moral connotations at all.

Conceptions of welfare as basis for moral status and moral obligations

Tom Regan defines harm as diminished welfare: 'Individuals are harmed when their welfare is seriously diminished' (Regan, 1983: 94). He also stresses the experiential nature of animal welfare:

> Mammalian animals have a welfare. They fare well or ill during the course of their life, and the life of some animals is, on balance experientially better than the life of others (Regan, 1983: 82).

However, Regan also argues that one can do harm to an animal without the animal noticing it (Regan, 1983: 97) and that this type of harming is also covered with the principle of harm, i.e. it is wrong. To get a clearer understanding of this tension we begin by analysing his understanding of welfare – placing it in the context of the animal welfare discussion, where three basic paradigms are used within animal welfare science: the feelings-based (Broom, 1991), functioning-based (Duncan, 2004) and natural-living-based (Rollin, 2004) paradigms, before analysing Regan's account of 'deprivational harm'.

Should we grant functioning-based, feelings-based or natural-living based welfare to animals according to Regan? D. Fraser *et al.* (1997) have suggested that Regan is amongst the authors who uses the feelings-based approach only as a basis for *moral status*, whereas the requirement that the proper biological functioning or natural living conditions should be granted to individuals, applies independently of whether the animal does feel the harm done to its biological or natural behaviour interests: 'for sentient animals, good biological functioning is important in and of itself, not merely as a means of affecting the animal's subjective experience' (Fraser *et al.*, 1997: 194). We take the phrase 'important in and of itself' as referring to moral importance here. According to Fraser *et al.* (1997), for Regan, concerns about animal welfare *arise because of* the capacity of animals for subjective experience and this differs from cases where concerns about animal welfare *are* concerns about the subjective experience of animals (Fraser *et al.*, 1997: 194-195).

We think other interpretations of Regan are more legitimate even though none are self-evident – but then interpretations would not be needed at all. So even though Regan does speak about the importance of needs for animal welfare, he bases the according prescription on the subjective feeling of satisfaction, and does not exclude that this could compensate for lack of good biological functioning. Thus, we propose and examine the following hypothesis:

> According to Regan, welfare is subjective feeling of welfare considered over time (i.e in the long run). It is experiential. Satisfaction of both preferences and needs contribute to its achievement. That the subjective feeling of satisfaction is (usually) ultimately preferred by the individual, also gives normative reason to accept the principle of harm.

To test our hypothesis we will discuss it in the light of Regan's account of deprivational harm. We will show that Regan is ambiguous in these questions but, if our hypothesis is accepted, his account can be interpreted in a more coherent way.

Tom Regan's account of welfare

Tom Regan presents his account of the nature of animal welfare in Chapter 3 in his influential book 'The case for animal rights'. At the beginning of his analysis, Regan indicates that animal welfare is related to the ability of the animal to act autonomously and defines autonomy as preference-autonomy: as having preferences and 'the ability to initiate action with a view to satisfying them' (Regan, 1983: 83-85). Regan then continues with further explanations of the nature of welfare to clarify what the role of autonomy in welfare is (Regan, 1983: 87). At first, he distinguishes what he calls preference-interests and welfare-interests. The first illustrates what the individual prefers, is *interested in*: the likes, desires, wants, or preferences of the individual. Welfare-interests, in contrast, describe what is *in the interests* of a being. These two may coincide but need not to. A being might prefer something that is not to its benefit from a health-perspective (Regan's example is the wilful usage of injurious drugs) or might not prefer something that is (Regan compares this situation with lack of interest in keeping in shape) (Regan, 1983: 87-88). For the sake of clarity in this paper we take Regan's terms 'preference-interests' and 'welfare-interests' as corresponding to 'preferences' and 'needs' respectively, and will use these, except when referring directly to Regan's text.

There are basically two ways to interpret how preferences and needs can contribute to welfare for Regan: (1) both the satisfaction of preferences and needs contribute to welfare independently; or (2) one is reduced to another: the satisfaction of needs is a contribution to welfare only as far as it satisfies the animal's subjective interest of experiencing the feeling of satisfaction. Even if animals do not have the respective preferences, they feel frustration when they cannot satisfy needs. Only (2) fits with our hypothesis. Needs are often associated to the functioning-based paradigm of welfare and preferences to feelings-based (e.g. Fraser *et al.*, 1997). However, it is not immediately clear which one Regan adheres to.

As the text proceeds, Regan first only concentrates on needs, amongst which he mentions not only biological, but also social and psychological needs. He also has an eye on preferences, but initially only as far as these cohere with needs. It could be argued then that Regan treats the issue of animal welfare from a functioning-based perspective. However, the best way to see whether Regan also allows an independent role to preferences is to look at the kind of preferences which are in conflict with needs. As mentioned above, animals may be interested in things they do not need. For example, dogs might like the taste of chocolate, but eating it is known to be against their welfare-interests from a health perspective. Does the satisfaction of such desires contribute to welfare? They might, according to Regan:

> An individual's welfare does not consist merely in having those benefits that make living well possible. Whether humans or animals, individuals must take satisfaction in living as they do if they are to have a good life, all considered. And this will require that they get what they want, prefer, desire and the like, more often than not, since without this their life will be characterized by *frustration* and all that this is heir to (Regan, 1983: 90-91, emphasis added).

Later, Regan even seems to marginalise the importance of physiological welfare and to stress that of satisfactions (again, as being inclined towards feelings-based account), as he says:

> Physical health (...) is a benefit for those animals who have it; it contributes to their welfare. But the major contribution it makes is to be understood in terms of what it makes possible, not in terms of the state or condition it is. Healthy animals can do more than sick animals, and because they can do more they have a greater range of possible sources of *satisfaction*. That is what makes health the benefit it is, both for these animals and for us (Regan, 1983: 93, emphasis added).

Regan's conclusive definition of welfare is that animals: 'live well relative to the degree to which: (1) they pursue and obtain what they prefer; (2) they take satisfaction in pursuing and getting what they prefer; and (3) what they pursue and obtain is in their interests' (Regan, 1983: 93).

The condition (3) seems to imply that animals cannot live well if their preferences are *not* in their interests. This would speak for the functioning-based paradigm and against our interpretation. However, this is so only if extracted from the rest of the text at those pages (90-94), where Regan explains that bare possession of a benefit does not yet enrich one's life.

This is why we think it is not the only possibility, to consider Regan as a thinker who holds that 'good biological functioning is important in and of itself, not merely as a means of affecting the animal's subjective welfare' (Fraser *et al.*, 1997: 194). In Regan's case, the animals' 'needs' are not simply associated to the biological functioning. 'Needs' are important as the means of satisfaction and 'satisfaction' is understood along to the subjective feeling paradigm. It is because what animals are – autonomous beings – that their welfare does not consist just in satisfaction of several kinds of needs (as is the case for plants):

> One aspect of animals' faring well or ill is the degree to which they are able to exercise their autonomy, since to thwart their will by denying them the opportunity to do what they prefer is to cause them frustration, while to allow them to do what they prefer is to make it possible for them not only to obtain what they want but to take satisfaction in obtaining it by their own devices (Regan, 1983: 116).

However, our hypothesis that for Regan welfare concerns are basically experiential could be undermined by his account of 'deprivational harm'. Thus, we will turn to this issue. More comprehensive discussion of other places in Regan's work that puts our interpretation into question will have to await a longer paper.

Regan's account of harm as deprivation

Whilst we have shown that initially, Regan's account of harm seems to lean on the feelings-based approach in some aspects, there is still an important aspect that contradicts this interpretation as Regan also states that 'individuals can be harmed in ways that do not involve causing them to suffer' (Regan, 1983: 96). To elucidate this, we have to look at Regan's account of harm as deprivation.

According to Regan, an individual can be harmed in two ways: by infliction and/or by deprivation. A typical case of infliction is to make an animal to suffer from intense and long-lasting pain. The other kind of harm – deprivational – does not involve suffering, as Regan claims (Regan, 1983: 94-96). Harms of this second type should be understood 'as deprivations or losses of those benefits that make possible or enlarge the sources of satisfaction in life' (Regan, 1983: 97). Although deprivation of something often causes suffering (e.g. grief about lost opportunities), to be harmed by deprivation, according to Regan: 'the victim does not have to be aware of these harms or to suffer physically or psychologically as a result' (Regan, 1983: 97).

An example to convince the reader that harm can come as deprivation is a 'happy' slave: a person who is contented with his situation, has no knowledge about what he has lost, and thus cannot even wish to be free (Regan, 1983: 97). Nevertheless, as Regan argues, intuitively we feel that the slave is harmed, although he does not suffer. Regan takes this intuition as true and draws parallels to animals, whose welfare is similar to that of humans in relevant respects. If such losses are seen as harms to humans, it must follow that parallel situations are also harms to animals.

However, Regan's examples do not actually show that the subjective feelings of the deprived individuals are not affected at all and that deprivational harm must mean unexperienced harm. In the example of 'happy' slave Regan puts 'happy' in quotation marks (Regan, 1983: 97). So isn't the slave really happy? If he isn't, then these examples do not prove that this type of harm can remain unexperienced. This speaks for our reading that feelings-based paradigm has more important place in Regan's account than it is typically thought to have.

It can be argued that it would be better for a man to be free, even if this does not bring more satisfaction. However, in arguing this, one is not addressing the question whether it is really better *from the individual's point of view*, but argues that the situation where a human being is free is better *from a moral point of view*. But how could one argue for this? Do we think that freedom is somehow good in itself and *should* be preferred, even if it is not enjoyed by the slave? If this, however, is a suspicious claim in the case of human beings, the more so in case of animals. We, humans, can have the imaginations of something being morally worthier for us, but animals hardly do. It is possible, that this is where Regan errs: he thinks that a slave cannot be *truly* happy, because his situation is not morally appreciatable. But is there something wrong with his condition from his own perspective? This is the question one should ask when arguing for direct duties towards either humans or animals.

The problem we wish to bring out is that Regan shifts from the evaluative usage of harm to the descriptive one without noticing. The reason could be that Regan presumes that certain acts, like depriving an individual of freedom, are wrong in themselves. And at the same time as a synonymous usage of 'harm' and 'wrong' can be found, he reverses the explanatory relationship between them. Instead of remaining to the argument that each harm is wrong (other things equal), he seems to imply that if something is wrong, it is harm. What happens in his argument can be opened up with following steps:

1. Deprivation of freedom is wrong in itself (hidden intuition).
2. All wrongs are harms ('harm' is used synonymously with 'wrong' here).
3. Thus: deprivation of freedom is harm.
4. According to the principle of harm, all harms are wrong ('harm' is descriptive, otherwise the principle would be question-begging).
5. Thus: deprivation of freedom, even if the subject is not aware of it, is wrong (but not in itself as in (1), but after concluding that it is 'harm' in the first place because of being wrong (1-3)).

If this is what happens, the argument that unexperienced harms are wrong does not succeed.

Most will probably agree that it is generally better for an animal to be free in the sense that it can satisfy more of its preferences. Freedom can thus be seen, *prima facie*, as a benefit to an animal and deprivation of it constitutes harm. However, according to our interpretation of Regan, an animal finds more satisfaction when it can fulfil more of its preferences and not because freedom is a benefit for an animal in and of itself. In each negative example of deprivation that Regan brings, we can imagine the animal actually to suffer in these conditions. Even if the animal does not have a conscious preference to be free, for example, this does not entail that it cannot have negative experiences in confinement.

Conclusion

Regan's account of welfare and of why it is wrong to diminish it is based on experiences – on satisfactions and frustrations that the animal may have. We showed that this is true of deprivational harm too, although it is easy to interpret Regan as saying that deprivation can be unexperienced and that the needs of animals must be granted, no matter whether the subjective feelings of individuals are affected or not. We also argued that the difficulty in interpretation arises from not keeping the concepts of 'harm' and 'wrong' apart. Especially when arguments for taking animals directly into consideration are in spotlight, there is a tendency to consider each and every wrong act as harm to animals themselves. In such cases, the alternative and – from our perspective – better reasons for considering the alleged acts of unexperienced and unknown deprivation wrong, will too easily be cast off. However, what these other reasons are, remains a topic of another paper.

References

Broom, D.M. (1991). Animal welfare: concepts and measurement. Journal of Animal Science 69: 4167-4175.

Duncan, I.J.H. (2004). A concept of welfare based on feelings. In: Benson, G.J. and Rollin, B.E. (eds.) The well-being of farm animals. Challenges and solutions. Blackwell Publishing, Oxford, UK, pp. 85-102.

Fraser D., Weary D.M., Pajor E.A. and Milligan B.N. (1997). A scientific conception of animal welfare that reflects ethical concerns. Animal Welfare 6: 187-205.

Regan, T. (1983). The case for animal rights. University of California Press, Berkeley, CA, USA, 425 pp.

Regan, T. (2004). Preface to the 2004 edition. In: The case for animal rights. University of California Press, Berkely, CA, USA, pp. xi-xlix.

Rollin, B. (2004). The Ethical imperative to control pain and suffering in farm animals. In: Benson, G.J. and Rollin, B.E. (eds.) The well-being of farm animals. Challenges and solutions. Blackwell Publishing, Oxford, UK, pp. 3-20.

Dignity of creature: beyond suffering and further

S. Camenzind
Messerli Research Institute, University of Veterinary Medicine Vienna, Medicine University of Vienna, University of Vienna; Veterinaerplatz 1, 1210 Vienna, Austria; samuel.camenzind@vetmeduni.ac.at

Abstract

Based on considerations on the criterion *excessive instrumentalisation,* the following paper argues that the concept of *dignity of animals,* as stated in the Swiss Animal Protection Act, leads to a paradigmatic change in human-animal interaction. Establishing the term *dignity of creature* in the Swiss Federal Constitution in 1992 has triggered an enduring interdisciplinary discourse about its meaning, extension and implementation. After specification by the Federal Ethics Committee on Non-Human Biotechnology (ECNH), the term *dignity of animals* was finally established in the Animal Protection Act in 2005. Its violation is prosecuted and causes legal effects. The concept of animals' dignity goes beyond common welfare approaches based on sentientistic grounds because besides avoiding suffering and pain it states that animals should also be protected from unjustified interventions on their appearance, from humiliation and from excessive instrumentalisation. Switzerland represents a worldwide unique biocentrism (the term 'creature' refers to all living being), which consequences aren't fully explored. The criterion *disproportionate instrumentalisation* has played a key role since the debate's origins. Nevertheless, different authors state their meanings and the impact is still not elaborated well to date. Starting from the general agreement that the concept of dignity is based on an inherent worth of nonhuman organisms, which has to be taken in account, this paper focuses on the specification of the instrumentalisation criterion in the sense of the Swiss law. According to the Swiss Animal Protection Act, dignity of animals can be violated if an animal is used merely for human ends. In contrast to human dignity in the sense of Kant, dignity of animals can still be respected if its violation is justified on the basis of a careful evaluation of certain interests specified in the Swiss Act of Genetic Engineering. On the other side, the animal's dignity is disregarded if the evaluation of interests shows that the animal's interests outweigh human interests. Based on a concrete example in the field of meat consumption, it will be shown that a consistent implementation of this understanding of animal dignity affects our fundamental position towards animals. Furthermore, the paper raises questions about responsibility in case of structural violation of animal dignity, which hasn't been discussed so far.

Keywords: instrumentalisation, structural violation, animal dignity, Swiss Animal Protection Act

Introduction

The current Swiss Animal Protection Act ('Tierschutzgestz', hereinafter: TSchG) is classified as one of the strictest Animal Welfare Acts worldwide. It protects dignity of animals and differs from other legislations insofar as it reaches further than the sentientistic principle of non-maleficenci, which is firmly established in common sense. Additional to the criteria 'pain', 'suffering' and 'harm', it includes the non-sentientistic criteria 'humiliation', 'major interference with appearance or abilities of an animal' and 'excessive instrumentalisation'. Although animal's dignity in law and ethics is not new to the debate (e.g. Balzer *et al.*, 1998; Baranzke, 2002; Kunzmann, 2007; Nussbaum, 2004; Teutsch, 1995), it has arrived at the point where the theoretical considerations have to be followed by concrete actions. Even though the Swiss concept of dignity has been put into more concrete terms by the expert's reports by 'Eidgenössische Ethikkommission für die Gentechnik im ausserhumanen Bereich' (Ethics Commitee on Non-human Biotechnology, ECNH) (1999), and the ECNH and the 'Eidgenössische Kommission für Tierversuche' (Federal Committee on Animal Experiments, FCAE) (2001), implementation – especially of the non-sentientistic criteria – seems to be difficult (see Friedli, 2009; Sigg, 2007). While problems of implementation affect mainly Swiss citizens and authorities, grounds for dignity of animals and its violation contributes to the philosophical discussion of animal dignity in general.

In the following, I will first give an overview of what are the specific features of the Swiss concept of animal dignity. Kirsten Schmidt's (2011) differentiation of three argumentation levels will give as an accurate tool to examine the main characteristics of the Swiss concept. Based on a concrete example about excessive instrumentalisation in the field of meat consumption, the issue of a structural violation will be discussed in a second step. It will be shown that even in evident cases of disregarding animal dignity punishment in terms of fines or imprisonment (see Art. 26 TSchG) is not always accomplishable. In the conclusion it is stated that the implementation of this understanding of animal dignity is a permanent, still ongoing process that affects our fundamental position towards animals.

On animal dignity in the Swiss Animal Protection Act

Switzerland incorporated 'dignity of creature' ('Würde der Kreatur') into the Swiss constitution in 1992 and 'dignity of animals' ('Würde des Tieres') into the Swiss Animal Protection Act in 2005. The purpose of the Swiss Animal Protection Act is to protect dignity and welfare of animals (see Art. 1 TSchG). It defines dignity as follows (Art. 3, lit. a TSchG):

> 'Inherent worth of the animal that has to be taken into account when handling it. If any stress imposed on the animal cannot be justified by overriding interests, this constitutes a disregard for the animal's dignity. Stress is deemed to be present in particular if pain, suffering or harm is inflicted on the animal, if it is exposed to anxiety or humiliation, if there is major interference with its appearance or its abilities or if it is excessively instrumentalised.'

Moral status

On the first level of argumentation, it will be discussed who belongs to the moral community according to the Swiss approach. According to 'dignity of creature' as it is stated in the Federal Constitution, every entity that has the ability to flourish and have its own good is morally considerable. Therefore, the term can be defined as a biocentric approach, including nonhuman animals such as mammals, vertebrates and invertebrates, but also plants, fungi and algae.

As the Swiss Animal Protection Act concedes dignity of animals only to sentient beings (see Art. 2 TSchG) such as mammals, vertebrates, cephalopods and crustaceans, the position can be described as 'limited' with reverence of the moral objects. Different authors state that if the criterion to be morally respected is flourishing, this limitation clearly conflicts with the concept of 'dignity of creature' as stated in the Federal Constitution (see Balzer *et al.*, 1999: 36 or Saladin, 1995: 366).

Specification of moral status

The second level of argumentation defines the weight of the moral status. To put it in different terms, the question is: Does Switzerland provide an egalitarian or an hierarchic ethical approach? To this day, the philosophical discussion about the relation between dignity of humans, nonhumans and plants isn't completely clarified. Based on practical consequences (see Balzer *et al.*, 1998: 49) and the anthropocentric orientation of the law, it is assumed that in case of a conflict, human interests count more than animal and plant interests. Arz de Falco and Mueller (2001) state the same opinion. In contrast to those views, Teutsch (1995: 55) holds an egaliltarian position and states that dignity of creature is the same in all living beings, but it can be violated in different ways.

To sum up, dignity of creature can be described as hierarchic, limited (moral objects) and extended (criteria of stress) biocentrism. It is a non-egalitarian biocentrism because humans overweigh animals and plants in a moral point of view. Further, it is extended biocentrism in case of the morally relevant criteria because other criteria are taken in to account additionally to flourishing.

Content of implementation of moral obligations

In contrast to other dignity concepts, the Swiss Animal Protection Act gives *prima facie* clear guidance how to respect the inherent worth of an animal: 'If any stress imposed on the animal cannot be justified by overriding interests, this constitutes a disregard for the animal's dignity' (Art. 3 lit. a TSchG). According to this article, the implementation level can be described as a two level concept: I will refer to the first level as 'level of stress' and to the second level as 'level of balancing of interests'.

On the first level, Art. 3 lit. a TSchG mentions 'stress' of animals that leads to disregard of animal dignity. 'Stress' is defined as pain, suffering, harm, humiliation, major interference with its appearance or its abilities and excessive instrumentalisation.

If stress is involved in an action with animals, its dignity is only disregarded when involved stress can't be justified by balancing of interests. Although not explicitly stated in the TSchG, it is common to speak of violation of animal dignity on the first level. If this violation can be justified by overweighing interests, dignity of animals is regarded. If this violation can't be justified and therefore the animal's interests outweigh human interests, animal dignity is disregarded. This is followed by legal actions such as imprisonment or fines. In contrast to utilitarian balancing, not every interest will be considered in the balancing. According to the Swiss Act of Genetic Engineering (ar. 8, para. 2, lit. b), only legitimate interests such as human and animal health, providing of sufficient nourishment, reduction of ecological degradation or scientific knowledge count.

Implementation and structural violation of animal dignity

There is the risk that dignity of animals embedded in an anthropocentric legal system will not be paid attention to in an appropriate, egalitarian way. In this case, animal interests will be considered, but they will always count less than human interests.

However, as Jörg Leimbacher (1997: 96) states, dignity of creature must not be seen as a nice gesture. If the concept of animal dignity is taken seriously as an attitude to respect the animal's inherent worth, animal and human interests have to be considered equally. Stress which occurred when balancing interests must never be taken lightly. It remains an important moral issue in every case. Violating animal dignity is in any case not a trivial offence that can generally be justified by overweighing human interests.

The Swiss concept of dignity, interpreted in this way, raises – especially concerning the criterion of instrumentalisation – questions about the implementation and responsibility in case of structural violation of animal dignity, which hasn't been discussed so far. What kind of implementation problems and what is meant by structural violation, will be explained in the next section based on a concrete example in the field of meat consumption.

The starting point is an animal husbandry company that raises and sells animals for food consumption. As the animals are kept on an economic base, their value corresponds with their economic value. If the value of an animal is only defined as the instrumental value for the company, dignity of the animal is violated. According to the Swiss Concept of animal dignity, violating dignity can be declared a moral problem, but it can still be justified by overweighing legitimate interests of humans and therefore doesn't cause legal actions. The Swiss Act of Genetic Engineering (art. 8, par. 2, lit. b) states that 'providing of sufficient nourishment' would be a legitimate human interest in the case of food production as stated in the. Therefore, stressing animals to fulfill human interests is legal in this case. The question if, and if yes, for whom and to which degree, meat consumption is necessary to stay healthy will not be discussed here. For the sake of argument it can be assumed that a certain amount of meat is necessary for mankind to stay healthy. But if the figures by V. Ahne (2012) are true, which state that one third of the produced meat is wasted in Europe, it is clear that this case is not only violation but also disregard of animal dignity. As the wasted meat cannot be justified by overweighing human interests, the disregard of animal dignity has to be sanctioned by imprisonment or fine. Another case is the acceptance of wasting animals in industrial egg production. While female chicken are kept as lay hens or used for meat production, most male chicken don't serve any purpose and are killed right after hatching. This case can be described as structural violation of animal dignity because it occurs in a system that doesn't intend to kill male chicken, but accepts it for economic reasons.

In both cases the question arises who can be held responsible in legal terms? Who has to pay the fine? The employee who kills redundant animals, the company, or even the government that subsidizes certain companies? Because household cause half of the wasted food (see WWF, 2012: 2) consumers are responsible for disregarding the dignity of animals too. The debate gets even more controversial, because violation of the Animals Protection Act is defined as public prosecution and therefore must be sanctioned by government. Therefore the problem of disregarding animal dignity applies on different levels. The conundrum lies in the fact that authorities must sanction violation of the Animal Protection Act, while the question of responsible party remains unanswered to date.

The main tensions regarding dignity of creature stem from the conflict between the biocentric approach of the Swiss legislation and the still widespread anthropocentric attitude in common sense. In an anthropocentric society, the implementation of the animal dignity is certainly limited and burdensome. It still serves as a tool, of permanent questioning human-animal interaction in every possible way. While the role of philosophy of law referring to dignity of animal is to clarify ethical terms and show it's range and consequences the implementation depends mainly on politics and executive authorities.

References

Ahne, V. (2012). Unser tägliches Fleisch gib uns heute..., 10.8.2012. Available at: www.greenpeace.org/switzerland/de/Kampagnen/Stromzukunft-Schweiz/Klima/Klimawandel/Unser-tagliches-Fleisch-gib-uns-heute/.

Arz de Falco, A. and Mueller, D. (2001). Wert und Würde von 'niederen Tieren' und Pflanzen. Ethische Überlegungen zum Verfassungsprinzip 'Würde der Kreatur' (Défis et dialogues, Bd. 17), Universitätsverlag, Freiburg, Germany.

Balzer, P., Rippe, P. and Schaber, P. (1998). Menschenwürde vs. Würde der Kreatur. Begriffsklärung, Gentechnik, Ethikkommissionen, Alber, Freiburg, Germany.

Baranzke, H. (2002). Würde der Kreatur? Die Idee der Würde im Horizont der Bioethik, Königshausen und Neumann, Würzburg, Germany.

Eidgenössische Ethikkommission für die Biotechnologie im ausserhumanbereich (EKAH) and Eidgenössische Kommission für Tierversuche (EKTV) (2001). Die Würde des Tieres. Eine gemeinsame Stellungnahme der Eidgenössischen Ethikkommission für die Biotechnologie im Ausserhumanbereich (EKAH) und der Eidgenössischen Kommission für Tierversuche (EKTV) zur Konkretisierung der Würde der Kreatur beim Tier, Bern, Swisterland.

Eidgenössische Ethikkommission für die Gentechnik im ausserhumanen Bereich (EKAH) (1999). Stellungnahme zur Konkretisierung der Würde der Kreatur im Rahmen der geplanten Revision des Tierschutzgesetzes, Bern, Swisterland.

Friedli, K. (2009). Die Würde des Tieres in der neuen Schweizer Tierschutzgesetzgebung. Journal für Verbraucherschutz und Lebensmittelsicherheit, 4: 387-391.

Harris, J. (1992). Clones, genes and immortality: ethics and the genetics revolution. Oxford University Press, Oxford, UK.

Kunzmann, P. (2007). Die Würde des Tieres – zwischen Leerformel und Prinzip, Alber, Freiburg, Germany.

Leimbacher, J. (1997). Zur rechtlichen Bedeutung der Würde der Kreatur. In: Bondofi, A., Lesch, W. and Pezzoli-Olgiati, D. (eds.) Würde der Kreatur. Essays zu einem kontroversen Thema. Pano Verlag, Zürich, Switzerland, p. 89-105.
Nussbaum, M.C. (2004). Frontiers of justice. Disability, nationality, species membership. Belknap Press, Cambridge, UK.
Saladin, P. (1995). 'Würde der Kreatur' als Rechtsbegriff. In: Nida-Rümelin, J., Von der Pfordten, D. (eds.) Ökologische Ethik und Rechtstheorie. Nomos, Baden-Baden, Germany, p. 365-369.
Schmidt, K. (2011). Concepts of animal welfare in relation to positions in animal ethics. Acta Biotheoretica 59: 153-171.
Sigg, H. (2007). Die Würde der Kreatur in der Tierschutzgesetzgebung. Talk at the conference 'Tierversuche und Würde der Kreatur', 8. 6. 2007, Botanisches Institut, Zürich. Available at: www.tierschutz.uzh.ch/links.html.
Swiss Animal Protection Act (2005). Tierschutzgesetz of 16. 12. 2005. Available at: www.bvet.admin.ch/themen/tierschutz/index.html?lang=en.
Teutsch, G. M. (1995). Die 'Würde der Kreatur'. Erläuterungen zu einem neuen Verfassungsbegriff am Beispiel des Tieres. Haupt, Bern, Switzerland.
WWF (2012). Lebensmittelverluste in der Schweiz – Ausmass und Handlungsoptionen. Available at: http://assets.wwf.ch/downloads/12_10_04_wwf_foodwaste_ch_final.pdf.

The moral status of animals: a relational approach

B.K. Myskja
NTNU Norwegian University of Science and Technology, Department of Philosophy, 7491 Trondheim, Norway; bjorn.myskja@ntnu.no

Abstract

There are many accounts ascribing moral status to animals, most of them departing from what we take to justify moral status for human beings and discussing similarities and differences. In order to frame our obligations for promoting the interests of others, we first have to understand the basis for moral obligation. Christine Korsgaard (2005) suggests that self-reflection on normative issues is the defining characteristic of human morality. Arguably, this sets human persons apart from other living beings inhabiting this planet, including some belonging to the human species, and is the basis for according moral worth. Allen Wood has argued that we have reason to include human beings who lack the capacity of self-reflection in the moral community as they have some part in rational nature (Wood and O'Neill, 2008). Apparently, this position implies that there is a moral gulf between humans and other animals. But neo-Kantians as Wood, O'Neill and Korsgaard have argued that several of the elements that we find to be of moral consideration for persons, are such that we also share with animals and are part of the basis for human normative reflection. We should therefore accord them some moral status and count animal desires and needs to be morally significant. Assuming this account gives an acceptable basis for moral consideration of animals, it does not give any specific indications of what kind of moral duties we have towards animals, beyond that we should avoid causing unnecessary suffering and restrictions on their life-space. I will suggest a relational extension of this approach to the moral status of animals, based on existing interactions between humans and animals. Animals form an inescapable part of human life-worlds, and belong to our value systems. Human beings form different kind of relationships with different animals – domestic and non-domestic. We express respect, awe, compassions, disdain and contempt for animals, thus emphasizing their likeness with and difference from humans. We regard them as having particular natures or ways of life. Some are more important for our understanding of our own life than others. All these relations provide basis for moral arguments regarding what we do owe animals, suggesting a middle way between the commodification witnessed in industrial husbandry and the moral status suggested in animal rights approaches.

Keywords: moral value, speciesism, rationality, community, domestication

Introduction

Well established human practices indicate that most of us regard animals to be of little or no moral value. Many – perhaps the majority – accept the killing of animals for human self-protection, for food and clothing – and we accept that animals are kept in slave-like conditions for our benefit, forced to work and unable to change employment. Yet other animals are kept for entertainment and company without being given the liberty to leave at choice, and few people object to that. Even the well-documented fact that a large number of animals are kept in inhumane conditions bound to make them suffer, is in practice accepted by large numbers of people throughout the world. We must assume either that people neglect their moral duties or that they believe that animals can be treated radically differently from the ways we think humans should be treated. The latter explanation is better because it assumes people act consistently. Although there are growing communities of animal liberators, most people seem to rest contently in the belief that there is a moral gulf between animals and humans. But is there one, and should they?

This picture indicates that the majority take this negligible moral value of animals in comparison with humans as premises for how we ought to behave towards animals. But that is not the case. Law regulation, such as the Norwegian Animal Welfare Act (2009), indicates that it is wrong to make animals suffer or to kill them for no reason, meaning that we have some duties towards animals. When we hear stories about animal abuse it is not only animal advocates that condemn the acts (Vollum *et al.*, 2004). Many, even those who gladly eat meat, seem to hold that it is sometimes worse maltreating animals than adult humans, when they are confronted with concrete cases. Likewise, many countries prohibit the killing of animals without reasons such that the animal is suffering or we need it for food or for research purposes (see e.g. German Animal Welfare Act, 2010). But these moral judgements also appear to be contingent on what kind of animal and which relation we have with it. A pet, an old 'faithful' horse or a mighty tiger is worse to kill than a wild rabbit or one of thousand sheep in a flock. So we appear to have at least three positions regarding our moral obligations for animals:

1. There is nothing special with being human and what matters are the morally relevant characteristics the being in question possesses – Singer's position.
2. There is a moral gulf between humans and other animals and if we have obligations, they are not to the animals but rather to ourselves – Kant's position
3. Our obligations towards animals are not absolute, but contextual – the common morality and ethics of care position (Donovan and Adams, 2007).

I will develop a neo-Kantian version of the third position, extending the Kantian framework with a relational account. But first I will discuss the two other positions that have dominated the ethical debate from the latter part of the Twentieth Century and onwards.

Speciesism and anthropocentrism

Those who insist on the significant difference between our duties towards humans and animals respectively, provide arguments that may support human commodification of animals in farming, hunting and leisure. One obvious reason supporting this view is the important distinction in that only humans can act morally, and we therefore have special duties of real or potential reciprocity. The problem with this approach is pointed out by many authors: not all humans can act morally, but most of us would not accept that they be kept as slaves, subjected to systematic suffering and killed for food or clothing as we do with animals. So there seems to be inconsistencies, giving argumentative ammunition to animal liberation groups.

Peter Singer (1990) argues strongly that the fact that only humans can act morally is not decisive for the question of who should be taken into account morally. He suggests that Bentham's criterion of ability to suffer and feel pleasure should be the decisive one when discussing moral considerability (Ibid.: 7-8). In these feelings we find the basis for interests, and we have no grounds for holding that the interests of one individual should count more than the interests of another. But his argument is convincing only if we accept his and Bentham's metaphysical assumption that we have moral responsibility for promoting the interests of others, whoever they may be. Then it makes sense asking what reasons we have for distinguishing between humans and other animals on analogy with the way the unfounded moral distinctions that once were made between whites and blacks were challenged. The first problem with this account is, as Singer (1990: v) points out, is that oppressed groups of humans could argue their own cause and thus contributed to their own liberation. They were able to explain why they should be treated equally with the oppressors, thus claiming their own right. So the analogy breaks down at the outset. A deeper problem with this account is that it fails to explain Singer's emphasis on interests. Why should I pay heed to the interests of others than myself or those I happen to take an interest in? In what sense do moral values and obligations actually exist? It is clear that pleasure and pain exists, as well as what we can loosely call subjective and objective interests, but we cannot deduce any obligations from that. Therefore it seems to be metaphysical assumptions underlying Singer's argument. In order to frame our obligations for promoting the interests of others, we first have to understand the basis for moral obligation.

Onora O'Neill, in response to Allen Wood, states that moral reasoning is by necessity anthropocentric, in the senses that it is only humans or beings with similar capacity to act on reasoned grounds that can exercise and understand morality (Wood and O'Neill, 1998: 217). This means that human beings have a special role in the system of morality, but not that only humans should be subject of moral consideration. But it is reasonable to hold that the implication is speciesist in the weak sense that morality should discriminate between humans and other animals. Singer has famously coined the term speciesism in analogy with racism and sexism to classify what he takes to be the last unfounded moral discrimination basis. The test for him is whether we say that we should treat infants and retarded humans better than animals with obvious higher cognitive capacity, for example for research purposes (Singer, 1990: 15-16). Then we discriminate due to species membership rather than morally relevant characteristics. But one could argue that if there are *other* morally relevant distinctions between humans and animals than ability to feel pleasure and pain and having self-consciousness, then discrimination can be legitimate. My suggestion is to look closer at the basis for moral judgements to understand why most people are right in holding on to the species-based distinction.

If we refrain from metaphysical assumption, the only possible source of morality and value systems in the world, are human beings. I do not exclude the possibility of more or less developed morality in other animals with high cognitive abilities, but as we cannot communicate with them in any advanced way, it is not meaningful to say that they can be the source of morality. Morality is a way of *human* thinking about the world and closely connected to human sociability. We are able to study moral or moral-like behaviour in animals just because we have the capacity to analyse and understand the distinctions involved. Animals cannot do the same with us. If they did, we would be morally obliged to include them as equal or superior members of our moral community. If there were no humans or other creatures with the characteristic human combination of sensibility and rationality, there would be no moral values and no obligations. There would be no one to make moral judgements or to experience duty (as long as we refrain from religious assumptions). Thus, as Christine Korsgaard (2005: 101) points out, morality is a human construction. It is not a construction in the sense of something intentionally brought about, but a construction of the world due to the particular way human beings live in the world. We have intentions and interests; we evaluate these and we decide whether they are good. And we discuss and defend our decisions in a community of beings who share this way of constructing the world. In this way we create value in the world, rather than responding to some values that exists independently of our valuing them. Humans belong to a species characterised by understanding the world in moral terms. According to this normative position, everybody within this community has personhood and moral worth. They are irreplaceable members of the community of those who bring value into the world – the community Kant (1785 [1965]: 438-439) calls the Kingdom of Ends.

Animals and the moral community

This speciesist conception of morality as connected to human communicative rationality and sociability is not particularly connected to Kantian frameworks. Aristotle (1996: 13) in *the Politics* I, 2 connects this way of judging right from wrong and just from unjust to human sociability exercised through language, and sees this as expression of the fundamental human *ergon*, the species-characteristic function distinguishing humanity from the other social animals. We do not need an Aristotelian metaphysics, either, to accept that we belong to a species constructing the world in moral terms. Singer's analysis is an example of the same assumptions; he takes morality for a given, social fact. But he is right that the implication of this fact need not be speciesism in deciding who should be considered morally.

On this account, fully rational, morally competent adults have a special status, belong to the Kingdom of Ends and should be treated according to categorical imperatives. But why should we include those who do not talk and make morally justifiable decisions on how to act, such as small children, dement people and the mentally retarded in this moral community? Wood argues that we should 'behave with respect towards non-rational beings if they bear the right relations to rational nature. Such relations, I will argue, include having rational nature only potentially, or virtually, or having had it in the past, or having parts of it or necessary conditions of it' (Wood and O'Neill, 1998: 197). We may in some cases have to prioritise these over fully rational humans, not because they belong to the same species as us, but because they take part in that which is the basis for moral worth, namely the capacity for making moral value judgements.

But also animals 'take part' in rational nature, that which is the basis for our moral perceptions and judgements. Animals have 'capacities which we should value as the infrastructure, so to speak, of rational nature' (Ibid.: 200). This infrastructure includes desires, pleasure and pain. Thus animals should be respected as taking part in the moral community. But Wood holds that animals are, like children and other human non-persons, not full-fledged members of the community. Furthermore, unlike these human non-persons, they have not been, will not and could never become such members. Thus their part in this moral infrastructure is weaker than that of children and other human non-persons. But we do still have duties towards them, because we share not only the same physical world, but also the same moral world. We recognise that animals have desires and feelings that are just like those desires and feelings that are integrated parts of our moral capacities for creating value. Without these capacities that we happen to share with animals, there would be no morals in the world.

Our moral duties towards other persons does not stop when we do not force or deceive them against their will, we must also refrain from hurting them physically or prevent them from fulfilling their physical desires even if we do not think these desires are morally good as this is contrary to treating them as ends in themselves. Thus we have a duty to respect the animal nature of other persons. We have similar duties towards non-persons, including animals, who obviously share our animal nature – as long as we have no other more significant grounds of moral obligations. This is the major difference in the duties we have towards persons and non-persons. The duties we have towards non-persons have no absolute ground, and may be overruled. But even if this gives an acceptable basis for moral consideration of animals, it does not give any specific indications of what kind of moral duties we have towards animals, beyond that we should not cause unnecessary suffering and restrictions on their life-space. How do we decide on the content and the limitations on our duties towards non-human animals?

Relational duties

My suggestion is that we build on Korsgaard's approach and determine our duties towards animals on the basis that humans bring value into the world, and that these values are validated in a community of moral reasoning. That means that it is through the way we relate to animals we accord them moral value, but any merely private relation cannot form the basis for such value. It is only when this value is such that it can be recognised and accepted as worthy of consideration in a moral community we can say that it is a moral value. Animals form an inescapable part of human life-worlds, and belong as such to our value systems. Humans form different kinds of relationships with different kinds of animals – domestic and non-domestic. These relations carry with them clear and distinct duties. This is analogous with the kind of special duties we have with humans we interact with. We owe every human respect and help in need, but my special relations with my wife, my children, my siblings and friends give rise to particular, more specified obligations (Kant 1797 [1996]: 469-473). In the same way we acquire special duties towards particular animals that we choose to get involved with. As Keith Burgess-Jackson (1998: 161) states, 'human beings have special responsibilities to the animals they bring into their lives – precisely *because* they bring them into their lives'. This is a relational account of responsibility. Our moral obligation is derived from the relation we choose to establish with the animal, not due to their properties, as for example Singer holds (Ibid.: 173-174).

I think there is something intuitively right about Burgess-Jackson's approach, but it rests on some implicit presuppositions that are not clearly explicated. The basis for the establishment of this particular relationship and its moral character is the human domestication of animals for thousands of years. By bringing wild animals into our life-sphere as companions and servants we establish a general relationship, and thereby acquire general responsibility for them. The special relationships are derived from these general ones. This means that even stray dogs without any owner must be treated well, and ideally we owe it to them to see to that they are taken care of – on weak analogy with the responsibility we have for human strangers who are in need of help. This obligation for pet animals can be considered a wide moral duty (Kant 1797 [1996]: 390), one that does not determine how far we should go in fulfilling the duty. Thus we have a general duty to treat any domesticated animal well since they belong to animals that we as humans have integrated in our moral sphere, and a special duty to take care of those that we establish special relationships with.

Kant holds that we have duties towards ourselves to treat animals well, lest we weaken our natural compassion that is useful for our moral sense (Ibid.: 443). I believe Kant is wrong about the reasons for the duty and the direction of it, but right about the content. I would rather suggest that it is because domestic animals take part in our moral community and we establish special relationships with them we have special duties towards them, as Kant writes, to refrain from violent and cruel treatment, not overstrain them and to kill them quickly and painlessly – when we have to. Likewise, unnecessary experiments (where do we draw the line of necessity?) are prohibited – and we should show 'gratitude for the service of an old horse and dog (just as if they were members of the household)' (Ibid.). The combined account of animals sharing our moral infrastructure, our incorporation of them in our moral community and our special relationships with particular animals can better explain the wide duties we have towards them, according to Kant.

There is nothing in this account that says it is wrong to eat animals. But it is certainly wrong if the practices turning them into food involve causing avoidable suffering. We know that large parts of modern food industry are systematically causing suffering, treating these beings that belong to our moral community as mere things without desires and feelings of pleasure and pain. These practices are plainly wrong, and we know about them, so we cannot eat meat produced under such conditions without becoming part in the wrong-doing. But on this relational account, it is not wrong to eat meat coming from an animal that has lived under good conditions, been given the opportunity to fulfil its natural desires with adequate food shelter and companionship. There are, however, other reasons to reduce the amount of food we eat, connected to environmental and food security concerns.

What about wild animals? Can a relational account give any indications of our duties towards them? We have the same general duty to refrain from interfering with their life unless necessary. This means that they should have sufficient free space and any kind of cruelty is unacceptable. There is no reason to assume that we should refrain from hunting them, as long as we use humane hunting practices. Many wild animals have humans as their main predator and they will suffer more from overpopulation if we refrained from regulated humane hunting. Thus, we have relational duties towards wild animals as well, and they may include humane hunting practices.

It is possible that we can use the relational approach to distinguish between duties towards different animal species. We express respect, awe, compassion, disdain and contempt for different kinds of animals, thus expressing their value for us. Some of them have more value for our understanding of our own life than others. This is not morally irrelevant. It is a greater loss for humankind if some of these iconic animals such as the wolf, the Bengal tiger or the African elephant disappeared, than if we lose others with less impact on our life-worlds. If I am right, we have a stronger duty to protect and preserve these animals that are important for our reflection on our own lives as a social animal with reason and the ability to bring moral value into the world. Some cultures have special symbolic relations with particular wild animals due to their close interactions with them. Even these particular symbolic relations carry moral weight, in a way similar to that of the special relationships that exists between ethnic groups and their herd animals, such as that between the Sami people and the reindeer.

The basis for our moral duties towards animals is found in the way we ascribe moral value to the world. Because animals take part in our moral infrastructure by sharing the sensible basis for moral perception and judgement, we owe them moral consideration. The particular content of these obligations can be further specified in regard to the kind of general and special relations we establish with groups of animals or particular animals.

References

Aristotle (1996). The politics and the constitution of Athens. Cambridge University Press, Cambridge, UK.

Burgess-Jackson, K. (1998). Doing right by our animal companions. The Journal of Ethics 2: 159-185.

Donovan, J. and Adams, C.J. (2007). The feminist care tradition in animal ethics. A reader. Columbia University Press, New York, NY, USA.

German Animal Welfare Act (2010). Available at: www.gesetze-im-internet.de/bundesrecht/tierschg/gesamt.pdf.

Kant, I. (1785 [1965]). Grundlegung zur Metaphysik der Sitten. Felix Meiner Verlag, Hamburg, Germany.

Kant, I. (1797 [1996]). The metaphysics of morals. Cambridge University press, Cambridge, UK.

Korsgaard, C.M. (2005). Fellow creatures: kantian ethics and our duties to animals. The Tanner Lecture on Human Values. Available at: http://tannerlectures.utah.edu/lectures/documents/volume25/korsgaard_2005.pdf.

Norwegian Animal Welfare Act (2009). Available at: http://www.regjeringen.no/en/doc/laws/Acts/animal-welfare-act.html?id=571188.

Singer, P. (1990). Animal liberation. New revised edition. Avon Books, New York, NY, USA.

Vollum, S., Buffington-Vollum, J. and Longmir, D. (2004). Moral disengagement and attitudes about violence toward animals. Society and Animals Volume 12(3): 209-235

Wood, A. and O'Neill, O. (2003). Kant on duties regarding nonrational nature. Aristotelian Society Supplementary Volume 72(1): 189-228.

Meat and the benefits of ambivalence

C.N. van der Weele
Department Communication, Philosophy and Technology, Wageningen University, P.O. Box 8130, 6700 EW Wageningen, the Netherlands; cor.vanderweele@wur.nl

Abstract

Meat is a troubling subject. On the one hand most humans love to eat it; the world is expected to double its meat consumption within a few decades. On the other hand, this increase also makes many meat related problems get out of hand. In societal debates on meat, meat eaters often appear to be on the wrong side of the moral line. Yet, below the surface of moral polarizations, many people – meat eaters as well as non-meat-eaters – are ambivalent about meat. Ambivalence is unpleasant and it comes with various (subconscious) mechanisms to reduce it, such as strategic ignorance. The result is that attitudes to meat look more unequivocal than they really are. I will discuss some mechanisms of strategic ignorance concerning meat. Attention for ambivalence may enable a more satisfactory understanding of present attitudes to meat than analyses in terms of straightforward attitudes. In addition, a greater appreciation of the plausibility and legitimacy of ambivalent attitudes may counter tendencies of denial and reveal similarities between meat eaters and vegetarians. This in turn may enable a less polarized and more creative societal space for the search for alternatives and solutions. One idea that may help to acknowledge ambivalences on meat is cultured meat, or in vitro meat.

Meat in moral decline

In her book *Anständiges Essen* ('Eating decently'), the German writer Karen Duve reports on her experiments and self-examination concerning her diet, with a strong focus on meat (Duve, 2011). At the beginning of the book she tells us what made her take up the project. On the hand, she loved animals and been living with pets for a long time, on the other hand she bought the cheapest chicken available in the supermarket, without thinking about it. A new vegetarian housemate criticizes her: doesn't she know what kind of life those chickens have been leading? Yes, says, Duve, somewhere at the periphery of my mind I knew, and I didn't like to think about it at all, because it meant I had to give up on the chicken. When she later reflects on the incident she suddenly realizes that she has never been thinking about her diet in moral terms. The more I thought about it, she writes, the more amazed I felt about the huge discrepancy between what I knew and what I had been buying so far. Filled with self-reproach and wonder she decides to start an experiment that involves trying different diets and to write a book about what she learns about food and about herself in the process.

Duve's thoughts and feelings illustrate the growing tensions concerning meat, at least in some Western societies. On the one hand, most people are fond of it; they are very attached to their meat eating habits. On the other hand, information on the dark sides of meat increases, as well as the number of dark sides: on top of animal suffering, which has been on the societal agenda from the 1970s onward, many other problems have been piling up, especially since *Livestock's long shadow* (Steinfeld *et al.*, 2006) drew wide attention to the environmental problems of husbandry. Greenhouse gas emission, land and water use, energy, pollution, animal diseases, the problems of antibiotics resistance and the prospects of global food scarcity are all expected to worsen further as a growing and richer human population is expected to double its global meat consumption in the coming decades. One prominent response to these problems is to argue that animal husbandry should become still more efficient and intensive, to feed all human mouths as sustainable as possible. But, apart from the question whether such a strategy can ever be sustainable enough, it ignores the growing moral uneasiness concerning animals. Announcements of welfare improvements within systems of intensive animal husbandry have so far not been able to take this uneasiness away, and it is doubtful if they ever will. It looks like human-animal relations are increasingly questioned in a more fundamental way. This may well be a symptom of the slow historical process observed by Darwin, in which human moral sensitivities gradually expand. Darwin thought that, due to our growing ability to attend to the more remote consequences of our actions in the course of history, our 'sympathies became more tender and widely diffused, extending to men of all races, to the imbecile, maimed, and other useless members of society, and finally to the lower animals' (Darwin, 2004/1871: 149). Though this conjecture may be too simple and too linear, Steven Pinker, in *The better angels of our nature* (Pinker, 2011), has forcefully argued and documented that violence has slowly been declining in the course of human history and that this process has come to include animals due to an expansion of moral circles – the metaphor is derived from Singer (1980).

Against this background, and given the present intensity of animal husbandry, it is not hard to understand that the moral reputation of meat has been in decline in recent decades, at least (again) in some Western countries. For example, Jonathan Safran Foer (2009) writes that when he started to study meat, everybody he encountered, himself included, expected that vegetarianism or even veganism would be the moral outcome of the project. The New York Times, observing that nowadays, 'ethically speaking, vegetables get all the glory' noted that meat eaters had had surprisingly little to say in return and held a contest, challenging its readers to submit essays on moral reasons for meat eating (Kaminer, 2012). Yet the outcome of such searches is not necessarily vegetarianism or veganism; eating modest amounts of meat from animals that have been having good lives may also do, according to both Foer and the winner of the New York Times contest. What is definitely wrong, however, according to a growing number of people, is factory farming. And the continuation of factory farming, according to Foer, must stem from indifference to the fate of animals, because, despite secrecy from the side of the meat industry, quite a lot is known about it now: 'We can't plead ignorance, only indifference' (Foer, 2009: 252). Yet he also tells us the book is filled with facts because they are 'a necessary starting point.' These convictions seem to be somewhat at odds with each other, which takes us to the paradoxes of strategic ignorance.

Ambivalence, meat and strategic ignorance

That people are indifferent about factory farming can increasingly be doubted. An ever growing number of meat eaters is very ambivalent about meat, especially because of what they know or suspect about factory farming. More than ten years ago, Holm and Møhl (2000) already reported that many (Danish) meat eaters voiced the same criticism as non-meat eaters. Ambivalence is not the same as indifference; it does not signal the absence of concern, it refers to tension between concerns, in this case between concern for animals and cherished habits. It thus involves a form of dissonance, not simply between two cognitions, but between cognitions associated with important actions. That dissonance may create psychological discomfort has been known ever since the theory of cognitive dissonance (Festinger, 1957). In answer to the question why and when dissonance and ambivalence are really troubling, it has since been found that it is especially unwelcome when it is connected with action (Harmon-Jones and Harmon-Jones, 2007, Van Harreveld *et al.*, 2009).

In the case of ambivalence about meat, cherished actions are clearly at stake, and people have been found to deal with the discomforts of ambivalence and dissonance in a variety of ways. Let me briefly list some examples.

- Holm and Møhl (2000) found that their respondents tended to buy meat that was less visibly part of an animal (for example minced meat) and that meat tended to disappear as the centre of the meal; instead, it became just one of the ingredients.
- People have been found to reduce dissonance by adjusting cognitions: they tend to ascribe lower mental capacities to animals we eat, such as cows, than to animals we do not eat, such as cats and dogs (Bastian *et al.*, 2012; Loughnan *et al.*, 2010).

- A further mechanism involves a compartmentalization of knowledge. For example, Korzen and Lassen (2010) found that when people talked about meat in the context of daily consumption, meat was associated with taste, price, convenience, health, additives and other consumption qualities. In the context of food production, quite different issues came up: animal welfare, sustainability, safety, trust and GMO's. Apart from the issue of additives, the researchers hardly found any overlap between the contexts. People tend to keep thoughts about meat eating apart from thoughts about animals, just like Karen Duve used to do.
- Stanley Cohen, in a book on the denial of human atrocities, describes the phenomenon of compassion fatigue, referring to his own experience. Much like other people throwing away an Amnesty leaflet, he says, his own filters go into automatic drive concerning animals; his attitude here is one of 'total denial'; 'this is not my responsibility, there are worse problems, there are plenty of other people looking after this' (Cohen, 2001: 289).

All these ways to deal with dissonance deny, obscure or deform threatening forms of knowledge in more or less automatic and subconscious ways, and they can all be seen as varieties of strategic ignorance. Such ignorance is in fact a paradoxical phenomenon, for in order not to want to know, one has to know enough to know that knowing more would be undesirable or dangerous. The phenomenon is primarily known from studies on the question what people did or did not know about the holocaust. In his book *The Germans and the final solution: public opinion under nazism*, David Bankier writes 'Clearly there was no scarcity of information, but some people could not or would not take it in. There is no doubt that those who wished to know had the means at their disposal to acquire such knowledge. Those who did not or could not believe reacted so because they did not want to believe. In one sentence: they knew enough to know that it was better not to know more' (Bankier, 1996: 115).

But strategic ignorance is not restricted to extreme situations. Behavioural economists found the same phenomenon in situations of far more ordinary decision making. Quite some people, confronted with a choice between a smaller and a larger sum of money, decide they do not want to know about the possible disadvantages for other people of the more attractive option, even when that knowledge is available with one mouse-click. They thus avoid a conflict of conscience, but the paradox of not knowing is acute here: in these cases it is a conscious decision not to want to know (Dana *et al.*, 2007).

In the context of a project on societal appreciation of the agricultural sector in the Netherlands, we recently conducted a series of interviews, asking people of different age and level of education what they do and don't want to know about food. We found that strategic ignorance about meat is widespread. People spontaneously say things like 'if you want to eat meat, you should not know too much about it'; 'I know but I don't want to know'; 'If I paid attention to every factory farmed chicken ['plofkip' is the recent Dutch term], shopping would be much more expensive'; I already have so many complicated things to think about'. The underlying ambivalence too is easily and spontaneously voiced: 'If I knew more, I would feel guilty'; 'If I am honest I must say I don't really pay attention' (Van der Weele, unpublished data). Clearly, while ambivalence about meat may not show up in these people's behaviour or their conscious considerations, it is not deeply hidden either; right under the surface it appears to cause some degree of psychological unrest.

Though strategic ignorance may help, it comes at a price. The more information is around, the more psychic energy it takes to remain ignorant. At the time of the interviews just mentioned, a media campaign was going on in the Netherlands against 'plofkippen'. All the respondents knew that term and they also knew to various degrees that something was wrong with the lives of these chickens. In that situation, to remain strategically ignorant may require hard (subconscious) work. Another price to pay is a certain loss in the feeling of being in contact with reality. Meanwhile, other motives receive reinforcement.

Given such shifts, the strategy to remain ignorant may collapse. One example shows up in this first sentence of the foreword of a book on food and ethics: 'I must confess I did not want to know much of what I learned by reading this book' (Hauerwas in Wirzba, 2011). Clearly, the author has overcome his 'desire to remain ignorant', as he describes his previous attitude, just like Karen Duve had overcome hers.

This is not to say that strategic ignorance is absent from then on. When you open up to the problems of meat, other information may become dissonant. For example, vegetarians may deny that they like meat, or that humans have evolved as omnivores, or that (some amount of) meat is healthy. They may also too eagerly embrace welcome information, for example that a vegetarian diet is always more healthy or that people who eat meat are morally or socially inferior. In other words, a conscientious stance in itself does not immunize against the tendency to respond to the discomforts of ambivalence and/or dissonance with strategic ignorance and firm unequivocal opinions.

Strategic ignorance, or denial, is not just an individual phenomenon. Kari Norgaard, in her book about denial of climate change, describes how denial is socially and culturally organized. People collectively protect themselves against threating knowledge, especially when they do not clearly see what to do with that knowledge. Traditions, local identities and established practices, while being sources of cultural meaning and virtue, are at the same time places to hide (Norgaard, 2011).

Rethinking ambivalence

Denial of ambivalence has unfortunate effects, as John Muller (2011) argues in a paper about disability, a theme on which ambivalence and denial of ambivalence are both pervasive. Through an analysis of attitudes towards disability in the contexts of selective abortion and the treatment of handicapped babies, he observes a widespread tendency to 'embrace a conscious view of disability that belies our mixed sentiments'. He also argues that this denial is harmful; first, it implies that our conscious views do not reflect our full range of sentiments and second, the suppressed sentiments may emerge in intensified, harmful and unanticipated forms. These harms look relevant in the context of meat as well.

From a psychological point of view, ambivalence is uncomfortable, especially when decisions must be made; that much probably belongs to the facts of life. But scientific and societal (lack of) attention to it is not cast in stone. It has been suggested that ambivalence is increasingly prevalent in modern societies (Van Harreveld *et al.*, 2011) and meat certainly seems to fit in with that observation. Attention for ambivalence may therefore create more insight in present attitudes to meat than analyses in terms of (only) unambiguous attitudes. A greater recognition of the presence and plausibility of ambivalence may also serve to counter denial, which in turn may increase the congruity between our conscious and our suppressed views, as Muller suggests. In addition, a greater acknowledgment of ambivalence may diminish societal polarization. When ambivalence on complex problems is denied, opposing parties spend part of their energy in fighting their own discarded views by fighting each other. Recognition of ambivalence inevitably reveals similarities between meat people who do and don't eat meat and thus enables a less polarized societal search space for solutions.

A greater scientific and societal appreciation for the ambivalences of meat may directly result from attention to ambivalence, but it may also result from unexpected new perspectives. Let me finish by briefly pointing out how to idea of in vitro meat, or cultured meat, is particularly promising for making ambivalences on meat explicit. The idea of cultured meat is to make meat from animal stem cells with the help of tissue engineering techniques and/or 3D printing. What results is not animals, but animal tissue, especially muscle tissue. The idea of cultured meat implies acceptance of (many) people's attachment to eating meat, while at the same time it promises to alleviate all the major problems associated with meat. While cultured meat is still in early stages of research and development, it is widely regarded as having great promise for animals as well as the environment (Hopkins and Dacey, 2008; Tuomisto and Teixeira de Mattos, 2011).

Although in vitro meat comes with ambivalences of its own (for example concerning food and technology), discussions on cultured meat also tend to highlight ambivalences of 'normal' meat. For example, thoughts about the alleged 'unnaturalness' of cultured meat invariably lead to the question how natural conventional meat really is (Welin and Van der Weele, 2012); the idea of cultured meat thus exemplifies 'world-disclosing' effects (Driessen and Korthals, 2012). Since it conspicuously undermines the inherent connections between meat, animal suffering and environmental problems, the idea of cultured meat also creates a common ground, however uneasy, between meat eaters and vegetarians, which can be seen in the support of animal rights organizations such as PETA for cultured meat.

It is plausible that attention to ambivalence is increasingly needed for understanding the valuation of meat. If a recognition of ambivalence also helps to loosen moral lines of division, it may help to liberate energy for cooperative inquiry in search of solutions for the many problems associated with meat.

References

Bankier, D. (1996). The Germans and the final solution: public opinion under Nazism. Wiley-Blackwell, Chichester, UK.

Bastian, B., Loughnan, S., Haslam, H. and Radke, H. (2012). Don't mind meat? The denial of mind to animals used for human consumption. Personality and Social Psychology Bulletin 38/2: 247-256.

Cohen, S. (2001). States of denial. Knowing about atrocities and suffering. Polity Press, Cambridge, UK.

Dana, J., Kuang, J. and Weber, R. (2007). Exploiting moral wiggle room: experiments demonstrating an illusory preference for fairness. Economic Theory 33(1): 67-80.

Darwin, C. (2004/1871). The descent of man, and selection in relation to sex. Penguin, London, UK.

Driessen, C. and Korthals, M. (2012). Pig towers and *in vitro* meat: disclosing moral worlds by design. Social Studies of Science 42(6): 797-820.

Duve, K. (2011). Anständiges Essen: ein Selbstversuch. Galinai, Berlin, Germany.

Festinger, L. (1957). A theory of cognitive dissonance. Stanford University Press, Stanford, CA, USA.

Foer, J.S. (2009). Eating animals. Little, Brown & Co, New York, NY, USA.

Harmon-Jones, E. and Harmon-Jones, C. (2007). Cognitive dissonance theory after 50 years of development. Zeitschrift für Sozialpsychologie 38(1): 7-16.

Holm, L. and Møhl, M. (2000). The role of meat eating in everyday food culture: an analysis of an interview study in Copenhagen. Appetite 34: 277-283.

Hopkins, P. and Dacey, A. (2008). Vegetarian meat: could technology save animals and satisfy meat eaters? Journal of Agricultural and Environmental Ethics 21: 579-596.

Kaminer, A. (2012). The meat you eat. The winner of our contest on the ethics of eating meat. New York Times, May 3. www.nytimes.com/2012/05/06/magazine/the-winner-of-our-contest-on-the-ethics-of-eating-meat.html?_r=0.

Korzen, S. and Lassen, J. (2010). Meat in context. On the relation between perceptions and context. Appetite 54: 274-281.

Loughnan, S., Haslam, N. and Bastian, B. (2010). The role of meat consumption in the denial of moral status and mind to meat animals. Appetite 55: 156-159.

Muller, J.F. (2011). Disability, ambivalence and the law. American Journal of Law and Medicine 37: 469-521.

Norgaard, K. (2011). Living in denial. Climate change, emotions and everyday life. MIT Press, Cambridge, MA, USA.

PETA (2008). *In vitro* meat press release. People for the Ethical Treatment of Animals www.peta.org/features/In-Vitro-Meat-Contest.aspx.

Pinker, S. (2011). The better angels of our nature. Why violence has declined. Viking, New York, NY, USA.

Singer, P. (2011/1981). The expanding circle. Ethics, evolution and moral progress. Princeton University Press, Princeton, NY, USA.

Steinfeld, H., Gerber,P., Wassenaar, T., Castel, V., Rosales, M. and De Haan, C. (2006). Livestock's long shadow, environmental issues and options. Food an Agricultural Organization of the United Nations. Rome, Italy.

Tuomisto, H. and Teixeira de Mattos, M. (2011). Environmental impacts of cultured meat production. Environmental Science and Technology 45(14): 6117-6123.
Van Harreveld, F., Van der Pligt, J. and De Liver, Y. (2009). The agony of ambivalence and ways to resolve it: introducing the MAID model. Personality and Social Psychology Review 13: 45-61
Welin, S. and Van der Weele, C. (2012). Cultured meat: will it separate us from nature? In: Potthast, T. and Meisch, S. (eds.) Climate change and sustainable development. Wageningen Academic Publishers, Wageningen, the Netherlands.
Wirzba, N. (2011). Food and faith. A theology of eating. Cambride University Press, Cambridge, UK.

Section 7. Sustainable aquaculture

AquAdvantage or disadvantage: social and legal pros and cons of genetically modified fish

E. Issatt
Centre for the Study of the Sciences and Humanities (SVT), University of Bergen, Allégaten 34, 5020 Bergen, Norway; issatt@inbox.com

Abstract

Aquaculture is now responsible for more than half of our seafood production. The industry is very successful, and has been increasing steadily since the 1970s. This may be expected to ease the strain upon capture fisheries, given that global fisheries are now fully- or over-exploited, and many fish stocks are depleted. A further solution is suggested in the development of genetically modified (GM) fish for aquaculture; they could have a significantly enhanced growth rate, lower feed intake and improved disease resistance; which combine to reduce the strain upon global fisheries. At present there are no GM fish licensed for consumption as foodstuff, but the *AquAdvantage* salmon, manufactured by AquaBounty Technologies, is under assessment for licence by the US Food and Drug Administration. The assessment process is accompanied by protest, both at grassroots and Senate level. The concept of genetically modified living animals is more negatively perceived than genetic crop modification and is in fact repugnant to many; the GM fish has been given the soubriquet 'Frankenfish'. There is also strongly differing scientific opinion of its value, and biosafety. This paper investigates the claims of the supporters and detractors of GM fish, and finds incompatibility; the safety claims for *AquAdvantage* do not counter the risk claims for GM fish in general.

Keywords: biotechnology, salmon, regulation

Introduction

In December 2012 the US Food and Drug Administration issued a Finding of No Significant Impact (FONSI) as a result of the environmental assessment conducted on the *AquAdvantage* salmon, to be grown at sites in Canada and Panama. The assessment is part of an on-going approval process for the first genetically modified (GM) animal to be licenced and produced for human consumption. The process has taken seventeen years; *AquAdvantage*'s developers AquaBounty Technologies (ABT) first applied in 1995.

Given that the US is significantly the main adopter of plant biotechnology, possessing nearly two-thirds of global GM crops, the slowness of the FDA's approval process raises questions about public perceptions: of genetic modification of animals, of GM animals as foodstuff, of regulators and government agencies, and of scientists and science itself. In tandem with distrust is the sense that non-GM aquaculture and modern fishing practices are accepted due to being perceived as traditional. There may also be an unwillingness to consider that genetic modification of animals may offer potential alternatives in meeting the food demands in the near future of a 9 billion global population, predicted for 2050.

Background to GM fish

The genetic modification of fish stems from the late 1980s, with the creation of a GM salmon in the early 1990s, which led to the commercial development of the *AquAdvantage* (www.aquabounty.com). The technology enables the organism's genetic structure to be altered to an extent which could not occur naturally, as the exchange of genetic material is inhibited by a common species barrier- only species of the same genus can produce hybrid offspring, eg. horse and zebra. Crossbreeding also naturally occurs when differing breeds of animals from the same species are bred, usually to enhance traits such as size; this selective breeding is a traditional agricultural process for improvement of stock fitness. By contrast, genetic modification introduces into the animal novel genes from DNA of animals outside the common species barrier.

Research into modification of fish has economic and practical benefits. The consumption and therefore production of fish as foodstuff has increased dramatically since the early days of aquaculture in the 1970s; aquaculture is estimated to account for half of the global fish consumption by humans, so the commercial incentive is clear (Food and Agricultural Organization, 2008). From a practical viewpoint it is considerably easier to modify fish than mammals. Livestock such as cows or sheep produce few eggs, which have to be removed, modified then replaced within the female in order to gestate. In addition to being less economically sound, the process is invasive and causes some stress and suffering to the animal. Fish eggs by contrast are multiple and can be painlessly removed.

The *AquAdvantage* salmon is an Atlantic salmon (*Salmo salar*) with genetic inserts from Chinook salmon (*Oncorhynchus tshawtscha*) and ocean pout (*Zoarces americanus*). The additional genetic information is a recombinant DNA sequence consisting of a growth hormone gene from the Chinook salmon and a promoter gene for cold tolerance from the ocean pout. Salmon only grows for part of the year, in response to water temperature; the ocean pout promoter gene codes for continual tolerance of cold waters, so its insertion in the salmon leads to continual growth. The result is that the modified fish is claimed to grow at least twice as quickly to 1 kg as conventional salmon, according to ABT (www.aquabounty.com). It does not grow any larger, so the animal at slaughter is comparable in size to conventional salmon, at 3-4 kg. The *AquAdvantage* salmon are all-female, and triploid; possessing an additional third set of chromosomes which renders them sterile. They are to be licenced for contained use only in land tanks as opposed to the open-net sea pens of traditional aquaculture.

GM fish: pros and cons

GM aquaculture conveys both actual and potential benefits, and potential detriments. As the *AquAdvantage* is not yet licenced, no actual detriment can be ascertained. Genetic modification can have a role to play in improving conventional aquaculture: although the industry has grown at a considerable rate in tandem with the decline of capture fisheries, it cannot be considered a solution to overfishing. The *AquAdvantage*'s accelerated growth is claimed to lead to more efficient food processing, with an alleged 10% better food conversion rate than conventional farmed salmon (www.aquabounty.com).

A further potential benefit may be found in fish modified to convert food to even greater efficiency. This may be a potential solution to food security issues.

A further environmental benefit is that the *AquAdvantage* is to be licenced only to be grown in land tanks: the wastes from aquaculture in conventional open water net pens drop to the sea bed below, whereas effluent from closed land tank systems can be collected and used as land crop fertiliser.

Modification could also convey benefits to the welfare of farmed fish, as they are susceptible to communal diseases and require dosage with pesticides. Resistance to disease and parasites is already a factor in selective breeding, but it may be possible that in the future, genetic modification may remove the need for dosage.

There are many more potential advances in biotechnology for fish with biomedical application, but this article considers GM fish as foodstuff only, in the context of the *AquAdvantage*.

The above environmental benefits also have economic worth: fish that grow twice as quickly are market-ready in half the time, leading to a doubled turnover. The increase in stock growth plus reduced fishmeal costs and possible future reduced pesticide costs, reduces the cost associated with an aquaculture start-up. They also mitigate the expense of land tank infrastructure, which in turn opens up fish farming as a viable inland economy, reducing transport costs and creating employment. Incidentally, this may also be applied in non-GM fish farming.

Genetic modification, whether of plants or animals, is a contentious issue for many people, who consider that the potential detriments far outweigh any benefits of the technology. This has led to delays in the FDA regulatory process. The issues are science-based concerns and ethical concerns, but also the regulatory process itself has been criticised as inadequate. The US Office of Science and Technology Policy decided on a Coordinated Framework of pre-existing legislation to regulate all genetically modified organisms (GMOs) on the principle that they are substantially equivalent to their conventional counterparts.

GM animals are regulated by the FDA under the Federal Food Drug and Cosmetic Act (FDCA). They are evaluated as New Animal Drugs, as the genetic alterations to the animal may be interpreted as a drug: 'an article intended to alter the structure or function of the body of man or animal' (21 USC §.321(g)(1)(C)) The evaluation process consists of a seven-step risk assessment. As part of the evaluation and as a requirement of the National Environmental Policy Act, an Environmental Assessment (EA) is conducted for the GM animal which considers its impact on the human environment. The results then lead to either the issue of a Finding of No Significant Impact (FONSI) or the commissioning of a more intensive Environmental Impact Statement (EIS). As stated above, at present the EA for *AquAdvantage* has been evaluated and a FONSI has been issued.

The concept of regulating GM fish for food as an animal drug seems unorthodox but has some advantages: New Animal Drug applications are subject to severe scrutiny, with a reversed burden of proof. Further to this, the FDCA gives the FDA powers to impose restrictions and recall post-market products. Post release, the applicant must provide data every six months for the first two years then yearly- so GM fish would be permanently under review (21 CFR §.514.80(a)(4)).

Despite the precautionary measures in place, concerns are still raised about the efficacy of the risk assessment process, because of fundamental uncertainty in the genetic modification of animals. Scientific uncertainty is defined as three types: statistical, where more data is needed; model, where the interactions are not fully known; and fundamental, where there is a risk of ignorance of knowledge deficit. For the *AquAdvantage*, this is most prevalent in the case of environmental risk. Although the fish are to be grown in land tanks with various confinement measures in place, the presumption that all fish farmers will adhere to the physical containment conditions is perhaps naïve; flouting regulations in order to save money is a key characteristic of our species. Given that GM fish may end up in net sea pens, and that triploidy is not 100% effective, it is possible that fertile fish may escape into open water. Therefore this is a possible detriment of GM fish with the most real concern and potential of serious ecological damage. The harm inherent in fish escape cannot be quantified with certainty as it depends upon three factors: the ability of the fish to escape and disperse into open water and interact with native wild fish populations, the effect of modification upon the fish's overall fitness, and the resilience of the native fish population. The Net Fitness methodology (Muir and Howard, 1999) considers various scenarios according to different genetic interactions: if the GM fish is fitter than its counterparts then it will successfully spread its genes into the local population ('spread' scenario). If the local fish are fitter, the GM fish will fail to establish, and its genes are purged ('purge' scenario). If the GM fish seems fitter but has a flaw, such as poor adult viability despite early sexual maturity, it will spread its genes to generations that die off early, leading to local extinction ('Trojan gene' scenario). The net fitness of *AquAdvantage* has been assessed as reduced, with a strong presumption that its gene flow would result in a purge scenario, described as a 'low probability of harm' (Van Eenennaam and Muir, 2011). Nevertheless, given that harm probability includes the condition of the receiving population, there may still be negative effects on an unstable low population community. Sterile fish are larger and stronger and able to compete better for resources (National Research Council, 2002) or may become predators on the local population. Their impact may be on a par with invasive exotic species, a contributory factor in recent animal extinctions. There is not just the local population to consider; studies have shown that the introduction of a new species to an ecosystem has a 'cascading impact' on the other species within that ecosystem, affecting its entire equilibrium (National Research Council, 2002).

Another scientific concern was food safety. Queries were raised over the presence of allergens, level of IGF-1 hormone and levels of polyunsaturated fatty acid. According to Van Eenennaam and Muir (2011) the results of all three food safety queries were either satisfactory or inconclusive and no food safety concerns were detected.

In addition to scientific concerns are the social concerns attached to the genetic modification of animals. The consideration of the welfare of animals is a concept dependent on the attitudes held toward animals by a culture, but certain values are generally agreed: that the animal should be free from physical or psychological distress, and that it should be free to behave according to its species (National Research Council, 2002). The modification of fish in earlier research caused physical distress with cranial deformities reported as standard, and other expressions of the random gene insertion such as affected swimming ability (National Research Council, 2002). Irrespective of phenotypic peculiarities, it may be possible that a fish that grows two to ten times faster than its peers would suffer pain due to rapid growth. Furthermore, the potential benefit of genetic modification to reduce disease and parasites may lead to a new welfare detriment in that more fish can be reared in the same space, causing crowding and psychological stress.

The ethical debate on modification of animals ranges from cultures where the subject is practically taboo, for example First Nations peoples in British Columbia, Canada, who view the salmon as a powerful spiritual source for their community (Tansey and Burgess, 2006) to cultures who welcome the technology as an encouraging next step in the selective breeding process, creating the possibility of greater efficiency in gene targeting. Somewhere in between these two extremes lies a majority opinion which appears to be a sense of the violation of animal integrity; 'a definite antipathy towards the transgression of [animal] species boundaries' (PABE, 2001).

Further concerns may be found in the legal issue of intellectual property. Apart from the ethics of animals as entities capable of being patented, the concern is that stock patenting creates an inequitable closed system, as fish farmers cannot breed the fish, and are reliant on continually buying eggs. The monopoly rights of ABT may cause economic disadvantage for developing countries, many of which have a thriving aquaculture economy. The GM aquaculture industry may have a detrimental effect on conventional aquaculture farmers, and fishers of wild salmon.

The scientific uncertainty inherent in the possibility of GM fish escape leads to legal concerns over the difficulty of establishing liability and redress in this scenario. The *AquAdvantage* EA states that the FDA consider global risk as part of its analysis, but concludes that as confinement measures are so secure, there are no risks to global commons. This does not of course apply to other GM fish in the future. Although many countries are signatories of the Cartagena Protocol on Biosafety, which regulates the transboundary movements of living GMOs (LMOs), the US and Canada are not signatories. The Protocol's Biosafety Clearing House provides information, shipment traceability and co-ordinated response in case of unintentional transboundary activity.

The new Nagoya Kuala Lumpur Supplementary Protocol addresses liability and redress for LMO transboundary damage. It states 'A causal link shall be established between the damage and the living modified organism in question in accordance with domestic law' (Nagoya Kuala Lumpur Supplementary Protocol, Art. 4). The major issue with liability systems for an unconfined and untraceable entity such as fish, is that proof of harm and causation may be impossible to demonstrate, particularly in a Trojan gene scenario where extinction may occur.

Should the Spread scenario occur, it may be impossible to prove whether the fish are the new generation and a product of gene contamination, or the original escapees. This is of particular relevance should GM fish be patented, as accidental infringement of the patent may occur when escaped fish are captured by local fishers. If the escapees are able to reproduce, do they also transmit the patent rights to the next generation? The difficulties inherent in patenting fish, and the equitable issues raised, are a topic for on-going discussion (Rosendal *et al.*, 2006; Tvedt, 2006). The novelty of the technology creates novel legal scenarios. It may transpire that it is impossible to establish a liability and redress system for transboundary GM fish damage.

Opinions about GM fish

The FDA and ABT have acknowledged the need for transparency and public consultation as key democratic governance strategies for controversial new technologies, by inviting public comment, both on the FDA Draft Guidance on Regulating GE Animals, and the *AquAdvantage* Briefing Packet and EA, but the tactic seems to have failed to gain public trust. The time period for public response to the FONSI has been extended to late April 2013, but so far the comments are overwhelmingly negative (Regulations.gov, 2013). At Senate level, Alaska's Congressman Young sponsored an amendment in 2011 to a farm spending bill to prohibit use of its funds on the FDA approval process of GM salmon, described by the Congressman as 'Frankenfish' (Don Young Press Release available at http://donyoung.house.gov/news).

Underlying some of the distrust of GM is the belief that 'natural is better'. As much of the developed world is urban and unaware of farming techniques, the belief may be romanticised. The industrial reality of, for example, chicken and fish farming, is broiler birds that have been selectively bred to such a size that they cannot walk normally; and spinal deformity is so commonplace in conventional farmed salmon that a classification of the most common deformities has been mooted (FDA, 2010). It might also be suggested that for aquaculture salmon, 'natural' is in fact highly 'unnatural'; salmon are migratory animals that are not best suited to being trapped in fish pens for the entirety of their lives, and the psychological stress must be considerable.

Scientific opinion is divided: the 2010 letter from various NGOs and concerned scientists to the Commissioner of the FDA stated the above concerns about food safety and environmental safety of the *AquAdvantage*, and requested a full Environmental Impact Statement to be conducted (Letter to FDA available at http://www.aquabounty.com/documents/misc/Env_FDA_). By contrast, other scientists are frustrated at the constant barrier to innovation that the FDA's slowness is causing; the 2012 letter from concerned scientists to President Obama describes the delay as 'inexplicable', and emphasises the need for the FDA to approve the *AquAdvantage* to be farmed in contained pens (Letter to President Obama 2012 available at http://aquacomgroup.com/wordpress). The strength of the language shows the extent to which this first GM fish as food is viewed as a testing ground for other research and development, a fact which was implicitly acknowledged in the award to ABT in 2011 of a research grant from the US Department of Agriculture. At present, ABT have sufficient funds to continue until early 2013, but their experience is discouraging to other industries seeking to innovate in GM animals.

Conclusion

Both sides in the debate hold strong opinions, but the opinions are not two sides of the same coin: protesters of *AquAdvantage* ignore the fact that the fish is to be safely confined and is safe to eat, preferring to focus broadly on their objections to GM fish in abstract. The supporters of *AquAdvantage* ignore the real concerns held about the future of GM fish and animals, preferring to focus narrowly on the approval process as a gateway to further innovation. Therefore the opposition to the *AquAdvantage* is really the opposition to genetic modification of living animals; the distrust of the new technology, moral objection, belief in 'naturalness' and the assigning to ABT the symbolism for an industry that may cause ecological harm. Detriments that can be stated with certainty are hard to find, and the benefit of a contribution to global food security is a serious consideration, but whether *AquAdvantage* will finally be served up for dinner remains to be seen.

References

Food and Agricultural Organization of the United Nations (2008). The State of World Fisheries and Aquaculture. FAO, Rome, Italy. Available at: http://www.fao.org.

Muir, W.M. and Howard, R.D. (1999). Possible ecological risks of transgenic organism release when transgenes affect mating success: sexual selection and the Trojan gene hypothesis. In: Proceedings of the National Academy of Sciences of the United States of America PNAS 96: 24.

National Research Council (2002). Animal biotechnology: science-based concerns. The National Academies Press, Washington, DC, USA, 182pp.

PABE (2001). Public perceptions of agricultural biotechnologies in Europe. PABE, Lancaster, UK. Available at: http://csec.lancs.ac.uk/archive/pabe/docs.htm.

Regulations.gov (2013). Public comments on the AquAdvantage FONSI. Available at: www.regulations.gov/#!docketDetail;D=FDA-2011-N-0899.

Rosendal, G.K., Olesen, I., Bentsen, H.B., Walløe Tvedt, M. and Bryde, M. (2006). Access to and legal protection of aquaculture genetic resources: Norwegian perspectives. Journal of World Intellectual Property 9: 392-212.

Tansey, J.D. and Burgess, M. (2006). Complexity of public interest in ethical analysis of genomics: ethical reflections on salmon genomics/aquaculture. Integrated Assessment 6: Tansey.

Tvedt, M.W. (2006). Elements for legislation in user countries to meet the fair and equitable benefit-sharing commitment. Journal of World Intellectual Property 9: 189-212.

Van Eenennaam, A.L. and Muir, W.M. (2011). Transgenic salmon: a final leap to the grocery shelf? Nature Biotechnology 29: 706-710.

Comparing the ethics of capture fisheries and aquaculture

M.E. Lam
University of British Columbia, Fisheries Centre, 2202 Main Mall, Vancouver, BC V6T 1Z4, Canada; University of New Mexico, Department of Biology, 167 Castetter Hall, MSC03 2020, 1 University of New Mexico, Albuquerque, NM87131-0001, USA; mimibethlam@gmail.com

Abstract

The fish sector (wild or capture fisheries and aquaculture) is an important global source of food, income, livelihood, and culture. Aquaculture currently supplies 42% of the world fish production and is predicted to soon eclipse capture fisheries. The balance between these two production systems in supporting global seafood consumption has serious implications for food security, income distribution, ecosystem services, and overall sustainability. Here, the ethics and sustainability of capture fisheries and aquaculture will be analyzed and compared. An innovative practical ethics approach will be presented which adapts the ethical matrix, a conceptual tool that analyzes the welfare, freedom, and justice of different interest groups, and *Rapfish*, a semi-quantitative, rapid appraisal technique used to evaluate the sustainability of fisheries. In analyzing the ethics of seafood production and consumption, the roles of the law, market, and citizen are emphasized. Based on rapid appraisal case studies of large- and small-scale capture fisheries and carnivorous and omnivorous finfish aquaculture, it is argued that all three are essential for creating ethical and sustainable seafood production systems.

Keywords: ethical matrix, food security, harm principle, *Rapfish*, sustainability

Introduction

The fish sector (wild or capture fisheries and aquaculture) is an important global source of food, income, livelihood, and culture. In 2010, it accounted for 148 million tonnes of fish, valued at US$217.5 billion, with 86% consumed as food and 19 kg per capita food fish supply (FAO, 2012). Demand for seafood is steadily increasing, due to both a growth in the global population and per capita consumption (FAO, 2012). Fish provide about 10% of the world's total caloric intake, including over half of the animal protein and minerals for some of the world's poorest people (UNEP, 2012). Globally, in 2009, fish provided 17% of the world population's animal protein and 6.5% of all protein consumed (FAO, 2012). The fish sector supports the livelihoods of 8% of the world's population and plays a particularly vital role in poverty alleviation for small-scale fisheries (UNEP, 2012). Aquaculture currently supplies 42% of the world fish production (Klinger and Naylor, 2012): it may soon eclipse capture fisheries, which have stagnated or declined since the 1990s (FAO, 2012; Pauly *et al.*, 2002). Thus, the balance between these two fish production systems in supporting global seafood consumption has serious implications for food security, income distribution, ecosystem services, and overall sustainability for the world's 7 billion people, particularly the poorest.

Here, the ethics and sustainability of capture fisheries and aquaculture will be compared. An innovative practical ethics approach will adapt the ethical matrix (Mepham, 2000, 2008, 2012a,b), a conceptual tool that analyzes the welfare, freedom, and justice of different interest groups. This theoretically-based method will be operationalized through *Rapfish* (Pitcher and Preikshot, 2001; Pitcher *et al.*, 2013), a semi-quantitative, rapid appraisal technique used to evaluate the sustainability of fisheries along six fields: *Ecological*, *Technological*, *Economic*, *Social*, *Ethical*, and *Institutional*. In particular, the ethical matrix of Lam and Pitcher (2012), modified from the analysis by the Food and Agriculture Organization of the United Nations (FAO, 2005), will be updated to evaluate the ethics and sustainability of capture fisheries and aquaculture (Appendix A). This work is motivated by Lam and Pitcher's conclusion (2012: 364) that 'ethical fisheries are also sustainable', based on a *Rapfish* analysis (Pitcher *et al.,* 2009) to evaluate compliance with the Code of Conduct for Responsible Fisheries (FAO, 1995), which showed that more compliant fisheries are also more ecologically sustainable (Coll *et al.,* 2012).

Capture fisheries

Capture fisheries account for approximately 90 million tonnes of fish annually (FAO, 2012). They are typically categorized either as large-scale, industrial fisheries, which use large specialized vessels, such as trawlers, purse seiners, long liners, and gill netters, with salaried crews, or small-scale, artisanal fisheries, relying on small owner- or family-operated crafts, some non-motorized, operated mostly inshore with more selective, multiple fishing gear technologies. The ethics of large- and small-scale fisheries are qualitatively assessed (Table 1) employing literature sources, such as information contrasting their benefits in terms of number of fishers employed in total and for each $1 million invested in vessels, capital cost of each job on fishing vessels, annual catches for human consumption and reduced to fishmeal and oils, annual fuel oil consumption, catch per tonne of fuel consumed, fish and other sealife discarded at sea (Pauly, 2006) and subsidies (Jacquet and Pauly, 2008). This represents the first ethical analysis of fisheries, using an updated version of the matrix of Lam and Pitcher (2012).

Large-scale fisheries

Large-scale fisheries include industrial commercial fishing fleets, which, generally of high volume and profit, have contributed to global overfishing and declining fish stocks (Pauly *et al.,* 2002). In the process, they have caused significant damage to ecosystems (Lam, 2012), which varies by gear types (Chuenpagdee *et al.,* 2003). This historical legacy of post-World War II economic expansion of national fishing industries reflects a time when environmental impacts of global fisheries were largely unrecognized and the industry was heavily subsidized (Sumaila *et al.*, 2010). A sense of entitlement persists today among fishers, who have 'captured' regulators in many developed countries (Lam and Pauly, 2010). Industrial-scale fisheries are considered largely unethical (Lam and Pauly, 2010), as private enterprises exploit public fishery resources without compensation to the public treasury, such as through 'pay-as-you-fish' policies or extraction fees paid for fish landed via a harm principle (Lam, 2012). Emerging ethical issues around the sustainability of seafood production and consumption have led to recent ethical analyses of fisheries (FAO, 2005, Lam and Pitcher, 2012), which I take one step further by applying the ethical matrix to both large- and small-scale fisheries.

Table 1. Comparing the ethics of large- and small-scale capture fisheries.

Capture fisheries large-scale	Welfare	Freedom	Justice	Capture fisheries small-scale	Welfare	Freedom	Justice
Ecosystem	Poor	Poor	Poor	Ecosystem	Average	Average	Poor
Fish populations	Poor	Poor	Average	Fish populations	Average	Average	Average
Society	Poor	Average	Poor	Society	Average	Average	Poor
Government agents	Poor	Poor	Poor	Government agents	Average	Average	Poor
Fishers	Average	Average	Average	Fishers	Poor	Poor	Poor
Consumers	Average	Average	Poor	Consumers	Average	Average	Average
Other stakeholders	Poor	Average	Poor	Other stakeholders	Average	Average	Average
Overall	Poor	Poor	Poor	Overall	Average	Average	Poor

Good	Average	Poor

Small-scale fisheries

Small-scale fisheries, which account for approximately 30 million tonnes of catch annually (Pauly, 2006), make important contributions to food security, nutrition, livelihoods, and poverty alleviation in many developing countries. Despite their importance, small-scale fishers and fish workers are often marginalized and not involved in resource decision-making affecting their livelihoods (Pauly, 2006). Human rights, including economic, social, cultural, political, and civil rights, are critical to achieving sustainable development for small-scale fisheries and the fishing communities they support (Allison *et al.,* 2011, 2012). Securing human rights for fishers to a decent standard of living, work, healthcare, and an education, for example, will involve changing governance institutions and power structures that determine resource allocation and access. An ethical matrix evaluating the welfare, freedoms, and justice of small- and large-scale fisheries is presented in Table 2. Problematic elements of small-scale fisheries include: income and asset poverty, vulnerability to climate and other variability, exclusion from decision-making, lack of recognition in planning, limited access to social services and infrastructure, and political marginalization (Bene *et al.,* 2007).

Aquaculture

Aquaculture refers to the farming or cultivation of aquatic animals or plants under controlled conditions and accounts for approximately 64 million tonnes of fish annually (FAO, 2012). 60% of all aquaculture (by tonnage) is produced in freshwater, 32% in seawater, and 8% in brackish water (Klinger and Naylor, 2012); freshwater aquaculture produces mostly finfish (55% of total), predominantly carp, while marine and brackish aquaculture is composed primarily of mollusks (25%), finfish (10%), and crustaceans (10%). Freshwater finfish are often integrated within agricultural and polyculture systems, causing less environmental damage than marine fish and crustaceans, which are typically reared near shore and in coastal ponds, respectively, modifying habitats and degrading ecosystem services. Aquaculture operations can be commercial farms that rely on intensive methods to produce commodities for regional and global markets or family and cooperative farms relying on less intensive practices and low-value species for household subsistence or local markets (Naylor *et al.* 2001). Aquaculture production has increased by an annual average of 8% since 1970 and is predicted to eclipse capture fisheries in its provision of global seafood through intensification of fish, shellfish and algae production (Klinger and Naylor, 2012).

Table 2. Comparing the ethics of carnivorous and omnivorous fish in aquaculture.

Aquaculture: carnivorous fish (Atlantic salmon)	Welfare	Freedom	Justice	Aquaculture: omnivorous fish (Nile tilapia)	Welfare	Freedom	Justice
Ecosystem	Poor	Poor	Poor	Ecosystem	Average	Average	Poor
Fish populations	Average	Poor	Poor	Fish populations	Good	Poor	Poor
Society	Poor	Average	Poor	Society	Average	Average	Poor
Government agents	Poor	Poor	Average	Government agents	Poor	Poor	Average
Aquaculture farmers	Good	Good	Good	Aquaculture farmers	Average	Good	Good
Consumers	Average	Poor	Poor	Consumers	Good	Average	Good
Other stakeholders	Poor	Average	Poor	Other stakeholders	Average	Average	Average
Overall	Poor	Poor	Poor	Overall	Average	Average	Average

Good	Average	Poor

Associated environmental and resource damage to the marine environment and wild fish populations threaten its sustainability, ocean health, and global food security, viz.: nutrient and chemical pollution; marine resource dependence via feeds; threats to wild species by farmed-fish escapes, parasites, and diseases; and limitations of freshwater and land resources to aquaculture growth (Klinger and Naylor, 2012). Solutions being investigated include novel culture systems, alternative feed strategies, and species choices (Klinger and Naylor, 2012).

Carnivorous finfish

The farming of high-value carnivorous species (e.g. salmon and shrimp) requires vast amounts of wild-caught fish for feed (Naylor *et al.,* 2001): Atlantic salmon (*Salmo salar*), for example, are fed fishmeal and fish oils from the reduction of forage fish at a ratio of 4:1 (Cressey, 2009), fish that might otherwise be consumed by human populations or provide essential ecosystem services (Pikitch *et al.,* 2012). As a byproduct of confining a large number of fish in ponds, tanks or cages in coastal waters, especially in mangroves and wetlands, aquaculture facilities often cause unmitigated environmental damage, viz.: nursery habitat destruction; eutrophication through excessive nutrients, wastes and antibiotics causing algal blooms and oxygen dead zones; sea lice and other diseases which can threaten wild fish; and escapes which can compete or hybridize with indigenous fish populations (Naylor *et al.,* 2001). Salmon are now being fed a mixture of soya bean and fishmeal and fish oils, making them 'more like pigs' (Cressy, 2009), which reduces the amount of healthy omega-3 fatty acids. Traits are being introduced, such as fast maturation and disease resistance, by selective breeding and genetic modification, respectively, which raise potential ecological and human health risks, often unknown to the consumer. Recirculation systems on land, fully monitored and controlled, avoid some of the potential hazards of genetically modified fish in open cages in the ocean, but this technology has added costs which open systems do not bear in the absence of regulations for environmental damage. Offshore farms are also an option over near-shore farms, as they have higher water quality and fewer conflicts with recreational water users, but the engineering and licensing requirements are more complex (Cressy, 2009).

Omnivorous or herbivorous fish

Omnivorous or herbivorous fish (e.g. tilapia or carp, respectively) and filter feeders (e.g. oysters, clams, and mussels) are lower in the food web and thus have less adverse impacts on ecosystems, as they consume primary producers (e.g. aquatic plants and plankton) and/or low trophic level fish. For example, an omnivorous fish, the Nile tilapia (*Oreochromis niloticus*), named the 'aquatic chicken' for its speedy and efficient growth, converts fishmeal and fish oils at a rate of 1:3, i.e. they are net fish producers (Cressy, 2009). They are less likely to build up toxins such as mercury in their flesh and have a sweet and inoffensive flavor, but are seen as bland and not favored in the west, being roughly a third less valuable than Atlantic salmon. Industrial-scale farming of omnivorous and herbivorous species still uses fishmeal and fish oil (about 15%) for compound feed, diverting food away from human consumption. Table 2 compares the ethics of farming Atlantic salmon versus the Nile tilapia.

Ethics and sustainability of capture fisheries

The ethical evaluation field of *Rapfish* was designed to evaluate ethical attributes of fisheries as they pertain to the overall sustainability of the fisheries (Pitcher *et al.,* 2013). The ethical attributes, issues scored, and justice addressed are listed in Table 3, along with a qualitative *Rapfish* ethical analysis for both large- and small-scale fisheries. Note that while the analysis is crude, averaging all fisheries, trends are revealed between the ethics and sustainability of large- and small-scale fisheries. Specifically, large-scale fisheries score poorly, while small-scale fisheries score average in terms of their sustainability evaluated along ethical criteria, which correlates with their ethical evaluation using the ethical matrix.

Table 3. Rapfish ethical analysis of capture fisheries (modified from Pitcher et al., 2013).

Ethical attribute	Issue scored	Justice addressed	Large-scale	Small-scale
Adjacency and reliance	Geographical proximity and historical connection with resource	Distributive		
Alternative livelihoods to fishery	Sources of support within the community	Distributive		
Equity in entry	Fishery entry based on traditional customs, historical access, harvest history	Distributive		
Just management	Inclusion of fishers and local community in management	Social		
Mitigation of habitat destruction	Attempts to mitigate damage to fish habitat	Ecosystem		
Mitigation of ecosystem depletion	Attempts to mitigate fisheries-induced ecosystem change to predators, prey or competing organisms	Ecosystem		
Illegal fishing	Illegal and unreported fish catches from poaching, trans-shipments or non-compliance with size, place, species or quota regulations	Retributive		
Discards and wastes	Discards and waste and/or bycatch of fish, birds, mammals, reptiles, invertebrates, including 'high grading' and discards made to comply with local legal regulations	Restorative		
Overall				

Conclusions

Based on rapid appraisal case studies of capture fisheries (large- and small-scale) and aquaculture (carnivorous and omnivorous finfish), I argue that the law, market, and citizens are essential for creating ethical and sustainable seafood production systems. First, drawing on the laws of damage and nuisance, I examined to what extent a harm principle has been incorporated, i.e. whether the harms or costs of environmental damages caused by the industry have been internalized (Lam, 2012). Second, ethical issues in the market were explored, such as eco-labeling, seafood certification and tracing, and corporate responsibility for fisheries (Lam and Pauly, 2010), and culture system improvements, alternative feed strategies, and species selection, including genetic modification, for aquaculture systems (Klinger and Naylor, 2012). Third, the roles and responsibilities of citizens and consumers were considered in the context of food security, consumer choice, and social justice (Pitcher and Lam, 2010). This preliminary analysis of the ethical dimensions of the fish sector and the ethics of seafood consumption highlights necessary changes in the law, market, and citizens.

References

Allison, E.H., Åsgård, B. and Willmann, R. (2011). Editorial: human rights approaches to governing fisheries. MAST 10(2): 5-13.

Allison, E.H., Ratner, B.D., Åsgård, B., Willmann, R., Pomeroy, R. and Kurien, J. (2012). Rights-based fisheries governance: from fishing rights to human rights. Fish and Fisheries 13(1): 14-29.

Bene, C., Macfadayen, G. and Allison, E.H. (2007). Increasing the contribution of small-scale fisheries to poverty alleviation and food security. Fisheries and Aquaculture Technical Papers 481. Food and Agriculture Organization of the United Nations, Rome, Italy, 141 pp.

Chuenpagdee, R., Morgan, L.E., Maxwell, S.M. Norse, E.A. and Pauly, D. (2003). Shifting gears: assessing collateral impacts of fishing methods in U.S. waters. Frontiers in Ecology and the Environment 1(10): 517-524.

Coll, M., Libralato, S., Pitcher, T.J., Solidoro, C. and Tudela, S. (2012). Sustainability implications of honouring the Code of Conduct for Responsible Fisheries. Global Environmental Change 23: 157-166.

FAO. (1995). Code of conduct for responsible fisheries. FAO, Rome, Italy, 41 pp.

FAO. (2005). Ethical issues in fisheries. FAO Ethics Series 4. FAO, Rome, Italy, 39 pp.

Klinger, D. and Naylor, R. (2012). Searching for solutions in aquaculture: charting a sustainable course. Annual Review of Environment and Resources 37: 247-276.

Lam, M.E. (2012). Of fish and fishermen: shifting societal baselines to reduce environmental harm in fisheries. Ecology and Society 17(4): 18, http://dx.doi.org/10.5751/ES-05113-170418.

Lam, M.E. and Pauly, D. (2010). Who is right to fish? Evolving a social contract for ethical fisheries. Ecology and Society 15(3): 16. http://www.ecologyandsociety.org/vol15/iss3/art16.

Lam, M.E. and Pitcher, T.J. (2012). The ethical dimensions of fisheries. Current Opinion in Environmental Sustainability 4: 364-373.

Mepham, B. (2000). A framework for the ethical analysis of novel foods: the ethical matrix. Journal of Agricultural and Environmental Ethics 12: 165-176.

Mepham, B. (2008). Bioethics: an introduction for the biosciences. 2nd Edition. Oxford University Press, Oxford, UK.

Mepham, B. (2012a). Agricultural ethics. In: Chadwick, R. (ed.) Encyclopedia of applied ethics. 2nd edition. Vol 1. Academic Press, San Diego, CA, USA, p. 86-96.

Mepham, B. (2012b). Food ethics. In: Chadwick, R. (ed.) Encyclopedia of applied ethics. 2nd edition. Vol 2. Academic Press, San Diego, CA, USA, p. 322-330.

Naylor, R.L., Goldburg, R.J., Primavera, J., Kautsky, N., Beveridge, M.C.M., Clay, J., Folke, C., Lubchenco, J., Mooney, H. and Troell, M. (2001). Effects of aquaculture on world fish supplies. Nature 405: 1017-1024.

Pauly, D. (2006). Major trends in small-scale marine fisheries, with emphasis on developing countries, and some implications for the social sciences. MAST 4(2): 7-22.

Pikitch, E., Boersma, P.D., Boyd, I.L., Conover, D.O., Cury, P., Essington, T., Heppell, S.S., Houde, E.D., Mangel, M., Pauly, D., Plaganyi, E., Sainsbury, K., and Steneck, R.S. (2012). Little fish, big impact: managing a crucial link in ocean food webs. Lenfest Ocean Program, Washington, DC, USA, 108 pp.

Pitcher T.J. and Lam, M.E. (2010). Fishful thinking: rhetoric, reality, and the sea before us. Ecology and Society 15(2): 12. http://www.ecologyandsociety.org/vol15/iss2/art12.

Pitcher, T.J. and Preikshot, D.B. (2001). *Rapfish*: a rapid appraisal technique to evaluate the sustainability status of fisheries. Fisheries Research 49(3): 255-270.

Pitcher, T.J., Kalikoski, D, Pramod, G. and Short, K. (2009). Not honouring the Code. Nature 457: 658-659.

Pitcher, T.J., Lam, M.E., Ainsworth, C., Martindale, A., Nakamura, K., Perry, R.I. and Ward, T. (2013). Improvements to the 'Rapfish' rapid evaluation technique for fisheries: integrating ecological and human dimensions. Journal of Fish Biology (in press). http://dx.doi.org/10.1111/jfb.12122.

Sumaila, U.R., Khan, A.S., Dyck, A.J., Watson, R., Munro, G., Tyedmers, P. and Pauly, D. (2010). A bottom-up re-estimation of global fisheries subsidies. Journal of Bioeconomics 12: 201-225.

United Nations Environment Programme (UNEP) (2012). Avoiding future famines: strengthening the ecological foundation of food security through sustainable food systems. UNEP, Nairobi, Kenya.

Appendix A. Ethical matrix for seafood production systems (updated from Lam and Pitcher, 2012).

System	Welfare	Freedom	Justice
Natural			Ecosystem
Ecosystem	Ecosystem integrity; protection of biodiversity, food webs and habitats	Preservation of adaptive capacity and resilience to anthropogenic perturbations (e.g. fisheries, pollution, etc.)	*Productive:* stewardship, conservation, protection *Restorative:* damaged ecosystems restored
Fish populations	Fish stock abundance and genetic conservation; fish welfare, including minimal genetic mixing or lice and diseases from farmed fish	Limited migration barriers and access to feeding or breeding habitats	*Productive & restorative:* policies for sustaining fish biomass, growth and reproduction
Human			Social
Society	Minimal environmental and social externalities (costs) from private enterprises; sustainable flow of aquatic resources for benefit of present and future generations; healthy economy and environment	Freedom to information and to express concerns about the management of resources to ensure it benefits all of society	*Distributive:* fees collected by management agencies distributed to social programs *Retributive:* compensated for ecological harm caused
Government agents	Alternative management and policy choices to serve public interests	Freedom to decide based on transparent information, open participation in public debates, and co-governance	*Social:* transparency; accountability; liability; access to information
Fishers and fishing communities/ aquaculture farmers	Work and social security: adequate income and safe working conditions; poverty eradication; health care, educational and other capacity-building opportunities; cultural diversity respected	Freedom to choose fishing and alternative livelihoods; empowerment of fishers, including women and ethnic minorities; distinct identities of indigenous communities and cultural rights to fish in traditional fishing grounds recognized; food sovereignty in fish production system	*Distributive & retributive:* participation in decision-making or co-management; equitable and secure access to and use of resources; fair treatment in entry, use, credit, market, trade, subsidies, regulations, policies, and law; compensation for equal work and social inequities
Consumers	Food security: access to safe, nutritious, affordable, and sufficient food	Food sovereignty: freedom to choose food through eco-labeled choices of responsibly harvested seafood and culturally appropriate foods	*Distributive:* equitable access to food; no trade barriers; balance low trophic level fish consumed for food and forage fish converted to fishmeal
Other stakeholders	Non-consumptive uses also valued in resource decisions	Freedom to compete for share of resources; participatory decision-making and collaborative governance	*Distributive & retributive:* equitable share of and access to resources for food, income, livelihood, culture, and recreation; dispute resolution for resource conflicts

Appendix A. Continued.

System	Welfare	Freedom	Justice
Interaction			Ecosystem & Social:
Overall system	Economic viability and stability; sustainable development; safety; competitive industry	Conditional freedom or privilege to fish (fishing rights with societal obligations); small-scale equal opportunity as industrial-scale enterprises	*Productive, restorative, distributive and retributive:* ecosystem-based management; historically based restoration; cross-sectoral equity in law and taxes; compensation for harm

Whose sustainability counts? Engaging with debates on the sustainability of Bangladeshi shrimp

S. Bremer, A.S. Haugen and M. Kaiser
Centre for the Study of the Sciences and the Humanities, University of Bergen, Allegaten 34, 5020 Bergen, Norway; scott.bremer@svt.uib.no

Abstract

Once a luxury food, shrimp today is both cheap and easily accessible for Europeans, owing largely to the rapid expansion of shrimp aquaculture in Asia, including Bangladesh. But, the economic benefits offered by this global shrimp trade are arguably accompanied by significant social and environmental costs in the producing countries, leading some groups to call for a complete boycott of shrimp from countries like Bangladesh. The European Union 7th Framework-funded research project 'Sustaining Ethical Aquaculture Trade' (SEAT) sought to engage with this political debate on the sustainability of the Bangladeshi shrimp trade. It attempted to draw attention to the complexity of the value-chain taking shrimp from farm to fork, and the plurality of actors along this chain, posing the politically significant question, 'whose sustainability counts'? This paper argues that scientists undertaking research on these themes cannot shy away from the political nature of these questions, and relates the experiences of the SEAT Project and its study of the highly politicised Bangladeshi shrimp value-chain. This political role demanded reflexivity from SEAT researchers relative to the role of the scientific community in political debates. While the paper acknowledges the potential conflict between objectivity on the one side and being a force in social change on the other, the value-laden nature of the term 'sustainability' seems to imply consequences for actual research and outreach transcending the traditional role of the scientific community. The paper has the joint aims of describing a particular method for undertaking research that contributed to political debates, and stimulating the broader dialogue on the role of science in political debates.

Keywords: aquaculture debates; science-in-society, reflexivity, dialogue, applied ethics

Introduction

European publics are engaging in politically-charged debates that question the sustainability and ethics of their food supply, with the stakes particularly high for seafood. With many European wild fishery stocks on the brink of collapse, today it is necessary to import more than half of the seafood demanded by the European market, increasingly from aquaculture in Asia. However, the long 'value-chains' that bring Asian aquaculture products to Europe have been criticised by some as unsustainable and unethical, with questions posed as to the safety of this food, the environmental impact of aquaculture, the welfare of the fish in the farms, and the living and working conditions of people in Asia. These debates play out in many different arenas at many different scales, from the international media to the local market place, and mobilise a plurality of voices on what constitutes 'ethical and sustainable' seafood; each drawing on its own constellation of values, principles and evidence. When discussing international value-chains – with seafood traversing political, economic and cultural boundaries – this only serves to compound the complexity and plurality characterising these debates. With so many competing ideas of what makes a sustainable trade, this introduces dilemmas about 'whose sustainability counts?' Some of the most stinging controversy in this respect has been attached to the trade of shrimp from Bangladesh.

The scientific community is increasingly recognising that it too participates in this complex lattice of political arenas, marking a departure from a 'demarcation model' of science towards a model of 'science-in-society.' By posing questions on the sustainability and ethics of seafood value-chains the scientific community is forced to confront the value-ladenness of these questions, and the inescapable implications for both undertaking their research and debating their findings. While recognising that there are some scientific endeavours that will be confined to 'the lab', many undertaking research on these topics today find themselves needing to participate in more political settings, whether this be with policy-makers, the industry or the popular media. This is translated into an expectation that science not only assumes a descriptive role, but also acts as a normative voice in political arenas.

There is a significant and expanding literature discussing the role of science-in-society. Much of this work is led by scholars from the field of science and technology studies, and the philosophy of science. Authors like Jasanoff (1987) describe a highly politicised boundary drawn around science as a means of controlling its role in society, with this boundary continuously renegotiated. A number of commentators have argued that this boundary should recognise scientists' active role in political debates, with Gibbons (1999) discussing this in terms of 'science's new social contract'. Gibbons and colleagues (1994) have argued for greater reflexivity amongst the scientific community to open their research to the critical scrutiny of scientific and other actors, while also critically appraising their own role as scientist *and* citizen. These ideas have found credence also within the field of fisheries and aquaculture governance, with Kooiman and Bavinck (2005) for instance, arguing for an active role for science in governance, including negotiating the role of science in political arenas.

This paper relates the experience of a European Union 7th Framework-funded research project titled 'Sustaining Ethical Aquaculture Trade' (or SEAT), as it engaged with the specific debate on the sustainability of the Bangladeshi shrimp trade. The paper has two aims. First, it describes how the SEAT consortium assumed its normative role in this debate, and the way that this framed their methods for research and outreach. Second, it argues that this can serve as one example for greater reflexivity in our scientific endeavours, and contribute to a wider dialogue on the role of science on politically charged questions.

Sustaining ethical aquaculture trade as a normative undertaking

The SEAT Project is a four-year EU-funded research project initiated in 2009 to explore the 'value-chains' bringing farmed seafood from Asia to Europe, and argue for how they can be more sustainable (see: seatglobal.eu). It brings together 13 institutional partners from across Europe and Asia to give effect to an interdisciplinary systems-approach drawing upon diverse disciplines ranging from aquaculture science and aquatic ecology, to life-cycle analysis, food safety, development studies, and applied ethics. SEAT is steered by four broad objectives: (1) to describe a sub-set of value-chains; in order to (2) normatively develop measures of their sustainability and ethicalness; while (3) actively seeking to influence practice along the value chains through 'action research'; and (4) enhance scientific, business and policy linkages between Asia and Europe. The latter three objectives reveal that at its inception SEAT was designed to contribute to on-going political debates on the sustainability of Asian aquaculture, and though the Project has in practice gone on to devote more energy to its descriptive goal, the normative aspirations have nonetheless found expression.

The normative role of the SEAT Project is strengthened through the inclusion of an ethics work package, led by the University of Bergen, which has developed an ethical framework for undertaking SEAT research. Importantly, applied ethics within the context of SEAT is not an exercise in prescriptively applying 'grand ethical theories' to the seafood value-chains to evaluate their performance. Rather, the framework begins from the heuristic of the value-chain to unpack the complex social and ecological systems that comprise this chain, and the significant uncertainties they imply. This framework brings to the fore the inescapable plurality of moral agents interacting with this notional value-chain, and their disparate perspectives on what constitutes an ethically good and sustainable seafood trade. It frames ethical enquiry as a social choice problem, and challenges SEAT to explore ways for bringing together these diverse systems of values, principles and knowledge in a rich picture of possible ethical and sustainable futures. In giving effect to this framework, the SEAT project has sought to extend moral agency to a wider group of actors than are usually engaged in these discussions. It has sought to enable dialogue across the different notions of 'the good', while preserving the integrity of each perspective to co-exist along-side others. To these ends, SEAT has engaged novel methods that combine research and dissemination roles, to present its own authentic voice in political debates.

Engaging with debates on the sustainability of Bangladeshi shrimp

Shrimp production is the second most important export sector in Bangladesh; directly employing more than 1.2 million people along the value chain to produce 170,000 million tonnes of shrimp per year, more than half of which is exported to Europe (Kruijssen *et al.*, 2012). This has seen a number of European enterprises working hard to facilitate this trade, and improve the quality and traceability of Bangladeshi shrimp. 'Naturland', a German association for organic agriculture, has been working to establish organic shrimp production in Bangladesh for example. However, there are a number of groups, significantly non-governmental organisations like the 'Swedish Society for Nature Conservation' that have engaged in campaigns to highlight the social and environmental damage they argue this trade causes, and call for an outright boycott.

The SEAT Project studied Bangladeshi shrimp and engaged with debates on its sustainability. In accordance with its ethical framework, SEAT sought to tell the story of a long and complex value-chain, linking diverse groups of actors with disparate conceptions of a sustainable future, demanding a reflexive question 'whose sustainability counts?' To this end, the Project adopted an approach that challenged the traditional boundary between conducting research, communicating the findings, and contributing to change.

The research began by mapping the diverse aspirations for the value-chain. From the European end, this took the form of a desktop review of completed quantitative surveys of European consumers' perceptions of seafood, and what attributes they look for in seafood products. From the Bangladeshi end, this took the form of a large quantitative survey of more than 400 shrimp farmers, followed up by a second round of qualitative interviews with a smaller set of actors from along the value-chain. The survey and interview results were interpreted as describing a rich 'ethical landscape' (Bremer *et al.*, 2012), revealing different values that were important to Bangladeshi producers underpinning principles and indicators of 'good' production. The challenge lay in bridging the European and Bangladeshi understandings of a sustainable shrimp trade, by facilitating communication along the value chain, and to heated political debates beyond, for a more nuanced appreciation of these diverse views.

Communicating European consumer values to Bangladeshi producers

The research used a combination of films and workshops to facilitate communication, and present the SEAT Project's perspective on the shrimp value-chain. This began with communicating the European consumers' perspective to Bangladeshi producers. The desktop review of European consumers was translated into a short film to portray one version of the European reality and ten key attributes that European consumers would like seafood producers to have consideration for. The film was filmed in English and uploaded for open-access on YouTube, before being subtitled in Bangladeshi and presented to more than 60 Bangladeshi stakeholders from along the regional shrimp value-chain in a workshop in Khulna. The workshop encouraged discussion amongst participants on European 'expectations' of shrimp that sparked an animated discussion, with most Bangladeshi unaware of how their products end up on European shelves.

Dialogue between Bangladeshi actors

A second purpose of the Khulna workshop was to facilitate dialogue between Bangladeshi value-chain stakeholders themselves. 'Categories' of stakeholders where grouped together, such that shrimp farmers could discuss their common aspirations for the future of the sector and voice them to other groups of input-suppliers, processing plant owners, and regulators for example. In this way it presented a uniquely inclusive and open forum; important for both conducting research and presenting the SEAT Project's authentic voice within political discussions.

A message from Bangladeshi producers to European consumers

A third purpose of the workshop was to 'close the loop' of indirect dialogue between Bangladeshi producers and European consumers, by asking workshop participants to send a message back to Europe. Participants were invited to collectively write a giant postcard to Europe, on which each included their own short individual message of what they would like European consumers to consider. This postcard was digitised after the workshop, and posted for dissemination on the SEAT website. At the same time, a film-crew from the University of Bergen was present in Khulna to capture the workshop, and collect images from along the value chain. This footage was compiled into a second short film, which aimed to communicate back to Europe the Bangladeshi realities, and how these stakeholders see the sector developing. This film was designed to work as a pair with the film on European realities, and was also placed on YouTube for open-access.

A platform for on-line debate

This 'indirect dialogue' described above culminated in an on-line platform for more direct debate among those with an interest in the Bangladeshi shrimp trade. An independent 'Facebook' group was created, with invitations to participate disseminated widely in partnership with seafood industry spokespeople from 'Seaweb'. The web-page presented participants with links to the two films on YouTube together with two 'infographic' posters depicting: (1) the postcard from Bangladesh to Europe; and (2) a diagrammatic representation of the shrimp value-chain. With access to these resources, participants were posed three open questions to stimulate an on-line debate. In this way, SEAT not only acted to engage with on-going debates in existing political arenas, but indeed strived to create new and novel arenas through workshops and on-line debating platforms, supported by new media.

The Bangladeshi shrimp 'narrative'

In parallel with these attempts at establishing direct and indirect dialogue, a separate SEAT initiative sought to describe the Bangladeshi shrimp value-chain in a 'narrative', which brought together the scientific outputs of the project in a story that emphasised complexity and plurality, while also being easily accessible to a wide audience. This narrative was written by an intern that was engaged on SEAT for three months, and as her own authentic voice it presented both a fresh perspective and brought the added legitimacy of being independent from the SEAT consortium, and other stakeholders along the chain. The internship began with three weeks exploring the value-chain in Bangladesh, before moving to Europe to investigate how the products reach consumers. It was actively informed by SEAT research, though the intern, with no background in aquaculture, had freedom to write her own narrative.

The internship and narrative proved to be successful mechanisms for presenting the SEAT Project and its research in support of on-going debates on Bangladeshi shrimp. While in Bangladesh the intern kept an on-line blog attached to the SEAT website, on which she posted photos and described her experiences. This blog was linked to other industry websites, and attracted a wide viewership. After following the value-chain from Bangladesh to Europe, the intern completed her narrative, which was disseminated via three channels. A summarised version of the narrative was distributed via emails to stakeholders within the seafood and aquaculture industry in Europe in partnership with industry advocates 'Seaweb'. At the same time, a version of the narrative featured in the March 2013 edition of the World Aquaculture magazine (Hensler and Bremer, 2013). Finally the intern presented her narrative in a session on organic seafood with industry partners at the Brussels Seafood Show, April 2013.

Some critical reflections on the SEAT Project approach

The SEAT Project's political engagement posed significant challenges to the consortium, not least related to their very identities and methods. In consortium meetings some scientists, from aquatic ecology for example, saw the Project as a challenge to their identity. These scientists described their role as documenting 'objective measurements', and discussed 'normative claims' as compromising the integrity of this endeavour. On the other hand, SEAT scientists regularly demonstrated loyalty to those actors along the value-chain that they studied, in the sense that they did not want SEAT research used to the disadvantage of poor Bangladeshi aquaculture producers. SEAT therefore demanded reflexivity and honesty from scientists on how they framed their identity, to recognise the human relationships that underpinned their science. Beyond identity, many SEAT scientists argued that they did not have the expertise, or relationships with media outlets, to meaningfully engage in political debates. This was a very real concern, with researchers at times pushed to engage with media like film that they had no experience with, with implications for the quality of these outputs. In reflecting on the experiences of the SEAT Project, one conclusion was that if we are, as a scientific community, to meaningfully engage in political debates through our research projects, then there is a need for a dedicated core of consortium members devoted to what has in the past been described as 'dissemination'. These consortium members ought to have a double-competency; professionally-trained in dissemination techniques and also knowledgeable of the study topic, aquaculture in SEAT's case. These professionals are needed to coordinate the often fragmented dissemination efforts in a large consortium, and target key messages to salient political arenas, while always evaluating their impacts. To this end, SEAT did engage aquaculture industry advocates 'Seaweb' in the last year of the project, though ideally they would have been engaged throughout the whole project.

Conclusions

This short paper purported to show by practical example that scientists exploring ethical and sustainable seafood can take on a role of responsible actors within on-going political debates on this topic. There were three key messages. First, politics need not be a 'dirty word'. Many scientists in SEAT were uneasy about engaging with political debates, but the reality is that much science is already, and has long been, used toward political ends; with aquaculture science a salient example. The scientific community, and the applied ethicist, can bring an appreciation for the complexity of seafood value-chains, and the plural perceptions of sustainability, such that we can reflexively pose 'whose sustainability counts?' Second, research on such controversial themes ought to be transparent, and transcend traditional boundaries between research, and the dissemination of findings. In this way we presented research methods that were able to be accessed and followed by an external audience like the Khulna workshop. At the same time, dissemination ought to move from a 'deficit' model of a uni-directional 'telling' of findings, to a model of dialogue. We can site the on-line deliberative forum, which not only disseminated findings via films and infographics but also used them as a basis for discussion, inviting critique. Third, it is important that scientific research engage with the full spectrum of media. As society changes the way it engages in political debates, so must the academy change also, and look increasingly to new methods in film and on-line deliberative fora alongside more conventional approaches like focus groups. Key to such an approach is creatively assembling a portfolio of various media, and also having a dedicated core of professionals with a double-competency in 'dissemination', and knowledge of the area of study; in this case aquaculture. Such political engagement, to be meaningful, must continue over the length of a project, and be subject to continuous evaluation itself.

References

Bremer, S., Haugen, A.S. and Kaiser, M. (2012). Mapping core values and ethical principles for livelihoods in Asia. In: Potthast, T. and Meisch, S. (eds.) Climate change and sustainable development – ethical perspectives on land use and food production. Wageningen Academic Publishers, Wageningen, the Netherlands, p. 419-424.

Gibbons, M. (1999). Science's new social contract with society. Nature 402: 81-84.

Gibbons, M., Limoges, C., Nowotny, H., Schwartzman, S., Scott, P. and Trow, M. (1994). The new production of knowledge. SAGE Publications, London, UK, 179 pp.

Hensler, L. and Bremer, S. (2013). An organic future for Bangladeshi shrimp? World Aquaculture Magazine(March): 28-35.

Jasanoff, S. (1987). Contested boundaries in policy-relevant science. Social Studies of Science 10: 195-230.

Kooiman, J. and Bavinck, M. (2005). The governance perspective. In: Kooiman, J., Bavinck, M., Jentoft, S. and Pullin, R. (eds.) Fish for life: interactive governance for fisheries. Amsterdam University Press, Amsterdam, the Netherlands, p. 11-24.

Kruijssen, F., Kelling, I., Hong, M.C., Jespersen, K. and Ponte, S. (2012). Value chains of selected aquatic products from four Asian countries. A review of literature and secondary data. World Fish, Penang, Malaysia.

Section 8. Food philosophy

Personalised nutrition and social justice: ethical considerations within four future scenarios from the perspective of Nussbaum's capabilities approach

K. Nordström[1] and J. Goossens[2]
[1]Jönköping University, Box 1026, 511 11 Jönköping, Sweden; karin.nordstrom@hlk.hj.se; [2]Bio-Sense, Elisabethlaan 68, 3200 Aarschot, Belgium; jgoossens@bio-sense.be

Personalised nutrition

The idea of personalised nutrition (PN) is to give dietary advice based on personal health data. Phenotype and lifestyle measures are used as a basis for individually tailored nutritional advice. Early attempts have been made for nutrigenomics-based PN. The scientific knowledge of the interaction between nutrients and the genome is still limited, but recent studies have increased the understanding of how individual genotypes modulate the responses to dietary factors (Görman *et al.*, 2013; Ronteltap *et al.*, 2012). PN may be seen as part of a general trend towards more individualized health care (Sahlin and Hermerén, 2012; Schumpelick and Vogel, 2011) and offerings of various business models are available (Ahlgren *et al.*, 2013; Ronteltap *et al.*, 2012). Facilitating improved individual health promotion through more precise information, PN has a potential of empowerment and increased motivation for healthy food intake. Yet, PN may also affect values related to our understanding of health and responsibility for health (Nordström *et al.*, 2013). This paper explores ethical aspects of PN within the contextual environment of four future scenarios of health and nutrition in Europe in relation to Martha Nussbaum's capability approach to social justice. Within each of these scenarios, PN can be seen to play a more or less important role in promoting health via tailoring of food consumption patterns in a society. Because the logic of each scenario is fundamentally different, varying from a control or gentle stewardship to an economical investment or consumer choice paradigm, the ethical concerns about the context of the PN approach are different too.

Scenario planning methodology and process

Future scenario planning, a widely accepted and applied methodology (Godet, 2006), aims to make uncertainty about the future explicit, by describing different future environments and to understand what is driving these developments. In the frame of the Food4Me project[28], four scenarios were developed to provide insight about possible future contexts in which PN might be embedded. These scenarios describe how European societies may evolve with regard to nutrition and health by 2030, a far enough future for significant societal changes to have taken place. The scenarios were created in a collaborative scenario planning process by a mixed group of industrial and societal stakeholders and experts. In three consecutive workshops, the group got acquainted with the subject through a background presentation on PN based on 26 stakeholder interviews. Participants then brainstormed an extensive set of driving forces that might affect the future of health and nutrition in Europe. A selection of critical uncertainties and driving forces with highest anticipated impact was made. In all of the scenarios one of the underlying driving forces for change is growing financial pressure on current curative health care systems. This resulted in the definition

[28] Food4Me is the acronym of the EU FP7 project: 'Personalised nutrition: an integrated analysis of opportunities and challenges' (Contract no. KBBE.2010.2.3-02, Project no. 265494). The parties involved in the project are listed on the project's Web site http://www.food4me.org. Project coordination was carried out at University College Dublin, Ireland, Institute of Food and Health; Project Coordinator: Professor Michael J Gibney, Project Manager: Dr. Marianne Walsh. For overall correspondence regarding the Food4Me project: Professor Michael J Gibney, UCD Institute of Food and Health, University College Dublin, Dublin, Ireland, Tel: +353 (1) 716 2824, e-mail: mike.gibney@ucd.ie

of a scenario space consisting of two dimensions (Figure 1). Horizontal dimension: conception of health. This dimension is about the extent to which an individual perceives health as a purely manageable issue (action/reaction) or as a personal conviction within a holistic perspective that balances reason and emotion. Vertical dimension: logic of health care. This dimension is about the extent to which health care systems are designed based on economic principles (which cost for which performance) or on quality principles (which effort for which benefit). Figure 1 illustrates the resulting scenario space and the four scenarios. Each of them was explored in terms of food, health care, business, regulatory, information and ethical environments and the narrative of how this scenario unfolds.

Scenario narratives

Smart sister

Scarcity of material and human resources triggers the need to improve resource efficiency to increase production, self-sufficiency and global economic competitiveness of the European region. Growing social unrest also undermines social cohesion and adds to the societal costs of unemployment, health and economic recession. As the system shows signs of spinning out of control, governments are pressed to re-allocate resources on the basis of best economic value return and to control counter-productive behaviour. To be democratically acceptable, decisions are based on evidence that is built through information systems – tracing problems back to the people. Health information is centralized in a coherent system: a Superbrain that collects, consolidates and analyses data and defines the feedback to society and citizens. Health control measures are thus directed to the right population groups and slowly the balance in the health care system is regained. The system gradually evolves towards more preventative measures to reduce future health risks. Superbrain also increasingly monitors food intake and advises on dietary behaviour, triggering food and supplement industries to adjust their offerings in line with the recommended lifestyle patterns. Health risks are evaluated early on in life and people are assisted to make the appropriate lifestyle choices to increase average productive working life. Non-responders face societal pressure and possibly financial penalties.

My health my home

Europe's limited budgets are rather directed to support the economy and to the most essential societal services leaving little money for the growing public health problems. When overall infant mortality starts to rise it is decided to roll out the Health Care Savings Account (HECSA) aimed at generalising the privatization of health care. By linking health to an economic value, HECSA makes health a tangible

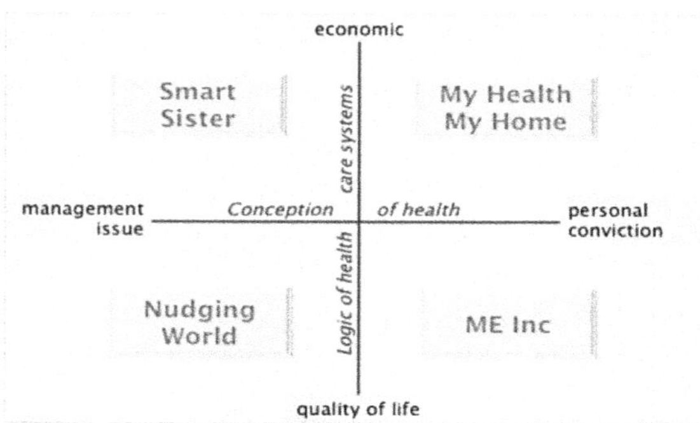

Figure 1. The scenario space: two dimensions opening up for 4 scenarios.

asset and rewards health-promoting behaviour. People start to invest in their health and, if well managed, get financial benefits for health promoting activities such as sport, wellness, fitness, health food, etc. It is a free market with regard to health choices; people choose a way of life that balances the desired quality of life with the cost of health care they can afford. There is a very basic safety net in case of hardship or bad luck in terms of health and for those struck by unemployment or income limitations. The most affluent may even have the budgetary freedom to indulge in unhealthy choices and pay for the cure. A wealth of new businesses, products and services emerge, all aiming to tap into the spending power the HECSAs generate and fuelling a new 'health economy'. The notion of healthy activity and body care becomes ingrained in society. Everyone is free to invest in HECSAs as they please but can also decide not to invest in health and to carry the high cost of curative interventions. The main thrust of the My Health My Home society is survival of the fittest.

Me Inc

Europe's highly regulated environment, which has led to a lack of individual responsibility and a wave of costly litigations, is increasingly contested by a growing share of the population. New consumer movements start to emerge through people who are taking responsibility for a healthy lifestyle and therefore disagree with taxation and limitations to distribution imposed on the population as a whole to curb unhealthy consumption patterns. As these new communities get organised, they start interacting with food, pharma and insurance companies to address their specific needs and desires. Embracing the opportunity of this direct interface with end-users, companies are eagerly developing new and specialised products, services and distribution channels. The success of this spreads rapidly, facilitated by the essentially virtual nature of these communities. Different communities emerge, each focusing on a different aspect of 'quality of life' (food, health, lifestyle, spiritual, art, etc.). Soon they represent large numbers of the most adept and able citizens and governments need to accommodate the desires of these new power blocks by developing suitable legal frameworks. Governments continue to assure basic societal values, such as civil and human rights, but health care services are increasingly privatised and communities broker the contracts between individuals and health care service actors. A basic public health care is available as a safety net for people who do not have the intellectual, physical, or financial capability to join a community. As community-driven systems require a relatively high level of intellectual engagement, it will take time and significant educational effort to establish the community-based approach throughout Europe. The core aspect of Me Inc is individual choice.

Nudging society

A growing societal awareness that current health problems require a long-term approach leads to the decision to 'start acting now for the current and future wellbeing'. The current curative health care system is found unacceptable and the new system that gradually emerges is centred on a democratically agreed 'Quality of life' view that encompasses all aspects of wellbeing. While the focus of health management is on prevention of illness, especially chronic diseases, the main driving value in society becomes 'Quality of Life' which can be measured as Quality Adjusted Healthy Life Years (QAHLY), a more broadly defined concept than the current 'free of disease' state. Because 'quality of life' means different things to different people, people with an identified health risk are given a choice between several courses of action to reduce their risk. People are also free to do nothing, but they are then solely responsible for the consequences and costs. Health caretakers and authorities 'nudge' people to accept the recommended options, and societal pressure starts to build on non-responders. The new system is democratically operated by citizens, designed to deliver a 'minimum standard QAHLY' that is universally applicable to everyone in Europe but is adjusted to accommodate for local differences, cultural or religious attitudes as well as special needs of different population groups. Authorities act as enablers; they provide resources to build the system, help citizens to set the rules and limitations, and provide education for users. A qualified service is available to assist those who are incapable of engaging with the system. Education and information campaigns drive the greater public understanding of the impact of behaviour on QAHLY and improve efficient use of resources and economic output. Gradually the effects become visible as a decline of disease-related absences from work, a beginning of reversal of the obesity ratio, as well as change in food consumption patterns and food companies improving nutritional quality. The new emerging affluence is invested by to extend the 'quality of life' vision with other important aspects e.g. mental health, life-long learning, etc. The key aspect of a Nudging Society is stewarding the public. Table 1 summarizes the key aspects of the four scenarios.

Within each scenario, PN can be seen to play a more or less important role in promoting health via tailoring of food consumption patterns in a society. Because the logic of each scenario is fundamentally different, varying from a control or gentle stewardship to an economical investment or consumer choice paradigm, ethical concerns about the context of PN are different too. After presenting Nussbaum's capability approach to social justice, PN is in the following evaluated ethically within each scenario in terms of its potential effects on social justice according to the capability approach.

Social justice within Martha Nussbaum's capabilities approach

Martha Nussbaum's capabilities approach consists of an account of justice based on human dignity. Being critical against social contract theories, she argues that they fail to give an inclusive account of justice. The idea of a social contract presupposes not only equal and free partners, but also mutuality concerning rights and duties. Groups with a significantly higher degree of dependency are not equal partners in a social contract. In the case of impairment or disability, for instance, contractarian approaches are according to Nussbaum not able to account for how special conditions and needs of disabled people could be integrated with a status of 'full equals to other citizens' (Nussbaum, 2007: 5, 7, 15). Nussbaum points thus at a vital feature of a social justice theory: the capacity to account for 'serious asymmetry of power and capacity' between different members of society (Nussbaum, 2007: 16). Her concern for asymmetries and her attempt to present an account of social justice, which is able to include them, is relevant in relation to PN and a general trend towards personalised medicine (Sahlin and Hermerén, 2012; Schumpelick and Vogel, 2011). Personalised information about health risks, as well as means to manage them, such as personalised nutritional advice, make variations within populations visible. Personalisation of health has implications for principles for the distribution of health care resources. As personalised information about people's health is available, randomized distribution of risks is not a rational principle of distribution anymore (Sahlin and Hermerén, 2012).

Table 1. Differences between the four scenarios as shown by comparing the key dynamics in terms of their dominant logic, key activity, governance type, and food consumption patterns.

Scenario	Dominant logic	Key activity	Governance type	Food consumption patterns
Smart Sister	Efficiency	Monitoring	Hierarchy	Monitoring, steering and controlling by a central authority
My Health My Home	Private health asset management	Investing	Responsible autonomy	Individual food choice balancing between health and economic affordability
Me Inc	Manifesting values	Choosing	Heterarchy	Hyper-individual choice but also responsibility for food intake
Nudging Society	Health commons management	Stewarding	Responsible collectiveness	Education and coaching for healthier food consumption but with limited choice

Nussbaum conceptualizes justice as 'entitlements to 'the capability' requirements of dignified flourishing' (Hailwood, 2012). Her approach contains a list of ten capabilities, all of which indicate 'core human entitlements' as a 'minimum of what respect for human dignity requires' (Nussbaum, 2007: 70). The capabilities with relevance to PN are: Life – being able to live a life of normal length; Bodily health – being able to have good health, to be adequately nourished; Practical reason – being able to form a conception of the good and to engage in critical reflection about the planning of one's life; Affiliation – Being able to live with and towards others, having a social basis of self-respect and non-humiliation; Control over one's environment – (political) being able to participate effectively in political choices and (material) to hold property on an equal basis with others.

Personalised nutrition and social justice within the four scenarios

Potential effects of PN on social justice are dependent on the function of PN within the different scenarios and affect health promotion and health care, as well as food consumption patterns. Nussbaum's capability approach focuses on a minimum standard for human dignity. This implies equality as a minimum of entitlements for everyone. Yet, since capabilities, such as the ability to form one's life plan and control over one's environment, are included in Nussbaum's account of social justice, there is also an idea of individual choice being essential for social justice. Personalisation of health, as strived for in PN, is therefore acceptable, if a minimum of entitlements is guaranteed, at the same time, as individual certain choice concerning own preferences is allowed for. Individual variation of how people might benefit from PN should therefore be consistent with the minimum situation defined by the list of capabilities.

The two dimensions of the scenario space (logic of health care system and conception of health; Figure 1) are significant for identifying ethical issues of social justice regarding the function of PN. On the vertical dimension (logic of health care system), PN's potential to facilitate empowerment in terms of citizens' control of health is challenged in scenarios with economic incitement for good health (Smart sister and My Health My Home). If PN is combined with economic incitements or sanctions that are based on common concepts of health, personal choice is limited. Furthermore, individual variations regarding premises for health, such as genetic, environmental, and socioeconomic aspects, are only in a very limited way controllable by the individual. Economic incitements or sanctions related to health, should therefore from the perspective of the capability approach be balanced with a fundamental minimum of entitlements for all regarding health promotion and health care. In contrast, PN might within a focus on health as quality of life (Nudging Society and Me Inc) facilitate people's possibilities to define and adhere to personal preferences in relation to health and wellbeing. This is consistent with Nussbaum's capability approach, because the ability to form one's own life plan is a fundamental aspect of dignity and social justice. Yet, a minimum of health promotion and care should also here be available for all on equal terms, in order to counterbalance with the potentially segregating effect of PN.

On the horizontal dimension (conception of health), within scenarios emphasizing health in terms of personal conviction (My Health My Home and Me Inc), PN may be beneficial for those taking interest, but may also have a potential of reinforcing social injustices. Everyone will not be able to benefit from personal freedom of choice and prospective responsibility in the same way. Again, within the capability approach, personalisation of health and personal freedom should be balanced with a minimum of equality in terms of entitlements for health care. In contrast, within scenarios emphasizing health as a management issue (Smart Sister and Nudging Society), citizens' personal autonomy is compromised for the sake of equality in society. Within these scenarios, common or societal interests moderate or counteract effects of personalisation. As Nussbaum's approach to social justice includes an idea of personal life plans and control as vital aspects of social justice, the more subtle societal control in the Nudging society scenario seems more adequate than the strong societal control in Smart sister.

With regard to food consumption patterns, the different key dynamics of the scenarios (Table 1) indicate that effects of PN might vary in different scenarios. In My Health My Home PN, as part of healthy food consumption is seen as an investment. Individual food choice is in this scenario balancing between health and economic affordability. While this allows for personal freedom, there is due to the economic dimension a potential of increasing social injustice, because people will not equally be able to benefit. In Me Inc food choice is highly individual, and so is responsibility for food intake. This is linked to consumer movements. Depending on the evolving social dynamics of the consumer communities, there is a risk for increasing inequalities here as well. Yet, both these scenarios stress personal conviction and with regard to Nussbaum, this is an important aspect of social justice. In contrast, the monitoring, steering and controlling of food intake by a central authority in Smart Sister seems to guarantee social justice in terms of equality. Yet, personal freedom is in this scenario very limited and social justice in terms of the ability to personal life plans and control over one's life seems therefore deprived. In Nudging Society, a balance between individual choice and societal interest seems to be strived for by education and coaching for healthier food intake with limited choice. Nudging Society comprises thus a combination of societal interests (by stewarding or nudging) and some toleration for individual preferences, which in relation to Nussbaum's concept of social justice seems more acceptable than the controlling situation in Smart Sister. Table 2 shows potential effects of PN on social justice in relation to Nussbaum's capabilities within the scenarios.

Table 2. Potential effects of personalised nutrition (PN) on social justice in relation to Nussbaum's capabilities within different scenarios.

Capabilities	Smart Sister	My Health My Home	Me Inc	Nudging Society
Life of normal length	PN contributes to efficiency, reinforces monitoring	PN beneficial for those taking interest in their health, contributing to segregation	PN beneficial for those taking interest in their health, contributing to segregation	PN contributes to attempted balance between common and private interests
Bodily health, nourishment	PN a means for all, enforced by authorities	PN a means for those interested if they can afford it	PN a means for those interested	PN a means for those
Practical reason, critical reflection, life plan	PN used as a means for societal interest, compromising autonomy	Availability of PN in various forms contributes to choice	Availability of PN in various forms contributes to choice	PN as a voluntary means to adhere to societal health principles
Affiliation, social basis of self-respect and non-humiliation	Refusing to use PN would probably result in sanctions and stigmatization	PN empowers those consumer groups that choose to use it	PN is a community-building factor	PN challenges responsible collectiveness, as some individuals will be predicted to have higher health risks
Control over environment, political choice	PN aimed at societal interests, not personal ones	PN embedded within economic incentives, as investment	PN regarded as an expression of personal responsibility	PN subjugated to democratic consensus

Conclusions – personalization of health and social justice

A core driving force for the development of the scenarios was growing financial pressure on current curative health care systems. The horizontal dimension (conception of health) of the scenario space indicates an ethical concern on both sides. Scenarios with a concept of health as manageable contain a risk of reducing personal autonomy. Scenarios with a health concept as expressed in personal conviction contain a risk of reinforced social injustices. Within a social justice perspective, two challenges may be identified. First, it is vital to discern between controllable and uncontrollable health risks and diseases. Ethically, it seems important to argue for a perspective, which combines incentives for healthier lifestyles with an understanding of health and disease as partly uncontrollable. Second, there is an ethical challenge (addressed differently by all scenarios) to find a balance between individual autonomy and societal interest to stimulate healthy lifestyles and food consumption patterns in order to manage controllable health risks. This latter challenge concerns the increasing scope of controllable or manageable health risks and diseases and can be expressed in terms of a dilemma of individualization (Nordström *et al.*, 2013). Personalised information on health risks may contribute to empowerment, as people are made aware of risks and means of managing them. At the same time, as health information is available on personal level, lifestyles or food consumption patterns, which are in opposition with nutritional advice, might appear increasingly intolerable within a societal perspective and especially in a context characterized by difficulties to finance health care. PN may within scenarios that incorporate strong economic incitement for good health contribute to an instrumental use of food as a means for health promotion at cost of food consumption as a matter of personal freedom and cultural expression.

References

Ahlgren, J., Nordgren, A., Perrudin, M., Ronteltap, A., Savigny, J., Van Trijp, H., Nordström, K. and Görman, U. (2013). Consumers on the Internet – ethical and legal aspects of commercialization of personalized nutrition. Genes & Nutrition 8(4): 349-355.

Godet, M. (2006). Creating futures: scenario planning as a strategic management tool. Economica-Brookings, London, UK.

Görman, U., Mathers J.C., Grimaldi, K.A., Ahlgren, J. and Nordström, K. (2013). Do we know enough? A scientific and ethical analysis of the basis for genetic based personalized nutrition. Genes & Nutrition 8(4): 373-381.

Hailwood, S. (2012). Bewildering nussbaum: capability justice and predation. The Journal of Political Philosophy 20(3): 293-313.

Nordström, K., Juth, N., Kjellström, S., Meijboom, F. and Görman, U. (2013). Values at stake: autonomy, responsibility, and trustworthiness in relation to genetic testing and personalized nutrition advice. Genes & Nutrition 8(4): 365-372.

Nussbaum, M.C. (2007). Frontiers of justice. Harvard University Press, Cambridge, MA, USA.

Ronteltap, A., Van Trijp, H., Berezowska, A. and Goossens, J. (2012). Nutrigenomics-based personalised nutritional advice: in search of a business model? Genes & Nutrition 8(2): 153-163.

Sahlin, N-E. and Hermerén, G. (2012). Personalised, predictive and preventive medicine: a decision-theoretic perspective Journal of Risk Research 15(5): 453-457.

Schumpelick, V. and Vogel, B. (eds.) (2011). Medizin nach Mass. Individualisierte Medizin – Wunsch und Wirklichkeit. Herder Verlag, Freiburg, Germany.

Individual animal welfare and the collective dimension of sustainability: the role of animal welfare in developments towards sustainable food production and consumption

F.L.B. Meijboom
Ethics Institute, Utrecht University, Janskerkhof 13a, 3512 BL Utrecht, the Netherlands; f.l.b.meijboom@uu.nl

Abstract

Sustainability is a broad and multi-layered concept that is not easy to define. In the practice of food production this becomes clear in a wide range of – sometimes mutually excluding – proposals to secure food production in a sustainable way. Currently there are a number of initiatives to make food production more sustainable. Animal welfare increasingly plays a role in the initiatives to come to a sustainable food production and consumption. However, in spite of the growing attention to animal welfare in sustainability debates, animal welfare quite often appears to conflict with the ideas on sustainability. On the one hand, problems occur as a result of the need to weigh different aspects of sustainability. These are questions, such as how to weigh the added value of giving animals the opportunity of free ranging against the related animal and public health risks. On the other hand, a recent proposal to make poultry meat production more sustainable shows an additional problem in the relation between animal welfare and sustainability. This is the potential conflict between the emphasis on the individual animal in the welfare debate and the orientation on collectives in the sustainability concept. An improvement of overall sustainability might still imply that the welfare problems of individual animals remain unaddressed. In this paper, I elaborate on the relation between animal welfare and sustainability. I use the debate between environmental and animal ethics – that is characterised by a similar gap – in order to look at the opportunities to deal with the tensions between individual animal welfare and the collective focus of sustainability. Finally, I propose that defining sustainability as a moral ideal is helpful to include individual animal welfare in the sustainability debate.

Keywords: sustainability, poultry farming, animal welfare, moral ideal

Sustainability as an ideal and the role of animal welfare

Sustainability is a broad and multi-layered concept that is not easy to define. In the practice of food production this becomes clear in a wide range of – sometimes mutually excluding – proposals to ensure sustainable food production. A lot of discussion remains, even if we take the broadly shared Brundlandt definition as a start for this paper, i.e. sustainable as the development that meets needs of the present without compromising the ability of future generations to meet their own needs (1987). What we define as a sustainable solution or strategy is still subject of genuine debate.

Elsewhere, I proposed that we had better approach sustainability as a moral ideal, rather than to search for one single definition of sustainability (Meijboom and Brom, 2012). Moral ideals are relevant in: (1) situations in which we are uncertain about what is morally desirable; and (2) situations in which one evaluates the common sense or traditional morality as no longer sufficient. As both appear to be applicable in the case of sustainability, moral ideals can be essential as they are in principle dynamic and open, but not empty. Sustainability as a moral ideal guides the normative practice, because it functions as a perspective or compass. Hence it gives direction and structures the deliberation about the desirable future states of affairs linked to the fields that are discussed in terms of sustainability (Van der Burg, 1997).

However, the question is what sustainability – understood in terms of a moral ideal -, can contribute to problems that surface if we look at the discussions on sustainable animal production. This is an interesting case, since a number of questions in the sustainable food discussion are related to the role of animal welfare. Currently there are a number of initiatives to make food production more sustainable. For instance, to reduce the carbon food print of primary food production or to use less natural resources in a processing of food. Animal welfare increasingly plays a role in the initiatives to come to a sustainable food production and consumption (e.g. Barbier and Elzen, 2012; EFFAB, 2009; SAID, 2009; Poppe *et al.*, 2009). This should not come as a surprise. Once we start to recognize animals as beings that are morally considerable for their own sake and therefore have their own interests, they equally have needs that have to be met and the needs of future generations can be harmed (cf. Brundlandt, 1987).

Animal welfare as source of moral problems

In spite of the growing attention to animal welfare in sustainability debates, animal welfare quite often appears to conflict with the ideal of sustainability. The classical case is the indoor and intensified housing systems of poultry. With the help of careful selection, broiler chickens have become the most efficient meat-producing animals. Furthermore, the current housing systems are claimed to be sustainable in a number of ways, such as human health, environmental impact and efficient use of natural resources. Yet, it raises many animal welfare problems (cf. Decuypere *et al.*, 2010), and it appears that reducing these welfare problems conflicts directly with other parameters for sustainability. For instance, allowing animals to walk outside implies risks at the level of animal and human health, and using breeds that grow slower may conflict with the aim to use natural resources as optimal as possible. Consequently, it has been argued that we end up in a deadlock position in which a low CO_2 footprint directly conflicts with genuine welfare problems that appear to be inherent to the high-production system (Bos *et al.*, 2010). This conflict between animal welfare and other parameters of sustainability equally holds for other practices of animal use and raise ethical problems that require a careful balance and assessment (Korthals, 2001; Meijboom, 2010).

Of course it might be difficult to find solutions that do justice to all aspects of sustainability and therefore one can argue that these dilemmas are part and parcel of the sustainability debate. It equally holds for debates on sustainable fashion or transportation. Similar to the problems with animal welfare, these contexts require a weighing of broad range of interests of many stakeholders. Nonetheless, a recent proposal shows an additional question in the relation between animal welfare and sustainability. This is the potential conflict between the emphasis on the individual animal in the welfare debate and the orientation on the collective aspects in the sustainability debates. In this paper, I elaborate on this tension along the lines of a recent initiative.

Mixing up to improve sustainability

Partly as a result of current public debates in the Netherlands on the acceptability fast growing broilers that are kept in intensified housing systems, new initiatives are developed and/ or are more explicitly put forward in the discussion. One of these initiatives is the idea to mix meat from conventional and more sustainable origin in order to increase the supply of sustainable meat production and to enhance sustainable consumption. The idea is similar to what has already been used in the context of green or renewable energy. If you buy 'green' electricity, it is not by definition that your refrigerator really works on electricity from wind or solar energy. The electricity company guarantees that they buy or produce the capacity of green energy you pay for, although it might be delivered at your neighbours. This is the result of having just one type of infrastructure that transports electricity from all kinds of origin. In the case of food production, we do have another system: we have different supply chains for the various ways food is produced. Organically raised chickens are mostly not mixed with hens raised in conventional housing systems.

A recent proposal, however, aims to show the relevance of using the same supply chain for product from different producing systems. The price differences between conventional and organically produced poultry meat are only to a very limited level related to the costs at the level of the farmer. The substantial price differences are mainly the result of the costs of processing and retail of relative small quantities of organic meat in contrast to the very efficient supply chain of conventional meat. Therefore, the 'mixing' concept has been introduced for sustainable poultry production (Nationale Denktank, 2012). The idea is to mix poultry meat from conventional farming systems with broilers that are raised in systems that pay more attention to animal welfare. The claim is that this idea of mixing is an efficient and effective approach to make real steps towards sustainable production and consumption. Rather than sustainability as a niche product, the concept of mixing provides farmers a better guarantee that they can sell their sustainable products. Furthermore, the additional costs to make the animal production more sustainable can be divided among more units and thus will lower the costs of the individual product. This idea makes poultry meat more animal-friendly and at the same time more affordable.

Leaving aside the discussion whether eating poultry can be sustainable or that we have to strive for meatless options, this proposal is an innovative idea to enhance sustainable production and consumption. Nonetheless, it poses questions at the level of animal welfare. These questions originate from the focus of many approaches of animal ethics: they tend to focus on the individual animal. If we look at the discussions of why animals matter morally for their own sake, and why we should care for their welfare, theories mostly refer to characteristics that belong to the individual animal and lead to direct duties towards the individual animal. For instance, the individual rather than the species is sentient, and therefore we should take the interests of the individual into account. Likewise, the individual rather than the group to which the animal belongs has inherent value, and consequently it's rights should be respected. As Broom states 'Welfare refers to a characteristic of the individual animal...' (2010: 86). This start in the individual animal, however, leads to questions with respect to the evaluation of the concept of 'mixing' that do not occur in the case of green energy. In the latter case, a 30% CO_2 reduction is mostly considered as a good step forward in our aim to reduce emissions. However, an improvement of the welfare by 30% of the animals might be an improvement, but still implies that the welfare of the other 70% is still under pressure. This is different from the energy case, for at least two reasons. The most obvious reason appears to be the assumption that, rather than energy, chickens are more than just resources for food. They can be harmed in themselves and thus the animals whose welfare remained unchanged are still harmed in their interests. Second, the mixing meat case shows a clash between the collective character of sustainability and the focus on the individual animal with respect to welfare. In terms of overall sustainability, the supply chain that contains meat from mixed origin might be the most optimal way to take as much parameters for sustainability into account including animal welfare. However, from an animal welfarist or an animal rights perspective it still does not do justice to the interests and value of the individual animal. To deal with this tension it is helpful to look at a similar debate between environmental ethicists and animal ethicists.

Learning from an ongoing debate and looking at moral ideals

Like the tension between a collective ideal of sustainable development and the focus on securing the welfare of individual animals, there is a gap between the more holist approach of environmental ethicists and the more individual based approach of those who work in the literature on animal welfare and animal rights. This is not to say that environmental ethicists would reject the relevance of animal welfare all together, but they focus on the welfare of ecosystems and species rather than on individual animals (Light, 2004: 125). In practice this implies that from an environmental perspective it can be justified that the management of ecosystems goes with harm to individual animals. This is the reason for animal liberationist to be sceptical to holistic approaches.

From a pragmatist perspective, however, Light shows that those tensions are not insurmountable. He proposes that, in spite of focusing on the meta-ethical differences one better should strive for convergence on ends. This does not imply a denial of the differences or evaluating the underlying theoretical assumptions as irrelevant, but Light argues that for the current debate it is more pressing to be after various reasonable arguments for shared common ends, such as the reduction of CO_2 emission or the rejection of factory farming (pp. 127-130).

I agree with his focus on the final aim rather than on the point of departure. Nonetheless, it feels as a way of begging the question. In formulating shared aims, the tensions will probably reoccur. To deal with this, the above-mentioned proposal to define sustainability as a moral ideal is helpful. Formulating moral ideals is not just about specific aims and goals, but define what is worthwhile striving for. It implies one step in between a mere theoretical discussion on fundamental assumptions and a rather pragmatic discussion on specific end points. Since moral ideals are about future states of affairs that are worthwhile striving for, they may not (yet) be the solution to the questions at stake. However, they are relevant as they structure the discussion on the preferred direction. Of course on the practical ways to achieve the aim there are discussions too. Nonetheless, by referring to moral ideals it is possible to get grip on what kind of discussion is necessary. Does a controversy starts from genuine differences on the moral ideal of sustainability, or on the general way to achieve that ideal, or on a specific proposal? Differentiating between these aspects helps the discussion.

Furthermore, a focus on ideals help to evaluate specific proposals to make food production sustainable, and to deal with the gap between the individual and the collective perspective. Related to the initiative of mixing meat from different production systems, the moral ideal approach helps to shift the focus of evaluation from the current proposal to the ideal that underlies this innovation. This can imply that – even if the current product is not as welfare-friendly as possible, it can be acceptable from an animal welfarist perspective if the moral ideal incorporates attention to individual animal welfare. In this case a reference to the precautionary principle can be a sufficient reason to agree. If the proposal is a step in the development to further improve animal welfare and the increase of respect for individual animals, then this will be beneficial for individual animals in the future. This is more than the result of negotiating or looking for compromises. With a shared moral ideal of sustainability that includes attention to individual animal welfare as a compass one has good arguments to endorse the idea.

Formulating the moral ideals that are worthwhile striving for will certainly not be easy, but as said before the sustainability discussion is complex by definition. However, defining sustainability in terms of moral ideals helps to find room for discussion and prevent that the debate on the relation between animal welfare and sustainability get stuck between either a theoretical debate or a discussion on specific proposals only.

References

Barbier, M. and Elzen, B.E. (eds.) (2012). System innovations, knowledge regimes, and design practices towards transitions for sustainable agriculture. INRA-SAD, Paris, France, p. 132-153.

Bos, A.P., Janssen, A.P.H.M., Leenstra, F.R. and Groot Koerkamp, P.W.G. (2010). Pluimvee met smaak: behoeften-inventarisaties, Project report on Sustainable broiler production. Wageningen UR, Wageningen, the Netherlands.

Broom, D.M. (2010). Animal welfare: an aspect of care, sustainability, and food quality required by the public. Animal welfare in education and research. Journal of veterinary medical education 37: 83-88.

Brundlandt, H. (1987). Our common future. Oxford University Press, Oxford, UK.

Decuypere, E., Bruggeman, V., Everaert, N. Li, Yue, Boonen, R. De Tavernier, J., Janssens, S. and Buys, N. (2010). The broiler breeder paradox: ethical, genetic and physiological perspectives, and suggestions for solutions. British Poultry Science 51(5): 569-579.

EFFAB, (2009). Code of good practice for farm animal breeding and reproduction organisations. EFFAB, Wageningen, the Netherlands.
Korthals, M. (2001). Ethical dilemmas in sustainable agriculture. International Journal of Food Science and Technology 36: 813-820.
Light, A. (2004). Methodological pragmatism, animal welfare, and hunting. In: McKenna, E. and Light, A. (eds.) Animal pragmatism: rethinking human-nonhuman relationships. Indiana University Press, Bloomington, IN, USA, p. 119-139.
Meijboom, F.L.B. (2010). Ethical reflection on a practical proposal of sustainable poultry farming. In: Romeo Casabona, C.M., Escajedo, L. and Emaldi Cirión, A. (eds.) Global food security: ethical and legal challenges. Wageningen Aacademic Publishers, Wageningen, the Netherlands, p. 87-90.
Meijboom, F.L.B. and Brom, F.W.A. (2012). Ethics and sustainability: guest or guide? On sustainability as a moral ideal. Journal of Agricultural and Environmental Ethics 25(2): 117-121.
Poppe, K.J., Termeer, C.J.A.M. and Slingerland, M.A. (eds.) (2009). Transitions toward sustainable agriculture and food chains and peri-urban areas, Wageningen Academic Publishers, Wageningen, the Netherlands.
SAID, (2009). Principles & practices for the sustainable dairy farming. Version 2009. SAI Platform, Brussels, Belgium.
Van der Burg, W. (1997). The importance of ideals. Journal of Value Inquiry 31: 23-37.

Ethics of authenticity of food: analogies from biodiversity protection

H. Siipi
Department of Behavioural Sciences and Philosophy, Assistentinkatu 7, 20014 University of Turku, Finland; helsii@utu.fi

Abstract

People commonly value and want to eat authentic food. In this paper I ask: (1) what different meanings 'authentic' has in the context of food; and (2) which, if any, of the identified meanings offer good justifications for individual food choices and for local and global food policies. The following three senses of authenticity are distinguished: authenticity as a contrast to copies and fakes, authenticity as a contrast to artificiality, and authenticity as being true to one's essence. Within the second sense, the analysis is strengthened by drawing analogies to the context of biodiversity protection, in which arguments for the value of authenticity are common. I argue that all three senses can, at least when combined with relevant beliefs, offer good justification for favouring authentic food.

Keywords: conceptual analysis, fake, artificial, essence

Background and the research question

According to James H. Gilmore and B. Joseph Pine II (2007), people increasingly see the world in terms of the authentic and the inauthentic and prefer the authentic offerings to inauthentic ones. This also concerns food. People value and want to eat authentic food, and they want to avoid inauthentic food products (Sagoff, 2001; Stiles *et al.*, 2011). This is realised in food marketing in which the claims concerning authenticity are very common (Gilmore and Pine, 2007).

The aim of this paper is to understand these appeals for authenticity. I ask: (1) what different meanings 'authentic' has in the context of food; and (2) which, if any, of the identified meanings offer good justifications for individual food choices and for local and global food policies. The point of view of the paper is philosophical, and the questions are answered by analysing the term 'authentic'. The analysis is strengthened by drawing analogies to biodiversity protection, in which authenticity is, at least sometimes and at least to some extent, acknowledged as a goal and a value behind conservation measures.

Meanings of authenticity

Original, real, and fake

The term 'authentic' is in food marketing often associated with the terms 'real' and 'original' (Gilmore and Pine, 2007). The term 'original' refers to something being firsts of its kind, something never seen (or eaten) before. Often repeated examples of authenticity in this sense are Coca-Cola, the very first cola drink in the beverage markets, and Kellogg's corn flakes, the very first cereal flakes (Gilmore and Pine, 2007). Contrasts to this kind of authenticity are copies and modifications. Copies are highly similar to the original products and are intended to be resemble them to a very high – sometimes the highest possible – degree (Gunn, 1991). Despite the high degree of similarity, copies (numerous other brands of cola drinks and corn flakes now available in markets) are often presented as distinct and different from the original. Modifications (for example Diet Coke and Kellogg's Frosted Flakes) claim to be the original product with some interesting modifications and extra qualities.

When a copy is falsely claimed to be the original (or an instance of the original), it becomes a fake. The intent to deceive is essential to and always present in fakes; fakes differ from other copies in pretending to be something that they are not (Gunn, 1991; Radford, 1978). At least two types of food fakes can be found. Brand fakes claim to be instances of a certain brand, Heinz Tomato Ketchup for example, but are actually produced by someone not allowed to use the brand. These kinds of fakes are common in some sectors of markets; everybody has heard about fake brand clothes, bags, sunglasses, and watches. Also food brand fakes can be found (McCluskey, 2012). Fakes of the second type deceive consumers regarding their ingredients. They are made of different matter or by a different method than presented to the consumers (in a menu or in a list of ingredients, for example). The second type fakes are often fakes in the first sense too (McCluskey, 2012), but the connection is not necessary. The recent horse meat lasagne and melamine milk cases are instances of the second type fakes (but not of the first type). Interestingly, less clear cases can be found, and there seems to be a large grey area between clear second type fakes and authentic food products. In *Den hemlige kocken* Mats-Eric Nilsson (2007) argues that even when food industry does not violate law it is constantly deceiving consumers regarding ingredients and production methods of food. Many food products are, according to him, presented to consumers in a way that leaves them little or no chance of detecting the inappropriate ingredients and undesirable ways of production. As examples Nilsson discusses vanilla aroma made from spruce, pistachio ice cream without any pistachio, and method of adding smoke aroma to meat by an artificial spray. I will further elaborate the authenticity of these grey area products below which concerns authenticity as being true to one's essence.[29]

As noted above, fakes always carry with them an intention to deceive. This dishonesty and insincerity imbedded in fakes makes them morally suspect and implies untrustworthiness of their producers. Since food affects our health, since food consumed may be integral to one's identity (Pascalev, 2003), and since we rely on others regarding our food, lack of trust is especially serious in food context. Thus, there are strong moral reasons against promotion of fake food products for example by buying or selling them.

Artificiality

What about copies and modifications? Are there any ethical reasons, beyond respecting copyright and honouring the creativity of the inventor, to favour original authentic products over copies? The question is answered by drawing an analogy to so-called restored ecosystems – ecosystems that have after human caused destruction been brought back to their former state by human beings. In his *Faking nature*, Robert Elliot (1997) and in his *The big lie: human restoration of nature*, Eric Katz (1997) argue against the equal value of original and restored ecosystems by noting that genesis and origin of entities may matter to their value, and that original authentic ecosystems have some value the restored ones lack. The crucial difference between original and restored ecosystems is in the former being natural whereas the latter are human created artefacts. In the words of William R. Jordan (1993), 'however skilfully restored and lovingly maintained, the *artificial* natural system is not and can never be fully *authentic*, or quite as real or valuable in some fundamental sense as its *natural* counterpart' (italics added).

I do not agree with Katz's, Elliot's, and Jordan's view that all restored ecosystems are artefacts (Siipi, 2003). Nevertheless, their view that artefacts and natural entities differ regarding their value is, for the following three reasons, relevant to questions concerning value of authenticity of food. First, the

[29] It might be suggested that there exists a third type of food fakes: look-alike products that are not meant to be eaten. Plastic grapes and wooden cheese used in different kinds of decorations are examples of these. These look-alike food products differ from two fake types described above in being honest about their inauthenticity. The element of deceiving is openly manifested in them, and there is no intention to really mislead anyone to eat the products. Thus, I do not include look-alikes to the class of food fakes.

distinction between artificial and natural is constantly made in food context. The food additives, for example, are commonly classified into natural and artificial ones. Second, the view of lesser value of artefacts is prevalent also in food context. People favour natural food and ingredients over artificial ones (Kaplan, 2012; Rozin, 2005), and naturalness claims are a common marketing strategy (Scott-Thomas, 2009). Third, as can be seen from the above quotation from Jordan, it is common to contrast authenticity and artificiality. In food context authentic vanilla, for example, is seen to come from a plant, whereas vanillin (imitation of vanilla) is an artificial synthetic product. That which is unmodified by human beings is then taken as natural and authentic, whereas at least certain forms of human influence are associated with inauthenticity (Gilmore and Pine, 2007).

But why should we value authenticity over artificiality? The case of ecosystem restoration forces us to spell out what we value in nature – human independent origin or certain physical properties. Katz and Elliot, of course, are arguing for the former. As Robert Gooding (1992) puts it, 'what it is that makes natural resources valuable is their very naturalness. That is to say, what imparts value to them is not any physical attributes or properties that they might display. Rather it is the history and process of their creation'. However, Katz's, Elliot's, and Gooding's view does not imply that natural origin is valuable also in food context (or any other contexts). Further elaboration is needed.

In addition to ecosystem restoration also other conservation approaches consisting of active human involvement have been suggested. One such approach is assisted migration, which means intentionally moving species to new more favourable areas in order to mitigate biodiversity losses caused by anthropogenic climate change. The outcome on assisted migration is not merely a saved species but also an anthropogenic change in the receiving ecosystem – the ecosystem to which the species is moved (Hewitt *et al.*, 2011; Minteer and Collins, 2010). Interestingly, alien species (species moved to new ecosystems by human beings) are usually seen as a major problem in biodiversity conservation. The crucial difference between assisted migration and undesirable introduction of alien species lies in the intentions behind them. Undesirable alien species are introduced unintentionally or at least without conservation purposes.[30] Assisted migration, on the other hand, requires by definition a conservational purpose behind the measures taken. At least in theory, the very same overt actions, with the very same results, might, depending on the intentions behind them, be considered either as assisted migration or as introduction of alien species.

Even though ethics of assisted migration is still under discussion, the mere possibility that it might be accepted as a conservation method implies that all artificiality is not equally undesirable. Rather undesirability of artificiality may depend on intentions behind it. It is worth asking whether this is the case also in food context. Academics (see for example Sagoff2001; Saher2006) are often keen to label consumers' desires to eat authentic and natural food as somehow mistaken or misguided, since natural ingredients cannot be said to be healthier or otherwise better than artificial ones. I suggest that from a consumer point of view, the acceptability of artificiality can be seen analogous to the distinction between alien species and assisted migration. It may well be that artificial ingredients, additives, and food products are much more willingly accepted when intentions behind them are perceived as sincere and consumer friendly, aiming to make food better for consumers. However, most consumers do not usually associate this kind of intention with the use of artificial ingredients. Rather artificial ingredients are commonly seen, as in Mats-Eric Nilsson's *Den hemlige kocken* (2007), as something that is low quality and less expensive than the authentic. The intention behind the use of artificial ingredients is then interpreted as cutting production costs and bringing more profit to food industry. If that kind of belief concerning the

[30] An example of an alien species is a Giant Hogweed in northern Europe. It was used as an ornamental plant in gardens. It escaped from domestication and is now spreading rapidly (Centre for Ecology and Hydrology, 2004.).

intentions can be detected behind the reluctance towards inauthentic and artificial, consumers' desire for authentic food becomes much more easily understandable.

Being true to one's essence

That certain artificial ingredient and additives are cheap and used with intention to gain more profit does not logically or even factually imply them being unhealthy, bad tasting or otherwise poor (Nilsson, 2007; Sagoff, 2001). Selfish intentions do not exclude good consequences to others. Why is replacing authentic ingredients with artificial ones, nevertheless, seen undesirable by many consumers? The answer may lie in the third sense of authenticity. Sometimes the term 'authentic' refers to being pure and true to one's essence. Authenticity of food then means that food is essentially normal in not containing material not belonging to it and in not lacking materials belonging to it. Inauthenticity analogously implies unsuitable mixing of elements – that is mixing that crosses the limits of some important categories (Bergin, 2009; Rozin, 2005). For example, as long as an egg is seen as an essential ingredient of mayonnaise and meringue, a food offering can be authentic mayonnaise or authentic meringue only if it contains egg. Analogously, authentic olive oil is not diluted with other cheaper oils, nor does authentic honey contain corn syrup, as the counterfeit products of these food staffs do (Sunshine, 2012).

Possibility to interpret authenticity as being true to one's essence implies that artificial ingredients may seem undesirable, not only because one is worried about their taste and healthiness, but also – and maybe mostly – because they move the food products away from their essences. Being made of egg is basically what it is about to be a real meringue. And containing avocado is necessary for something to be authentic guacamole. Lacking the true essence of meringue or lacking the true essence of guacamole is, of course, not a problem as such, but it becomes a problem when the product is presented as a meringue or as guacamole. And this is exactly what Mats-Eric Nilsson (2007) argues the food industry to do. He discusses numerous cases in which a food product lacks ingredients its claimed identity implies it to contain. He presents even more cases in which food contains ingredients that do not essentially belong to its. Consumers' accusations of inauthenticity can then be understood as accusations of being an unreal or a very poor instance of the presented type of food. Meringue without egg is not authentic meringue. Rather it something that merely resembles the real authentic product. Guacamole without any avocado is, at its best, a very poor guacamole.

As can be seen from the above examples, authenticity as being true to one's essence has two different meanings: being real (instance of something) and being good (representative of one's class). Both rely on the idea of an ideal instance of the food product type in question. This ideal product, a pizza for example, is a perfect and authentic representative of its class. When we claim that a certain product is not an authentic pizza, we may want say that it is not a real pizza at all, but rather a fake pizza, something that is falsely claimed to be pizza. Alternatively we may want to refer to the low quality of the product. Even though we still regard it as pizza, it is a very lousy one, poor representative of pizzas. The first type of authenticity is associated with fakery, the second with low quality.

As noted by Lisa Bergin (2009), authenticity as being true to one's essence is strongly associated with the ideas of purity and belonging. Bergin argues against this kind 'purity-valuing metaphysics' and 'ethics of belonging' by pointing out how the very same type of thinking can be found behind sexist and racist practices aiming for racial purity and strict gender roles. According to her, 'from feminist perspective, there is nothing unnatural per se about mixed beings' (Bergin, 2009). In other words, according to her, referring to something being (or not being) true to its essence can never form a strong argument, since essences of entities are social conventions, that are often used as forms of social control.

The view that essences are social conventions is easy to accept in food context. Authentic meatballs are different in Sweden and Greece, and the ideas concerning essences of different types of food have evolved through social and cultural practices. Moreover, ideas of real and good food are not foreign to power relationships. Nevertheless, these objections are not relevant to the point that a food product may appear inauthentic exactly because its actual ingredients are not compatible with claims concerning its identity as a certain type of food. Food industry actively participates in the social and cultural process of defining essences of different types of food. At the same time, it identifies (through package labels and other marketing strategies) the type and identity of its products. When an actual product fails to meet the essence of its claimed type (which according to Nilsson (2007) is common) the product can be said to be inauthentic in a sense that offers a good justification for rejecting it.

In extreme cases, what is eaten may not only fail to be a claimed type of food but also food at all. In cases of severe food crises people eat material (i.e. bark of pine tree) that is not usually considered as food. Authenticity of food may then refer to eaten material being real and good food, something suitable for fulfilling eater's nutritional needs.

Conclusion

Appeals for authenticity and desires to eat authentic food can be interpreted at least in three ways that make them sensible and understandable. First, authenticity may be contrasted with being a fake. The intention to deceive present in fakes offers a reason against promoting them. Second, it is common to contrast authenticity with artificiality. Artificiality as such is in food context not sufficient for moral undesirability. Nevertheless, when consumers' views regarding intentions behind use of artificial ingredients are taken into account, their desires for authentic food become more sensible and easier to understand. Third, authenticity can be understood as being true to one's essence. Accusations of inauthenticity can then be understood as claims concerning incompatibility of the claimed product type and the actual ingredients of the product. This kind of incompatibility may well justify decision not to promote the product in question. To conclude, 'authentic' is an ambiguous term. In some of its senses it is irrelevant to food choices, in other of its senses it can – at least when combined with relevant beliefs – offer good justifications for individual food choices and for local and global food policies.

References

Bergin, L.A. (2009). Latina feminist metaphysics and genetically engineered foods. Journal of Agricultural and Environmental Ethics 22: 257-271.

Centre for Ecology and Hydrology (2004). Information sheet 4: giant hogweed. Available at: www.ceh.ac.uk/sci_programmes/documents/GiantHogweed.pdf.

Elliot, R. (1997). Faking nature. Routledge, London, UK, 177 pp.

Gilmore, J.H. and Pine, B.J. (2007). Authenticity. Harvard Business Review Press, Boston, MA, USA, 299 pp.

Goodin, R. (1992). Green political theory. Polity Press, Cambridge, UK, 250 pp.

Gunn, A.S. (1991). The restoration of species and natural environments. Environmental Values 13: 291-310.

Hewitt, N., Klenk, N., Smith, A.L., Bazely, D.R., Yan, N., Wood, S., MacLellan, J.I., Lipsig_Mumme, C. and Henriques, I. (2011). Taking stock of the assisted migration debate. Biological Conservation 144: 2560-2572.

Jordan, W.R. (1993). The ghosts in the forest. Restoration and Management Notes 11: 3-4.

Kaplan, D. (ed.) (2012). The philosophy of food. University of California Press, Berkeley, CA, USA, 312 pp.

Katz, E. (1997). The big lie: human restoration of nature. In Katz, E. (ed.) Nature as subject: human obligation and natural community. Rowman & Littlefield Publishers, Lanham, MD, USA, 257 pp.

McCluskey, M. (2012). Counterfeit foods: are you eating the real thing?. Daily Finance, 31 July 2012. Available at: http://www.dailyfinance.com/2012/07/31/counterfeit-foods-are-you-eating-the-real-thing/.

Miteer, B.A. and Collins, J.P. (2010). Move it or lose it? The ecological ethics of relocating species under climate change. Ecological Applications 20: 1801-1804.

Nilsson, M.-E. (2007). Den hemlige kocken. Ordfront, Stockholm, Sweden, 309 pp.

Pascalev, A. (2003). You are what you eat: genetically modified foods, integrity, and society. Journal of Agricultural and Environmental Ethics 16: 583-594.

Radford, C. (1978). Fakes. Mind 87: 66-76.

Rozin, P. (2005). The meaning of 'natural': process more important than content. Psychological Science 16: 652-658.

Sagoff, M. (2001). Genetic engineering and the concept of the natural. Philosophy and Public Policy Quarterly 21: 2-10.

Saher, M. (2006). Everyday beliefs about food and health. Yliopistopaino, Helsinki, Finland, 64 pp.

Scott-Thomas, C. (2009). US consumers think that natural is greener than organic, says survey. Food Navigator-USA, July 6th 2009. Available at: http://www.foodnavigator-usa.com/Financial-Industry/US-consumers-think-natural-is-greener-than-organic-says-survey.

Siipi, H. (2003). Artefact and living artefacts. Environmental Values 12: 413-430.

Stiles, K., Altiok, Ö. and Bell, M.M. (2011). The ghosts of taste: food and cultural policies of authenticity. Agriculture and Human Values 28: 225-236.

Sunshine, A. (2012). Fake food trying to make its way into the U.S. Consumer, 18 June 2012. Available at: http://miami.cbslocal.com/2012/06/18/fake-food-making-its-way-into-the-u-s/.

Section 9. Technology, society and ethics

Justifying pro-poor innovation in the life sciences: a brief overview of the ethical landscape

C. Timmermann
Wageningen University, Department of Social Sciences, Applied Philosophy Group, Hollandseweg 1, 6706 KN Wageningen, the Netherlands; Centre for Society and the Life Sciences, P.O. Box 9010, 6500 GL Nijmegen, the Netherlands; cristian.timmermann@wur.nl

Abstract

An idea is a public good. The use of an idea by one person does not hinder others to benefit from the same idea. However in order to generate new life-saving ideas, e.g. inventions in the life sciences, a huge amount of human and material resources are needed. Powerful, but highly criticized tools to speed up the rate of innovation are exclusive rights, most prominently the use of patents and plant breeders' rights. Exclusive rights leave by nature a number of people empty-handed, with starvation, stuntedness, prevalence of disease and death as preventable and quotidian consequences. To stimulate a human rights compatible use of exclusive rights a wide range of moral frameworks have been developed to condemn current praxes. Most prominent in the debate are theories building on: (1) utilitarian calculations of weighing benefits with Peter Singer as a prominent advocate; (2) Pogge's vindication of compensatory duties for institutional harms; (3) a comprehensive analysis on how the current innovation incentive system fails to secure human rights and human capabilities; and lastly (4) showing how the status quo nurtures misrecognition. With help of those theories modest targets as well as a thorough restructure of the innovation incentive system can be justified. Those theories have the mammoth task of restraining well-established ideas supporting the permissibility of a reckless use of property rights that are deeply anchored in the property law discourse. Life sciences raise a range of special problems when justifying pro-poor innovation. Healthy people living in a society with a good sanitary infrastructure need far less resources to tackle health problems than people in places with a poor infrastructure. Patents that involve gene sequences (or part thereof) make inventing around impossible, making the seeking of licenses mandatory for investigators wanting to make follow-up research with the molecule. Speedy sharing of data concerning public health hazards or threats to food security are vital to maintain living standards.

Keywords: global justice, intellectual property, benefiting from science, human rights

Introduction

The principle of progressive realization of all human right is one of the most fundamental principles in human rights law. A steady rise in life expectancy has been observed all around the world since World War II. Multiple factors have contributed to this remarkable achievement, the rapid development of the life sciences one of them. However, since the beginning of the 21st century and for the first time in decades, not war but disease has lowered the average life expectancy in a number of countries (cf. World Health Organization, 2012). The AIDS pandemic in the present and most likely climate change in the future are serious hurdles for continuing increasing average life expectancy. Some states are in such a dire situation that the label 'developing country' does not reflect current realities. The term 'retrogressive society' would be more suitable to describe those nations.

Despite those negative scenarios, we should keep in mind that the last 70 years have also been an epoch of highly laudable achievements, both in medicine and agriculture. Yields in agriculture have increased by over 130% in the last 50 years (Baulcombe *et al.*, 2009). Even taking into consideration the horrendous levels of hunger and malnutrition, we can proudly say that the earth is feeding many more people than ever before in history. Some diseases have been eradicated completely – polio a well-known example. A continued homage to this optimism demands an acceleration of the innovation rate coupled with a strong orientation towards the needs of the poor to counter current global problems.

The reduction of average life expectancy in some countries is a clear sign that one of the most sacred principles of human rights law, the principle of progressive realization, has been violated, demanding urgent response. Scientific and technological innovation, being a key element in advancing welfare in the past cannot exempt itself from this call for action.

Innovation and cosmopolitanism

Science has been bound to serve human needs from its very beginning. The freedom scientific curiosity had to blossom has been in various degrees undermined or encouraged, often but not necessarily reflecting the level of societal welfare. Global justice demands a wider inclusion of the group of people science has to be at service at and thus, in a world of limited resources and dire needs, also undermines the freedom to practice science following mere curiosity or to develop products that allow lavish exploitation. Putting science at the service of the poor has to be however justified. Therefore I will explore some prominent ethical justifications to conscript science to social welfare, those are:

Weighing benefits

The extra euro given to a person *x*, will benefit the person who earns an euro a day substantially more, than the one given to a person who earns 100 euros a day. In this simplest form, this fact is reason enough for a thinker like Peter Singer (1993) to favour the poorer person as the recipient. The most efficient use of resources in the pursuit of welfare maximization would demand that resources are given to those who can transform them into well-being most efficiently. A person who reached a certain level of welfare is increasingly, one could even argue exponentially, less efficient in transforming resources into well-being. Under this line of thought Peter Singer condemns the research and development needed to produce an additional shaving cream for an already saturated market while some dread diseases hardly find research money at all. Hereby Singer has developed a strong argument for shifting research money to investigate problems that affect the poor, as well as the necessity of research having an effect on the promotion of social welfare.

Compensatory duties

The global institutional design lacks democratic legitimacy and brings substantial disadvantages to the global poor while benefiting the wealthiest inhabitants. This, argues Pogge (2008), is enough to amount as a violation of the negative duty of not imposing a harmful regime on others. Such a violation leads to compensatory duties that create the obligation to establish institutions whose positive effect outbalances the negative effects of the current institutional order. Pogge's most prominent example of one of such institutions that should be in place (among others) is the Health Impact Fund (Hollis and Pogge, 2008). While focusing on medicines, the mechanism advocated by Pogge aims at targeting both the availability of medicines for the poor as well as securing accessibility.

Basic needs

The doctrine of basic rights, prominently discussed by Henry Shue (1996) recognizes that some needs are essential to be able to enjoy other rights. The freedom from hunger and disease are two examples. This doctrine aims at securing those fundamental entitlements, standards that are still well below of what is protected by international human rights law. The level of health targeted by article 12.1 of the

International Covenant of Social Economic and Cultural Rights (1966) and the official commentaries thereafter clearly state that there is not only a right to health in the sense of freedom from disease, but that everyone has the right to the highest attainable standard of physical and mental health – a standard well above that which is protected under the doctrine of basic rights. Similarly, the right to food is not solely seen as freedom from hunger, but as a right to adequate food, specifying herewith that the available food has to be also culturally acceptable. The threshold line of when basic needs are met is clear: the moment you are able to play a constructive role in society. If the assigned role is the one the individual had in mind or not, is irrelevant for this doctrine. Important is the fact that she can physically do so, if she has the desire to fulfil the task. Clearly this is an ethic of extreme scarcity or in situations of emergency. Especially in relation to food, as Amartya Sen (1981) famously demonstrated, we do not live in a world that has to abide to such rough principles. Not scarcity but misdistribution is the main cause of hunger. In sum, according to this doctrine, objects of innovation that are vital to ensure the very basic needs should be made widely available.

Human rights and capabilities

Securing basic needs alone is not seen as sufficient, for either the human rights or capabilities discourse. The latter two have similarities; objects secured by human rights are often also targeted by capabilities theorists. Martha Nussbaum (1997) has shown that many of the central human capabilities she has identified have their correlates in human rights law. Those two discourses demand a series of positive and negative duties to ensure the minimum needs and basic securities to be able to lead a good life, and both guarantee those rights to all human beings in virtue of their human nature. Arguably, we can say that both discourses seek to secure the fundamental freedoms for people to be able to pursue their image of a good life. The catalogue of freedoms guaranteed is drafted (or should be drafted) according to democratic principles in order to give shape to the global society as agreed upon. Signatories to both international treaties, the International Covenant of Civil and Political Rights and the International Covenant Economic, Social and Cultural Rights, as well as capability theorists, acknowledge that if everyone should be able to work towards her ideal of the good life, individuals and states have to not only refrain from certain actions (i.e. abide to series of negative duties) but are also required to contribute to securing other people's needs (i.e. fulfil positive duties).

Allowing certain freedoms to pursue one's image of a good life and at the same time granting some powers to shape one's environment has a far stronger impact on the liberty inventors enjoy in using their exclusive rights. Here a much wider set of rights, e.g. the rights to participate in scientific and cultural life, to food, to health, to self-determination, to education, and to enjoy the benefits of science, have to be balanced with intellectual property entitlements.

Recognition theories

A key aspect of recognition theories is stated in Hegel's memorable words 'they recognize each other as mutually recognizing one another' (Hegel, 1807/1970). Not surprisingly, this passage has led to countless interpretative scholarship, identifying in this statement that action leading to recognition has to be simultaneous, reciprocal, transitive, reflexive and symmetrical (Limmer, 2005). Coming out of this tradition, contemporary recognition theorists restate the importance of being able to mutually influence each other and condemn relations of unnecessary dependency. Thereby also advocating measures that lift people from this situation of dependency, e.g. by building-up scientific capacities and making research enterprises more inclusive. This tradition also objects failing to acknowledge contributions of a particular origin (i.e. traditional knowledge) while other types of inventions and discoveries are given full credit by the scientific community.

(Re)claiming the commons

The argument made in this discourse is predominantly based on principles of fairness. Inventions rarely come out of thin air, typically one relies on what previous researchers have observed, catalogued, invented and discovered. Further innovation builds up on what is freely available for use. When knowledge, or nowadays biomaterial, is enclosed by exclusive rights, innovation becomes more expensive and for almost every individual also restricted. This has led a series of scholars to defend the public domain and common pool resources, to ensure that future researchers (but also artists) have the building blocks to continue to innovate (Boyle, 2008), hereby stating both future social welfare and creative liberty as objects in need of protection.

Maintaining the commons is justified under different philosophical traditions. The Lockean proviso of leaving enough and as good to others is one of them. Fairness demands from researchers an adequate social return for the infrastructure and scientific development opportunities provided by the taxpayer.

Cooperative justice

It might be safe to say that if every individual in the developing world completely ignores intellectual property rights in every single aspect of daily life, the global enforcement of such rights by the developed world alone would be prohibitively expensive. Even though intellectual property rights as they are currently in force disproportionally advantage industrial countries, we rely on the enforcement and recognition of those rights by people who hardly or don't benefit from their global implementation. Principles of fairness demand thus a fairer distribution of burdens by having this global incentive scheme in force (cf. Timmermann, 2013).

Denied access to innovation

Research and development in the life sciences is predominantly incentivized by using intellectual property rights; particularly patents. As society grants innovators temporary exclusive rights, it makes the recouping of research and development costs possible, provided those costs were reasonable and the product developed can be sold and finds a large enough market.

Since the second half of the last century however, those exclusive rights are less referred to as privileges and more seen as property entitlements. This change in terminology resulted in exclusive rights being perceived as stronger than previously acknowledged. The aforementioned ethical theories have now not only to combat rights conceded for instrumental reasons, but also deeply anchored notions of property rights. Under the latter, we have Lockean labour theories that justify property as a form of desert and Hegelian conceptions that link property to personality ties.

What makes life sciences special?

Exclusive rights over uses of genetic resources have a different effect on innovation and accessibility than patents in other fields of research. Some peculiarities are:

Impossibility to invent around

Patents also foster innovation by encouraging inventing around a patented invention in order to not having to rely on a license granted by the patent holder. The idea is that a patent should not block further innovation but encourage other researchers to offer products that are similar but different enough to count as novel, and thus increase competition for the benefit of the public. Patents that are closely linked to biological material interfere with this mechanism as the possibility to invent around is impeded or greatly hampered due to the uniqueness of genetic material.

Patent thickets

Particularly for innovations that count as research tools, patent thickets are a major problem. Patent thickets may occur when patents owned by a large number of individuals with diverging interests cover an object of innovation. A classic example of how complex such negotiations can get is the 'golden rice' case (cf. Van den Belt, 2003). Here researchers aimed at producing a rice variety that was rich in beta-carotene destined for poorer people with deficiencies of vitamin A.

Temporality of delays

Agricultural innovation can have a significant role in securing the right to food. Medicines have to be taken at a certain time in order to be effective. While exclusion from the benefits of science is temporary for society at large, exclusion is permanent for those which the fruits of science arrive too late to save their lives or prevent life-long damages to their health.

Biodiversity

Much of biodiversity is secured by refraining to pollute, destroy or invade certain habitats. However in many instances, biodiversity is secured by enormous active conservation efforts. In the case of agriculture continuous work is done to enlarge the diversity of crops available – mainly through informal seed exchange programs. This work is often misrecognized and rarely remunerated.

Resilience

The philosophical, economic and legal discussion assumes that knowledge is a good of non-rivalrous consumption and that knowledge is still useful for society once the patent expires. Climate change, as well as resistance to herbicides, antibiotics, antifungi or pesticides, makes many technologies useless in the long run. If the knowledge encompassed in a lucrative innovation is of rivalrous consumption, a self-interest to sell it to a small number of high-paying customers or to overexploit it without regard to future generations raises. Incentives to conserve those resources are needed.

Self-multiplication

The vast majority of plant varieties that are protected by exclusive rights reproduce themselves naturally. Obviously this is not a characteristic shared with innovations in other fields of research. Who is responsible for the reproductive behaviour of plant varieties? The much-debated *Monsanto Canada Inc. v. Schmeiser* court case is a remarkable example of this complexity.

Speedy sharing of data and samples in global emergencies

Health and food security hazards show a capacity to cross borders unnoticed, that even the most talented smuggler in history would envy. This incapacity to seal borders hermetically makes it mandatory to have a global cooperative spirit that favours easy flow of data and biological samples in times of distress. Intellectual property and the extreme inequalities among rich and poor countries seriously hamper this possibility.

Biosafety

Inventions affect not only the users of a given technology but often society at large. When those effects are negative, regulations have to be arranged in order to limit unsolicited side effects. However, intellectual property can make taking risks (or being risk adverse) lucrative. Societies who do not have their basic needs secured are often more willing to take considerable risks into account. Regional harmonization of safety standards often impedes such options (cf. Toft, 2012).

Conclusion

The interaction between intellectual property and pro-poor innovation in the realm of life sciences is complex as demonstrated. Which ethical approach (or combination) to take will depend on the concept of justice one can most eagerly defend.

Acknowledgement

This article was written as a visiting researcher at the Brocher Foundation and is result of a research project of the Centre for Society and the Life Sciences in the Netherlands, funded by the Netherlands Genomics Initiative.

References

Baulcombe, D., Crute, I., Davies, B., Dunwell, J., Gale, M., Jones, J., Pretty, J., Sutherland, W., Toulmin, C., and Green, N. (2009). Reaping the benefits: science and the sustainable intensification of global agriculture. The Royal Society, London, UK.

Boyle, J. (2008). The public domain. Enclosing the commons of the mind. Yale University Press, New Haven, CT, USA.

Hegel, G.F.W. (1807/1970). Phänomenologie des Geistes. Suhrkamp, Frankfurt am Main, Germany.

Hollis, A. and Pogge, T.W. (2008). The health impact fund: making new medicines accessible for all. Incentives for global health. Oslo, Norway.

Limmer, R.M. (2005). Der Begriff der Anerkennung. Philosophisch-psychologische Untersuchungen, Ludwig-Maximillians-Universität, München, Germany.

Nussbaum, M.C. (1997). Capabilities and human rights. Fordham Law Review 66: 273-300.

Pogge, T.W. (2008). World poverty and human rights: cosmopolitan responsibilities and reforms. 2nd ed. Polity Press, Cambridge, UK.

Sen, A. (1981). Poverty and famines. An essay on entitlement and deprivation. Oxford University Press, Oxford, UK.

Shue, H. (1996). Basic rights: subsistance, affluence, and U.S. foreign policy. 2nd ed. Princeton University Press, Princeton, NJ, USA

Singer, P. (1993). Practical ethics. 2nd ed. Cambridge University Press, Cambridge, UK.

Timmermann, C. (2013). Sharing in or benefiting from scientific advancement? Science and Engineering Ethics (in press) doi:10.1007/s11948-013-9438-3.

Toft, K.H. (2012). GMOs and global justice: applying global justice theory to the case of genetically modified crops and food. Journal of Agricultural & Environmental Ethics 25: 223-237.

Van den Belt, H. (2003). Enclosing the genetic commons: biopatenting on a global scale. In: Baumgartner, C. and Mieth, D. (eds.) Patente am Leben? Ethische, rechtliche und politische Aspekte der Biopatentierung. mentis, Paderborn, Germany, p. 229-243.

World Health Organization (2012). World health statistics 2012. World Health Organization, Geneva, Switzerland.

Scaling values: a perspective from philosophy of technology

A.H. Kiran
Researcher, Department of Philosophy, Norwegian University of Science and Technology, 7491 Trondheim, Norway; asle.kiran@ntnu.no

Introduction

Natural foods, a classification that includes organic food and locally produced small scale food, have increasingly come into focus in the last few years. 'Today, the fastest growing segment of the food market is not fast foods, convenience foods, or inexpensive industrial foods but instead is "natural foods"' (Ikerd, 2008). Part of the reason for this is of course that these types of food taste better and more 'natural' than their industrially produced counterparts. However, that might not be the sole reason and maybe not even the most important reason for the upsurge in interest in more naturally grown foods. Opposition to industrialised forms of food production is as old as industrialised food itself; alternative ways of growing foods and farming animals have been linked to a spectrum of concerns and values about the earth, human health and animal welfare at least since the 1960s (Ikerd, 2008). In recent years, in the context of global warming and other environmental issues (itself nothing new), the tendency to link natural foods the broader issue of sustainability has become reinforced: 'the natural food movement is being driven by deeper political, ethical, and philosophical issues than the concerns that have driven the natural food movement of the past' (Ikerd, 2008). Similarly, in an initiative to increase the production and consumption of organic products by the Norwegian Ministry of Agriculture and Food (LMD, 2009), the focus is on issues other than taste; sustainability, biological diversity, and gaining new knowledge and new techniques that can be transferred to ordinary agriculture are among the reasons the Norwegian ministry cites for prioritising organic food production in the years to come (LMD, 2009: 8f). In other words, besides being 'good food', there are a range of political, practical and ethical issues that are associated with natural foods systems that should make both policymakers, producers and consumers avoid industrialised foods.

When pushing these aspects of organic production systems, the Norwegian government seeks to increase the market share of organic food; the ambition in Norway is that organic production and consumption should be 15% of all food production and consumption by 2020 (LMD, 2009)[31]. Despite some huge leaps the last few years, 10% increase from 2010 to 2011 and 17% from 2011 to 2012, the market share of organic food was still only 1.7% in 2011, so that ambition is some distance away. Paradoxically, farmlands being classified as organic saw a reduction in 2012, albeit only slight: 0.1%. Furthermore, in a report made by Centre for Rural Research in Norway, 91% of farmers say that they will not consider going organic (Logstein, 2012). This means that if the goal set by the Norwegian government is to be fulfilled, more effort must be put in by those already doing organic food production. Of course, organic food consumption can be, and is, supplemented by imported food, but the goal is, to be sure, that also organic food production should be 15% by the year 2020. Imported or not, a dramatic increase in organic food consumption requires that there is a corresponding increase in organic food production from today's level. In this paper, I shall discuss what such a scaling-up of market share means to the values associated with organic food (and other natural food systems). As mentioned, much of the rationale behind 'going organic' is related to a number of political and ethical issues, that is, to a set of values that is associated with natural foods, but not to industrialised food types.

[31] Numbers relating to organic farming in Norway are from the Norwegian Ministry of Agriculture and Food: www.regjeringen.no/nb/dep/lmd/tema/okologisk.html?id=1272.

Food and values

In discussing local food systems (LFS), Phil Mount (2012) asks if scaling-up is possible for food types that are strongly associated with values. Rather than concluding unambiguously one way or another, he identifies the main hindrances in scaling values when the production within a niche type of farming increases. Mount claims that there is a danger of undermining the very existence of LFS because 'scale developments may affect the perception and legitimacy of LFS' (2012: 117). How come? LFS (and organic food) are fringe activities, and part of the attractiveness of it is closely linked to being a fringe activity. That is to say, small scale production invites *transparency* between producer and consumer. Consumers often buy the products directly from the people that have produced it, or from small-scale agricultural organisations that distribute the products from a mere handful of farms. A larger production system (including packing, transport and storing) means that to maintain transparency becomes more complicated; farmers need to spend more time in the fields or has to hire more help, transporting and storing food require new and more complex solutions, etc.

However, overcoming barriers relating to infrastructure and logistics is not in itself the problem for LFS; the main asset of transparency is what it might bring in terms of a *trusted relationship* between the producer and the consumer. 'A full assessment of the local food premium must account for the possibility that local food is not simply another familiar value-added niche opportunity requiring only a distinct commodity, and strict control of the commodity process' (Mount, 2012: 109). This way of doing business, then, generates an 'added value that is difficult to quantify' (p. 109) that results from the encounter between producer and consumer. More precisely, the 'potential for interaction and transparency in this relationship delivers accountability and trust, and, through this, security and confidence – qualities that can only be assumed, or accepted as given, without this interaction' (p. 109). Also, 'authenticity' is a value that a non-transparent relation between producer and consumer do not invite (p. 117). Mount concludes that these types of values will not simply follow if 'the form of exchange is altered to accommodate scale' (p. 109). Can it be that local food inherently is an activity that only can occupy a very small niche of the market?

For organic food, the situation is different. Regular organic products are not handed over to the consumers in direct encounters, with the added bonus of establishing a transparent relationship with the producer, but are sold among other goods in a regular store, and are often produced by companies that also deliver non-organic products. However, for Norway and the EU, organic products are subject to strict regulations governing the production process; regulations that grant a kind of transparency that non-organic products do not have (cf. the recent horse-meat scandal that disclosed, to many consumer's surprise, the short cuts that parts of the food industry sometimes take in order to keep prices down). Furthermore, the values that Mount lists for LFS – 'trust, authenticity, safety and confidence' (p. 117) – are values that also the organic consumer can relate to. Ikerd (2008) categorises the values associated with natural food into three principles that is recognisable in the action plan of the Norwegian Ministry of Agriculture and Food: (1) ecological integrity (matters of cultural and biological diversity); (2) social integrity (includes attitudes important to resist production short cuts, such as trust, kindness, and courage); and (3) economic integrity (includes a balancing between input and output to keep prices competitive).

It makes good sense, then, to also ask if the added values of organic food production can be upheld when goals like the one of the Norwegian ministry are to be fulfilled. After all, the regulations for what farmlands are categorised as organic and what animals that have been raised organically are very strict. For this reason, an expansion of the organic assortment will necessarily be slow. One reason for this is that organically raised animals require a sufficient supply of organic feed, so that an expansion in the production of organic meat, eggs and dairy put even more pressure on other organic products. One bottleneck is that the price for organic feed are very high (LMD, 2009: 11). Moreover, as mentioned, the expansion of farmlands categorised as organic is less than for organic products, making demand higher than supply, possibly influencing prices to become even higher. Will the strict regulations governing small scale organic production not hamper the desired increase in market share? Will it be possible to maintain the values associated with organic products if production increases; can the consumers *trust* that all the rules in all aspects of production, feeding, transporting, storing, etc. are being maintained?

Although it is premature to conclude, there is *no reason to take for granted* that the values embedding natural foods today will do so in the case that the market share increases dramatically. If we are to expect consumers to actually start buying organic and local foods in larger quantities, they must be assured that the food is produced in accordance with the principles that govern these types of food; that organic *is* organic, that local *is* local, etc. For small scale production, this trustworthiness comes 'innately. But what happens when the demand for organic and local food increases; how can this type of transparency be retained?

Technological mediation

In the following I shall ask if *technology* might be a bridging element that enables values to follow the scaling-up of production. That is, I am going to make an argument that innovation and deliberate use of technology can help sustain transparency in a market where organic and local foods make up a larger part than what is the case today. At first glance, this might seem like a strange take on Mount's dilemma, for two reasons. First, are not technologies merely instruments that help us organise a more efficient production-chain; are values not an 'added extra' to the transactions between producers and consumers, and not something that comes about because of the instruments that help transactions happen? Second, if we talk about technology in relation to values; does not technology imply a value system and rationalist system in direct opposition to the kind of 'natural' values being advocated by organic and local food systems; would it not be more sensible to develop production routines that place *less* importance to technology instead of more?

In order to reply to both objections we will need to consider one concept; *technological mediation*. This concept indicate that when we use technology – whether it is a single person listening to music on an iPod or scientists using heavy technology to gain knowledge about some very abstract features of the world – the relation between us and our surroundings changes in manners that are technology-specific; we perceive and act differently in the world because our relation to it is mediated by technology. For us, the world itself is different because it is a world enhanced with technology. We do not see the world mediated by technology as world + technology, but as constituting a different untity altogether. We relate to this unity *as the world* even though it would have been a very different world had not the technology mediated out relation to it (Ihde, 1990; Verbeek, 2005). Furthermore, through the technologically mediated relation to the world we obtain new potentialities for how to live in the world, which means that we come to develop a different self-understanding. New technological possibilities means that we can act differently and that we can be different persons; understanding *what we are* is intimately related to how we see ourselves as being able to act in the world and towards other human beings (Kiran, 2012b). In view of their mediating role, then, technologies are not merely helpful instruments that we pick up when we want to accomplish something, but they contribute to a re-arrangement of the structure of the world and the human-human and human-world relations within it.

The view that technologies can innocently be put to use within a given practice with only a restricted function of increasing the efficiency of logistics cannot be upheld. Technology's impact will transcend its mere functional role; new technology means that the practice itself becomes a new type of practice, with new material and social arrangements, necessitating new ways of relating to this arrangement for all actors involved in making this practice operate. For instance, the emergence of telecare technology in healthcare has resulted in different professional identities for both nurses and patients, and has contributed to shift the meaning of 'care' (Oudshoorn, 2011; Kiran, 2012c).

The re-structuring of practices that technology contributes to often evades prediction. In designing new technologies or in implementing a new technology into a practice, the focus and effort will most often be on the functional aspects. Unexpected consequences of the new technology will to a certain degree be calculated in, in risk-analysis and estimates of how the technology will impact on issues of health, safety, employment, pollution, etc. However, such estimates will usually be of quantifiable consequences of technology use, and as such, they deal with so-called hard impacts of innovation. Given how new technology often contributes to shape attitudes, identities, roles and values, engineers and designers of new technology should also pay attention to the so-called 'soft impacts' of innovation (Boenink *et al.*, 2010; Swierstra and Te Molder, 2012). With an understanding of *how* technologies impact on their designated practices, it should be possible to anticipate some of the soft impacts of a certain technology and maybe even – to a certain degree – proactively shape impact the technology has on norms and values.

It is important to stress 'to a certain degree' here, because we cannot *use* technology to shape a context, not its functional aspects, nor its norms and values. That would also be to treat technologies like instruments for our intentions (Kiran, 2012c). As has been shown by historians and sociologists of technology, the impact of technology often escape their intended impact, either by having unforeseen consequences – as happened when power consumption went up instead of down in the Netherlands because energy saving light bulbs resulted in people starting to lit up dark areas like their backyards (Verbeek, 2009), or by being used in ways that differs from their designed use – philosopher of technology Don Ihde claims that he cannot think on a single use of technology that strictly has followed the design (Ihde, 2008). Technologies are *multistable*; their material and symbolic properties encourage users to both adopt and adapt to technology in ways that are suited to their situation (Ihde, 1990, 2008).

Neither does technological mediation imply any form of technological essentialism. The impact that technologies have on a context is not due to an inherent meaning, or essence, that makes users conform to specific ways of dealing with them. But through their material and symbolic capacities (sometimes called their *affordances*), users relate to technologies in manners that are restricted. However, the uses of technology are also restricted for a range of reasons; user's knowledge and competencies, other user's preferences and motivations, norms and values, availability of other technology, legal and ethical aspects, etc. In this sense, the way a technology is adopted into a practice is *interdependent* on what it was designed to do, how it is received by users, what it offers in terms of material and symbolic possibilities (Kiran, 2012a; Latour, 1994). In this sense, there are no essential 'rationalist' values in technology; instead, new technologies enter a relation with an established practice to interdependently shape how they are put to use.

The value-laden, *soft*, impact technologies have, then, is a result of a contextual and practice-bound shaping. The values associated with organic and local food systems should not be seen to be about (or requiring) a nostalgic-romantic return to a pre- or early technological time, but is about having a specific relation to the world, to environmental issues, to animal welfare, to our fellow human beings, etc. There is nothing anti-technological in that (although some technologies can certainly be contradictory to 'natural food' values). Because technologies are no mere instruments, but is a contributing factor to how values within a practice is shaped, careful and knowledgeable use of technological solutions in organic and local food systems might retain and reinforce values like trust and authenticity rather than go against them.

Being proactive

Once it is established that technologies transcend their instrumental role, this knowledge can be used *proactively*, although, as we saw, not perfectly. We cannot shape a use context perfectly by design, but we can influence it through realising that technology is not just about executing specific functions. Of course, this is not a recent realisation. Technology assessment used to be an after-the-fact assessment of (usually negative) consequences of new technologies. The manner in which one tried to limit negative consequences was through political and legal measures (Palm and Hansson, 2006). However, these measures had limited success, because technologies with time are so tangled up in socio-economic networks that unhappy impacts of technology use (think pollution, unemployment, etc.) cannot easily be remedied through governance. Instead, technology assessment started going 'upstream' in the innovation process, aspiring to shape how technologies impact on society by selecting technical solutions that were seen as more desirable for society prior to the technology being 'unleashed' on society ('downstream'). *Participatory design* (Schuler and Namioka, 1993) and *Constructive Technology Assessment* (Rip *et al.*, 1995) are but two proactive approaches.

In the last few years, because of a growing realisation that technologies have soft impacts, there has also been an effort in assessing the ethical and normative impacts of technology 'midstream' and then try to influence, or at least making explicit to engineers and designers. The many ELSA-initiatives in Europe and North-America are a testimony to that, and various methodologies for how ethicists can influence innovation has emerged. *Midstream Modulation*, *Ethical Technology Assessment*, *New and Emerging Science and Technology ethics*, *Embedded* and *Parallel Ethical Research* are all fuelled by the conviction that undesired ethical consequences of innovation can be constrained if ethicists and philosophers enter into a dialogue with scientists and innovators during the innovation process (Kiran (2012b). Doing ethics, then, should not any longer be seen as being passive, withdrawn, and assessing, but as being active, present and proactive.

Technology for scaling food values?

Is it possible to apply this approach to technology also in the question of whether the values attached to natural food systems can be scaled if production and consumption increase? In my opinion, yes, and in fact, it is already being done.

ICTs are today being used to track various foods. The tracking technologies help identifying the origin of a piece of meat or a carton of milk. This is *useful* when a batch of damaged food appear, as it points to the sources and possible reasons for the bad food. However, besides this functional aspect, tracking technologies also have a value-aspect, namely in facilitating a relation of trust between consumers and producers. Tracking technologies disclose the entire chain that food goes through from its origins to the shelf in the store (in principle, if paid by card, all the way to the consumer). It is a type of technology that enables transparency between producer and consumer, and therefore has the potential to establish trust even in cases where producer and consumer do not meet. This potential is already being acknowledged (Sintef, 2011), and recently the Norwegian Ministry of Agriculture and Food called for a report to be made about the current traceability of Norwegian food (at the time of writing, this report has not yet been published). Potentially, natural food producers can rely on such technolog to make sure that consumers can trust the production and logistical methods even in larger scale operations.

Another example of technology that has a positive potential for values associated with natural foods is the milking robot. According to farmers, this very high-tech automatic milking device has increased the welfare of the cows; it for instance enables the cows to decide for themselves when to be milked (Storstad, 2009). As mentioned, there is no *essential* contradiction between technology and food values.

But can technology be used proactively, or will values only be 'added extras' to the functional aspects of technology? Well, why not? For instance, in LMD (2009), the main cause for turning farmers to organic production forms is *motivation*; most farmers find the rigorous rules governing organic production to be an obstacle to go organic. Without offering a solution to this problem, let me just suggest that there might be technologies that has the potential to reduce that kind of resistance. In any case, I hope that this paper has contributed to a line of thinking where pursuing such strategies can be seen as productive.

References

Boenink, M., Swierstra, T.E. and Stemerding, D. (2010). Anticipating the interaction between technology and morality. Studies in Ethics, Law, and Technology 4(2): 1-38.

Ihde, D. (1990). Technology and the lifeworld. From garden to earth. Indiana University Press, Bloomington, IN, USA.

Ihde, D. (2008). The designer fallacy and technological imagination. In: Vermaas, P.E., Kroes, P.A., Light, A. and Moore, S. (eds.) Philosophy and design. From engineering to architecture. Springer, Dordrecht, the Netherlands, p. 51-59.

Ikerd, J. (2008). The relocalization of food: vales-added agriculture. Paper at The 10th Annual Midwest Value Added Agriculture Conference. http://web.missouri.edu/ikerdj/papers/WI%20Eau%20Claire%20--%20Local%20Values-Added.htm.

Kiran, A.H. (2012a). Responsible design. A conceptual look at interdependent design-use dynamics. Philosophy and Technology 25: 179-198.

Kiran, A.H. (2012b). Technological presence. Actuality and potentiality in subject constitution. Human Studies 35: 77-93.

Kiran, A.H. (2012c). Does responsible innovation presuppose design instrumentalism? Examining the case of telecare at home in the Netherlands. Technology in Society 34: 216-226.

Latour, B. (1994). On technical mediation – philosophy, sociology, genealogy. Common Knowledge 3(2): 29-64.

LMD (Ministry of Agriculture and Food) (2009). Økonomisk, agronomisk – økologisk! http://www.regjeringen.no/upload/LMD/Vedlegg/Brosjyrer_veiledere_rapporter/Handlingsplan_okologisk_200109.pdf.

Logstein, B. (2012). Trender i norsk landbruk 2012. Centre for Rural Research, Norway.

Mount, P. (2012). Growing local food: scale and local food systems governance. Agriculture and Human Values 29: 107-121.

Oudshoorn, N. (2011). Telecare technologies and the transformation of healthcare. Palgrave Macmillan, Basingstoke, UK

Palm, E. and Hansson, S.O. (2006). The case for ethical technology assessment (eTA). Technological Forecasting and Social Change 73: 543-558.

Rip, A., Misa, T.J. and Schot, J. (eds.) (1995). Managing technology in society. The approach of constructive technology assessment. Pinter Publishers, London, UK.

Schuler, D. and Namioka, A. (eds.) (1993). Participatory design: principles and practices. Lawrence Erlbaum, Hillsdale, NJ, USA.

Sintef (2011). Sporing av mat gir økt tillit www.sintef.no/Presserom/Forskningsaktuelt/Sporing-av-mat-gir-okt-tillit/.

Storstad, O. (2009). Robot og dyrevelferd. Nationen. 12. juni. http://oddveigstorstad.blogspot.no/2009/06/robot-og-dyrevelferd.html#!/2009/06/robot-og-dyrevelferd.html.

Swierstra, T. and Te Molder, H. (2012). Risk and soft impacts. In: Roeser, S., Hillerbrand, R., Sandin, P. and Peterson, M. (eds.) Handbook of risk theory. Springer, Dordrecht, the Netherlands.

Verbeek, P.P. (2005). What things do. Pennsylvania State University Press, University Park, PA, USA.

Verbeek, P.P. (2009). Ambient intelligence and persuasive technology: the blurring boundaries between human and technology. Nanoethics 3: 231-242.

The knowledge society as pleonasm: towards mobilisation of social intelligence in the agricultural and food domain

V. Beekman and H. Dagevos
Agricultural Economics Research Institute (LEI), P.O. Box 29703, 2502 LS The Hague, the Netherlands; volkert.beekman@wur.nl

Abstract

This paper presents a vision on the future of research and innovation. The kernel of this vision is that the knowledge society of the twenty-first century evolves towards an integration of science and society. Therefore, the word pair 'knowledge society' becomes a pleonasm. Times pass for the situation in which knowledge and society are distanced from each other, the knowledge community is only populated by scientists, and the fruits of these bright minds serve the interests of the academic world. This aristocratic episode gets competition from a perspective on research and innovation in which knowledge is not a freestanding goal but a means towards economic growth and corporate competitive power. This 'American' approach is present in European research and innovation policies that focus on public-private partnerships in science funding. However, this 'American' approach is only an intermediate episode towards a situation in which the distance between science and society is bridged at a more fundamental level. Bridging this distance between science and society is particularly pivotal in the domain of agriculture and food, since food scandals, ethical consumer concerns and the gap between production and consumption call upon socially integrated research and innovation. This urgency is fuelled by the profound nutritional, emotional and ethical values of food for experts and lay people as the co-creators of ethical food consumption. Attention economics calls upon research and innovation to aim at improving present and prospective chances and choices of people and society at large. In the attention economy research and innovation not only serve societal issues but are co-created by societal beneficiaries. This democratisation of research and innovation implies that the knowledge community is fully opened up. Hence, the mobilisation of social intelligence is fundamental to responsible research and innovation in the knowledge society as a learning society. This implies that reducing government intervention in research and innovation processes is a pretty unwise idea.

Keywords: attention economics, responsible research and innovation, science in society

Introduction

The knowledge society is a word pair. It will become more and more self-evident that knowledge and society are inextricably connected. The knowledge society as a word pair will thus become a pleonasm. How this merger is happening, what its merits are and which open issues need to be addressed, these are the topics of the current paper.

Aristocratic episode

Knowledge as the core of society is at first sight less recognisable than the opposite image of a gap between knowledge and society. This latter image is easily drawn. How easy is it to imagine scientists in ivory towers focused on research topics of their interest and expertise without paying much attention to broader societal issues. Drawing up such an image of scientists as alienated from real life is not merely drawing up a straw man.

Tunnel visions in the academic community, directed at publication of scientific papers in international journals, contribute to this image of a gap between science and society. Who wants to make a career in science interacts first and foremost with fellow scientists of the same disciplinary denomination. Status and success are to a large extent determined by scientific peers.

Specialisation is far more important for status in the academic community than broader societal relevance. No surprise that societal concerns and wishes do not have primacy within science. No surprise either that the academic community is confronted with lowering esteem and attention from society at large. After the priest and the politician also the professor has lost much of its societal standing (United Nations, 2005).

However, other organisation forms (input) and other performance criteria (output) are possible. In fact, sketches of such an alternative relation between science and society are developing right now. The dominance of what we would like to call the aristocratic episode in science and society relations has gained competition from a quite different input-output model. This model partly leaves the monopolistic production of knowledge within research and innovation projects. We would like to coin this competing approach the 'American' episode in science and society relations.

American episode

The 'American' approach has gained importance with the emphasis on thinking in terms of a knowledge economy. Knowledge qualified as instrumental to economic progress and competitive power has fuelled a vision that emphasises the contribution of knowledge to economics and entrepreneurship. Knowledge production has thus developed organisation forms in which other actors are welcomed within the research aristocracy. Apart from academic representatives also civil servants and corporate representatives gain a role in setting the research and innovation agenda now.

Both in terms of input and in terms of output knowledge production in the 'American' episode differs from the aristocratic model with its emphasis on scientific autonomy as the core value. The pursuit of knowledge is no longer a goal in itself but becomes a means towards economic growth. The difference with the situation in which the research and innovation agenda is set by and for academic experts is obvious (European Commission, 2007).

Although the 'American' approach to research and innovation deviates from the aristocratic approach, we expect that looking at knowledge production from the perspective of the knowledge economy is nothing more than a transition period to the next stage. This stage will close the gap between knowledge and society on a more fundamental level than the currently popular 'American' approach.

Limitations of the 'American' model are easy to identify. The exclusive focus of research and innovation on economic profit is hardly less reductionist that a focus on academic publications. Measuring the performance of research and innovation merely in terms of its contribution to cost reduction in production processes, market growth for products or profitability on the foreign market is hardly less one sighted than measuring performance in terms of numbers of publications and citations.

Attention episode

The vision on knowledge and knowledge production developing after the dominance of the aristocratic and 'American' approaches we would like to call the attention approach. The core idea of this approach is that attention means appreciation and that paying attention means social growth. The attention approach thus corresponds to the idea of an attention economy in which behaviour is no longer solely motivated by self-interest but increasingly also by its meaning for others (Stolze, 2011). Characteristic for this approach is that social profit is the most important parameter in measuring performance: research and innovation should aim at contributing to social needs and ambitions. Maximising the social benefits of knowledge replaces the maximisation of economic profit in the 'American' approach.

With respect to input the number of participants in research agenda setting multiplies quantitatively and qualitatively with a resulting organisation form that is better classified as democratic than aristocratic. More than in the 'American' episode the attention approach engages NGOs and citizens in research programming and innovation processes. The basis of public participation is broad and concerns society at large. After and next to research agenda setting by and for the academic community and corporate representatives respectively, society gets centre stage now. The attention approach thus emphasises societal participation in knowledge production. Much attention is paid to the mobilisation of social intelligence. Influences, interests and ideas from outsiders get access to decision-making processes about knowledge production, implying that these outsiders (with varying organisation degree and professionalism, from well-informed NGOs to lay people) are also engaged and used as knowledge providers. Attention for research agendas and projects is no longer just an issue for academic specialists but embedded in societal relations, responsibilities and revisions.

With respect to the output of the attention approach the aim of knowledge and its production is oriented at the good society. It is obvious that this performance measurement criterion differs from the specific and quantitative publication and citation criteria in the aristocratic approach to science and society relations. The attention approach does not aim at spelling out such overly simplified performance criteria. Of course, scientists cannot provide answers as to what counts as the good society, since science is not in the business of selecting and legitimising moral choices. At the same time scientists cannot retreat to the safe haven of value neutrality whilst leaving moral choices to politics and society. The inconvenient truth is that in the knowledge society no neat distinction exists between the domain of facts and the domain of values.

When the attention approach to science and society relations becomes dominant and the knowledge society thus becomes a pleonasm, societal relevance and desirability will no longer be the topping on research agenda setting but the very basis of this process. Any idea that research and innovation are better off in splendid social isolation is definitely history now. The rather different output objectives of the attention approach may also be illustrated by flagging that it becomes crucial that research and innovation contribute to maintaining or improving chances and choices of people and society. In sum, the kernel of the input-output model in the attention approach is to fully engage customers, the public at large, in the formulation of the research and innovation agenda (input), which is aimed and judged at the contribution of science and technology development to the good society (output). We witness, in other words, a shift from the head (aristocratic episode) via the hands ('American' episode) to the heart (attention episode) (Figure 1).

Responsible research and innovation

Already in the 1970s and 1980s next to the dominant aristocratic approach an initially strongly Marxist and Habermasian undercurrent existed within the tradition of 'science and society' studies. In the 1990s this undercurrent seemed to evaporate, with the rise of the 'American' approach, in a postmodern anything goes and the erection of a separate social constructivist ivory tower that looked at science as the product of market and power relations.

However, in Brussels a small group of people resisted under the banner of 'science in society'. Here, the seeds were sown for what initially was called ELSA (ethical, legal and social aspects) of science and

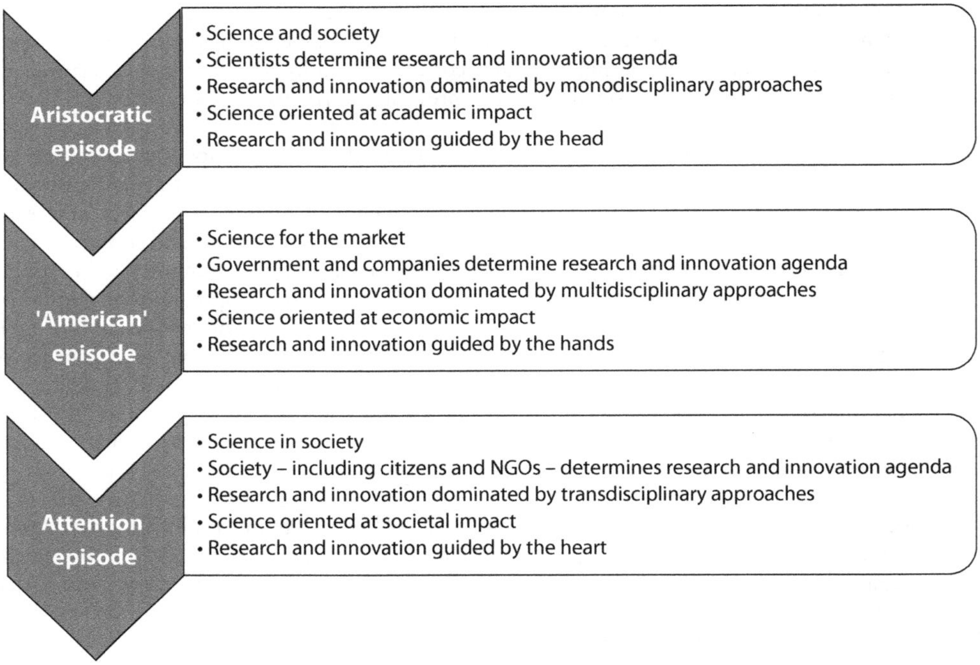

Figure 1. Episodes in relations between science and society.

technology development. ELSA strongly emphasised the importance of social acceptance studies and stakeholder participation in FP6 and FP7 of the European Commission. And now we witness the revival of science and/in society under the banner of 'responsible research and innovation' (RRI) as one of the cross-cutting themes in Horizon 2020.

Transdisciplinarity is one of the key words in RRI. This implies that RRI inherently builds on the idea that social intelligence is a source of knowledge, both in the formulation of questions (agenda setting) and in the formulation of answers (knowledge production). Transdisciplinarity means that organisation, execution and justification of research and innovation are done without predetermined or reified boundaries. It implies cooperation between scientists from various disciplinary backgrounds, with heterogeneous expertise and experience and from a variety of knowledge centres. The context of both formulation of research projects and application and interpretation of research outcomes is thus formed by a broadening societal setting (Sutcliffe, 2011; Von Schomberg, 2013).

Social learning platforms

We need to be very precise about the concrete novelty of RRI as the heart of the attention approach to the knowledge society. The difference that RRI makes is that we start thinking and doing in terms of social learning platforms. These learning platforms aim at equally ambitious and realistic changes, and pursue a radically centred approach between piecemeal engineering and system innovation. This radically centred approach should allow entrepreneurs, (local) governments and (social) organisations to work together and thus build a basis for changes in their own practices. A critical issue becomes who is included and excluded – and through which explicit and implicit mechanisms – as participants on these learning platforms.

RRI does not advocate a linear view on research and innovation processes but does distinguish between three basic stages in these processes: reflection, action and observation. The reflection stage deals with defining the principles of social responsibility for research and innovation. On a high level of abstraction 3P (people, planet, profit) interpretations of sustainable development set the cornerstones of the RRI principles. However, on a more practical level these sustainability dimensions need to be operationalized into more specific guiding principles. The action stage deals with facilitating the process of socially responsive research and innovation. This is a participatory process that includes process monitoring and evaluation. The observation stage deals with measuring the performance of socially responsible research and innovation. This is the stage of temporarily stocktaking in terms of for instance sustainability performance measurements.

This social learning process is demanding and not so much asks for building bridges across the gap between science and society but rather to change these now inextricably connected worlds. It is more about organising interaction than integration. It is more about generating new knowledge and meaning than executing predetermined plans. It is more about stimulating learning through offering space for experiments and reflection than about the formulation of criteria and rules. In short, this learning process calls for formulating the right questions and objectives rather than answers and means, since this is a situation where no single party could claim unique access to cognitive or moral rightness (Regeer, 2010).

Implications for the agricultural and food domain

The attention approach and the responsible research and innovation agenda are particularly relevant for moral issues in the agricultural and food domain. When experiences of citizens and societal interests are to provide guidance for science and technology development in terms of both input and output, ethics and the social sciences are natural allies for these societal drivers of research and innovation. Knowledge development by and for society at large – the core idea of the attention approach – demands knowledge of what happens in society. Responsible government and governance in the knowledge society thus need to be fed with monitoring of societal trends and interactions. The social sciences are pivotal in identifying and monitoring societal interests and in flagging what counts as responsible research and innovation. Ethics is pivotal in reflecting on food production and consumption as activities with deeper moral meanings than meeting nutritional needs, on sustainable development as a social innovation process that transcends mere resource efficiency, on bridging the gap between food producers and consumers in a movement from alienation towards connection and interaction.

In other words, the attention approach is embodied in the agricultural and food domain by consumer concerns, short supply chains, slow food and other alternative food styles, chains and processes. The attention approach thus enriches images of agriculture and food production that merely focus on economies of scale and cost efficiencies, which in turn opens new directions in addressing emergent and emerging issues in the agricultural and food domain. These new directions might even bracket current divisions of roles between producers and consumers of food. The attention approach builds on creative and innovative forces within society at large. The sum of these forces is indeed needed for a sustainable development of current and prospective food systems. In short, the attention approach would bring a heart to liberalisation trends in the agricultural and food domain.

Conclusions and recommendations

The attention approach to the knowledge society as pleonasm, and its culmination in the agenda of responsible research and innovation, is the temporary finalisation of a much longer democratisation process. This attention approach is demanding. With respect to its output the attention approach argues that research and innovation need to be ethically acceptable, sustainable, societally desirable and in the interest of public values. With respect to its input the attention approach radically changes the organisation of the knowledge production process. Participation does not remain restricted to a few actors but concerns passionate interaction with society at large.

One of the storylines in the development from the aristocratic via the 'American' to the attention approach is that more and more actors are engaged in designing and executing the research and innovation agenda. Governments would be out of touch with reality and their social responsibility, if they opted out as one of the co-creators of science and technology development. The challenge for governments is not to outsource science and technology policies to other social actors but to proactively pursue – perhaps sometimes even unexpected – roles in the societal co-creation of the research and innovation agenda. This implies that reducing government intervention in research and innovation processes is a pretty unwise idea.

References

European Commission (2007). Taking European knowledge society seriously: report of the expert group on science and governance to the Science, Economy and Society Directorate. Directorate General for Research, European Commission, Brussels, Belgium.

Regeer, B.J. (2010). Making the invisible visible: analysing the development of strategies and changes in knowledge production to deal with persistent problems in sustainable development. Vrije Universiteit, Amsterdam, the Netherlands.

Stolze, J. (2011). Uitverkocht!: Welkom in de aandachtseconomie. Bruna, Utrecht, the Netherlands.

Sutcliffe, H. (2011). A report on responsible research and innovation. European Commission, Brussels, Belgium.

United Nations (2005). Understanding knowledge societies: In twenty questions and answers with the index of knowledge societies. Development of Economic and Social Affairs, United Nations, New York, NY, United States.

Von Schomberg, R. (2013). A vision of responsible research and innovation. In: Owen, R., Heintz, M. and Bessant, J. (eds.) Responsible innovation. John Wiley, London, UK.

Section 10. Animal welfare

Ethical concerns beyond the border: how European animal welfare policies reach Brazil

C. Maciel[1] *and B. Bock*[2]
[1]*Wageningen University, Environmental Policy Group and Law and Governance Group, Hollandseweg 1, 6707 KN Wageningen, the Netherlands;* [2]*Wageningen University, Rural Sociology Group, Hollandseweg 1, 6707 KN Wageningen, the Netherlands; carolina.maciel@wur.nl*

Abstract

To accommodate the growing ethical concern regarding the way farm animals are treated, a comprehensive body of legislation has been developed to ensure the welfare of animals from birth to slaughter. Alongside governmental measures, the last decades have witnessed a proliferation of voluntary animal welfare standards developed by supermarkets and non-governmental organisations. Thus, within the European context, significant advancement has been reached in the implementation of good animal practices. But how to address ethical consideration in countries that export meat to Europe? Ethical demands for foreigner food production are most likely to revolve into an international trade conflict as it is regarded as an extra jurisdictional measure. By looking at the relation between European and Brazilian actors in the field of animal welfare policies, this article seeks to capture the transnational governance of animal welfare. Drawn upon the literature of policy transfer and stakeholders' interview in Brazil and Europe, this study identified three main channels through which European animal welfare policies and practices reach farms in Brazil. Accordingly with the set of actors, instruments and procedural differences underlying the functioning of each channel, we have named them as: governmental, commercial and civil. We observed that in the context of EU-Brazil relation, ethical claims for animal welfare prospered mostly through commercial and civil channels. The outperformance of governmental channels by non-governmental channels indicates a shift from 'hard transfer' to 'soft transfer' in cross-border animal welfare dialogue. The finding that measures developed by governmental actors are surpassed by private measures poses questions from the perspective of the international trade regulation. Will the current welfare governance evolve into a shift from 'hard trade barrier' to 'soft trade barriers'? And if so, how does it relate to the multilateral trade system since private standards and animal welfare are both controversial topics for Members of the World Trade Organisation?

Keywords: policy transfer, governance shift, soft barriers

Introduction

Over the last 40 years, a comprehensive governance structure has been put in place to address the growing concern for the welfare of farmed animals. Consequently, animal welfare has become one of the most rapidly expanding policy areas in Europe, with public and private rules and standards regulating rearing, transport and slaughter practices for a range of species. Still, European regulation alone is insufficient to address the new questions that result from the increasing globalisation of animal production and trade– and as a result of which many animal products reach European consumers that do not comply with the strict regulation that European producers have to meet. Having European products sold alongside foreign products (which might have been produced under lower welfare conditions) is considered not only detrimental for the competition of producers in European Union (EU) but also a potential source of confusion and discouragement for European consumers and producers. Therefore, European Members of Parliament, farmers' organization (Copa-Cogeca, 2010) and the European non- governmental organization of animal protection (Eurogroup, 2000) stressed the need for applying equivalent animal welfare regulations to imported products. This is, however, quite a difficult task that most likely conflicts with the rules of the World Trade Organization (WTO). Animal welfare measures are highly controversial within the WTO. While some WTO-Members, like the EU, consider animal welfare as a legitimate non-trade concern that should be 'globally addressed in a consistent manner within the WTO' (EC, 2000), other Members see it as a disguised restriction to trade. Since achieving consensus through the WTO may still take a long time due to ethical, cultural, economic and political divergences, it is important to consider which other avenues there may be to take care of global farm animal welfare concerns (EC, 2002).

Within Europe, market-based initiatives play an increasingly important role in meeting consumers' concerns about the treatment of farm animals (e.g. Veissier *et al.*, 2008; Maciel and Bock, unpublished data). So far, however, little is known about the outreach of such private regulation to non-European countries and their role in the transnational governance of animal welfare. This article looks at the relation between European Union and Brazil in order to better understand how animal welfare is governed transnationally. The objective is to identify how European animal welfare practices and policies reach one of the most important EU trading partners: Brazil. The next section presents the research design and methods. We then present a summary of our findings followed by a brief discussion.

Study design and methods

The objective of this study is to explore the different ways in which European private and public animal welfare policies and practices reach out and affect the incorporation of animal welfare policies and practices in Brazilian institutions. The study followed a qualitative design that combined semi-structured interviews with relevant experts in the field of livestock production and international trade, with the review of key documents. Interviews were conducted in Brazil during January to April 2012 with representatives of the Ministry of Agriculture and Livestock, Ministry of Foreigner Relation, Agricultural Research Federal Institute, supermarkets, farmer's trade union, slaughter houses, certification organisms and academics. In Europe, interviews took place during September to December 2012 with representatives of European Commission, food retailer's outlets and associations, non-governmental organisations for animal protection, association of European farmers and meat industry. The data were analysed with the help of policy transfer theory. In short, policy transfer is defined as the processes through which knowledge of policies, administrative arrangements, institutions and ideas in one political system are used for the development of policies, administrative arrangements, institutions and ideas in another political system (Dolowitz and Marsh, 1996).

Summary of findings

The data collected reveals that European animal welfare policies and practices reach Brazil in different ways – under the influence of different actors and through the employment of different instruments. In analogy to the three subsystem of governance (state, market and civil society), we clustered the activity of transference in three channels depending on whether governmental, commercial or civil actors take the lead in introducing European policies and practices into Brazilian territory. We refer to the channels as: governmental, commercial and civil.

Governmental

At the governmental level, Europe and Brazil maintain relations at three levels: (1) bilateral; (2) regional (in negotiations such as the free trade agreement between EU-Mercosur); and (3) multilateral (in forums such WTO). The bilateral procedure of 'establishment approval' seems to account for the greatest transfer of policy and practices from European Union to Brazil. According to article 11 and 12 of Regulation (EC) No 854/2004, for a food products of animal origin to enter the European

Union market the corresponding foreigner establishment needs to be approved by the European Food and Veterinary Office (FVO). Thus, Brazilian farmers and slaughterhouses aiming to export to EU's market need to fulfil the requirements of the EU legislation regarding animal health and food safety. In doing so, some European animal welfare practices are transfer to Brazil. Many of the respondents referred, for instance, to regulation EC n. 1099/2009 on slaughter as an illustration of European animal welfare requirements that entered the Brazilian production meat chain. In order to Brazilian establishments to obtain EU pre-export approval they needed to adapt their production methods to the European regulation and change the electrical parameters for the stunning water bath of poultry as well as employ an animal welfare officer. Another situation for policy transfer at the bilateral level came with the recent memorandum of understanding for technical cooperation in the area of animal welfare signed by the Brazilian Ministry of Agriculture, Livestock and Food Supply (MAPA) and the Directorate General of Health and Consumers of the European Commission (DGSANCO). With this memorandum the parties agreed to exchange scientific knowledge and technical information on the welfare of farm animals.

The negotiations at the regional and multilateral level have, in contrast, been less expressive for policy transfer. Since 1995 EU is negotiating a trade agreement with the South American trade bloc Mercosur (whose members are Argentina, Brazil, Paraguay, Venezuela and Uruguay). Nine negotiation rounds have taken place (the last one from 22 to 26 October 2012) but no agreement has been reached yet. Animal welfare is one among the issues that hinder the conclusion of an agreement between the EU and Mercosur. While EU argued for animal welfare to be part of the Sanitary and Phytosanitary chapter of the agreement, Mercosur countries insisted on placing it on the Cooperation chapter (interview Brazil and Brussels, 2012). Negotiations are also stalled at multilateral level of the Doha Development Round. Launched in Qatar in November 2001, this latest round of multilateral discussions on trade has not yet seen a consensus among WTO members in major issues such as agriculture, industrial tariffs and non-tariff barriers (like animal welfare).

Commercial

European food retailers increasingly demand compliance with scheme standards such as GlobalGAP, BRC Food or retailers' own codes of practices from both domestic and foreigner suppliers. Such standards specify guidelines for production, transport and slaughter of farmed animal in line with European consumers' preferences and EU legislation. Hence, besides the transference that comes within the establishment approval procedures, animal welfare policies and practices reach Brazil through the certification process against European private scheme standards. Observations in the field, moreover, revealed that the influence of European supermarkets in Brazil is not restricted to certification within import operations. The scheme 'Filière Qualité' developed in France in 1991 by Carrefour was brought to Brazil in 1999 under the label 'Guarantee of Origin'. The scheme consists of a set of food standards with labour, safety, environment and animal welfare measures that are above the local legal requirements. The scheme, for instance, introduced an extended loading platform[32] to enhance the protection of animals during transport of cattle (interviews). Similar to the French scheme, 'Guarantee of Origin' requires the identification of suppliers as well as regular audits of producers and slaughterhouses by Carrefour's own veterinary staff. Carrefour collaborates in this scheme with the Braford's producer association and Frigorifico Silva slaughterhouse, which was chosen because of its status as EU approved establishment (interview Brazil, 2012).

[32] Consists of a surface of approximately two meter that is placed on vehicles to help the lifting and lowering of animals.

Civil

The World Society for Animal Protection (WSPA) is a non-governmental organisation for animal protection founded in 1981 in United Kingdom. Along the years, WSPA has expanded its operation to over 50 countries. In Brazil this organisation started in 1989 and today stands out as the most active civil agent transferring animal welfare policies and practices from Europe to Brazil. Almost all the respondents mentioned the participation of WSPA in Brazilian public and private initiatives of animal-friendly practices. For instance, WSPA is involved in the training of federal inspectors and slaughterhouse's personnel within the National Program of Humane Slaughter, named STEPS. The programme is the outcome of an agreement signed in 2008 between WSPA and the Brazilian Ministry of Agriculture, Livestock and Food Supply (MAPA) with the aim to improve food safety and animal welfare in Brazilian production chain of beef, pork and chicken. The content of the training incorporates provisions regarding animal welfare issues during the loading, unloading, and pre-slaughter handling stipulated by Brazilian law, World Animal Health (OIE) and European Union directives The programme started in April 2009 in the southern part of Brazil (Santa Catarina); by 2011 already more than 1,128 professional from 250 processing plants have received training. In addition, WSPA closely collaborates with producers' associations (e.g. UBABEF) and slaughterhouses (e.g. Marfrig) in the development of their own animal welfare programmes. Another European non-organisation involved in initiatives that encourages good animal welfare practices is the World Wide Fund for Nature (WWF). Originally from Switzerland, WWF began to work in Brazil in 1961 with the aim to conserve nature and promote sustainable development. Since 2004, WWF has actively participated in supporting Organic Beef Certification in Pantanal in collaboration with the slaughterhouse JBS-Friboi. EMBRAPA-Pantanal, a governmental research institute, supports the Organic Beef project by developing and disseminating good agricultural and livestock practices that can enhance environmental protection and animal welfare.

Discussion of research findings

The above presented research findings indicate that animal welfare concepts, policies and practices, flow from Europe to Brazil through governmental, commercial and civil channels. It is, hence, not only public legislation which is transferred; quite on the contrary it seems to be that private regulations reach Brazil most easily through the engagement of commercial and civil actors, which reflects the on-going governance shift in the (inter)national political landscape. The shift to governance is usually described as a move away from the classic command-and-control mode of policy making towards collaborative governing arrangement between public and private actors (Kooiman, 1993). A similar tendency is observed in the transnational governance of animal welfare where one observes that the international transfer of practices and policies is no longer an exclusive activity of governmental representatives of the EU but a shared responsibility of state officials, civil society organisations and multinational business.

With the traditional multilateral governmental negotiations taking too much time to advance, as the experience of several WTO rounds has proven, new instruments that doesn't rely on the authority and sanctions of governments are advancing. What we see is that non-governmental actors have been much more successful in transferring quite a range of animal welfare policies and practices. However, the academic debate so far gives little attention to the role of non-state actors in policy transfer. According to Stone (2004) policy transfer literature is still state-centrist. But as the experience in Brazil demonstrates, it is worthwhile to study the new policy transfer arrangements more in detail as the entrance of non-public actors testifies more than just state withdrawal. The increasing reliance on market-based instruments and voluntary agreements for the transfer of policies and practices constitutes not only a shift from hard (legislator and obligatory) modes towards softer voluntary modes of transfer. It also implies a profound change in the kind of trade barriers organising international trade. The shift from hard to soft transfer modes is precipitating the proliferation of private scheme standards that function as soft trade barriers.

This is of crucial importance because although of a voluntary legal nature, these standards have become *de facto* requirements for accessing the European market. Many European trading partners have seen the rise of soft trade barriers such as private standards as protectionist measures that violate the multilateral trade regime. But how these soft trade measures interrelate with WTO legislation is not yet clear.

References

COPA – COGECA (2010). The reaction of European farmers and European agri-cooperatives to EU strategy for the protection and welfare of animals 2012-2015. Available at: www.copa-cogeca.be/Download.ashx?ID=938718.

Dolowitz, D.P. and Marsh, D. (1996). Who learns what from whom: a review of the policy transfer literature. Political studies 44(2): 343-357.

European Commission (2000). European Community proposal for animal welfare and trade in agriculture – submitted to WTO committee on agriculture (WTO document n. G/AG/NG/W/19). Available at www.wto.org/english/tratop_e/agric_e/negs_bkgrnd02_props1_e.htm.

European Commission (2002). Communication from the Commission to the Council and the European Parliament on animal welfare legislation on farmed animals in third countries and the implications for the EU. Available at: http://ec.europa.eu/food/animal/welfare/references/2002_0626_en.pdf.

Eurogroup (2000). Discussion paper on farm animal welfare and the WTO. Paper prepared for EC consultation meeting on agriculture. Available at: http://trade.ec.europa.eu/doclib/docs/2005/april/tradoc_122238.pdf.

Kooiman, J. (ed) (1993). Modern governance. Sage, London, UK.

Stone, D. (2004). Transfer agents and global network in the 'transnationalization' of policy. Journal of European Public Policy 11(3): 545-566.

Veissier, I., Butterworth, A, Bock, B. and Roe, E. (2008). European approaches to ensure good animal welfare. Applied Animal Behaviour Science 113(4): 279-297.

Reasoning rejection of factory farming: the importance of aesthetic and eudaimonistic arguments

L. Voget-Kleschin[1] *and M. Langanke*[2]
[1]*Kiel University, Institute of Philosophy, Leibnizstr. 6, 24118 Kiel, Germany;* [2]*Greifswald University, Institut of Theology, Am Rubenowplatz 2-3, 17489 Greifswald, Germany; voget-kleschin@philsem.uni-kiel.de*

Abstract

While confinement methods constitute the predominant form of chicken and pig husbandry in industrialized countries since several decades, in the last few years several countries have encountered a considerable increase in scale of production units. The proposed contribution analyzes arguments against such large-scale production units. First, we demonstrate that animal welfare concerns and environmental concerns are less able to specifically criticize units of large scale than appears at first view. On the one hand, some environmental issues such as nutrient leakage and non-GHG-emissions are more pertinent in large-scale units. On the other hand, other issues such as GHG-emissions, issues related to arable-based animal feed and need-based arguments for animal welfare do not allow for specific criticism of large-scale units. Second, we therefore analyze alternative arguments, namely care based and aesthetic arguments. In regard to the former we show that a broader interpretation of the notion of care that includes issues such as emergency evacuation or even a personal relationship to individual animals results in specific criticism of large-scale units in so far as these do not allow for such care. In regard to the latter, we argue that drawing on the notion of monument allows for intersubjectively reasoning aesthetic objections against certain large-scale units. In conclusion, our contribution employs the example of reasoning rejection of factory farming for arguing that an encompassing approach towards agricultural and food ethics necessarily encompasses both genuinely normative as well as eudaimonistic and aesthetic issues. As such, we understand our contribution as an argument for exploring and strengthening the role of eudaimonistc and aesthetic arguments in food and agricultural ethics.

Keywords: industrial livestock-farming, animal welfare, care-based arguments, nutrient-leaching, greenhouse-gas emissions

Introduction

While confinement methods constitute the predominant form of chicken and pig husbandry in industrialized countries since several decades, in the last few years several countries have encountered a considerable increase in scale of production units.

The proposed contribution analyzes arguments against such large-scale production units. To this end, we first analyze common arguments against factory farming, that is, arguments drawing on animal welfare concerns as well as environmental concerns. We show that the ability to specifically criticize units of large scale is less distinct than appears at first view. Second, we therefore turn to alternative arguments. In this regard we discuss different versions of care-based arguments as well as aesthetic arguments. We show that and in how far these arguments allow for specific criticism of units of a certain scale. Accordingly, we conclude by pointing to the possibilities of widening the ethical discourse about food and agricultural ethics by including eudaimonistic as well as aesthetic arguments.

Common arguments against factory farming

Animal welfare concerns

Animal welfare concerns are typically based on claims for a way of animal husbandry which allows animals to satisfy their needs. This claim does not only encompass basic needs such as those for food, water and sleep, but also social and behavior related needs, such as communicative interaction, scratching (poultry) and rummaging (pigs).

Large-scale units do not as such preclude compliance with such claims. Though the owner/keeper is not able to look after each specific animal, technical monitoring systems allow for a central 24-hour monitoring of vital signs and behavior. Accordingly, it is not a priori clear that large-scale rearing systems cannot adequately account for the animals' needs. Rather, animal welfare can be secured by designing production units with a suitable internal structure and providing artificial enrichment.

By contrast, small-scale animal husbandry does not necessarily allow for adequate welfare orientation just in virtue of being small-scale. Rather, small scale production might as well go along with inadequate housing and/or inadequate handling of livestock. Thus, the number of animals kept in one facility does as such not allow drawing conclusions regarding animal welfare issues.

However, it needs to be emphasized that though it is theoretically possible to create large-scale units that allow for adequate animal welfare, this possibility is de facto not used by operators managing such facilities today. The reason for this is that given current marked-conditions creating such facilities would result in loss of income for operators.

Environmental concerns

The most important environmental concerns raised in regard to large-scale units encompass:

- Nutrient leaching into ground- and surface-water bodies: nutrient leaching primarily results from excessive application of manure to agricultural surfaces (FAO, 2006). As such, it is not directly linked to the number of animals on a farm, but rather to the animal/area-ratio. In so far as the manure resulting from large-scale facilities is spread over a sufficient large area, excessive nutrient leaching does not necessarily occur (Jering *et al.*, 2012). It should be noted, though, that nutrient leaching is also a function of the form of manure applied and the way it is applied. While poultry farming generally produces solid manure, pig farming can result in both solid and liquid manure, depending on the type of housing. Large-scale units typically produce liquid manure which is linked to a higher risk of nutrient leaching as well as negative effects on biodiversity. By contrast, the way manure is applied is in theory not specific to the scale of units. However, de facto, large-scale units often result in excessive manure being applied close to the production units. Furthermore, in so far as larger production units necessitate larger facilities for storing manure, there is an increased risk related to leakage resulting in large-scale nutrient leaching.
- Greenhouse-gas emissions: GHG-emissions related to livestock rearing encompass carbon dioxide-emissions (CO_2) from land use change for arable production of animal feed, methan-emissions (CH_4) related to enteric fermentation from cattle and emissions of nitrous oxides (N_2O) from soil (in this order of magnitude)(cf. Foresight, 2011: 134, with further references). In regard to chicken and pig rearing, only the first and the last are of importance. Both are primarily related to the absolute number of animals kept rather than to the number of animals kept in a specific production unit. Thus reducing livestock-related GHG-emissions necessitates reducing the number of livestock kept globally. Furthermore, emission of nitrous oxides is also related to the form of manure produced. In general, emissions are higher in systems that produce solid manure than in those producing liquid

manure (Monteny *et al.*, 2006). Large-scale units for pigs typically produce liquid manure, thus resulting in lower emissions of nitrous oxide in comparison to alternative types of housing.

- Non-GHG-emissions encompass ammonia (NH_3), odor and particulate matter. Ammonia-emissions result in excessive nutrient loads in water bodies, forests and other ecosystems. Odour annoys local residents. Finally, aerosols negatively affect human health (Melse *et al.*, 2009). In contrast to GHG-emissions the negative effects of non-GHG-emissions are related to their concentration. Thus in so far as large-scale units constitute point sources, their negative effects resulting from non-GHG-emissions are actually higher than corresponding negative effects of smaller units.
- Arable based feed production: a significant part of negative environmental consequences related to livestock-farming results from arable-based feed production, especially in so far as this in turn results in direct and indirect land use change. Direct land use change describes the transformation of non-arable land (both forested and unforested) to arable land. Indirect land use changes describes a process in which a change of use of a certain area of arable land (e.g. from food production to feed production or from food production to biofuel-production) results in displacement of the former use to formerly non-arable and often marginal land. Direct and indirect land use change are inter alia related to CO_2-emissions from soil and biomass, biodiversity loss, soil erosion, etc. (FAO, 2006) However, as in regard to GHG-emissions these effects are not linked to a specific way of keeping animals (i.e. to a certain size of production unit) but rather to the absolute amount of animals reared.

Alternative arguments that specifically apply to large-scale production

Care-based arguments

One way of reasoning claims for a way of animal husbandry that allows animals to satisfy their needs consists referring to the notion of care. Given the additional premise that facilities and circumstances which do not allow for such satisfaction of needs jeopardizes animals' health and wellbeing, care-based arguments allow for reasoning such claims.

Yet claims for allowance of need satisfaction do not logically exhaust care-based arguments; i.e. the two arguments do not coincide. Rather, care-based arguments allow reasoning additional claims not linked to the concept of needs. By way of example, care-based arguments justify claims for facilities that allow for emergency evacuation, e.g. in case of fire. One could argue that the possibility for emergency evacuation draws on a need for health and wellbeing. However, needs are defined as those claims which if not met result in negative impacts on animals' health and wellbeing. Therefore, claims for health and wellbeing cannot themselves be framed as needs without lapsing into circular reasoning.

It seems plausible that possibilities for adequately guaranteeing emergency evacuation are negatively correlated to the size of units. Given this premise, a broader understanding of care that includes the necessity of providing possibilities for emergency evacuation allows specifically criticizing large-scale units.

Finally, care-based arguments can be broadened even further. Thus, a rich notion of care is linked to caring for animals as individuals and thus developing an individual relationship to each animal. Such a notion can build on ideas from virtue ethics (cf. Gjerris *et al.*, 2011) and agrarianism (cf. Thompson, 2010) respectively. In contrast to the former two interpretations of care which can be reasoned from a deontological as well as from a eudaimonistic perspective, such a broad notion is closely linked to ideas about what it means to live a good human live and to ideas about *eupraxia* respectively. As such, it rather tilts towards the eudaimonistic realm. Obviously, a notion of care that asks for individual relationships to specific animals is not compatible with large-scale units. Note, however, that leastwise in regard to poultry rearing such a notion is neither compatible with a scale of units that today qualifies as small or medium scale.

Aesthetic arguments

Aesthetic criteria do not feature prominently in the debate about factory farming. This is probably due to the general uneasiness regarding the possibility to intersubjectively reason aesthetic judgements. Rather, such judgments are regularly understood as mere expressions of individual taste. Furthermore, there is a general impression that given today's tight economic constraints we cannot afford to take soft issues such as aesthetics into account.

Contra such concerns, we argue that certain aesthetic arguments can indeed be employed in arguing against large-scale units. This is so because such units encompass buildings whose sheer scale results in optical prominence that exceeds the effects of small-scale enterprises. These effects can be questioned from an aesthetic point of view.

From a methodological perspective it does obviously not suffice to claim that a certain landscape or villagescape is 'beautiful'. Rather, to allow for intersubjective claims, one needs to draw on less subjective categories. We propose that the concept of monument can act as such a category. This is so because the classification as monument can be operationalized by linking it to more or less intersubjective criteria (cf. Table 1, cf. also UNESCO 1972, Art. 2 as well as codes like DSchG –MV 2013, Art. 1-2.).

Accordingly, if some landscape or villagescape is classified as monument, optical changes resulting from large-scale production units can be addressed from an aesthetic perspective. As a matter of course, case-specific evaluation needs to take into account the architectural design, specific location and extend of aesthetic 'erosion' already in place due to existing buildings.

In so far as impacts resulting from large-scale units are more severe than those related to small-scale enterprises, arguments in this line of thought allow for specific criticism of units of a certain (large) scale.

Table 1. Qualifying a landscape/villagescape as monument.

A landscape...	A villagescape...
...is important from a natural history perspective in so far as it bears witness to certain geological processes	
...or part of it is characteristic for a certain type of historic use	...is characteristic for a certain (regional) architectural style of a certain period of cultural history
	...features distinguished buildings from a perspective of history of architecture
...characteristically illustrates a certain type of landscape (either in terms of average or in terms of particularity)	
...is exceptional in comparison to the region of which it forms a part	
...is important from a history of arts perspective	

Conclusion

In this contribution we have demonstrated that animal welfare concerns and environmental concerns are less able to specifically criticize units of large scale than appears at first view. On the one hand, some environmental issues such as nutrient leakage and non-GHG-emissions are more pertinent in large-scale units. On the other hand, other issues such as GHG-emissions, issues related to arable-based animal feed and need-based arguments for animal welfare do not allow for specific criticism of large-scale units. Accordingly, we turned to an analysis of alternative arguments, namely care based and aesthetic arguments. In regard to the former we have demonstrated that a broader interpretation of the notion of care that includes issues such as emergency evacuation or even a personal relationship to individual animals results in specific criticism of large-scale units in so far as these do not allow for such care. In regard to the latter, we have argued that drawing on the notion of monument allows for intersubjectively reasoning aesthetic objections against certain large-scale units.

In conclusion, our contribution employs the example of reasoning rejection of factory farming for arguing that an encompassing approach towards agricultural and food ethics necessarily encompasses both genuinely normative as well as eudaimonistic and aesthetic issues. As such, we understand our contribution as an argument for exploring and strengthening the role of eudaimonistc and aesthetic arguments in food and agricultural ethics.

References

Denkmalschutzgesetz (DSchG M-V)(2013). Denkmalschutzgesetz. In der Fassung der Bekanntmachung vom 6. Januar 1998, Zuletzt geändert durch Gesetz vom 23.5.2006. Available at: www.landesrecht-mv.de/jportal/portal/page/bsmvprod.psml?showdoccase=1&doc.id=jlr-DSchGMVpP1&doc.part=X&doc.origin=bs.

Food and Agricultural Organisation (FAO) (2006). Livestock's long shadow. FAO, Rome, Italy. Available at: ftp://ftp.fao.org/docrep/fao/010/a0701e/a0701e.pdf, checked on 8/12/2010.

Foresight (2011). The future of food and farming. Challenges and choices for global sustainability. Final project report. Available at: www.bis.gov.uk/assets/bispartners/foresight/docs/food-and-farming/11-546-future-of-food-and-farming-report.

Gjerris, M., Gamborg, C., Röcklinsberg, H. and Anthony, R. (2011). The price of responsibility: ethics of animal husbandry in a time of climate change. Journal of Agricultural and Environmental Ethics 24(4): 331-350.

Jering, A., Klatt, A., Seven, J., Ehlers, K., Günther, J., Ostermeier, A. and Mönch, L. (2012). Global Landflächen und Biomasse nachhaltig und ressourcenschonend nutzen. Available at: www.umweltdaten.de/publikationen/fpdf-l/4321.pdf.

Melse, R.W., Ogink, N.W.M. and Rulkens, W.H. (2009). Air treatment techniques for abatement of emissions from intensive livestock production. The Open Agriculture Journal 3: 6-12.

Monteny, G., Bannink, A. and Chadwick, D. (2006). Greenhouse gas abatement strategies for animal husbandry. Mitigation of greenhouse gas emissions from livestock production. Agriculture, Ecosystems & Environment 112(2-3): 163-170.

Thompson, P.B. (2010). The agrarian vision: sustainability and environmental ethics. The University Press of Kentucky, Lexington, KY, USA.

United Nations Educational, Scientific and Cultural Organisation (UNESCO) (1972). Convention concerning the protection of the world cultural and natural heritage, Adopted by the General Conference at its seventeenth session, Paris, 16 November 1972. Available at: http://whc.unesco.org/en/conventiontext.

Hunting for food in environmental ethics

G. van Wieren
Michigan State University, USA; vanwie12@msu.edu

Hunting has long been a hotly contested issue in environmental ethics. On the one hand ecocentric scholars such as Holmes Rolston have heralded hunting as a positive type of land management and nature-based experience. On the other hand animal rights authors such as Marti Kheel have opposed hunting as a detrimental form of anthropocentrism and domination of the natural world. Some scholars such as J. Baird Callicott have attempted to bridge the divide in such debates, suggesting that holistic and animal rights approaches represent a triangular, rather than oppositional affair.

Yet hunting continues to be a highly polarizing topic in the field and in society more generally. Similar to debates over nature's non/anthropocentric value, positions have become so entrenched that scholars and activists have either agreed to disagree or otherwise simply ignored the topic. This is too bad for hunting spotlights fundamental aspects of the human relationship to the natural world that have been largely overlooked in modern environmental thought, including, most notably, the intrinsically troublesome dimensions of this relationship (Jordan *et al.*, 2012).

Recently the issue of hunting has reemerged in ways that make it ripe for fresh environmental ethical analysis. In the popular press, for instance, numerous new essays and books have appeared, touting hunting as an ethical, even *the* ethical, model of food procurement and consumption. The *New York Times* book review featured a series of memoirs written by 'a new breed of hunter' who 'shoots, eats, and tells' about it (2012). These hunters are not, as the article states, your typical macho, pick-up truck driving set. Rather the new breed of huntin writers 'have loaded their rifles and shotguns for complicated reasons, including culinary one-upmanship'.

This paper attempts to remedy the oversight of hunting in environmental and food ethics by examining several recent hunting memoirs. It asks: are the justifying reasons for killing wild animals sufficient in locavore hunting accounts? Does eating ones own hunted food really represent the 'perfect meal' (Pollan)? Is hunting a legitimate meaning making activity? Should it be considered an important tool in wildlife management and land conservation? If so, is this an adequate justification for hunting food? If not, why is hunting as conservation method an outdated or limited viewpoint? In order to explore these questions I draw on ecocentric and ecofeminist perspectives within environmental ethics.

I begin by analyzing Tovar Cerulli's *The mindful carnivore: a vegetarian's hunt for sustenance.* Cerulli tells the story of the author's conversion from New York City vegan to rural Vermont deer slayer (2012). It begins with Cerulli juxtaposing two discombobulating personal experiences: catching and filleting a trout and attending a retreat with the world reknowned Buddhist teacher Thich Nhat Hanh. 'Awareness, aliveness, being awake', Cerulli writes. 'I severed the trout's head and felt a sharp twinge as the blade bit' (2012: 1-2).

The second memoir I analyze is Lily Raff McCaulou's *Call of the mild: learning to hunt my own dinner.* In it Raff McCaulou narrates a transformation from city slicker non-hunter to country dwelling hunter similar to Cerulli's (2012). A journalist, McCaulou surprised herself by taking a job at a small town newspaper in Oregon. She further surprised herself by developing an interest in hunting. 'You would be hard-pressed to find an unlikelier hunter than me', confesses McCaulou. 'I'm a woman, and married to a man who does not hunt. I grew up in a city, terrified of guns. I love animals and even entered college on track to become a veterinarian. Yet, at the age of twenty-six, I made the strange decision to pick up a gun and learn to hunt. It was a complicated choice, but it started with one simple thing that almost all of us – hunters and non-hunters, women and men, city dwellers and country bumpkins – have in common: dinner' (2012: 7).

Steven Rinella's and Georgia Pellegrini's books – *Meat eater: adventures from the life of an American hunter* (2012) and *Girl hunter: revolutionizing the way we eat, one hunt at a time* (2011) respectively – most emphasize the culinary upshot of hunting and slaughtering ones own meat. Rinella, a lifelong hunter, grew-up hunting, fishing, and trapping in northern Michigan where he ate his way through college based on the meat he and his friends and family hunted. Pellegrini, who grew up on her family farm in upstate New York, came to hunting later in life. Training to be a French chef after leaving her job as a Wall Street investment banker she was forced to learn to slaughter and clean a chicken. From there, she thought, why not learn to hunt wild turkey – then it was javelina, pheasant and grouse, doves, hogs, deer, and squirrels.

Both Rinella's and Pellegrini's volumes concludes each chapter with instructions for cleaning and preparing the particular type of wild meat, along with favorite (in Pellegrini's case gourmet) recipes: Wild Turkey Schnitzel, Adobo Javelina Backstrap, Whole Pheasant Poached in Juniper Sauce, Curried Pigeon, and Balsamic Deer Heart, to name a few. Rinella claims his favorite meat in the world is squirrel, which his mother prepared in a slow cooker with a can of cream of mushroom soup. He gets so hungry for squirrel meat while cooped up in his Brooklyn, New York apartment that at times he sets snap-type rat traps in his backyard vegetable garden, and then cleans them over the tiny kitchen sink. To cook the squirrel he mashes 'coarse salt, fresh thyme, and garlic into a pulp in a mortar and pestle, and then wet[s] that mixture with olive oil and the juice of a lemon', pierces the meat, marinades, and grills until cooked well enough to 'pop the rear ball joints with a light twist' (2012: 41).

Rinella and Pellegrini are not the only new breed of hunters living in urban locales. A piece in *Slate* magazine highlighted 'hipsters who hunt', tracing the 'evolution of the new lefty urban hunter'. 'More liberals are shooting their own supper', the essay's subtitle reads. Now it's 'the bearded, bicycle-riding, locavore set' who have also taken the pledge to eat only, or at least mostly, meat they have hunted themselves. According to the article, the emergence of the new lefty urban hunter evolves something like this:

- 2006: Reads Michael Pollan's *the omnivore's dilemma,* about the ickyness of the industrial food complex. Starts shopping at a farmer's market.
- 2008: Puts in own vegetable garden. Tries to go vegetarian but falls off the wagon.
- 2009: Decides to only eat 'happy meat' that has been treated humanely.
- 2010: Gets a chicken coop and a flock of chickens.
- 2011: Dabbles in backyard butchery of chickens. Reads that Facebook founder Mark Zuckerberg decided to only eat meat he killed himself for a year.
- 2012: Gets a hunting permit, thinking 'how hard can it be? I already totally dominate Big Buck Hunter at the bar' (2012: 1).

The article postulates that that this new breed of urban locavore hunter may be responsible for the recent increase (9% from 2006-2011) in the number of hunters in America.

It also suggests that hunting is a necessary conservation tool for promoting ecosystem and species health, especially in urban areas. Citing Jim Sterba's book *Nature wars*, the essay notes the increase in population of certain wild animal species such as whitetail deer, turkey, and Canadian geese. This, accompanied by urban sprawl, which often creates sanctuary type setting for these animals, and a sentimental view of wildlife have promoted an anti-hunting approach to urban wildlife management, Sterba argues. 'Most modern North Americans have no idea what to do with these species', the article states based on Sterba's analysis. 'We gawk and gape; we feed them doughnuts; we run into them with our cars; we are surprised and alarmed by their messy habits and occasional aggressiveness; we manage them all wrong; we want them gone from our neighborhoods, but we abhor the idea of killing them' (2012: 1).

In order to address this problem, Sterba recommends wildlife underpasses and overpasses for roads and freeways. But he also recommends the introduction of sharpshooter hunters in urban areas in order to manage wildlife populations – and further, that the hunters bring the wild game to sell at local farmers markets. Hunting, however, should not only be taken up for ecological reasons, according to Sterba. It should also be adopted for social and experiential purposes. Even modern people today need connections with the natural world, he writes, 'in ways that, to put it bluntly, get dirt under their fingernails, blood on their hands, and even a wood splinter or two under their kneecaps and butts' (2012: 1).

Ethical consumption, good conservation, deep connection to the natural world, each of these are elements that the new breed of hunter incites when describing the value and meaning of hunting for food in today's modern age. But questions remain, as I have already stated. This paper addresses some of these questions, concluding that hunting serves as a critical context both for examining central issues related to consumption in environmental ethics, and for experientially and emotionally connecting humans to the natural world.

References

Cerulli, T. (2012). The mindful carnivore: a vegetarian's hunt for sustenance. Pegasus Books, New York, NY, USA.

Jordan, W.R., Barrett, N.F. Curtis, K., Heneghan, L., Honold, R., LeVasseur, T., Peterson, P., Thiele, L.P. and Van Wieren, G. (2012). Foundations of conduct: a theory of values and its implications for environmentalism. Environmental Ethics 34(3). Available at: http://dx.doi.org/10.5840/enviroethics201234324.

McCaulou, L.R. (2012). Call of the mild: learning to hunt my own dinner. Grand Central Publishing, New York, NY, USA.

New York Times (2012). A new breed of hunter shoots, eats, and tells. October 1, 2012. Available at: www.nytimes.com/2012/10/02/books/new-breed-of-hunter-shoots-eats-and-writes.html?_r=0.

Pellegrini, G. (2011). Girl hunter: revolutionizing the way we eat, one hunt at a time. DeCapo Press, Boston, MA, USA.

Rinella, S. (2012). Meat eater: adventures from the life of an american hunter. Spiegel & Grau, New York, NY, USA.

Tail docking in the EU: a case of routine violation of an EU Directive

H. Lerner[1] and B. Algers[2]
[1]Linköping University, Department of Culture and Communication, 581 83 Linköping, Sweden; [2]Swedish University of Agricultural Sciences, Department of Animal Environment and Health, P.O. Box 234, 53223 Skara, Sweden; henrik.lerner@liu.se; bo.algers@slu.se

Abstract

The question of tail docking in pigs is an ongoing problem despite the fact that it should have been solved long ago. In the Council Directive 2008/120/EC it is clearly stated that routine tail docking in pigs are prohibited and enrichment materials for the pigs must be provided, which is in line with the high animal welfare standards that the European Union aim for. This directive is in force in all member states. The habit of tail docking is widespread as a simple comparison by two reports by EFSA shows. We present these results together with results showing that some countries, like Sweden, Finland and Lithuania manage to still keep their production without tail docking routinely. We therefore suggest that the gap between the strong intentions of prohibiting tail docking in the directive and the weak (or non-existent) enforcement of it in most countries in the EU needs to be closed. Of the arguments saying that this will be a troublesome task, we will here focus on two of them. The first is that the directive is unclear or actually allows tail docking. The second is that the habit of routine tail docking is economically profitable. Both these arguments will not hold. There are three ways to bridge the gap. The first is to lower the threshold, lowering the animal welfare level in the directive. We believe strongly that this solution is contradictory to the trend in today's legislation about animals and not in line with the Lisbon treaty. The second is to demand stronger enforcement which is in line with the EU Strategy for the Protection and Welfare of Animals 2012-2015. The third is to accept that different countries will not enforce the directive, then leaving it to the consumer to choose between more or less animal friendly pork. EU seems to adopt this way in contrast to the EU AW Strategy. To properly inform consumers about animal welfare is a good help although it demands a lot of resources and is a rather slow process. Therefore, in order to have a rapid solution to the gap one need to have a stronger enforcement of the law.

Keywords: formal justice, enforcement of legislation, pig welfare, EU pig directive

The legislative background

The habit of routinely tail docking in pigs is an ongoing problem in the European Union (EU) despite the fact that the EU pig directive prohibits it. In the Council Directive 2008/120/EC it is clearly stated that routine tail docking in pigs is prohibited and enrichment materials for the pigs must be provided, which is in line with the high animal welfare standards that the European Union aim for (Camm and Bowles, 2000). We will here present the two passages from the directive that show this statement clearly.

Annex 1, Chapter 1, Paragraph 4, states that:

> … pigs must have permanent access to a sufficient quantity of material to enable proper investigation and manipulation activities, such as straw, hay, wood, sawdust, mushroom compost, peat or a mixture of such, which does not compromise the health of the animals (Council Directive 2008/120/EC).

Annex 1, Chapter 1, Paragraph 8 regulates tail docking, only allowing it as a final solution when no other solution is possible:

> All procedures intended as an intervention carried out for other than therapeutic or diagnostic purposes or for the identification of the pigs in accordance with relevant legislation and resulting in damage to or the loss of a sensitive part of the body or the alteration of bone structure shall be prohibited with the following exceptions:
> ... docking of a part of the tail, ...
>
> Neither tail-docking nor reduction of corner teeth must be carried out routinely but only where there is evidence that injuries to sows' teats or to other pigs' ears or tails have occurred. Before carrying out these procedures, other measures shall be taken to prevent tail-biting and other vices, taking into account environment and stocking densities. For this reason inadequate environmental conditions or management systems must be changed (Council Directive, 2008/120/EC).

This directive has been in force in all member states since a decade ago. Still, tail docking seems to be a regular routine in most European countries within the European Union (EFSA, 2007b) although this particular part of the directive has been in force since 2003.

The reason for tail docking is that it is supposed to prevent tail biting. Tail docking leads to the formation of amputation neuromas which make the end of the docked tail more sensitive and the pigs charged with a tail biter will protect their tail more.

There are several risk factors for tailbiting in pigs of which lack of space, lack of space at feeder and lack of manipulable material to satisfy needs for exploration and foraging are the dominant (EFSA, 2007b). Instead of changing the current industrial housing and management practices of pigs, where these obstacles occur, the simple, although painful, solution has been to tail dock the pigs to minimize the risks.

The habit of tail docking is widespread. A simple comparison by two reports presented by EFSA (2007a,b) shows clearly that tail docking is performed in countries where the use of slatted flooring and no use of straw is prevalent. Table 1 shows that some countries, like Belgium, Denmark, France, Germany, Greece, Ireland, Italy and Spain the use of straw or similar material is very limited and that, in these countries, tail docking is used for most pigs. Sweden, Finland and Lithuania has, however, managed to keep their production despite that their pigs are not tail docked. This, in a situation when the Commission has failed to enforce the EU Directive (Vryonides, 2012).

The attempts to induce enforcement

The Food and Veterinary Office (FVO) inspections in the member states have reported difficulties in compliance with the directive. Since January 2010, there have been 17 FVO missions that have investigated the compliance with the 2008/120/EC environmental enrichment or tail docking requirements (EU, 2013). Out of these, 12 reports included a specific recommendation concerning inadequate implementation of the requirements (Austria, Belgium, Bulgaria, Czech Republic, Denmark, France, Hungary, Italy (2 inspections), Portugal, Romania, Slovakia). Only 2 missions reported full compliance (Sweden and Luxembourg). The remaining 3 missions did not explicitly comment on the compliance or non-compliance with the directive requirements in the conclusion or recommendations.

The gap and the solution

We therefore suggest that the gap between the strong intentions of prohibiting tail docking in the directive and the weak (or non-existent) enforcement of it in most countries in the European Union needs to be closed. One could here launch several arguments that this will be a troublesome task and

Table 1. Tail docking, slatted flooring and use of straw in different countries (EFSA, 2007a,b and for finishers Finland; Valros pers. comm. 2013).

	Percentage of undocked pigs	Percentage of piglets		Percentage of finishers	
		Without or restricted straw	With straw	Without or restricted straw	With straw
Austria	1				
Belgium	<1	93	5	96	4
Cyprus	1.5				
Denmark	<1	90	10	92	8
Estonia	10				
Finland	95			<10	>90
France	<1	86	12	97	3
Germany	<1	83	9	90	10
Greece	15	90	0	100	0
Ireland	5	99	0	95	5
Italy	8	86	4	100	0
Latvia	25				
Lithuania	100				
Netherlands	1	86	0		
Norway	100				
Portugal	5	93	2		
Slovenia	<1				
Spain	8	90	5	100	0
Sweden	100				
Switzerland	100			80	20
United Kingdom	19	60	40	75	

we will here focus on two of these arguments. The first is that the directive is unclear or actually allows tail docking. The second is that the habit of routine tail docking is economically profitable.

The first argument claiming that the intention of the directive is unclear would not hold. The wording of the directive is easy to understand and would in theory be easy to follow. Even if some might argue that the directive actually allows tail docking when no other solution to the problem is available (see Annex 1, Chapter 1, Paragraph 8 quoted above), this is a weak argument. Enough evidence exists that indicates that farming methods with undocked pigs is possible without increasing the risk of tail biting. If sufficient space and enrichment materials are provided tail biting is prohibited in pigs that have their tail intact (Algers, 2012). Adding straw is a simple solution which enriches the environment and enables the pigs to have their exploratory behavior minimizing the risk of tail biting. However, it is associated with additional costs.

The second argument, therefore, is that the directive would not be economically applicable if applied. All production would cease due to the open market. As seen in Table 1, however, there are good examples of countries which are able to continue their production while enforcing the directive (i.e. Finland,

Lithuania and Sweden) although competing in an unfair market. Their possibility to compete would be even better if the market was fair.

The fact that there is this kind of discrepancy between the enforcement of law and the intention of law is sometimes highlighted. Recently a similar discrepancy was found in the Finnish legislation when the law was compared to cases of enforcement concerning animal welfare in broilers and turkeys (Wahlberg, 2011).

One could meet this discrepancy in at least three ways. The first is to lower the threshold, lowering the animal welfare level in the directive. We believe strongly that this solution is contradictory to the trend in today's legislation about animals (see Lerner, 2008) and not in line with the Lisbon treaty, 13§ (see Tjärnström, 2010). The second is to demand stronger enforcement which is in line with the EU Strategy for the Protection and Welfare of Animals 2012-2015. We believe this is the way forward because the continuing practice of non-enforcement is counterproductive to the legislation of the EU. To be recognized as a law in force, citizens in the member states need to experience enforcements for that piece of legislation. Therefore to meet the demand of the directive all member states need to enforce the prohibition of routinely tail docking of pigs. In the long run, the strength of EU as a legislator might be questioned if the directive would not be followed. If the directive is a weak one when it comes to enforcement, the third way is to accept that different countries will not enforce the directive, then leaving it to the consumer to choose between different kinds of pork, some more animal friendly, some less. EU seems to adopt this way (see Vryonides, 2012) in contrast to the EU AW Strategy. To be able to manage such an approach implies that a lot of resources are invested so that proper and easily accessible information about the welfare of the animals is present for the consumers.

As Algers (2011) previously suggested, there might be a gap between the views of the producers and the views of the consumers on what is good animal welfare practices, which makes a tension in society when it comes to understanding these issues. One could close this gap both by informing consumers and producers so that expectations of what is good animal welfare will overlap. Consumers would be informed about the conditions where production takes place, for example how much animal welfare that is provided for the pigs by the producers in different countries. Producers on the other hand need to be informed about how the consumers think about meat production. The producers also need to get tools to inform the consumers. In a study of attitudes toward climate change among organizations that influence the meat production in Sweden (Lerner *et al.*, 2012), those producer organizations participating stated that they believed consumers to have too little information on what meat they buy. Still, they also believed that animal welfare values were hard to commercialize (see also Larceneux *et al.* (2012) have discussed this in depth for organic labels). In the discussion of the results, Lerner *et al.* (2012) also claim that the way of informing consumers to voluntarily change to a more climate friendly diet might be a slower process to end the problem than direct governmental steering. We believe that the same holds for animal welfare. To properly inform consumers about animal welfare is a good help. However, in order to have a rapid solution to the gap one need to have a stronger enforcement of the law.

Summary

In summary we have, by the case of tail docking, highlighted the problem when high standards given through legislation are not enforced. We have tried to analyze some of the arguments that are put forward in order to solve the discrepancy between the intention of the legislation and the actual enforcement of the legislation. Our solution is that one should primarily stronger enforce the particular directive. Secondarily consumers should be informed so that they could choose the meat they want.

References

Algers, B. (2011). Animal welfare – recent developments in the field. CAB Reviews: Perspectives in Agriculture, Veterinary Science, Nutrition and Natural Resources 6: 10. Available at: http://dx.doi.org/10.1079/PAVSNNR20116010.

Algers, B. (2012). Contemporary issues in farm animal housing and management – swine. In: Jakobsson, C. (ed.) Sustainable agriculture. Ecosystem Health and Sustainable Agriculture 1. Baltic University Press, Uppsala, Sweden, p. 329-337.

Camm, T. and Bowles, D. (2000). Animal welfare and the treaty of Rome – a legal analysis of the protocol on animal welfare and welfare standards in the European union. Journal of Environmental Law 12(2): 197-205.

Council Directive 2008/120/EC. Laying down minimum standards for the protection of pigs (codified version). Official Journal of the European Union 2009/L 47/5-13.

EFSA (2007a). Scientific Report on animal health and welfare in fattening pigs in relation to housing and husbandry. The EFSA Journal 564: 1-100.

EFSA (2007b). Scientific report on the risks associated with tailbiting in pigs and possible means to reduce the need for taildocking considering the different housing and husbandry systems. The EFSA Journal 611: 1-98.

EU (2013). http://ec.europa.eu/food/fvo/ir_search_en.cfm.

Larceneux, F., Benoit-Moreau, F. and Renaudin, V. (2012). Why might organic labels fail to influence consumer choices? Marginal labelling and brand equity effects. Journal of Consumer Policy 35: 85-104.

Lerner, H. (2008). The concepts of health, well-being and welfare as applied to animals. A philosophical analysis of the concepts with regard to the differences between animals. Ph.D. thesis. Linköping Studies in Arts and Science No. 438, Linköping University, Linköping, Sweden.

Lerner, H., Algers, B., Gunnarsson, S. and Nordgren, A. (2012). Stakeholders on meat production, meat consumption and mitigation of climate change: Sweden as a case. Journal of Agricultural and Environmental Ethics 26: 663-678.

Tjärnström, E. (2010). Ethical impact on EU animal welfare policies: the example of Article 13. Student report. Swedish University of Agricultural Sciences, Skara, Sweden, 25 pp.

Vryonides, D. (2012). Letter to Dyrenes Beskyttelse concerning 'Deres klage af 29. februar 2012 og brev af 1. august 2012 (Ref.: 9.2.3.41), mit brev af 5. juli 2012 (Ref. Ares(2012)815277) – (Vores réf.:CHAP(2012)00655)'. European Commission Ref. Ares (2012) 1023736 – 03/09/2012.

Wahlberg, B. (2011). Reglering och förvaltning av produktions- och slaktdjurs välbefinnande. En offentligrättslig undersökning. Ph. D. Thesis. Åbo Akademi, Åbo, Finland.

The reintegration of animals and slaughter into discourses of meat eating

J. Gutjahr
Group for Society and Animals Studies, Institute of Sociology, University of Hamburg, Allende Platz 1, 20146 Hamburg, Germany; julia.gutjahr@gsa-hamburg.org

Abstract

In my paper I analyze, how the slaughter of nonhuman animals is legitimated and framed in recent media discourses on 'responsible' meat consumption, and which role these phenomena play concerning ambivalences in the human-animal relationship. In modernity, specific strategies and social techniques have rendered the animals, which are killed within the system of meat production, invisible, and have served to cover up the violence against them. Recently, however, there is a new movement towards re-visualization of animals within the meat-production process. Here, contrary to regular discourses on meat-eating, the animal is very much present, and consumers actively acknowledge the fact, that animals have to be killed for their food. Through analyzing the content of depictions of German media and popular culture dealing with the process of ('do-it-yourself') slaughter and 'responsible' meat consumption, I will show, that the new visualization of animals and slaughter is embedded in a set of complex and ideological strategies, and that those phenomena can be interpreted as a reaction to current debates on the issues of intensive animal husbandry and meat consumption.

Keywords: meat consumption, 'happy meat', ambivalences in the human-animal relationship

Introduction and theoretical background

Ambivalences in the human-animal relationship

In modern Western societies, human-animal relations are characterized by an ambivalence, which comprises personal relationships to individualized 'companion animals' on the one hand and the institutionalized and violent termination of animal life, e.g. in the area of food production, on the other. In the latter, objectification, desubjectification and commodification are the predominant types of relationship to animals.

Some sociologists have proposed initial theoretical approaches on how to conceptualize these ambivalences. Morgan and Cole (2011) have developed a typology of the material and discursive positioning of animals (Figure 1) to draw attention to these ambivalences and the different constructions of animals, by focusing on the dimensions of visibility-invisibility and subjectification-objectification. 'Farmed Animals' and 'Dead Meat' therefore have different constructed positions as 'Pets' for example. The scheme also provides an explanatory framework for how the violent treatment of animals in certain fields is enabled by corresponding mechanisms such as invisibility and objectification.

The normality of meat consumption and the objectification of animals

With the increase in meat consumption in the modern era (Fiddes, 1991: 20f.) meat has increasingly become a staple food, and for most consumers, the consumption of meat is not perceived as a kind of political or ethical decision. This is similarly described by the social psychologist Melanie Joy (2010), who identifies the normalization mechanisms that reproduce the violent system of meat consumption, as a perception and belief scheme, which is deeply internalized in the subjects (similar to an ideology),

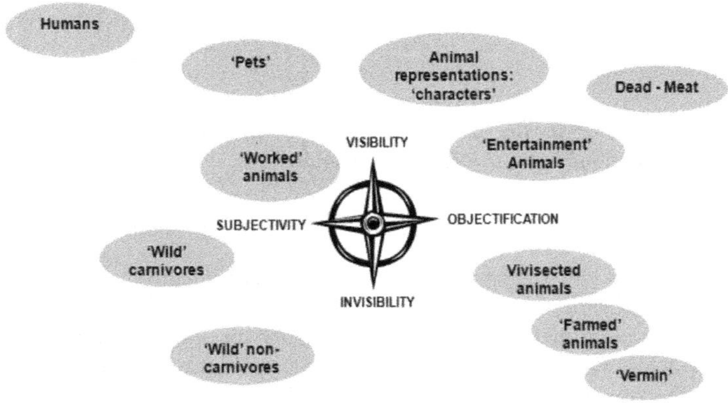

Figure 1. Material/discursive positioning of animals (Morgan and Cole, 2011: 113).

which she calls Carnism. Through the analysis of various socio-psychological (defense-)mechanisms to explain why the consumption of meat of certain animals appears as opportune. The constellation of these mechanisms is a modern phenomenon. To hide the violence in factory farms and slaughterhouses or to define it at least as unproblematic, certain mechanisms have been developed in modern times in, such as techniques of normalization, emotional and spatial distancing and rationalization (Buschka *et al.*, 2013). The spatial segregation of the violence, the division of labor in the production process, the objectification of animals in modern production systems, the linguistic-discursive level (Adams, 2010: 40f; Fiddes, 1991: 97; Heinz andLee, 1998: 94), the fragmentation and anonymity of animal bodies and their body parts (Fiddes, 1991: 119) can be mentioned here as key aspects. The 'commodity meat' is nowadays hardly reminiscent of the once-living individual animal (c.f. Adams, 2010).

Recent developments

Currently, the social norm of meat-eating seems to be losing its rigidity. In recent years, a development has begun in various media-public discourses, which addresses issues surrounding meat consumption and production, and increasingly this topic has become a field of social conflict. The number of vegetarians and vegans is constantly on the rise. My thesis is, that in response, some other developments have appeared, which criticize modern animal production methods as well, but chose different solutions and affirm a more conscious and 'sustainable' meat consumption. An additional thesis is, that within those meat eating-practices, some of the previously described mechanisms are more salient to the consumers and are partly criticized. Jovian Parry has referred to this development as the New Carnivore Movement (2009, 2010). Here, in contrast to mainstream discourses on meat-eating, the animal and production process is very much present, and consumers actively acknowledge the fact that animals have to be killed for their food, as long as they lived and died 'humanely'.

Methodology

I analyzed the content of depictions of German media (newspaper articles, blog-entries, etc.) dealing with the process of ('do-it-yourself') slaughter and 'responsible' meat consumption. Material was analyzed with computer programme Maxqda and Grounded Theory (Strauss and Corbin, 1990). Most of the texts are dealing with 2 phenomena that have received broad media attention:

1. My little farm ('Meine kleine Farm') is the project of student Dennis Buchmann, who works together with an organic pig farmer. Customers can vote on the Internet which pig will be slaughtered next and then purchase meat products from this pig. On the packaging of the meat-products there is a picture of the particular pig.
2. Porkcamp is a 'DIY-slaughter'-meeting, which took place two times in 2011. Different people met on a farm together to slaughter pigs themselves (with guidance from a butcher), process the meat and then consume the products.

Results

Problem framing

These phenomena can each be understood as a direct response to current debates about meat consumption or conscious consumption, and are received by media as such.

The protagonists question the normality of the consumption, or high consumption, of meat, and criticize the fact that consumers themselves do not worry about how their meat is produced and what consequences the production methods have. Consumers therefore are accused of acting irresponsibly.

Topics that are taken up especially are the ecological and ethical outcomes of intensive animal husbandry. High meat consumption and the desire for cheap meat could only be supported with factory farming, which is conceived to be ethically problematic. The circumstances in the factory farms and also partially at slaughter are seen as not corresponding to the needs of animals. They are merely understood as means of production; their needs are considered secondary to economic constraints.

Criticism is aimed especially at the alienation from the animal and the production process. Thus, some of the mechanisms, which were previously described in the theoretical part, are reflected. Consumers would just buy anonymous mass-meat, and thereby forget that it is a dead animal. This is described as a form of 'alienation' from the production process and from the once living animal. Some of the protagonists describe situations in which the question of the origin of the meat served as a light-bulb moment, and the desire arose to change their own consumption or to kill an animal themselves:

> I have [...] started to buy meat at a very good butcher. They were still hanging real pig heads that nowadays hardly any butcher does. I thought that was strange and at the time realized how I was actually alienated from the meat that I've eaten. This gave me the idea that you should actually participate in slaughter and should produce sausage in a traditional way' the initiator of Pork Camp tells (Erk, 2011).

Strategies for solution

The intention is to 'improve the world' (Erk, 2011) and to provide potential solutions for an alternative, reflexive meat consumption. The products of 'My Little Farm' are therefore called 'meat on a mission' (anonymous, 2011). Dennis Buchmann intends to directly make a change in attitudes and consumer behavior. The goal is a conscious, responsible use of food and a new appreciation for meat. The suggestion is to consume less meat, and buy organically produced, high quality meat. This approach is described on several occasions as a kind of 'middle ground' between vegetarianism /veganism on the one hand, and consumption of meat from factory farming on the other hand, which mediates the own interests of pleasure and other external interests: 'The most logical way: going vegan. But because this is, to my opinion, a subjective decision, and a balance between 'selfish' enjoyment of life and responsibility 'for the world', conscious meat consumption can be a middle ground' says Dennis Buchmann (Anonymous, 2012a).

The visibility of the formerly living animals, discursively and materially, is therefore a determining factor. The products of My Little Farm are called 'meat with a face'. The alienation or lack of connection to the anonymous animal which has been killed for supermarket-meat is supposed to end. Consumers are meant to overcome the distancing from food animals through transparency in the production process and being aware of the living animals, that are used for their food.

In the course of the recovery of the reference to the animal, these are occasionally even individualized, for example described with their specific character traits. They are granted specific needs, and the capacity to experience emotions such as fear during slaughter.

With the practice of visualizing, Dennis Buchmann intends to provoke thoughts in meat eaters: 'Yes, I want to achieve the effect that the customer [...] mentally stumbles when he looks into the eyes of the sausage, and asks himself: '[...] such a piece of meat, there is also a life behind it, and maybe it's better to just eat a little less meat' (Anonymous, 2012b). The awareness of the necessity of death to obtain meat in part leads to insights into those social mechanisms that normally hide this fact for most people. Thus, two participants of the Pork Camp talk about the English terms 'pig and pork': 'pork comes from a pig, and the words we use should not cover this too much' (Melican, 2010).

Justifications, morality and coping with violence

The death of animals is not concealed, but highlighted. Thereby also affective states are described, which appear when one is witnessing a slaughter. Facing slaughter can result in concern, and the whole experience can be moving or even depressing. The personal concern and the exact witnessing of the slaughter process is almost an act of moral catharsis: 'As intense as the moment was, I thought it was ultimately good, because it has made the production of meat transparent and accountable. That was the exact opposite of factory farming and meat from the supermarket. My relationship with food and animals has been strongly affected by this', one pork camp participant reports (Erik, 2011).

Pigs for example are indeed depicted as individuals, who do not want to suffer, and a treatment of animals as mere objects is clearly rejected by the protagonists in the texts. Nevertheless, certain objectification strategies are required to construct their killing as less problematic. It is emphasized that the pigs are still 'livestock' (Grossarth, 2012). This is reflected in the fact that in My Little Farm, the animals are identified on the basis of numbers: 'We don`t give the pigs names, only numbers, because there are livestock animals' says Dennis Buchmann (Grossarth, 2012).

A participant of the Camps Pork says: 'I've seen how a pig is killed, so we can eat its meat. I am reminded of this. But what I am also reminded of is that this doesn't have to be a barbaric act, if it is not slaughtered industrially and animals are treated as livestock, not as raw material' (Breithut, 2010).

The terms of a 'good life' and of 'respect' play a crucial role. The term 'respect' appears in nearly all of the examined texts and is considered a topmost maxim. Thus, the participants of Pork Camp claim that 'respect' was THE word of the weekend (Dickhaut, 2010). Also, another motto of My Little Farm is: 'Less meat, more respect' (Augustin, 2012).

As Buchmann describes it: 'I have been at slaughter. I had no pity, but a lot of respect. I very much appreciated the sausage I later had in my hand from this pig. Well, the pig died for me' (Augustin, 2012).The concept of respect is removed from the context of human society, but its content is adapted to the hierarchical human-animal relationship, and no further explanations are made, as to what respect towards animals actually means. Respect for animals is more an one-sided affair, which also evident in the following quote from the Pork Camp: ''No pictures please', was said when the animal was lifted by a crane from the box to give it a stab in the jugular vein to drain the blood, 'out of respect' (Dickhaut, 2010). Respect for animals or livestock is defined from an animal welfare point of view as a win-win-situation, but includes only the recognition that they have to die for consumption, as well as enabling them a suffering-free life before death, and a stress-free killing.

Discussion and conclusion

The presented practices and discourses can be interpreted as a result of (critical) discussion of meat consumption and can be understood as a -reaction to the burgeoning importance of diets as vegetarianism and veganism (c.f. Joy, 2011: 1), as they explicitly refer to those discourses Melanie Joy describes these new forms of arguments in the consumption of meat as a 'carnistic backlash' (ibid.). The carnist ideology is partly seen through, such that mechanisms which normally conceal the suffering of animals in factory farms or maks consumers distance themselves from certain practices are criticized.

Fundamental maxims of carnism though are affirmed, e.g. that we may use and eat certain animals. By the reference to the animal, by raising awareness of the process of killing, and by ensuring certain production conditions, the protagonists legitimate using and killing animals for human purposes, although meat consumption previously has been increasingly socially questioned. It is thus mainly the *form* of treatment of animals that they perceive as ethically problematic. Eating meat becomes, in the context of this development, a conscious decision, and not only functions as unquestioned normality. The people have dealt with the issue of meat consumption, have rationally weighed arguments, and have then deliberately chosen in favor of eating specific types of meat.

The concepts of the animals in the studied examples oscillate constantly between individualization and objectification. This can be seen as an expression of a change in the way a part of society conceives their relations with 'livestock' animals. The dilemma, that animals are recognized one the hand as individuals, capable of suffering, but their (apparently fulfilled) existence is violently ended, is solved by the ideologically charged concepts of 'respect' and a 'good' or 'happy' life. Although animals apparently are granted specific needs by propagating 'humane' animal husbandry conditions and therefore they are not constructed as insensitive objects, which can be unconditionally disregarded, not all the animals' needs are given equal consideration. Animals are not granted a right to exist for their own sake and so finally their individuality and their existence is ended by killing them. Klaus Petrus speaks in this context of a fragmented subjectivity: 'Because sentient beings are not merely half of, a third of or a tenth of subjects, but as a whole – or not at all' (Petrus, 2011: 8).

When interpreted within the framework of Morgan and Cole (Figure 2) the 'happy meat' animals in the analyzed texts are constructed with both a high visibility and a middle level subjectification. This is a difference in the way 'normal meat animals' are seen. But, in any case, the animals are definitely not given the same subject status as e.g. pets or people. They are still conceived of as 'livestock animals', function as means to an end and thus have a short life. Finally, the subject status is completely negated and the living beings are transformed into objects as they become a piece of meat.

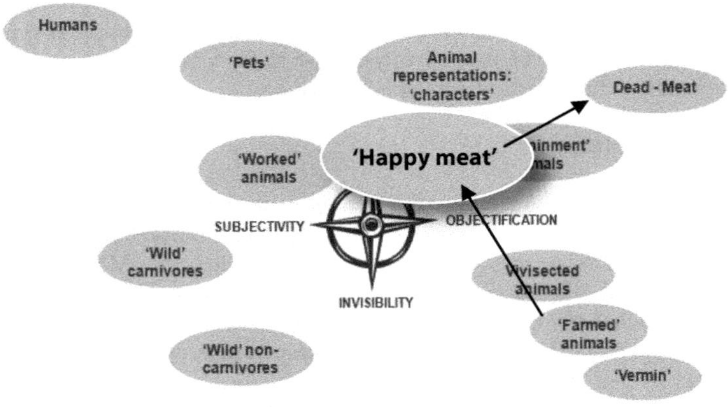

Figure 2. Own modification of the scheme by Morgan and Cole (2011: 113).

References

Adams, C.J. (2010). The sexual politics of meat. A feminist-vegetarian critical theory. Continuum, New York/London, USA/UK, 339 pp.

Anonymous (2011). Meinekleinefarm.org. Available at https://www.youtube.com/watch?v=H38M4swNCcE.

Anonymous (2012a). Pro vs. Contra: Mit gutem Gewissen Tiere essen. Available at: http://deinfreund.org/?p=3338.

Anonymous (2012b). Deutschland – Internetmetzger. Available at www.youtube.com/watch?v=8-6DQCDUiOw.

Augustin, M. (2012). 'Der Wurst in die Augen schauen'. Dennis Buchmann verkauft Fleisch von Schweinen, die man sich im Internet und auf der Packung ansehen kann. Available at www.schwaebische.de/politik/politik-aktuell_artikel,-Der-Wurst-in-die-Augen-schauen-_arid,5306084.html.

Breithut, J. (2010). Fleischkonsum oder Vegetarismus. Immer noch Lust auf Wurst? Available at www.stuttgarter-zeitung.de/inhalt.fleischkonsum-oder-vegetarismus-immer-noch-lust-auf-wurst.ab472c56-0add-41d4-bc28-5b5546cf8e05.html.

Buschka, S., Gutjahr, J. and Sebastian, M. (2013). Gewalt an Tieren. In: Gudehus, C. and Christ, M. (eds.) Gewalt. Ein interdisziplinäres Handbuch. Verlag J.B. Metzler, Stuttgart/Weimar, Germany.

Dickhaut, S. (2010). Wir sind dann mal Schwein – ein Wochenende im Porkcamp. Available at www.kuechengoetter.de/rezept-blog/wir-sind-dann-mal-schwein---ein-Wochenende-im-Porkcamp-beitrag-1466410.html.

Erk, D. (2011). Pork Camp 'Das Gegenteil von Massentierhaltung'. Available at: www.zeit.de/lebensart/essen-trinken/2011-01/pork-camp-neuruppin.

Fiddes, N. (1991). Meat. A natural symbol. Routledge, New York, NY, USA, 272 pp.

Grossarth, J. (2012). Internet-Metzger Buchmann im Gespräch ‚Der Wurst ein Gesicht geben'. Available at: www.faz.net/aktuell/gesellschaft/internet-metzger-buchmann-im-gespraech-der-wurst-ein-gesicht-geben-11621867.html.

Heinz, B. and Lee, R. (1998). Getting down to the meat: the symbolic construction of meat consumption. Communication Studies 49(1): 86-99.

Joy, M. (2009). Why we love dogs, eat pigs, and wear cows: an introduction to carnism. Conari Press, San Francisco, CA, USA, 208 pp.

Joy, M. (2011). Understanding neocarnism: how vegan advocates can appreciate and respond to 'happy meat', locavorism, and 'paleo dieting'. Available at: www.onegreenplanet.org/lifestyle/understanding-neocarnism.

Melican, B. (2010). Porkcamp. Available at: http://melicanskitchen.wordpress.com/2010/01/26/ porkcamp-2/.

Morgan, K. and Cole, M. (2011). The discursive representation of nonhuman animals in a culture of denial. In: Carter, R. and Charles, N. (ed.) Humans and other animals: critical perspectives. Palgrave, London, UK, pp. 112-132.

Parry, J. (2009). Oryx and crake and the new nostalgia for meat. Society & Animals 17(3): 241-256.

Parry, J. (2010). Gender and slaughter in popular gastronomy. Feminism & Psychology 20(3): 381-396.
Petrus, K. (2013). Tiere als fragmentierte Subjekte. Available at www.tier-im-fokus.ch/mensch_und_tier/fragmentierte_subjekte/.
Strauss, A.L. and Corbin, J. (1990). Basics of qualitative research. Sage, Thousand Oaks, CA, USA, 312 pp.

Does the society perceive its own responsibility for modern pig production?

D. Weible, I.B. Christoph-Schulz and P. Salamon
Thünen Institute of Market Analysis, Bundesallee 50, 38116 Braunschweig, Germany; daniela.weible@ti.bund.de

Abstract

Lately, the present system for agriculture and food production has been topic in many public discussions; especially modern animal husbandry does no longer match consumers' expectations or societal needs. This paper concentrates on the society's perspective regarding intensive pig production. By combining focus groups with a quantitative survey, a mixed method approach is pursued. Focus groups carried out in three German cities are to capture a variety of opinions and concerns among the population: perception, assessment, responsibility and expectation of the participants regarding pig production. Results showed participants' attention to the following topics: space available per pig was considered as insufficient and not species-appropriate, frequency of medications as too high, and in particular the prophylactic use of antibiotics as problematic. Also interaction among the lack of space, the higher use of medication and the behavioural disorders (e.g. pigs bite each other's tails) were also discussed. Participants often criticized that animals are only seen as a technical product in a production system and there is no 'real caring' due to fact that pigs are means of generating profit. Regarding the question of 'who is responsible for modern animal husbandry?' it is striking, that respondents have also mentioned consumers' responsibility as well as the role of the state. None of the participants expressed missing knowledge as reason for criticizing and simultaneously consuming meat from modern animal production. Based on these focus groups a quantitative online survey was developed to examine those topics in more detail and to quantify the outcome representatively. Based on 1,519 participants, the survey confirmed many of the critical views gained in the focus groups. Besides a very engaged group which is characterised by a strong criticism in general and a strong critical perception of current production systems, also a considerable group accepting modern animal husbandry was identified. The question about responsibility showed that more than 80% of the respondents stated to accept increasing prices as consequences of regulations which gain at animal friendly husbandry.

Keywords: modern pig production, societal expectations, responsibility, mixed method approach

Introduction

Today's agriculture and food production has been a topic in public discussions and the media during the last years. On one hand, modern animal husbandry does not match consumers' or societal expectations. And on the other hand, animal production is an important sector in German agriculture and in whole food processing chain. And what is more, it is a growing sector comprising about 60% of farmers' sales, but often suffering by small margins. There is no indication of a declining gap between consumers' expectations and their perception of animal husbandry which might affect consumption patterns in the long-run due to changes in the basic narratives. Objective of the overall study[33] is to compile information to enabling the development of strategies for all involved agents including farmers, politicians, processors, and distributors to shape agriculture production better towards society's expectation. In this context the following paper concentrates on society's views and expectations concerning intensive pig husbandry. Based on a literature review guidelines for focus groups are developed to capture and discuss society's

[33] This research has been funded by "Stiftung Westfälische Landschaft".

views and expectations on pig husbandry in detail. Outcomes of these focus groups are used, in a second step, to setup an online survey to quantify the results of the focus groups. In a last step strategies will be deduced to allow farmers and other actors to fit agriculture better to society's needs.

Literature review

Although expectations concerning animal husbandry in agriculture had already been touched in several studies their main topics lay elsewhere e.g. on animal welfare, hampered food quality, or on agriculture's role for the society.

Importance of animal welfare is rated by European population with 8.1 on a scale of 10, but most people (85%) stated that they know little or even less about husbandry and 77% think that improvements are required (Eurobarometer, 2007). Animal welfare in husbandry is seen as the third important objective of European agriculture (Eurobarometer, 2010). 55% of the respondents stated that agriculture policy does not give enough prominence to animal welfare (Eurobarometer, 2005). In the survey of TNS Emind (2012), only 46% of 1000 European respondents are interested in agricultural topics in general, while 85% mentioned good governance in animal husbandry as a desired property of German agriculture which is only partly achieved. But spontaneously, animal welfare is not mentioned as criteria of food quality. Only when questioned directly, concerns against modern animal husbandry are raised and often identified as reason for changes in individual consumption pattern (Harper and Henson, 2002) supported by Lassen *et al.* (2006). Kayser *et al.* (2012) evaluate attitudes on intensive animal husbandry in Germany indicating that animal welfare deficits in industrialized animal husbandry are perceived due to a limited available space per animal. The DLG (2009) determined a deep emotional relationship of the German population towards animal husbandry based on farmsteads in children's picture books. As a consequence, despite expected prices increases, society (79% of respondents) called for animal friendly husbandry; however, respondents perceived a partly improvement in animal welfare. Biggest room for further improvement is seen in pig production (Eurobarometer, 2005). Although negative impressions about animal welfare lead to latent unease, other criteria affect buying decisions (Alvensleben, 2002). Although currently animal welfare has little relevant for buying decisions it is gaining on importance (Alvensleben, 2002). Consumers presume they can improve animal welfare by buying respective products and state they would be prepared to pay a higher price for products of a higher animal welfare standard. However, when it comes to the buying decisions inadequate label [and awareness towards other criteria] hamper consumers' decision making (Eurobarometer, 2005). Nestle (2012) found 58% of respondents mentioning higher animal welfare standards among other criteria like sustainability as relevant for buying decisions. In the SGS Fresenius survey (2011) 69% of the population regarded animal welfare as an important factor of quality outscoring regional origin and ecological production. In general (83%), Germans are against the application of human antibiotics in animal husbandry (Forsa, 2012). In Lassen *et al.* (2006), Krystallis *et al.* (2009), and Boogard *et al.* (2011), also topics defining animal welfare from a society's perspective were analyzed like physical integrity and sound conditions or access to open land. 'Industrialized' animal husbandry was refused. Heid and Hamm (2012) drilled deeper to aspects of physical integrity.

Methodology

Based on the above mentioned findings in the literature focus groups and a representative online survey were used to generate necessary information developing strategies to abate the gap between society's perception of animal husbandry and the society needs.

Focus groups are moderated discussions of several people that are focussed on one topic such as animal welfare or like in our case pig production. It is 'a method for eliciting respondents' perceptions, attitudes and opinions' (Wilson, 1997: 209) and takes advantage of group interactions to determine participant's motives, which cause their behaviour. Participants are confronted with other participants' opinions, attitudes or perceptions, and may have to justify the own opinion, attitude or perception. Hence, 'individual response becomes sharpened and refined, and moves to a deeper and more considered level' (Finch *et al.*, 2003: 171). Usually, a guideline is written in advance to structure the discussion.

Six focus groups were carried out in September 2012 in the cities of Rheine (high concentration of pig farms), Mainz (low concentration of pig farms and Leipzig (close to an area with Germany's largest pig farms). Groups involved up to 11 participants, lasted about 90 minutes and were documented in audio and video format. Participants had no professional background. Focus groups included in a first part aspects like perception and knowledge of modern pig production. Main point of interest was to evaluate how much knowledge already exists regarding pig production among German society. After a more or less unstructured part, the perception of the stable, the pigs and the farmer were discussed in detail. An assessment of modern pig production followed in a second part. Positive as well as negative aspects were discussed in detail. In some groups, it was possible to go one step beyond. Here, conflicts of interest were regarded as well. In the end, discussants expectations' towards modern pig production and the question of responsibility were discussed.

Knowledge gained from the focus groups was used to develop the quantitative survey. An online survey was carried out with 1,519 adults in Germany in spring 2013. Participants were equally distributed regarding gender, region and income. But people older than 65 years were underrepresented. Respondents faced several seven-point likert scales about their attitudes, perception and expectations towards modern pig production. A special point surveyed was the question about responsibility for modern pig production. Respondents got specific actions on how pig production could be improved by the state, retailing companies or the consumers themselves. All of these actions were described with specific consequences like increasing prices or migration of producers abroad. Sociodemografic variables were questioned as well. Items regarding attitudes, perceptions or expectations were pretested using the Kaiser-Meyer-Olkin test, the measure of sampling adequacy test and the Bartlett test of sphericity. The complete questionnaire was tested using cognitive pretesting.

Results

Focus groups

An enormous amount of information was gained from the focus groups. This makes it necessary to concentrate on a few outstanding results. Discussants had a quite good understanding of modern pig production. Although not being told that the focus should be on intensive pig production discussants started to express their perception and opinion about it. During the discussion the opinion emerged that the lack of space in modern stables is the root for the main problems. Pigs were described as 'depressive', 'frightful' and 'maladjusted'. It was mentioned that 'they become somehow mentally deteriorated due to the way they are kept'. It is assumed that pigs start fighting for space and biting into other pigs tails ('that they become aggressive to each other or even against themselves, that they hurt each other. Because they are not together in their natural groups and they have a certain social behaviour that is not respected', 'If they have to eat a lot to become big on the one hand and on the other they have no possibility to move, then I think that they have energy what they cannot [discharge]..., they cannot run, maybe they also become aggressive').

Most of the discussants agreed that due to that lack of space pigs cannot behave naturally. But some respondents assumed that the pigs might have lost their natural behaviour ('born in captivity, the animal never got to learn something else. ... That is the question, because maybe that the mother was also born in captivity and cannot impart natural behaviour ... '). In most of the focus groups participants also established the link between farming practice and use of antibiotics ('If I have so many animals living on one spot, then I need antibiotics, because otherwise they all become ill', 'If they could move more freely [they did not need] the use of antibiotics', 'vicious circle').

Discussants argued much differentiated. Especially the relative low price of pork in Germany was seen as positive and negative as well. Again, a vicious circle was described but not mentioned explicitly. It was argued that people eat more and more pork due to the cheap price and therefore farmers will produce even more pork ('pork is relatively (...) cheap', 'giveaway prices for pork', 'meat for all', 'Price sensibility of consumers, because Germany is one of the countries that pay the less for their food. Problems are homemade', 'It is produced more and more meat for a cheap price and therefore it is consumed more because one can afford it').

Consumers' own responsibility was raised several times in all groups. The two main aspects were consumers' responsibility for the actual way pigs are produced and their responsibility for improving modern pig production. The former aspect was described with the conflict of interest between immoderately meat consumption and the price for pork. One respondent argued 'actually we consumers are the problem; also because we need to eat one kilo of meat per day, exaggerated...'. The same person mentioned '... [I should spent] few more Euros to get something of high-quality and then (...) [I would] moderate my meat consumption'. And another respondent argued that '(...) if people started to eat meat only on Sundays, there would be eaten less meat and then agricultural industry would be not so important. You have to look [first] at the consumer (...) and [then] agriculture is adapting'.

Regarding their expectations towards future pig production discussants also argue that all consumers have to change habits. Respondents were confident that these new habits would be retained. One discussant argued that 'humans are creatures of habit. They will pay more for pork und will go on eating it. We currently have it with coffee [explanation: the price for coffee increased by 15-20% during the last months but people still buy it]..... And we would also buy more expensive pork because it tastes better.' And another mentioned that 'we should not deal for this very cheap meat. Maybe with a higher price, meat will have a higher quality and then [we will] adjust our shopping habits'.

Asked who would be mainly responsible for the current situation most of the discussants stated the consumers are responsible for the pig meat he/she is buying ('the customer determines the price and in the end also the production system' or 'that the consumer can determine [the system] based on the fact that I buy or that I do not buy'). Some mentioned that more information would help to change consumers' habits ('one should try to change one's view based on more information'). The government is seen as responsible for the implementation of 'strict guidelines, to receive subsidies' and the 'monitoring of the guidelines'. And it was mentioned that monitoring has to be more continually.

Online survey

The analysis of the online survey confirmed many results of the focus groups. Based on the outcomes of the focus groups ten possible improvements in modern pig production were quantified in the online survey. More available space per pig and more materials available for pigs (65%), more outdoor access (60%) and no prophylactic use of antibiotics (48%) were considered as the most important aspects. Further, for 45% of the respondents no genetically modified feed was considered as one of the most important aspects and no surgical treatments without anesthesia such as piglet castration, tail-docking and/or grinding canines for almost 40%. The other possible improvements derived from focus groups, however, appeared less important in the quantitative survey[34].

[34] Respondents had to indicate the three improvements that were most important for them.

Using the data of 19 statements regarding attitudes, perception and expectations towards modern pig production (carried out with a seven-point likert scale), an exploratory factor analysis was used to define the underlying structure in the data matrix (Hair *et al.*, 1998: 90) and to describe society's perception towards modern pig production in more detail.

A principal component analysis was carried out[35]. Based upon these findings of the factor analysis, a cluster analysis was applied (Churchill and Nielsen, 1995: 985).

As shown in Figure 1, three clusters with enormous differences regarding the four extracted factors were found. The first cluster, the 'opponents', can be described as very engaged people since every single factor is strongly above or below the average. General criticism towards modern pig production and a critical perception of the farmers are (very) strong in this cluster, whereas the acceptance of the current system is very weak. Further, the opponents depicted strong claims regarding the behaviour of others. The second cluster, the 'tolerating' segment, is complementary to the opponents and thus, this cluster also shows a high engagement. They have a very high acceptance of the current system and no criticism towards modern pig production at all. Thus, the behaviour of others is not an issue. The third cluster can be described as the moderates or the indifferent persons who have no real opinions towards modern pig production. This interpretation appears from factor 1 (general criticism) and factor 3 (acceptance) which are both on average. Even factor 4 (behaviour of others) which is above the average, confirms the indifferent attitude as they have no claims regarding the current system. Nevertheless, the critical perception of this cluster is above the average. Finally, all three clusters are of comparable sizes.

Furthermore, the question whether respondents perceive their own responsibility for an improved pig production due to their purchase behaviour or whether they expect the state or retailers to be responsible was part of the online survey. All of these actions were described with specific consequences like increasing prices or migration of producers abroad.

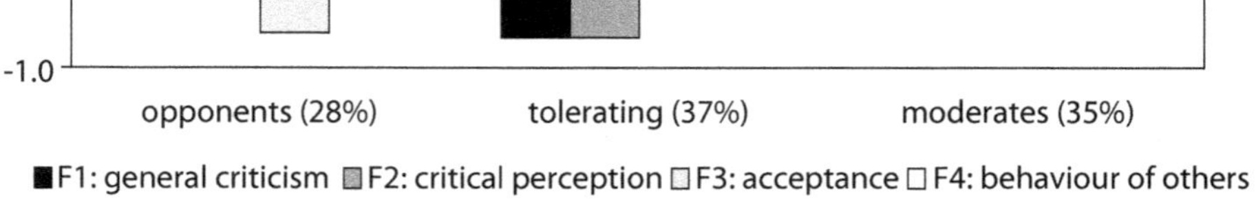

Figure 1. The affiliation of the clusters according to the extracted factors (mean deviation of each cluster from the grand factor mean of the overall sample).

[35] A table with estimation results of the factor analysis is available from the author on request.

83% of the respondents stated to accept increasing prices as consequence of stronger regulations or other similar measures to push animal friendly production systems. In contrast, only 38% of the respondents would accept it if several farms go abroad because of stronger regulations and 60% of the respondents would accept it if a couple of farms had to close. Thus, higher prices are only for few persons (17%) a reason to refuse improvements in modern husbandry, whereas migrating or closing farms seem to be more problematic. However, it is not clear to us, whether respondents realized their own responsibility in so far that the possibility to support animal friendly production systems is also a purchase decision. Moreover, the final question about the main responsibility for modern pig production resulted rather different. Here, only 13% of the respondents answered that consumers are mainly responsible and they see farmers (47%) and the state (31%) carrying responsibility for improving modern husbandry.

Discussion and conclusion

The findings partly support the outcome of older studies: respondents often identified limited space in modern pig keepings as a major source of their concerns. Related to this insufficient room other problems may arise according to their views, such as the need of tail docking or the use of antibiotics (legal and illegal) in pig production. Respondents displayed ambivalent attitudes towards the pricing of pork. German pork prices were perceived as relatively low in general, and price levels were regarded as inadequate to cover the cost of good governed pig production with sufficient space. In this respect also the responsibility of the consumers for the current pig production system and its improvement was very clearly stressed. The train of thought was that people need too much meat and thus requiring too much space respectively resources or that the prices paid do not cover costs of animal friendly production systems. In general, there was a feeling that consumers determine the handling of pig production systems to a greater extent. In addition, the government is seen in charge for the implementation of strict guidelines the monitoring hereof which to the mind of the respondents needs to be conducted more frequently.

Concerning further developments of the system several strategies can be followed; however, the picture is far from being clear cut und strategies need to be underpinned and clarified by new and additional results:

1. As the society expresses a very strong concern towards a more animal friendly production process the government may introduce obligations without compensation for the farmers towards animal welfare. Due to the fact that German farmers are integrated in an EU Common Market this strategy requires a common introduction in all EU Member States to avoid a reallocation of production into other countries without those obligations. Due to the cost of compliance market prices will increase.
2. As the consumers see their responsibility for the pig production systems they need to adjust their buying decisions. This is only possible if they have a choice to separate animal friendly produced meat form others which would require a labelling. However, when asked people showed weariness towards labels and in particular, the credibility of labels is regarded as low. Thus a special emphasis needs to be put on the compliance with label rules very frequently controlled by a really trustworthy organisation.
3. Respondents displayed a very limited knowledge of production and tend to reveal a retrospective, nostalgic view on pig production based on a picture book farmstead, but when questioned they were self conscious to this fact. To overcome this shortage group specific communication strategies will be required. A possible strategy may include a wide range of measures which might cover the opening farm gates to the public by e.g. frequent visits on farms, better representation on TV respectively the internet and inclusion of farms into TV series, better representation of farms in picture books, etc.

Based on these results there is no high road to overcome the problem in pig production, but a mixed of the different strategies might lead to a step by step improvement.

References

Boogaard, B.K., Boekhorst, L.J.S., Oosting, S.J. and Sorensen, J.T. (2011). Socio-cultural sustainability of pig production: citizen perceptions in the Netherlands and Denmark. Livestock Science 140: 189-200.

Churchill, G.A. and Nielsen, A.C. (1995). Marketing research methodological foundations. The Dryden Pree, Fort Worth, TX, USA.

DLG (2009). Image der Landwirte deutlich verbessert, Pressemitteilung 15.01.2009. Available at: http://presse.dlg.org/publicArtikelDetail.do;jsessionid=3909BAD40BD82154B4BB17DF72CA7570?artikelId=2189&spracheId=1&kategorieId=1&bereich=Aktuelles&type=pdf.

Eurobarometer (2005). Attitudes of consumers towards the welfare of farmed animals, Eurobarometer Spezial 229. Available at: http://ec.europa.eu/food/animal/welfare/euro_barometer25_en.pdf.

Eurobarometer (2007). Attitudes of EU citizens towards animal welfare, Eurobarometer Spezial 270. Available at: http://ec.europa.eu/public_opinion/archives/ebs/ebs_270_en.pdf.

Eurobarometer (2010). Europäer, Landwirtschaft und Gemeinsame Agrarpolitik, Eurobarometer Spezial 336. Available at: http://ec.europa.eu/public_opinion/archives/ebs/ebs_336_de.pdf.

Finch, H. and Lewis, J. (2003). Focus groups. In: Ritchie, J. and Lewis, J. (eds.) Qualitative research practice, Sage Publications Ltd., London, UK, p. 170-197.

Forsa (2012). Meinung zum Einsatz von Humanantibiotika in der Tierhaltung. Available at: http://bund.net/fileadmin/bundnet/pdfs/landwirtschaft/120508_bund_landwirtschaft_forsa_humanantibiotika_umfrage.pdf.

Hair, J.F., Anderson, R.E. and Tatham, R.L. (1998). Multivariate data analysis. Prentice-Hall International, Upper Saddle River, NJ, USA.

Harper, G. and Henson, S. (2001). Consumer concerns about animal welfare and the impacton food choice. EU FAIR CT98-3678. Final Report, 38 p.

Heid, A. and Hamm. U. (2012). Consumer attitudes towards alternatives to piglet castration without pain relief in organic farming: qualitative results from Germany. Journal of Agricultural and Environmental Ethics 25: 687-706.

Kayser, M., Schlieker, K. und Spiller, A. (2012). Die Wahrnehmung des Begriffs 'Massentierhaltung' aus Sicht der Gesellschaft. In: Berichte über Landwirtschaft 90(3): 417-428.

Krystallis, A., De Barcellos, M.D., Kügler, J.O., W. Verbeke and K.G. Grunert, (2009). Attitudes of European citizens towards pig production systems. Livestock Science 126: 46-56.

Lassen, J., Sandøe, P. and Forkman, B. (2006). Happy pigs are dirty! – conflicting perspectives on animal welfare. Livestock Science 103(3): 221-230.

Nestlé Deutschland AG (2012). Die 'Quality Eater' erobern den Supermarkt, Pressemitteilung 'Das is(s)t Qualität'. Available at: http://www.nestle.de/Unternehmen/Nestle-Studie/Nestle-Studie-2012/Documents/Pressemitteilung_Studie_Das_isst_Qualitaet.pdf.

SGS Fresenius (2011). Verbraucherstudie 2011 „Lebensmittelqualität & Verbrauchermacht.

TNS Emnid (2012). Das Image der deutschen Landwirtschaft: Ergebnisse einer Repräsentativbefragung, März 2012. Available at: http://www.ima-agrar.de/fileadmin/redaktion/download/image-studie/2012/ima-imagestudie landwirtschaft_charts-2012.pdf.

Von Alvensleben, R. (2002). Neue Wege in der Tierhaltung – Verbraucheransichten und -einsichten. In: KTBL (ed.) Neue Wege in der Tierhaltung. KTBL-Schrift 408: S. 25-32.

Wilson, V. (1997). Focus groups: a useful qualitative method for educational research? British Educational Research Journal 23(2): 209-224.

Section 11. Food in the public sphere

Public sector food procurement in UK local authorities: ethics and sustainability

M. Stein
Salford University Business School, Maxwell Building, The Crescent, Manchester M5 4WT, United Kingdom; markstein2010@live.co.uk

Abstract

The paper will consider national policies aimed at promoting health and sustainability in public sector food procurement for UK schools and their interpretation by UK local authorities. In particular it will examine the work of the Food for Life Partnership (FFLP) which has been working with schools and municipal catering services to enable children to eat good food, learn where it comes from, how it is produced and how to grow and cook it themselves. It involves children, teachers, parents and school managers and promotes greater consumption of local and organic food, with 4,200 schools enrolled in the programme and 300,000 children eating accredited meals every day. The research question is: what have been the factors promoting FFLP's relative success following the limited achievements of the preceding Public Sector Food Procurement Initiative? This paper is based upon news and academic reports, policy documents and websites and discussions with individuals. My analysis has been informed by the discussion of success of sustainability initatives in Booth and Skelton (2011). Deloitte's evaluation of the Public Sector Food Procurement Initiative shows that this initiative achieved limited success during the years 2003 to 2007. Independent evaluation of FFLP's has found more substantial evidence of success which includes: (1) improved eating habits among primary school age children and parents; (2) increased take-up of free school meals among low-income children; (3) benefit to local economies from increased spending on local food; (4) increased consumption of organic food; (5) increased job satisfaction among kitchen staff; and (6) food citizenship education among both primary and secondary school children.

Keywords: municipalities, catering, environmental, organic, local

Background to the development of new policies for UK school food

The British public sector buys substantial quantities of food, principally for schools and hospitals. Total annual spend is around £1.8 billion. Quality has often been unsatisfactory (Morgan, 2008: 91-92]. In 1984 prime minister Thatcher abolished the nutritional standards put in place in 1944 (Morgan, 2008: 91-92). This has had costly consequences. The UK has the worst child obesity problem in Europe. In recent years Britain's agricultural sector has been in difficulty, with increasing import penetration and concern about long term food security, amid volatile international prices.

New approach to school catering

The years 2005 to 2007 saw a radical policy shift towards school catering in the UK. Jamie Oliver's campaign in 2005 obtained a high profile for its criticisms of poor quality school catering. The government responded by introducing new mandatory school meal standards between 2006 and 2009. These required nutritious foods and excluded junk foods.

Takeup of meals initially fell and then recovered. New nutrition standards led some kids to stop eating school meals because they preferred the accustomed unhealthy foods. This threatened the financial viability of the school meals service. It was an unanticipated negative response of a sort which can lead to failure of sustainability initiatives (Booth and Skelton, 2011: 67). School caterers were obliged to develop a strategy for adverse public reactions. They made a strong effort to to improve school meal takeup, with considerable additional financial resources provided by the national government.

Public sector food procurement initiative launched 2003

The public sector food procurement initiative (PSPFI) was launched in August 2003. Its aimed to encourage public bodies to procure food in a manner that considers the principles of sustainable development. This included: increasing consumption of healthy and nutritious food, improving sustainability of production, processing and distribution, increasing tenders from small and local producers, promoting organic food, and animal welfare (Deloitte, 2009: 3). Local authorities which did most to pursue PSPFI were rural ones with substantial local food production – with clear local economic benefits. The international consultancy Deloitte produced an evaluation of PSPFI which acknowledged that it was difficult to quantify PSFPI's success because it has not been measured or monitored regularly during its lifetime and quantifiable targets were not established at the outset. Deloitte concluded that there is evidence of limited progress towards PSFPI goals – a very slight increase in the proportion of UK food used by government departments and increased use of seasonal produce and small producers. Deloitte also reported that buyers lacked necessary skills and the initiative relied on the heroic efforts of enthusiastic individuals for success – when these people move on what they have achieved may be lost. There was a perceived cost barrier – although Deloitte's research showed that PSPFI had often been implemented with reduced or stable costs. The initiative lacked political backing and there was no clear definition of 'sustainability' or 'local food' (Deloitte, 2009: 14-17; 22-40; 43-55).

Food for life partnership 2007 to 2012

The most important development in sustainable food procurement was the launch of the Food for Life Partnership (FFLP) early in 2007. FFLP is an alliance of four English NGOs – Soil Association, Focus on Food, Garden Organic and the Health Education Trust. The core idea has been to enable children to eat good food, learn where it comes from, how it is produced and how to grow and cook it themselves. FFLP received generous funding – £17 million over five years – from the Big Lottery Fund. This has enabled it to work with schools in every English region employing a total of around eighty staff. In terms of money this gave FFLP vastly greater resources than PSFPI – who had a small central team and a single civil servant in each English region. A great strength of FFLP has been the grassroots approach – how it develops a community in and around each school. Children who learn about food consumption and production, have cookery lessons, grow food on the school grounds and visit farmers and food producers. Parents who come into school and share cooking skills with children including the rich variety of ethnic food cultures to be found in England. Teachers who integrate food into the overall school curriculum. Head teachers who lead the process and receive the awards. School catering staff who prepare the food from fresh local ingredients. local farmers and food producers. FFLP was designed from the outset that school caterers could win a hierarchy of awards and progress would be publicly recognized at each level:

- Bronze Award criteria — 75% of dishes on menu freshly prepared; minimising additives; no hydrogenated fat; seasonal menus; meat farm assured as minimum; eggs cage free.
- Silver Award Criteria — Range of local items; range of certified organic items; poultry eggs pork Freedom Food; only sustainable fish; at least one Fair Trade product.
- Gold Award Criteria — At least 30% of ingredients are certified organic or MSC; at least 50% of ingredients local; emphasis on animal welfare; increased vegetarian food.

This is a much more structured approach to promoting best practice than was the case with the Public Sector Food Procurement Initiative, where people worked largely in isolation, with little public recognition of their achievements. It also accords with Booth and Skelton's advice to 'be realistic about what can be accomplished with the available time, resources and personnel...gaining acceptance with smaller proposals is crucial to building towards more ambitious ones' (Booth and Skelton, 2011: 67). A measure of FFLP's success is that by summer 2011 over 3,800 schools were enrolled in the program and over 300,000 children eating FFLP accredited meals every day. Detailed evaluations of FFLP's work was done by three groups of researchers, whose studies were published in June 2011 (Food for Life Partnership, 2011). The studies showed that FFLP had led to improved eating habits – eating more vegetables – among primary school age children. There was increased take-up of free school meals among children from low-income families – a group which suffers above-average levels of obesity. Inspector's ratings of overall performance of the schools involved had improved. Increased spending on local food has benefitted local economies. The emphasis under FFLP Gold on promoting organic food has important environmental benefits.

Since the change of government (May 2010)

Since the new government was formed in May 2010 the overriding priority has been to reduce public spending. The ring-fenced school meals grant has been removed. This has put pressure on school caterers to increase prices and reduce costs.

At local authority level new public health bodies are being set up as part of the government's NHS restructuring and these will have to decide their spending priorities – which will in some cases include work with children along the lines pioneered by Food for Life Partnership.

FFLP has continued to expand over the last three years. The way FFLP was designed makes it possible for individual schools to follow FFLP. This flexibility means that any break up of local catering services resulting from the government's new policies need not necessarily prevent the expansion of Food for Life. FFLP has had marked success in London. A November 2011 report by Sustain says that seventeen of thirty three boroughs have achieved at least FFLP Bronze Standard and eight have been awarded Silver – or have a catering contract stipulating that the Silver Standard must be attained (Sustain, 2011). Success of FFLP in London is partly attributable to the enthusiasm shown by ethnic minority communities bringing their food traditions into school. Better quality meals have led to increased uptake and revenue which has more than made up for increased costs due to better ingredients. FFLP has also developed well in a number of rural counties and urban local authorities FFLP has also been bolstered by increased public interest in food quality – as seen in the public outcry over adulteration of beef with illegal horsemeat. FFLP's five year funding expired at the end of March 2012. However the Big Lottery Fund has now provided funding for a further year which will maintain a core team and during this period it is hoped that funding for local projects will be provided by new public health bodies.

FFLP can be compared with sustainable food initiatives in Italy and Finland. Sonnino (2012) praises the achievements of enlightened city politicians and officers in achieving a school food revolution in Rome but notes that the election of a right wing administration in 2008 led to abrupt policy change – with abolition of ethnic menus and 'defensive localism' and asks 'how can we ensure that the gains of school food reform survive the vicissitudes of electoral cycles'. FFLP's success arguably reflects the way in which it has embedded itself in the wider community of the English schools, cities and towns where it has been operating. Risku-Norja & Mikkola (2010) comment on sustainable food initiatives in Finnish schools: 'Both among the catering personnel and in schools the means to influence the decisions regarding food are experienced as meager or non-existent: the decisions are made beyond the reach of catering and school personnel....[by] the municipal authorities'. FFLP by contrast involves catering and school personnel as well as children and parents in choices regarding school food.

Promoting ethical and sustainable food

- FFLP promotes animal welfare in a variety of ways: cage free eggs; Freedom Food pork and eggs. Meat is Farm Assured as minimum and organic food is encouraged which promotes better animal welfare (Ricke *et al.*, 2012: 91).
- FFLP has promoted local economic benefits through local food sourcing (Kersley, 2011).
- FFLP promotes consumption of sustainable fish – from fisheries approved by the Marine Stewardship Council.
- FFLP's growth has helped stem the decline in organic food consumption in the UK (Soil Association 2013:3]. Organic food is arguably more sustainable (Lynch *et al.*, 2011).
- Implementation of FFLP increased job satisfaction among kitchen staff (Kimberlee *et al.*, 2012).

Academic evaluations of FFLP have shown that it provides a range of opportunities for primary school pupils to learn about food production and sustainability – provided questions regarding effective, equitable and on-going implementation are addressed (Weitkamp *et al.*, 2013). Evaluation in 24 secondary schools shows successful implementation of food citizenship education but less evidence of positive student behavioural change – highlighting the organizational challenges of delivering a complex and ambitious program in the secondary school setting (Jones *et al.*, 2012).

References

Booth, A. and Skelton, N. (2011). Anatomy of a failed sustainability initiative. Sustainability: Science, Practice and Policy 7(1): 56-68.

Deloitte (2009). DEFRA- Public sector food procurement initiative: an evaluation-final report April 2009. London, UK. Available at: www.defra.gov.uk/farm/policy/sustain/procurement.

Food for Life Partnership (2011). Good food for all: the impact of the food for life partnership. Available at: www.foodforlife.org.uk.

Jones, M., Dailami, N., Weitkamp, E., Kimberlee, R., Salmon, D. and Orme, J. (2012). Engaging secondary school students in food-related citizenship. Education Sciences 2: 77-90.

Kersley, H. (2011). The benefits of procuring school meals through the food for life partnership. New Economics Foundation, London, UK. Available at: www.foodforlife.org.uk.

Lynch, D., MacRae, R. and Martin, R.C. (2011). The carbon and global warming potential impacts of organic farming. Sustainability 3: 322-362.

Morgan, K. and Sonnino, R. (2008). The school food revolution: public food and the challenge of sustainable development. Earthscan, London, UK, 231 pp.

Ricke, S.C., Van Loo, E.J., Johnson, M.G. and O'Bryan, C.A. (2012). Organic meat production and processing. Wiley-Blackwell, Oxford, UK, 444 pp.

Risku-Norja, H. and Mikkola, M. (2010). Towards sustainable food systems through innovative networks in public catering. 9th European IFSA symposium 4-7 July 2010. Available at: http://ifsa.boku.ac.at/cms/fileadmin/Proceeding2010/2010_WS4.2_Risku-Norja.pdf.

Soil Association (2013). Organic market report 2013. Available at: http://www.soilassociation.org/marketreport.

Sonnino, R. (2012). Up to the ground and from the ground up: Presentation 27th January 2012. Available at: http://www.scp-responder.eu/community/documentation/KB_Food_2012.

Sustain (2011). Good Food for London. Available at: http://www.sustainweb.org/londonfoodlink/good_food_for_london.

Weitkamp, E., Jones, M., Salmon, D., Kimberlee, R. and Orme, J. (2013). Creating a learning environment to promote food sustainability issues in primary schools. Sustainability 5: 1128-1140.

The scientifically motivated regulation of food: a discursive analysis of the EU health claims regulation on omega-3 fatty acids

M. Döring[1] and S. Stirn[2]
[1] University of Hamburg, Institute for Geography, Bundesstraße 55, 20146 Hamburg, Germany; [2] University of Hamburg, FSP BIOGUM, Biozentrum Klein-Flottbek, Ohnhorststraße 18, 22609 Hamburg, Germany; doering@metaphorik.de

Abstract

During the last decade, an increasing number of foods promoted with nutrition or health claims appeared on European supermarket shelves. These foods are perceived by a large number of European consumers as bearing a considerable health-promoting impact over comparable products not holding a health claim. To harmonize the internal market and to ensure a high level of consumer protection by banning misleading claims, the EU health claims regulation (Regulation EC/1924/2006 on nutrition and health claims on foods) has been adopted in 2006. One of the basic principles of it is that the use of nutrition and health claims should not suggest that consumption of functional foods (FFs) can replace a balanced diet. Omega-3 fatty acids represent in this context a special case as a variety of positive health outcomes are attributed to them ranging from healthy brain and eye development in infants to the prevention of neurodegenerative disease in adults. Applications for health claims related to these benefits have recently been authorised by the EU-Commission. A discursive analysis of the processes underlying the authorisation of these health claims forms the starting point of this paper in which we assessed their underlying conceptual structure. For this to be done, we studied interpretative repertoires in written evidence (scientific reviews, legal commentaries, policy documents, etc.) on omega-3 fatty acids. We focused on health claims promoting visual and brain development in infants and children. First, analytical emphasis was put on the semantic structure of scientific evidence that informed EU-legislation. The second step consisted in the analysis of interpretative repertoires apparent in the documents that were highlighted the processes of authorisation. Here, emphasis was put on embedded normative assumptions in the emerging interpretative repertoires. The aim of the paper is twofold: It analyses the conceptual structure of scientific evidence that informs regulative procedures and it critically assesses its normative assumptions that inform current food labelling practices on omega-3 fatty acids within the EU.

Keywords: functional foods, study of food and society, discourse analysis, childhood- and infant images

Introduction: the emergence of functional foods

The idea of functional foods emerged during the 1990s in the context of a variety of developments such as the newly emerging food sciences, their impact on industrial application and new strategies of marketing healthy foods. The growth of so-called nutritional knowledge stemming from a variety of scientific disciplines, boosted by new scientific insights and propelled by interdisciplinary fields of research such as public health nutrition, deepened the understanding about the relation of nutrition and health. This development generated interest in the food industry which saw the opportunity to produce a positive self-image of a health concerned and health improving industry. The challenge consisted in creating a new and profitable market which aimed at developing novel and profitable foodstuffs that were supposed to hold health improving benefits or even aimed at preventing major common diseases. These so-called functional foods (FFs), enriched for example with cholesterol-lowering substances or health improving fatty acids, were conceptualised as foods that hold a beneficiary impact on one or more body functions. The health improving impact, however, has to be demonstrated by appropriate studies and indicate that the health benefits go well beyond normal nutritional effects. In brief, improved health, well-being and the reduction of risk of illness have to be achieved by the technically enriched or fortified foodstuffs. This broad definition creates difficulties as it does not determine what it means to surpass normal nutritional effects and it does not outline measures how improved health and well-being could be measured. This ambiguity led to a lack of defining what precisely the positive health effects are and it is mirrored in a variety of FFs on the food market that range from relatively precisely defined and tested cholesterol-lowering margarines to generic definitions such as an improved resistance to illness via the consumption of probiotic bacteria in yogurts. The positive health effects exerted by conventional foods such as berries or different kinds of herbs are even harder to measure. These problems revolving around the definition and assessment of positive health effects led to a variety of regulative approaches in the European Union (Regulation EC/1924/2006 on Nutrition and Health Claims on Foods), the United States, Australia and elsewhere. These were designed to assess and legally regulate health claims used in the promotion and marketing of FFs. Their main aim consists in protecting consumers from being misled while industrial representatives constantly complain about the strict scientific assessment of health and nutrition claims which, following their line of reasoning, could be compared to the assessment of pharmaceuticals. This becomes particularly apparent if one looks at the economic expectations raised during the advent of FFs: representatives of the food industry estimated the development of a new market with a considerable increase in economic growth, turnover and profitability. Closely connected to these economic aspects were hopes envisioned by public health scientists such as a decrease in future health expenditures, an increase in public health, new possibilities for consumers to make health-promoting choices or simply citizens enjoying considerably better health than former generations. These hopes have been challenged right from the start by critics who underlined that a balanced diet in itself does not require any health improving food supplements whatsoever. Furthermore, attention was directed towards the problem of techno-foods which might erode people's understanding and awareness of a healthy diet while, on a general level, the creeping medicalisation of food via FFs was conceived to support a reductionist if not science-driven concept of eating and health which neglects the cultural and social dimension of food and nutrition. To sum up: FFs represent a special kind of 'borderline foodstuff' that could best be described by the neologism 'nutraceutical', a composition of 'nutrition' and 'pharmaceutical', and whose health improving aspects are managed in the European context by the Regulation EC/1924/2006 on nutrition and health claims on foods.

This regulation forms the starting point of our empirical investigation in which we analyse the scientifically informed decision processes underlying the authorisation of health claims revolving around omega-3 fatty acids and their positive impact on brain and eye development in infants and children. A variety of positive health outcomes are attributed to omega-3 fatty acids and applications for health and nutrition claims related to these benefits have recently been authorised by the EU-Commission (European Commission, 2011). Documents stemming from this review process were analysed from a discourse analytical point of view to unravel and tackle embedded normative assumptions (Komduur *et al.,* 2009) inherent in underlying images of infancy and childhood. The aim of the paper is to show and analyse which infant-images and images of child development taken from scientific evidence inform regulative procedures and governance processes and how these appeared in the area of food legislation.

Analysing omega-3 fortified foodstuffs for children: context, theory and method

Omega-3 fatty acids were chosen as a starting point of this paper due to the continued industrial interest in producing omega-3 fatty acids enriched foods such as fish fingers, eggs or breads. The interest in omega-3 fatty acids as a health promoting ingredient in food goes back to research undertaken in the 1930s by Dyerberg and Jorgensen. They discovered the paradox that members of an Inuit tribe in the North of Canada consumed high amounts of fatty sea food but were not prone to cardiovascular disease. Follow-up studies supported the hypothesis that omega-3 fatty acids played a vital role: They considerably contributed to reducing triglycerides, managing blood pressure and prevent atherosclerosis despite the usual Inuit diet which only contains small amounts of fibre and fruit. Since then, research on omega-3 fatty acids gathered momentum as evidence supported that they hold a multivalent therapeutic spectrum comprising the prevention of cancer, cardiovascular disease, neurodegenerative disease plus some other positive impacts on eyesight, neuronal development and mental disorders. Not surprisingly, the food industry – above all Mead Johnson Nutraceuticals, Merck and Unilever – developed an interest in omega-3 fatty acids which led to the fortification of all kinds of foods and opened-up a market for follow-up products addressing age groups such as elderly people, adults, children or infants. Especially parents are targeted by the advertisement of omega-3 fortified foods such as baby mashes or milk powders. These present the advantages of an omega-3 rich diet during pregnancy and afterwards promoting their positive impact on neuronal development of infants, the possible improvement of psychological disorders in children and the prevention of neurological diseases from childhood onwards throughout the whole lifespan.

The European Council and Parliament reacted to this development by adopting the Regulation on Nutrition and Health Claims Made on Foods (European Community, 2006) in December 2006. For the first time, this regulation laid down harmonised rules across the EU for the use of nutrition claims such as 'low fat' or health claims such as 'reducing blood cholesterol'. In accordance with the Health Claims Regulation, claims made on foods were only permitted if they were either listed in the register of permitted health claims or authorised individually in a Commission Regulation. One of the key objectives of this regulative attempt was to ensure that any claim made on a food label is clear, not misleading and substantiated by scientific evidence. The aim of this procedure was to improve the consumers' ability to make informed and meaningful choices.

Three categories of health claims are defined as follows:

- the 'function' or Article 13 health claims, describing the role of a nutrient in growth, development or the functions of the body;
- the 'risk reduction claims' or Article 14 (1) (a) claims on reducing a risk factor in the development of a disease; and
- health claims referring to children's development and health (Article 14 (1) (b) claims.

Later, in May 2012, the EU–Commission established the EU list of permitted 'function' claims ('Art. 13 list') (European Commission, 2012). Several decisions on health claims referring to children's development or health under Article 14(1) (b) have been published since then (European Commission, 2011). We took this ongoing regulative process as a starting point for our paper in which we first analysed underlying interpretative repertoires (IRs) (Wetherell and Potter, 1988: 172) of infancy and childhood in the scientific literature and documents taken from the review process. We generally build on insights provided by critical discourse analytical approaches combining them with insights provided by discursive psychology. Both approaches conceptually converge in the fact that they understand discourses as based on a structured set of IRs or schemas, which in many cases start with a linguistically motivated signifying activity. These signifying processes and their continuing repetition establish overarching patterns of meaning and interpretation called IRs which construct, connect and arrange the entity omega-3 fatty acids and brain and eye development in infants and children. This approach thus stresses the relevance of IRs as an outcome of processes of an underlying meaning construction and opens-up possibilities for a tentative analysis of implicit normative assumptions. A first analytical step consisted in the study of IRs in scientific documents. We then analysed documents taken from the review process to tackle surfacing IRs in the process of authorisation. As a result, a variety of IRs emerged during analysis which was then compared with the IRs tackled in the scientific literature. In doing so, an IR-catalogue of both discursive and intermingled strands became apparent which enabled us to track to a certain amount the underlying meaning structures informing European health claims regulation on eye sight and brain development in infants.

Acknowledging the multifaceted nature of our research object, we gathered a set of scientific articles and reviews, books, newspaper articles and internet resources covering the time span between 2000 and 2012. The juxtaposition of the texts offered a rich insight into the discourses and contexts surrounding the topic of omega-3 fatty acids and their positive health impacts. After tackling general developments providing an overview over contemporary scientific discourses, we decided to concentrate on the role of omega-3 fatty acids in brain and eye development in infants and children as they only recently attracted increasing scientific, industrial and regulative interest. The next step consisted in developing two consistent databases: One on the scientific discourse and the other on the process of authorisation by the European Food Safety Authority (EFSA). The corpus of scientific texts was reduced to scientific reviews as this type of text provides deep insight into current trends in research, the shaping of research agendas and important future research avenues. Publications from ten leading authors were taken from the Medline database and checked on the ISI-citation index regarding their intra scientific dispersion. Both datasets were analysed according to the systematic requirements outlined in Grounded Theory. This procedure enabled us to trace general developments in both databases and unravel their underlying semantic and conceptual structure. Questions addressed were the following: What IRs could be found? What models of infancy and childhood with regard to brain and eye development surface? What are their normative assumptions? The next section displays preliminary IR-oriented results taken from our analysis.

Omega-3 fatty acids and child development: from scientific evidence to legal regulation

The analysis of omega-3 related scientific reviews taken from our database provided a rich insight into current research. The grounded approach was chosen as it puts emphasis on the empirical development of analytical categories from data. Once main themes or topics emerge during the process of analysis, texts segments were grouped under these emergent headings. Data subsumed under these headings were re-read and analysed with the main aim to track and compare linguistics structures such as lexical items, metaphors or narratives which are supposed to exert a meaning making impact on the discourse. In the case of sufficient evidence, the status of a topic was transferred to an IR due to its distinctive and substantiated meaning structure. These IRs were then conceived as organising devices which enabled us to assess and compare to a certain extent prevailing images and connected normative assumptions of the impact of omega-3 fatty acids on eyesight and brain development in infants and children.

Taking a look at the main narratives of the scientific literature, omega-3 fatty acids are nowadays conceived as a very important and disease preventing nutritional ingredient. Omega-3 fatty acids are envisioned as crucial macromolecules which determine the brain's integrity throughout life due to the fact that nearly 60% of the substance of the human brain is made of fat. A problem, however, exists in the fact that essential fatty acids such as omega-3 cannot be synthesised by the body, they have to be obtained from dietary sources such as oily cold water fish, walnut or peanut oil. Thus, the human brain requires out-of-body-resources to build and maintain its structural and functional capacities. Contrary to this insight provided by scientific research runs the fact that the amount of omega-3 consumption has decreased over the last decades. Several factors seem to be responsible of which the most important are changes in modern working life (unstable working situation) and nutritional adaptation strategies (lack of home cooking and increase in the use of frozen meals) to these working conditions. These changes are supposed to hold a problematic impact with regard to visual and brain development for children as malnutrition with omega-3 fatty acids is supposed to cause ADHS or cognitive impairment.

Seven IRs materialised in the course of our analysis: the western-lifestyle-IR (normative assumption on the deficiency of omega-3 intake due to changes in modern working life), the feeding-mother-IR (normative assumption on nutritional responsibility from fertilisation to the end of childhood), the neonate-IR (normative assumption about the positive impact of omega-3 fatty acids on neurological development), the child-IR (normative assumption about the positive impact of omega-3 on neurological improvement), the disease-IR (conceptual focus on impairments and cognitive syndromes) and the prevention-IR (normative assumption that breast feeding or feeding omega-3 fortified milk helps to prevent disease and holds positive developmental impacts). This conceptual structure shapes the scientific discourse on the functional relation of omega-3 fatty acids and visual and brain development in infants and children and highlights the responsibility of the mother while the role of the father in this context remains unasked. The western-lifestyle-IR, for example, puts emphasis in the fact that so-called modern lifestyles are fine-tuned to working hours and that subsequent dietary habits do not provide relevant daily doses of omega-3 fatty acids. This IR links-up with the feeding-mother-IR which depicts the mother or women as being in charge of a healthy diet. Neonates and children are mechanistically portrayed as endangered brains under construction in the neonate- and child-IR while possible dangers are delineated in the disease-IR. Remedies are outlined in the prevention-IR which neatly links-up with the feeding-mother-IR and distributes responsibility to women or the maternal provision of fortified foodstuffs. Considered as a whole, it becomes apparent that the IR-structure of the scientific discourse is quite tight and puts mothers/women at the centre of responsibility. Infants and children are represented as human beings under construction who need the relevant nutritional building bricks to construct healthy brains and eyesight. This seems to be tricky in the current nutritional contexts and the only remedies available are breast or formula milk to be provided by the mother or by industry. One side effect of this development is that breastfeeding mothers do not trust their own milk and consume omega-3 fortified foods or use improved formula milk.

We compared the IR-structure found in the scientific publications with selected applications on health claims related to the beneficial role of omega-3 fatty acids referring to children's development and health. The process of authorisation in the administrative framework of the EFSA was informed and contextualised by an assessment of the scientific discourse which provided relevant grounds for the final decision and its justification. Thus, the IR-structure of the scientific discourse could be assumed to be known and the discussion mainly revolved around relevant daily doses which might exert a measureable impact in brain- and eye development. Interestingly enough, the administrative documents exhibit some structural aspects found in the scientific discourse, namely the neonate-IR and the child-IR with their normative assumptions about the positive impacts of neurological development and improvement. These aspects, however, were reframed and critically considered during assessment expressed in one claim underlining that Docosahexaenoic Acid (DHA) contributes to the visual development of infants (European Food Safety Authority, 2009). In another case, the prevention-IR explicitly addressed in the wording by the applicant 'DHA is important for early development of the brain in the foetus (unborn child) and infant. Maternal DHA supply contributes to the child's cognitive development' was finally changed to 'Docosahexaenoic acid (DHA) has a structural and functional role in the brain and maternal DHA intake contributes to normal brain development of the foetus and breastfed infants (European Commission, 2010). The removal of the phrase 'child's cognitive development' and its substitution by 'normal brain development' puts emphasis on physiological aspects although implicitly fostering the neonate-IR and the child-IR. In conclusion, three IRs stemming from the scientific discourse could be tackled in our provisional analysis of submissions on health claims made to the EFSA. The analytical tool of interpretative repertoires provided a structural insight into the discourse on omega-3 fortified foods and helped to uncover the conceptual and semantic structure which highlighted physiological images of neonate and child development which were coupled with images of preventing disease. Bearing in mind that the positive impact of omega-3 fatty acids is far from clear and was conceptually coupled with an apparent image of the feeding mother, the scientifically informed regulation seem to promote a mixture of physiological child- and traditional mother-images. The regulation is thus based on a physiologically motivated image of child development and motherly care.

Conclusion: the scientifically informed regulation of omega-3 fatty acids

By analysing written evidence taken from two sources, our analysis built on the still lacking sociological interest in the structure, organisation and dispersion of scientific evidence that informed the regulation of foods in the EU. We argued that the efforts made to conceptualise the positive impacts of omega-3 fatty acids on the development of infants and children is based on a variety of shared IRs that semantically structure and discursively organise the scientific assessment of health and nutrition claims. Our analysis thus provides a provisional insight into the still under-researched areas of functional foods from a science and technology perspective. It furthermore explored underlying images of child development and motherly care and uncovered the conceptual structure of scientific evidence that informs regulative procedures of current EU food labelling procedures. In doing so, it aimed at a critical assessment of current intervention tools that shape the food market.

References

European Commission (2010). Draft commission regulation of on the authorisation and refusal of authorisation of certain health claims made on foods and referring to children's development and health. SANCO/13017/2010 Rev. 2.

European Commission (2011). Commission regulation (EU) No 440/2011 of 6 May 2011 on the authorisation and refusal of authorisation of certain health claims made on foods and referring to children's development and health. Official Journal of the European Union L 119: 4-9.

European Community (2006). Regulation (EC) No 1924/2006 of the european parliament and of the council of 20 December 2006 on nutrition and health claims made on foods. Official Journal of the European Union L 404: 9-25.

European Food Safety Authority (2009). Scientific opinion of the panel on dietetic products, nutrition and allergies on a request from Mead Johnson Nutritionals on DHA and ARA and visual development. The EFSA Journal 941: 1-14.

Komduur, R., Korthals, M. and Te Molder, H. (2009). The good life: living for ealth and a life without risks? On a prominent script of nutrigenomics. British Journal of Nutrition 101: 307-316.

Wetherell, M. and Potter, J. (1988). Discourse analysis and the identification of interpretative repertoires. In: Antaki, C. (ed.) Analysing everyday explanation: a casebook of methods. Sage, London, UK, p. 168-183.

Obesity and costs of low energy density foods: a case for state against consumer responsibility

V. Sodano and F. Verneau
Department of Agriculture, University of Naples Federico II, via Università 96, Portici, Italy; vsodano@unina.it

Abstract

The paper, after a short review of the different (economic, ethical and political) pros and cons of state interventions against obesity, presents the results of a survey aimed at testing the inverse relationship between energy density of foods and their energy cost. The econometric model confirms the presence of an inverse relation between food product prices and energy value, and hence represents further confirmation of the importance of factors of strict economic relevance in leading to the emergence and progressive increase in obesity worldwide. Results bring evidence of socio-economic causes of obesity, with the poorest swathe of society at higher risk, and add arguments in favour of state interventions, according to the stewardship perspective.

Keywords: obesity, energy density, diet cost, state intervention

Introduction

Obesity rates have doubled over the past 30 years. Globally, 1.4 billion people are overweight and 500 million obese; by 2030, 50-60% of the population in many countries is on target to be classified as obese (WHO, 2007). Obesity is the fifth-highest global risk for death, accounting for at least 2.8 million adult deaths a year (Merrill Lynch, 2012). Unaffordable and potentially overwhelming costs of Health National Systems (HNS) have prompted governments to intervene. The debate on government's role in fighting the obesity epidemic opposes the supporters of state intervention (the stewardship model) against the advocates of consumers' freedom of choice and the individual's autonomy. The paper introduces to the debate on the legitimation of the wide range of interventions against obesity and focuses on the necessity of removing some important constraints which hinder healthy food consumption choice, such as the lower price of high energy food compared to that of low energy. In particular, the paper presents the results of a survey carried out in Southern Italy aimed at testing the relationship between price, food energy density and nutritional claims for a selected group of food items.

The main conclusion of the paper is that the consumer communication policies alone are not sufficient to stop the obesity epidemic and that regulatory and market based instruments need to be used as well. In particular, there is a need for regulatory interventions aimed at reversing relative prices between obesogenic and healthy foods.

Legitimating policies to tackle obesity epidemics: ethical and economic issues

Over the last 25 years, in the USA, the growing cost burden of obesity – especially on NHS- has spurred a strong state action in different fields of intervention (Gostin *et al.*, 2009). Also the European Union (EU), although more recently, has defined a set of actions to stop the obesity epidemic (European Commission, 2007). Designing policies to fight obesity is a challenging task because obesity has multiple causes -ranging from individual lifestyle factors to general socioeconomic and environmental conditions- which occur in conjunction (Faulkner *et al.*, 2011; Sacks *et al.*, 2008). Effective policies should be able to deal simultaneously with all of them. As a consequence, when assessing the costs and benefits of interventions taken individually, benefits appear to be limited and uncertain. Moreover, because many kinds of intervention are aimed either at changing individuals' behaviour, or at regulating firms' activity, they raise concerns over an excessive state power which might threaten personal freedom.

In order to better clarify the problem of legitimation (at economic, political and ethical level) of policies against obesity, it is useful to classify the possible interventions with respect, first, to their sphere of action and, secondly, to the objectives that they may help pursue.

With respect to the sphere of action, three fields of intervention may be identified:

1. Socioeconomic environment, where policy actions are aimed at turning the obesogenic environment into a healthy one. Interventions in this field include; restriction on junk food (i.e. food rich in calories and with low nutritional value) advertising targeted to children; restriction on marketing of unhealthy food; rebalancing agricultural policy in such a way as to support the production of fruits and vegetables; mandatory nutrition labelling information for restaurants and retail establishments; urban planning aimed at providing public transport, walking and cycling environment and facilities for physical activity; facilitating the distribution of fresh produce; regulating food production, as for example banning excessive fatty and sweet foods, prohibiting particular food ingredients and imposing limits on the amount of fat and salt in certain products.
2. Lifestyle and eating behaviour, with policies aimed at directly influencing the behaviour of individuals in such a way as to reduce energy intake and increase physical activity. Interventions: beverage and food taxes, such as caloric sweetened beverage and fatty food tax; fruit and vegetables subsidies; income transfer healthy food; fitness tax credit; social marketing programs; school nutrition education programs.
3. Health services, with policies aimed at providing obese people with adequate health services and clinical interventions. Interventions: providing obese patients with adequate treatment; creating multidisciplinary teams made up of physicians, surgeons, nurses and other health professionals able to deal with the complex obesity related syndromes.

In order to understand the kind of legitimation (economic, political, ethical), which underpins these various kinds of intervention, it is useful to identify what further objectives, besides the mere objective of tackling obesity, interventions may help achieve (Table 1).

The suggestions presented in Table 1 may help address the problem of the acceptability/legitimation of obesity policy in liberal democracies. As long as interventions against obesity are geared towards the goal of economic efficiency, as in the case of the majority of interventions concerning the socioeconomic environment, they are easily accepted by the public and do not raise legitimation concerns. This is true also when a strict neoliberal perspective is endorsed. Interventions that contribute to a fairer income distribution may be opposed by neoliberals, but are still supported when an (at least minimal) welfare state is accepted. Instead, when dealing with interventions whose legitimization rests on explicit political and ethical considerations, doubts may be advanced as to their consistency with principles of a liberal democracy, and a case of paternalism may be alleged. When obesity interventions are seen as an important part of the general food safety and public health policies (as in the case of interventions number 5, 6, 12 and 13, in Table 1) they are legitimated on the same basis as such policies. Since a reasonable level of health is regarded as a conditio sine qua non for the fulfilment of the right to life, liberal states (Rajcz, 2008) usually foresee some form of public health policy (and food safety policy as well). Nevertheless, when ranging from one extreme form to another of liberal state, with one endorsing the libertarian and the other the collectivist perspective, the measure of intervention may vary considerably. According to the libertarian perspective the state ought to limit its power to guaranteeing the basic human natural rights: life, liberty and property. The collectivist perspective instead acknowledges the role of the state as guarantee of a wider range of human rights, from classical

Table 1. Obesity interventions: goals and legitimation.

	Objective besides obesity	Legitimation
Socioeconomic environment		
1. Restriction on advertising targeted to children	Ethical objective: state commitment to safeguard vulnerable people	Ethical
2. Restriction on marketing of unhealthy food	Economic efficiency: correcting market failure due to excess of market power	Economical/ efficiency
3. Rebalancing of agricultural policy	Economic objectives: correcting policy bias	Economical/ efficiency
4. Mandatory nutrition labelling information	Economic efficiency: correcting market failure due to imperfect information	Economical/ efficiency
5. Regulating food production	Food safety policy goals	Political/ human rights
6. Urban planning	Health policy goals	Political/ human rights
Lifestyle and eating behaviour		
7. Beverage and food taxes	Correcting free riding (obese people increase the costs of HNS and should pay for it)	Economical/ efficiency?
8. Fruit and vegetables subsidies	Economic Social justice -income redistribution	Economical/ equity
9. Income transfer healthy food	Economic Social justice -income redistribution	Economical/ equity
10. Fitness tax credit		Ethical?
11. Social marketing programs		Ethical?
12. School nutrition educ. programs	Education policy goals	Political/ human rights
Health services		
13. Health services for obese people	Health policy goals	Political/ human rights

civil and political rights to the second (economic, social and cultural rights) and third (solidarity rights, right to peace, right to clean environment) generation of human rights. When interventions cannot be qualified as food safety and health policy and therefore do not deal directly with the issue of human rights, as in the case of interventions number 1, 10 and 11 in Table 1, their legitimation is even more difficult to promote. In this case, there is a need for specific ethical frameworks, which may be supportive of public interventions (Have *et al.*, 2010; Mepham, 2010).

To solve ethical and political concerns raised by obesity interventions, which cannot be legitimated on the basis of solely economic considerations, a framework has been suggested under the name of stewardship model (Nuffield Council, 2007). The stewardship perspective, set forth by the Nuffield Report, sketches the boundaries of the acceptable level of obesity intervention for the liberal states of western tradition, identifying the following goals and constraints (Kersh *et al.*, 2011). Concerning goals, public intervention should: attempt to reduce risks for obesity that populations might impose on each other; reduce causes of obesity through legislation or regulation that create environmental conditions which sustain good health (e.g. access to healthy foods and opportunities to be physically active); emphasize attention to the health of children and other vulnerable sub-groups (e.g. those with disabilities); promote health not only by providing information but also with programs that help populations maintain exercise and healthy diets; make leading a healthy life easy; ensure that populations have access to services; strive for justice in health. In terms of constraints, interventions should: -not attempt to coerce adults to lead healthy lives; -seek to minimize interventions that affect important areas of personal life; develop and introduce interventions without the consent or participation of those affected.

When endorsing the stewardship framework all the interventions outlined in Table 1 turn out to be acceptable in a liberal democracy. Nevertheless, when the state exhibits a strong neoliberal attitude, interventions affecting the socioeconomic sphere may be preferred to those affecting personal behaviours or entailing high public health expenses. One goal, which may reach a wide consensus in society, is the availability and affordability of healthy food for everyone. Affordability is ensured by reasonable prices for food rich in nutrients but low in calories (fruit and vegetable, lean meats and fish) compared with cheap energy-dense foods (junk foods). It is worth noticing that the achievement of affordability is conditional to the success of the other interventions, and especially to informative and educational interventions. There may be many reasons why healthy food is expensive (high production costs, scarcity, high distribution costs, market power, and so on) and many policies available to make it more affordable. An important step in obesity policy design is to determine whether energy-dense food options are actually cheaper than less caloric options (Drewnowski and Darmon, 2005; Monsivais and Drewnowski, 2007). In the next section a case study is presented in which we aim at assessing the relationship between energy density and energy cost for a sample of food products in a region of southern Italy.

Case study: energy density and energy cost

The survey was carried out in April 2012 in the Naples area at Italy's three main food retailers (Auchan, Carrefour and IperCoop). The following product categories were investigated: yogurt (n=236), frozen vegetables (n=204), savoury snacks (n=75), sweet snacks (n=146), breakfast biscuits (n=236), fresh-cut vegetables (n=91), fresh-cooked vegetables (n=13), ready-to-eat dishes (n=35). Among the surveyed items private labels ranged from a minimum of 19% for savoury snacks to a maximum of 68% for fresh-cut vegetables. For each product category prices and label information were recorded for all the items found on the shelf, with the exclusion of products affected by sale promotion. Data were collected on a total of 967 items.

The information regarding some indicators is reported in Table 2. The calorie content per 100 g represents the energy value index, while standard deviation shows the dispersion from the average calorie content in each category. A small standard deviation means that the eight product categories are fairly homogeneous with respect to the calorie. With reference to the six food categories for which a sufficient number of observations were available, an independent samples t-test was performed in order to test differences in the mean energy density between products with and without nutritional claims which explicitly or implicitly referred to the concept of 'lightness' and low energy density. The null hypothesis (with a p-value <1%) of the same energy density was rejected only in the case of yogurt

Table 2. Products cost and energy cost.

	Kcal/100 g	Standard deviation	P/100 g	P/100 Kcal
Fresh-cut vegetables	30.2	24.2	1.00	4.36
Frozen vegetables	50.4	36.1	0.54	1.52
Fresh-cooked vegetables	55.9	31.2	0.63	1.47
Yogurt	82.9	23.6	0.58	0.61
Ready-to-eat	145.2	27.6	1.20	0.88
Sweet snacks	410.0	36.2	0.67	0.17
Biscuits	457.4	25.5	0.47	0.10
Savoury snacks	522.7	28.0	1.16	0.19

and biscuits. Importantly, in the case of the two categories of sweet and savoury snacks, no statistically significant difference was found in the energy content with respect to the presence of nutritional claims.

In order to assess the existence of a relation between food price and energy value, a hedonic econometric model of price was used. This model determines the selling price of a commodity as a function of the material and non-material attributes that comprise it.

Starting from the complete dataset of 967 observations concerning eight distinct retail categories, our empirical survey selected the 367 observations relative to all the products in the dataset sold with the retailer's brand. Analysis was thus limited to private label products. The choice was dictated by the need to arrive at a more homogeneous set of items with respect to the large number of variables (such as, for example, the presence of organic products or products targeted with specific price strategies) which could mask the relation between food price and calorie content. Starting from the set of data available, we obtained the matrix of potential regressors, each of which containing information on a characteristic of the commodity which was directly obtained from the label or from the observation of the product (nutritional information, format, presence of nutritional claims, and so on).

In the hedonic model proposed, the dependent variable *P* was defined as the mean price in Euros per 100 gram for each article. The independent variables which proved statistically significant were the following:

- TW = overall product weight in 100 g of each item.
- NutClaim = dichotomous variable which identifies whether the product makes claims explicitly or implicitly connected with the idea of well-being, lightness and physical shape (1 = claim; 0 = no claim).
- D_{Funz}= dichotomous variable which identifies whether the product has added nutritional elements that might be termed *functional*, such as vitamins, mineral salts, fibre or probiotics, given the value 1, and 0 otherwise.
- Kcal/100 g = indicates the content in kilocalories per 100 gram of product.

The hedonic model, estimated with the ordinary least squares (OLS) method, was expressed as follows:

$$P_{100g} = \beta_0 + \beta_1 TW + \beta_2 Kcal/100g + \beta_3 Func + \beta_4 NutClaim + \varepsilon$$

The model was initially estimated by using Box-Cox transformation to test alternative functional forms. As the likelihood test ratios led us to reject both the double-logarithmic and semi-logarithmic form, we opted for a linear functional form. The latter also displayed some problems which were highlighted by White's test, which is discussed below. However, it allowed us to interpret the coefficients directly as implicit prices and obtain information concerning the influence of each individual characteristic (Table 3).

The model explained overall about 37% of total variance (corrected R^2=0.3709) and the signs of the coefficients were consistent with expectations. Submitted to the White test, the model showed the persistence of inherent heteroskedasticity in the data. The test was clearly run in the unknown variance mode, and to reach a consistent estimate of the coefficient values and the standard errors we proposed White's corrected matrix of variance and covariance. The model was thus estimated with standard errors robust to heteroskedasticity.

The dummy associated with the presence of functional elements showed a positive coefficient, indicating that the functional nature of food results in a price increase. The impact on the sale price, about 0.11 €/100 g, suggests that the functional choice represents a successful strategy for firms able to guarantee a significant price premium. The variable Kcal/g showed a negative sign. This result was

Table 3. Hedonic Model (OLS); Dependent Var.: Pg Robust standard errors for heteroskedasticity.

	Coefficients	Std. error	t-ratio	*P*-value
Const	0.56035	0.0623686	13.7946	<0.00001
TW	-0.0834	0.061829	-13.7924	<0.00001
NutClaim	0.114545	0.0392555	2.9179	0.00374
Kcal/100g	-0.0310517	0.00692907	-4.4814	<0.00001
Func	0.11196	0.0500429	2.2373	0.02587

Mean dependent var.	0.6434	SQM dependent var.	0.3705
Sum of squared residuals	0.3135	SE regression	0.2939
R^2-	0.3777	Corrected R^2	0.3709
F(4, 363)	48.88795	*P*-value(F)	<0.001

obtained by looking overall at the items belonging to all eight categories in question. Although the coefficient is very low, the result confirms the inverse relation between food energy value and cost per unit of kilocalorie. The coefficient of the product format also resulted in a negative sign. In other words, the greater the weight of the package sold, the lower the unit price of the product. The dummy associated with the presence of nutritional claims that evoke well-being and lightness had a positive coefficient. Hence the products which have such claims receive a price premium from the market.

Conclusion

In liberal societies, there is a strong resistance towards those obesity interventions deemed to limit personal and economic freedom. The paper demonstrated that interventions aimed at removing economic restraints to healthy diets might be accepted also in liberal contexts, especially when a stewardship perspective is endorsed. The case study brought evidence of some economic factors responsible for the obesogenic environment, which is responsible for the progressive increase in obesity worldwide. In particular, the econometric model showed the higher prices of healthy food choices, described in terms of low energy density, small servings and disclosed healthy characteristics. A general conclusion is that, when designing obesity policies, a priority should be given to those interventions able to lowering the price of healthy foods, such as: an adequate competition policy, for the removal of market power which enables firms to receive a price premium for healthy foods; the rebalancing of agricultural policy, in such a way as to redirect public resources towards non obesogenic sectors; adequate income supports for the poorest swathe of society.

References

Commission of the European Communities (2007). White paper on: strategy for Europe on nutrition, overweight and obesity related health issues. Available at: http://ec.europa.eu/health/archive.

Drewnowski, A. and Darmon, N. (2005). The economics of obesity: dietary energy density and energy cost. Amercan Journal of Clinical Nutrition 82: 265-273.

Faulkner, G.E., Grootendorst, P., Nguyen, V.H., Andreyeva, T., Arbour-Nicitopoulos, K., Auld, M.C., Cash, S.B., Cawley, J., Donnelly, P., Drewnowski, A., Dubé, L., Ferrence, R., Janssen, I., Lafrance, J., Lakdawalla, D., Mendelsen, R., Powell, L.M., Traill, W.B. and Windmeijer, F. (2011). Economic instruments for obesity prevention: results of a scoping review and modified Delphi survey. International Journal of Behavioural Nutrition and Physical Activity 8: 2-14.

Gostin, L.O., Pomeranz, J.L., Jacobson, P.L. and Gottfield, R.N. (2009). Assessing laws and legal authorities for obesity prevention and control. Journal of Law Medicine and Ethics 37: 28-36.

Have, M., De Beaufort, I.D., Mackenbach, J.P. and Van der Heide, A. (2010). An overview of ethical frameworks in public health: can they be supportive in the evaluation of programs to prevent overweight. BMC Public Health 10(638): 1-11.

Kersh, R., Stroup, D.F. and Taylor, W.C. (2011). Childhood obesity: a framework for policy approaches and ethical considerations. Preventing Chronic Diseases 8: 1-5.

Mepham, B. (2010). The ethical matrix as a tool in policy interventions: the obesity crisis. In: Gottwald, F.T., Ingensiep, H.W. and meinhardt, M. (eds.) Food Ethics. Springer Science – Business Media, Dordrecht, the Netherlands, pp. 17-29.

Merril Lynch Bank of America (2012). Globesity: the global fight against obesity. Equity | Global | ESG & Sustainability 21 June 2012. Available at: http://www.foodpolitics.com/wp-content/uploads/Globesity-Report_12.pdf.

Monsivais, P. and Drewnowski, A. (2007). The rising cost of low-energy-density foods. Journal of American Dietetic Association 107: 2071-2076.

Nuffield Council on Bioethics (2007). Public health: ethical issues. London (GB): Nuffield Council. Available at: www.nuffieldbioethics.org.

Rajcz, A. (2008). A liberal approach to the obesity epidemic. Public Affairs Quarterly 22: 269-288.

Sacks, G., Swinburn, B. and Lawrence, M. (2008). Obesity policy action framework and analysis grids for a comprehensive policy approach to reducing obesity. Obesity Reviews 10: 76-86.

World Health Organization (2007). The challenge of obesity in the WHO European Region and the strategies for response. Available at: www.euro.who.int.

Can healthy eating at school be considered a human right?

B.E. Mikkelsen
Meal Science & Public Health Nutrition, AAU-MENU, Aalborg University, Department of Development and Planning, A.C. Meyers Vænge 15, 2450 Copenhagen SV, Denmark; bemi@plan.aau.dk

Abstract

The increasing focus on unhealthy eating and the growing interest in providing a foodservice at school have fuelled a debate on its normativity. Should children enjoy the right to a nutritious meal, should they be protected against offers that might be harmful to their health or threaten their possibilities of enjoying a balanced diet? Could this potential right be compared to the well established basic right to enjoying education? This paper explores whether the broad framework of human rights can be invoked to ensure healthy eating in school. It gives a brief account on provisions of food and health from an international human rights framework and discusses the opportunities they offer for taking action to improve eating at school. In addition, the paper presents a short overview of provisions on healthy eating at school that can be found in intergovernmental policy documents. The paper concludes that there is broad support in both intergovernmental policy documents and provisions in the broad framework of human rights on the necessity for providing opportunities for healthy eating at school. It concludes that since children should enjoy, according to human rights framework, the right to adequate nutritious food and the highest attainable standards of health and reach their full physical potential, this framework should be invoked and translated into concrete strategies, policies and regulation at national, regional, local and school levels. The paper provides recommendations on how this could be set about to secure healthy school food environments.

Keywords: school meal programs, obesity, unhealthy eating, young people

Introduction

There is growing evidence suggesting that changes in the school food environment might positively impact the eating patterns of young people (French *et al.*, 2004; Hendy *et al.*, 2005) as well as influence obesity prevalence among this age group (Campbell *et al.*, 2001; Doak *et al.*, 2006). As such, improved school meal programmes are increasingly recognized as an important contribution to the promotion of good heath among children and adolescents.

Young people spend long hours in school for up to 10 years of their life and in many countries across the world, including the United Kingdom, Brazil, India, and the United States (Drezel and Goal, 2003; Gatenby, 2007; Hirschman and Chriqui, 2012; Sidaner *et al.*, 2012) school meal programmes have traditionally been an important part of school life. But faced with the increasing prevalence of obesity and overweight among children and adolescents in many countries it has become obvious that it should not only be about providing food but also about providing food of high sensory and nutritional quality. As such, the call for healthier eating in school has fuelled a debate on normativity and the quality of food served at school. For instance: Should children at school enjoy the right to a nutritious meal and should they expect to be protected against obesogenic food environments when at school? These types of questions are being asked increasingly.

Although school meal programmes offered by the public can be found as an integral part of school life in many countries it is important to note that traditions and school food cultures vary. In countries like Norway, Denmark, Iceland, Austria, Netherlands and some parts of Germany, food at school has not traditionally been a part of school life. Although in some public schools meal programs are emerging in these countries they tend to depend on a *consumership model* where school meals are purchased from a point of sale and not provided free as an integral part of the school day. In such cases, choosing the food provided by the school is seen as the choice of the consumer as in any other market environment. In other countries such as Sweden and Finland school meal programs are provided as part of a 'free meals for all' model. In these schools meal provision is seen as an integrated part of the welfare provision in an approach that can be referred to as a *citizenship model*. A number of intermediate approaches are also found in between these approaches. Transnational food provision programs in addition to national programmes are in effect, which is the case for the school milk scheme and more recently the school fruit scheme that are promoting these products at school as part of an European community aid program.

With longer school days being one of the reasons for the on-going attempts to reform school systems across Europe and with the on-going debate on unhealthy eating patterns of young people a new debate on what food at school should be about has begun to emerge. This paper explores whether the broad framework of human rights and international policy documents can be invoked to ensure healthy eating in school. It presents some of the provisions in international policy documents and in the human rights framework and discusses how they might be applied to improve eating at school.

Food, nutrition and human rights in the context of school

The idea that food and nutrition can be viewed from a rights perspective and that the protection of the right to good nutrition as well as food security can be regarded as a human right have gained increasing momentum over the past decade (FAO, 2004; Kondrup, 2004; Oshaug *et al.*, 1994). In Sweden a movement on the right of all children to good and nutritious food has begun recently (http://allebarnsratten.se/om-oss).

Traditionally, human rights in relation to nutrition and young people were thought in terms of the right to protection against undernourishment. However, with the on going nutrition transition where under nutrition and over nutrition seem to coexist the human right framework is being applied and thought about in relation to the right to protection against obesogenic environments.

Historically, school food programs were started to address inadequate nutrition, improve school attendance and enhance children's ability to learn. However, today, most countries are faced with the full spectrum of malnutrition ranging from hunger to over-consumption (Ashe and Sonnino, 2012), and are realizing that these conditions are taking place in tandem among young people. As a result there is growing recognition that the broad framework of human rights should be applied more systematically to eating at school and that schools should be regarded as 'protected places' where children should expect to enjoy certain standards that are supportive of good health, including access to nutritious foods.

There is a comprehensive legal basis in international human rights systems that can be invoked in the case of healthy eating at school. The framework of human rights began to evolve after the second World War as part of the attempt to secure and ensure citizens their basic rights as human beings. The Universal Declaration of Human Rights (United Nations, 1948), the European Convention on Human Rights (COE, 1950) and the Convention on the Rights of the Child (United Nations, 1989) are all examples of provisions that define the framework of human rights. In many sectors of high societal importance such as occupational safety, protection of the environment and women's rights the framework, despite the non-binding nature of its provisions, has played a significant role in turning these provisions into binding national regulations.

It is important to note that there is substantial philosophical disagreement on what the reach and status of human rights should be. For instance, should they be regarded as being universal and always existing or should they be regarded as socially constructed and thus dependant on their social and cultural contexts? In the context of this paper it will simply be understood as a common but abstract moral language for food at school – a language that ideally should inform public decision making simply because there are a large number of examples and experiences in practising such forms of morality.

The Conventions on the Rights of the Child (CRC) is of particular interest with regard to the rights of children and adolescents, since the parties of the convention are obliged to respect their rights to nutrition, health and education. For instance, according to the CRC, children have the right to the 'highest attainable standard of health' and 'adequate nutritious foods and clean drinking water'. Regarding responsibility, the CRC refers to 'institutions, services and facilities responsible for the care or protection of children'. In addition, the European Social Charter (ESC) is important since it refers to the right to protection of health. The ESC also states that there should be health policies for preventing illnesses and that there should be a healthy environment for all citizens.

In addition to the provision in texts from the international human rights system there is a growing number of policy documents adopted by intergovernmental bodies. Over the past .... (indicate number of decades here, if not drop the s in the word decades) decades these policy documents have increasingly been looking into the potential of settings such as the school to play a more important role in the provision of opportunities for healthy eating. These international policy documents include the WHO Istanbul charter (2006), the EU white paper on diet and physical activity (2007), the resolution of the Council of Europe on healthy eating at school and the action plan adopted by the Nordic Council of Ministers in 2007 on diet and exercise.

Application of human rights to food and nutrition at school

Although the human rights framework does have implications for the work on improving the quality and nutrition of meal programmes, it could be assumed that the pathway is a rather indirect one. As such, human rights serve as a foundation – an abstract common moral language – that can be referred to and used in the on-going debate on the future of school meal programs as well as in the discussion of future strategies for promotion of healthier eating among young people. As such, human rights are broad in scope and need to be translated and applied in the concrete work that takes places at national, regional, municipal and school levels. School food and nutrition policies and action plans are good examples of how human rights can be translated into action at school level. The literature provides good evidence on how to develop and maintain such policies and how to make them a concrete instrument in the day to day work with improvement of school meal programs. And in many countries the understanding of school as a protected place can be found to be mirrored in concrete legal provisions such as nutritional guidelines and regulations for school meals.

Application of a human rights perspective to eating at school should not limit itself to only looking at food provision. Food and eating is not only important from a physiological point of view but also play an important role in the formation of life skills of young people. It is important to see school meal programs in a broader perspective as part of the health promoting efforts taken in many schools. The Whole School Approach (Wang and Stewart, 2012) and the closely related framework of the Health Promoting School (Lee, 2009) offer a comprehensive framework on how the work with food and eating can be got about in practical terms at school level. More importantly the WSA underlines the importance of transdisciplinary cooperation and commitment from teachers, food service staff, school administration, parents and students in order to maintain optimal school meal programs.

Conclusion

The right to adequate nutritious food, the highest attainable standard of health and the right to be able to reach full physical potential could be considered basic human rights for children and these human rights are firmly established in international and European human rights instruments. As public places and as societal cornerstones school has a special obligation to make sure that these rights are respected and reflected in the way school meal programmes are implemented.

School has a tradition for being a protected place and to possess an ethos that includes certain levels of quality in the services it provides. Studies have shown that the nutritional quality of school lunches is often significantly better when it is provided by school than alternatives such as lunch packs brought from home (Sabinsky *et al.*, 2010; Stevens and Nelson, 2011; Clark, 2009).

The intergovernmental policy document and recommendations as well as the human rights documents presented here provide a good case for strengthening the efforts to make schools a better place in terms of providing opportunities for healthier eating. More recently the United Nations has appointed a special rapporteur on this issue and, in general, the debate revolves around not only availability and accessability of food but also around the notion of adequacy of food. Adequacy implies that foods should meet current nutritional recommendations.

The broad framework of human rights has been acting as a foundation for policies and regulations in many fields such as gender equality and occupational safety and health. Within these policy areas the foundation in the human rights documents has been functioning as one of the drivers that eventually lead to translation into policies at national level. This transfer of ideas from the realms of human rights to concrete legal provisions and strategies at national level should be replicated in the case of the right to healthy school food environments and school meal programs.

References

Ashe, L.M. and Sonnino, R. (2012). At the crossroads: new paradigms of food security, public health nutrition and school food. Public Health Nutrition 16(6): 1020-1027.

Campbell, K., Waters, E., Meara, S.O. and Summerbell, C. (2001). Interventions for preventing obesity in childhood. A systematic review. Obesity Reviews 2: 149-157-

Clark, M.A. (2009). Nutritional quality of the diets of US public school children and the role of the school meal programs. Journal of the American Dietetic Association 109,2(Suppl.): S44–S56.

Doak,C.M., Visscher, T.L.S., Renders, C.M. and Seidell, J.C. (2006). The prevention of overweight and obesity in children and adolescents: a review of interventions and programmes. Obesity Reviews 7: 111-136.

Dreze, J. and Goyal, A. (2003). Future of mid-day meals. Economic and Political Weekly 38(44): 4673-4683.

European Social Charter (revised) (1996). CETS No. 163. Available at: http://conventions.coe.int/Treaty/Commun/QueVoulezVous.asp?NT=163&CL=ENG.

FAO (2004). Voluntary guidelines to support the progressive realization of the right to adequate food in the context of national food security. (Adopted by the 127th Session of the FAO Council. 22-11-2004).

French, S.A., Story, M., Fulkerson, J.A. and Hannan, P. (2004). An environmental intervention to promote lower-fat food choices in secondary schools: outcomes of the TACOS study. American Journal of Public Health 94: 1507-1512.

Gatenby, L.A. (2007). Nutritional content of school meals in Hull and the East Riding of Yorkshire: a comparison of two schools. Journal of Human Nutrition and Dietetics 20: 538-548.

Hendy, H.M., Williams, K.M. and Camise, T.S. (2005). 'Kids choice' school lunch program increases children's fruit and vegetable acceptance. Appetite 45: 250-263.

Hirschman, J. and Chriqui, J.F. (2012). School food and nutrition policy, monitoring and evaluation in the USA. Public Health Nutrition, Public Health Nutrition 16(6):982-8.

Kondrup, J. (2004). Proper hospital nutrition as a human right. Clinical Nutrition 23: 135-137.

Lee, A. (2009). Health-promoting schools: evidence for a holistic approach to promoting health and improving health literacy. Appl Health Econ Health Policy 7(1): 11-17.

Oshaug, A., Eide, W.B. and Eide, A. (1994). Human rights: a normative basis for food and nutrition-relevant policies. Food Policy 19: 491-516.

Sabinsky, M., Toft, U., Andersen, K.K., Mikkelsen, B.E. and Tetens, I. (2010). Ernæringsmæssig evaluering af skolemads betydning for elevers kostindtag til frokost (projekt EVIUS – effektvurdering af Interventioner omkring frokost for børn og unge i skoler) [Nutritional evaluation of school administrations' impact on students' dietary intake at lunch (project EVIUS – effect evaluation of interventions around lunch for children and young people in schools)]. Aalborg University, Aalborg, Denmark.

Sidaner, E., Balaban, D. and Burlandy, L. (2012). The Brazilian school feeding programme: an example of an integrated programme in support of food and nutrition security. Public Health Nutrition 16: 989-994.

Stevens, L. and Nelson, M. (2011). The contribution of school meals and packed lunch to food consumption and nutrient intakes in UK primary school children from a low income population. Journal of Human Nutrition and Dietetics 24(3): 223-232.

United Nations (1948). Universal declaration of human rights. Adopted on December 10 by the General Assembly of the United Nations.

United Nations (1989). CRC – convention on the rights of the child. Adopted by United Nations. Available at: www.ohchr.org/EN/ProfessionalInterest/Pages/CRC.aspx.

Wang, D. and Stewart, D. (2012). The implementation and effectiveness of school-based nutrition promotion programmes using a health-promoting schools approach: a systematic review. Public Health Nutrition 16(6): 1082-1100.

Potential of transformative consumer learning for governance for animal welfare by public catering?

M. Mikkola[1] *and A. Morley*[2]
[1]*University of Helsinki, Ruralia Institute, Lönnrotinkatu 7, 50100 Mikkeli, Finland;* [2]*Cardiff University, Sustainable Places Research Institute, 33 Park Place, Cardiff CF10 3BA, United Kingdom; minna.mikkola@helsinki.fi; morleya@cardiff.ac.uk*

Abstract

Public catering, defined as the institutional provision of food by state, or quasi-state institutions, carries a political responsibility for progressing sustainable food systems. Enhancing animal welfare in conventional production is an often ignored strategy for public procurement however, sidelined by strategies such as increasing vegetable and organic food consumption. However, if public caterers are to represent their customers' societal views, what kind of approach could inform their procurement strategy regarding animal welfare? This paper uses a qualitative research approach to investigate consumers' advanced everyday transformative learning potential regarding animal welfare issues, highlighting consumer dispositions as defensive, critical, self-interested and change oriented. The framing of consumer views by transformative learning offers a dynamic view for developing demand for increased animal welfare within supply chains, although resistant consumer positions remain a challenge for proponents of transformative education. Advanced ethical consumer dispositions are claimed to legitimate the potential for the governance for animal welfare by public caterers and to represent a possible future orientation for sustainable food procurement practices.

Keywords: animal ethics, markets for sustainability, advancing policy goals, social learning

Introduction

'As we use animals for our food we are also responsible for them' concludes a young adult reflecting about consumers' need to know about animal welfare in the food chain. While consumers often express relative concern for animal welfare, neither their level of understanding nor consumption behaviour regarding animal husbandry practices seem to unfold respectively (Barnes *et al.*, 2009; Ingenbleek and Immink, 2011). The same approach appears to be shared by institutional consumers such as public caterers, deploying public funding to provide wholesome meals in schools, hospitals and governmental offices (Mikkola, 2009), despite increasingly being considered as having a political responsibility to promote sustainable development (Meadowcroft, 2007) and sustainable food systems (ICLEI, 2008; Morley *et al.*, 2012). While recommended vegetable consumption can be seen as a strategy of withdrawal from animal-based food chains, and organic meat consumption as a strategy for animal welfare (ICLEI, 2008; Lappalainen, 2012), neither of these strategies addresses animal welfare issues in predominant conventional animal-based food supply chains. Whilst animal husbandry improvements are being promoted through European research and regulations (Miele, 2010), a greater intensity and leadership appears possible through the interactive clarification of goals by societal self-steering (Meadowcroft, 2007) for animal welfare. When looking for government and industry alliances to promote animal welfare (Ingenbleek and Immink, 2011), public caterers represent by their possible intervention activities the political meta-object of sustainability, and thus have the potential to trade as one of the governing centres of commercial power within the market (Meadowcroft, 2007). As sustainable development is essentially a democratic process, implying social learning for sustainability, it benefits from societal interactions to help clarify its goals through social and scientific discourse (Meadowcroft, 2007; Wals, 2010).

This paper aims to understand how public caterers may promote advanced ethical consumer positions in terms of animal welfare by mapping their customers' perceptions and behaviour as legitimizing grounding for formulations of market governance. Rather than investigating consumer decision-making patterns (Ingenbleek and Immink, 2011), we explore consumer discourses about farm animal welfare through the enabling frame of everyday transformative learning (Illeris,2007; Kegan, 2009; Mezirow, 2009; Wals, 2010). The paper deploys the theoretically advanced notion of transformative learning to shed light on a complex issue of learning about animal welfare as a societally binding development. Our study describes a continuum of consumer positions, from non-transformational to ethical change oriented, as represented among higher education students in Finland, which offer a grounding for democratic and progressive ethical public procurement.

Methodology

The notion of transformative learning inherently signifies change and is understood as a process entailing a critical reflection of (more justified) beliefs, actions and dispositions (Mezirow, 2009). Transformative learning entails demanding conceptual, ethical and emotional reflections by the learner, whereby meaning perspectives are reviewed (Illeris, 2007). It is not only the cognitive content which changes but also the framing of this content (Kegan, 2009; Mezirow, 2009). Importantly, these kinds of 'learning leaps' or slowly proceeding learning processes imply the possibility of changing the learner's position in terms of his/her aims and activities. While transformative learning seems possible in everyday life contexts (Illeris, 2007), its conceptual character as an amorphous and contingent on-going process may be interpreted as an ante-narrative (Boje, 2001). Analytically the participants were classified within their process of transformative learning about farm animal welfare through levels from self-distancing non-reflection to self-inclusive and engaged reflection.

A methodology for group interviews was employed for the study (Kvale, 1996). Morgan (1997, 20) emphasises the importance of understanding about what, how and why participants think about an issue as they do. Focused interviews with groups (Morgan, 1988, in Kvale, 1996, 70-72) can also be seen to represent wider social reality, as '...in real life people talk most of all about what others talk about...' (Bakhtin, 1981: 338).

Six group interviews, with a total of eleven female and eight male students, were conducted in university canteens in Helsinki during 2008 and 2009. The discussions were recorded and transcribed verbatim. The groups' discourse dealt with (organic) food consumption in general and within their educational contexts, whereby the notion of organic was used as a clue proxy to animal welfare. The qualitative analysis condensed the core discursive meaning (Kvale, 1996; Parker, 1992) of the consumption of animal products. Organic consumption was interpreted as a successful transformation, while non-consumption signified a transformative process not yet started or still partial. These two discursive approaches were further divided into two subgroups each to address the different reasons for particular kind of consumption behaviour.

Results

The analysis identifies non-consumers as expressing either delimiting or critical ambivalence towards organic animal product consumption. Organic consumers indicated either self-seeking interests or change-oriented disposition towards animal welfare through their animal product consumption. The discourse is shortly illustrated by quotes of the groups' members.

Non-organic consumers: delimiting and critical ambivalence towards animal products

Delimiting ambivalence was exhibited by some participants who tended to say little about organic food and were clear about not consuming it. They framed organic food as an issue of human health, costly food and excessive economic benefits for businesses. They also articulated a narrow discourse about organic animal welfare while simultaneously invalidating that as a non-issue. In spite of the stated belief about health quality of organic animal products they were not consumed due to lacking counterevidence about the hazards of the conventional diet. Whilst the young respondents emphasised to become organic consumers on their own initiative they also reported to resist the prompting to organic consumption. They explained their organic consumption decisions as conditioned by wealth and framed themselves as 'free to choose', with no reflection about animal welfare practices.

> Businesses can sell better to their customers if they call the food organic so this can be contested because businesses are profit focused...I don't have personally very much sympathy for animals, I have a materialistic world view, but they're biological machines just like humans...I'm not very familiar with this, and I buy very little organic food, if any at all...If I won the lottery, I could become an organic consumer but only for selfish reasons.(G5,2).

Critical ambivalence was openly expressed by the participants who reflected their food consumption more profoundly while stating not to consume organic animal products. There seemed to be no 'victorious' evidence about organic superiority in terms of human health, the environmental or animal welfare. In principle, purchases were to support the humane treatment of animals. However, as humans and animals were not considered equal, ultimately for a meat eater the matter was deemed not of that great a concern. In fact, maltreatment of animals was considered a very generic condition and therefore seemingly redundant. Despite this, campaigns for animal rights were worth supporting, as intensive producers were not viewed sympathetically. However, participants expressed interested to learn more about the issue in a critical and factual manner allowing them to construct their own views about organic food instead of 'flatly' relying on labels. While viewing organic animal products in the canteens positively, there was no willingness to pay more. Finally, participants criticized the status of the organic issue among other more important ones.

> Obviously I'm an animal killer too so it's hypocritical to...want happy chickens...it was terrible to see how animals drowned in their own excrement in those pens... it's not humane... (G 2,2).

Organic consumers: self-centred and change-oriented interests

Self-centred interests in organic animal foods were expressed by participants, who afforded to buy organic animal products and who accounted for their organic consumption as a way of improving their own wellbeing. Simultaneously they remained somewhat detached from animal welfare and other more global food issues. The participants based their views on the scientific understanding of organic food, which meant primarily personal health and self-gratified motives. To justify purchases like expensive organic milk, the beneficial features of organic food should be made clear. These students criticised animal treatment in intensive production because the industrialised food system seemed to risk potentially disastrous trajectories such as resistant strains of human pathogens. Here genetic engineering was viewed as a human rights issue, contesting the organic position. Some of these students had also lost their commitment to make the world a better place, as it was deemed 'hopeless'; in other words they returned to their self-interest. Furthermore, for these students it seemed plausible that the meals in the canteen contained local, organic and fairly traded food.

> Organic milk is healthier and better...animals are given antibiotics which get into me and give bacteria resistance... When I buy organic it slows down the development of genetic modification... If they would serve here organic milk I'd like to have that... I don't look at the receipts in that way, I just pay the bill (G3,1).

Change-oriented dispositions were adopted by participants basing their accounts on a societally motivated scientific worldview. Politically oriented activity against the industrial and intensive food system, with the aim of making a better world, were expressed through organic consumption choices. Organic animal products, highest up in the food chain, were seen as a way to consume clean, chemical-free food that limits animal suffering and induces satisfaction in consumers. If organic food had a wider market share, possibly its health effects could become visible at the population level. Although organic agriculture and the environment had rather blurred connections, it was rather clear that in organic farming animal welfare and biodiversity were superior to conventional systems. However, it was recognised that the reference base of organic labelling was not quite understood. Ecology was regarded as a fun 'can do' and community matter to promote a 'happy future'. Although organic food in general was perceived as expensive, it was suggested to share income more equally across the chain. In terms of global problems, the collateral development of genetic modification and organic agriculture seemed desirable. Organic animal products in catering demonstrated ethical approaches and were claimed to induce appreciation among customers. However, not many choices of this nature seemed to be available in the student canteens.

> ...knowledge increases the pain...that I don't eat meat is a political statement against that for my sake, in a way the poor animals don't need to be raised and produced for food...if it says organic do we just trust in it or what is it actually...the reality of animals is hardly the one depicted on the milk cartons... (G4,1).

Discussion and conclusions

The transformative learning framework provides a continuum from non-transformative, even change resistant consumer dispositions to transformed and ethically more advanced ones. This framing may potentially explain how consumers expressing delimiting ambivalence to organic animal products are not interested in animal welfare; they actually dislike the information and refuse its significance (Illeris, 2007; Ingenbleek and Immink, 2011; Lappalainen, 2012; Miele, 2010). The framing also sheds light on how critically ambivalent consumers may learn about animal maltreatment and express their strong regrets about it while simply positioning themselves as meat eaters. Moreover, these consumers accept even scandalous events through reframing them as ordinary and disparaging their significance as issues in need of their attention and repositioning. Furthermore, self-centred consumers may mainly focus on their own wellbeing while understanding about and distancing themselves from systemic global problems and their contradictory and contingent solutions. Furthermore, there is the option for a consumer to refocus one's interest in one's own wellbeing after having made disappointing efforts for the betterment of the world. Finally, the change oriented consumers accepted their responsibilities for the animal welfare within the systemic structures of conventional and organic animal husbandry in terms of the market, although they recognised their limited expertise and possibly compromising practices in evolving organic animal husbandry practices. The notion of transformative learning obviously contains potential to understand consumer behaviour in everyday contexts. However, as the transformations are on-going and contingent, they may unfold in many ways, including previous behavioural practices.

Intriguingly, these ordinary consumer dispositions were not typically based on a detailed understanding of animal husbandry practices but on assumptions of different levels of animal welfare, mainly learned through the media and labelling schemes (Ingenbleek and Immink, 2011; Lappalainen, 2012; Miele,

2010). The basic approach seemed to be the inclusion or exclusion of oneself within the food chain (Wals, 2010); except the final change oriented consumer disposition, all other dispositions made efforts to different extents to detach the consumer's responsibility from the food chain and move that to businesses and wider society.

This type of framing of consumer understanding of ethical animal husbandry lends itself to the generation of developmental options in terms of consumer behaviour. Instead of 'fixed' attitudes, a more dynamic and positive relational view for future change is offered by the notion of everyday transformative learning (Illeris, 2007). However, the difficult part implied by this approach is the implication for learning by defensive, critical and self-centred organic consumers. Finally, even those with change orientation may benefit from more advanced market communication about animal welfare (Lappalainen, 2012).

This approach, however, suggests that public caterers have a legitimate potential to govern the development of advanced ethical consumer positions (Meadowcroft, 2007) regarding animal products. It is also possible that this kind of ethical position by public catering will be to some extent appreciated by consumers from the resistant to change end of the continuum, although currently the willingness to pay just for animal welfare seems more likely to rest upon the self-centred and change oriented consumers. The institutional canteen can help move animal welfare issues from the responsibility of the closed supply chain to the open market, as an issue committed to by reflective consumers and guided by the state.

References

Bakhtin, M. (1981). Discourse in the novel. In: Holquist, M. (ed.) The dialogic imagination. University of Texas Press, Austin, TX, USA, p. 259-422.

Barnes, A., Vergunst, P. and Topp, K. (2009). Assessing the consumer perception of the term 'organic': a citizens' jury approach. British Food Journal 111(2): 155-164.

Boje, D.M. (2001). Narrative methods for organizational and communication research. Sage Publications, London, UK, 152 pp.

ICLEI – Local Governments for Sustainability & Ecoinstitut Barcelona (2008). European Commission Green Public Procurement (GPP) training toolkit. Module 3: purchasing recommendations. Food and catering services. Background Product Report. Available at: http://ec.europa.eu/environment/gpp/pdf/toolkit/food_GPP_background_report.pdf.

Illeris, K. (2007). How we learn. Learning and non-learning in school and beyond. Routledge, London, UK, 289 pp.

Ingenbleek, P.T.M. and Immink, V.M. (2011). Consumer decision-making for animal-friendly products: synthesis and implications. Animal Welfare 20: 11-19.

Kegan, R. (2009). What 'form' transforms? A constructive-developmental approach to transformative learning. In: Illeris, K. (ed.) Contemporary theories of learning. Learning theorists...in their own words. Routledge, London, UK, p. 35-52.

Kvale, S. (1996). Interviews. An introduction to qualitative research interviewing. Sage Publications, Thousand Oaks, CA, USA, 326 pp.

Lappalainen, E. (2012). Syötäväksi kasvatetut. Miten ruokasi eli elämänsä. Atena, Jyväskylä, 355 s.[Grown to be eaten. How your food lived its life] In Finnish. Atena, Jyväskylä, Finland, 355 pp.

Meadowcroft, J. (2007). Who is in Charge here? Governance for sustainable development in a complex world. Journal of Environmental Policy and Planning 9(3-4): 299-314.

Mezirow, J. (2009). An overview on transformative learning. In: Illeris, K. (ed.) Contemporary theories of learning. Learning theorists ... in their own words. Routledge, London, UK, pp. 90-105.

Miele, M. (2010). Report concerning consumer perceptions towards animal farm welfare, Official Experts Report EWAP, Uppsala University, Uppsala, Sweden.

Mikkola, M. (2009). Shaping professional identity for sustainability. Evidence in Finnish public catering. Appetite 53(1): 56-65.

Morgan, D.L. (1997). Focus groups as qualitative research. Qualitative research methods series 16. 2nd Ed. Sage Publications, Thousand Oaks, CA, USA, 80 pp.

Morley, A., Sonnino, R. and Smith, A. (2012). Leading by procuring: the power of public sector purchasing. In: Gallagher, D.R. (ed.) Environmental leadership: a reference handbook. SAGE Publications, Inc., Thousand Oaks, CA, USA, p. 113-122.

Parker, I. (1992). Discource dynamics. Critical analysis for social and individual psychology. Routledge, London, UK, 169 pp.

Wals, A.E.J. (2010). Message in a bottle: learning our way out of unsustainability. Prof. dr. ir. Arjen E.J. Wals. Inaugural lecture upon taking up the posts of Professor of Social Learning and Sustainable Development, and UNESCO Chair at Wageningen University on May 27th 2010. Wageningen University, Wageningen, the Netherlands.

School meals: bridging the gap between citizen expectations, procurement skills and legislation

N. Lindström[1] and H. Röcklinsberg[2]
[1]Equal Value Project, Luleå University of Technology, 971 87 Luleå, Sweden; [2]Department of Animal Environment and Health, Swedish University of Agricultural Sciences, P.O. Box 7068, 750 07 Uppsala, Sweden; niclas.lindstrom@ltu.se

Abstract

Among others, Swedish citizens trust responsible authorities and legislation to ensure that producers manufacture food under reasonable conditions for both sentient beings and landscapes. Nevertheless many consumers are prepared to go much further and buy goods with e.g. fair-trade or ecological certificates. We believe that this trust is becoming increasingly compromised, and a closer study of what values influence purchase of food served in official institutions such as hospitals, preschools, and schools is important as individual choice according to certain certificates is very limited. This forthcoming study has two main aims. One is to examine inherent values in consumer expectations, policies and legislation regarding school meals through an ethical analysis of different aspects such as e.g. individual choice, public health, global justice, animal welfare and environmental sustainability. Another aim is to suggest a model for municipalities and the county councils to weigh these, sometimes conflicting, factors to one another in order to take reasonable ethical responsibility, live up to citizen expectations and meet national and international legislation regarding e.g. global justice, sustainability and animal welfare.

Keywords: animal welfare, decision-making, ecological sustainability, ethical analyses

Theme and hypothesis

Swedish consumers expect legislation to cover 'legitimate claims' and responsible authorities to ensure producers implement these regulations (Keeling *et al.*, 2012) i.e. expect the food on the market to be manufactured under reasonable conditions for both sentient beings and landscapes. These expectations traditionally include also the food served at the country's preschools and schools, but recent initiatives for higher standards are showing a new picture (Ingenbleek and Immink, 2011; Olofsson, 2013). This is parallel to many consumers being prepared to go much further and buy their goods with e.g. fair-trade or ecological certificates. Citizens seem on the one hand to expect a lot of authorities and producers and on the other seek alternative quality assurances. This can be seen as an indication that the trust in both authorities and industry has become increasingly compromised. Such a description of the current situation is taken as a point of departure in this project as it reveals a gap between three spheres: (1) increased interest in 'ethical food' in society; (2) food served at official institutions at lowest possible cost; and (3) policies and legislation regarding food at official institutions. Based on this we have the following two hypothesises: (1) potential conflicting values can both be overcome and remain in content regarding meals served at schools, i.e. meeting demand for e.g. animal welfare and environmental sustainability; (2) based on this analysis useful tools can be developed serving as guidelines for improving procurement skills and contribute to better inclusion of consumer interests for 'ethical food' also at schools or other official institutions.

This multidimensional study has two main aims.[36] One is to 'examine inherent values in consumer expectations, policies regarding school meals at Swedish schools and legislation on school meals through

[36] An application for funding of this study is currently under review.

an ethical analysis' of different aspects such as e.g. individual choice, public health, global justice, animal welfare and environmental sustainability. Another aim is to 'suggest a method for how we are to weigh these, sometimes conflicting, factors to one another in order to take reasonable ethical responsibility, live up to citizen expectations and meet national and international legislation regarding e.g. global justice, sustainability and animal welfare for food served at schools'. This will further provide tools and guidelines for municipalities and the county councils responsible for the food provision when writing policies, and facilitate balancing health, taste and respect to global justice, animal welfare and environmental sustainability at purchase.

In order to meet the main aims, we see the following process tasks:

- to examine whether or not there is a discrepancy between the actual school meals with regard to 'ethical food' in Sweden on the one hand and relevant policies and legislation on the other;
- to investigate where the main value conflicts lie in an ideal situation of ensuring public health and individual freedom while aiming at schools meals produced with considerations to e.g. animal welfare, climate change and global justice;
- to investigate how a balance for handling such 'added values' could be constructed, and, based on that;
- to suggest decision-making tools and a guideline for policy makers and those responsible for purchasing food to schools.

In order to investigate these issues, we will structure the scientific work according to four steps, including three workshops covering the themes in a, b, c and d for input from stakeholders, reference persons and collaboration with the Centre of Excellence in Animal Welfare Science.

Overview of the area

Food inevitably concerns us all on an everyday basis and involves much more than only our individual taste preferences. What we eat affect as diverse matters as our individual performance, health status and life expectancy (Dahl and Jensberg, 2011) and our eating habits are associated with social interaction (Delormier *et al.*, 2009). In many parts of the world consumption of food is in turn closely related to modern, large scale, production of food which also involves much debated issues of animal welfare, social responsibility, extensive land use and climate change. The 2006 FAO report 'Livestock's long shadow' evoked global awareness and a discussion about the level of impact from animal produce on climate and environment, as well as the interaction between food choice and ecologic footprint (Steinfeld, *et al.*, 2006). To improve sustainability consumers are encouraged to buy with awareness, which has generated initiatives to improve for instance the treatment of animals in extensive farming, (e.g. labelling non-stunned slaughter) and promote agricultural development in a more climate friendly direction (Belasco, 2008). Considering the multidimensionality and level of impact food consumption have it is of special interest to stress the importance of meals served at public institutions such as hospitals, day-care centres, and schools. Only in Sweden approximately 460 million meals are served at preschools and schools every year. What children eat at school is thereby governed by public policies and national legislation decided within a democratic framework by majority rule (Educational act, SFS 2010:100 § 10; Vernon, 2005). This begs the question; do these official institutions serve the best interests of everybody involved? The relation between food, health, and performance is often taken for granted in research within the area.

Ilkay and Wallace (2012) has identified balance between risk of obesity or diabetes 2 and 'tasty food' as an ethical dilemma for school administrators and stakeholders where economic cost is weighed against heath and taste preferences of schoolchildren. The bottom line is to serve nutritious food contributing to good health, which needs to be combined with taste preferences, but, as we argue, above this 'added values' like animal welfare and climate change are to be taken into account to meet consumer awareness and citizen expectations of 'ethical food' at official institutions.

Above this, habits matter for what children eat. According to Delormier *et al.* (2009), the development of an overall relationship to food, is especially important during our formative years. Persson Osowski's study (2012) is another rare attempt to map children's eating patterns, both home and at school, in order to analyse pupils' understanding of the significance of food and eating. Both studies show that it is of utmost importance to introduce a healthy and sustainable life style early on in life and can perhaps also help to explain why eating patterns are hard to break, regardless whether they consist of self-starvation, over intake or arrogance towards animal welfare, climate aspects and global justice. However, even if there is research conducted on eating habits and lifestyle issues in relation health and performance in general, other aspects of publicly provided school meals are seldom taken into consideration. If we take these questions seriously they are to be considered on an individual, social and in a wide sense environmental level. How can social patterns be established, within a learning environment, in order to promote a sensible relation to food and a healthy lifestyle? What measures are taken to grant that the food provided is produced in a way that takes reasonable considerations to animal welfare, environmental sustainability and climate change? Although national guidelines mention that school meals shall ideally be a 'learning occasion', neither juridical, sociological nor ethical studies have been performed to investigate whether or how this is implemented or what values are to be expressed in such a learning situation (see e.g. the Food Agency's advice).

We argue that despite changes in attitudes and increasing awareness of highly compromised animal welfare and severity of climate impact in today's industrialised food production, policies and legislation are lagging behind. It has also proven difficult for official institutions to defend other interests or take other aspects, than price alone, in account during the process of public procurement of food (Olofsson, 2013). Although a democratic legislative process is bound to be slow, thorough changes are called for in order to mirror public perception (Gavinelli and Ferrara, 2009). As the difference concerned individual consumers can make is, still, rather limited it is important to shed light on the following issue: *what is possible to achieve in areas such as public health, animal welfare and environmental sustainability to name a few, even by small alterations of public policies on food*? How to weigh these, sometimes conflicting, factors to one another in order to follow, not only national legislation and international conventions, but also reasonable ethical requirements, is not at all obvious, and hence in need of investigation.

Theory and method

The relationship between empirical and ethical studies is especially relevant in an interdisciplinary framework and can be described with help of three main alternatives:

1. Traditionally empirical and ethical studies have been considered as separate fields, with specific methods used in order to examine different aspects of the world. According to this view it isn't possible to, on the basis of what *is* draw conclusions about something *ought* to be. In this case the results are not perceived as compatible to one another but as valid within what, perhaps, best can be described as separate research paradigms (Hume, 1739).
2. Another alternative is a collaborative relationship, where results from both empirical and ethical research, can exercise mutual influence on each other. Studies can be conducted by representatives from separate fields, in several different ways, but the results in one area of investigation can still be relevant to the other.
3. Yet another way to view the relationship between empirical and ethical studies is in terms of a synthesis, where empirical and ethical studies are fused into one diverse area of investigation. This is closely related to a rejection of a strict dichotomy, between *is* and *ought* or *fact* and *value*, where different sorts of inquiry cannot be incorporated in a common field (Lindström, 2012; Weaver and Trevino, 1994).

A multidimensional study can thus be interpreted as collaboration across traditional disciplinary boundaries. This is consistent with representatives of different fields performing their studies in methodologically different ways but also requires that the results are compatible across traditional disciplinary boundaries. Only then, it is possible for different research fields, to some extent, exert reciprocal influence on each other. Empirical fact about what the Swedish school meal consists in and the quantity of products with, for instance, dairy or meat, is especially relevant for an inquiry regarding animals' situation in the food production. The same idea applies for empirical fact about animals' natural behaviour, how they are treated in large scale farming or industrial agriculture, and how this in turn affects them and the environment at large. These concerns are relevant for an ethical analysis that can help to highlight features that we need to know more about in order to make well informed judgements (Lindström, 2012; Röcklinsberg, 2006).

The underlying idea of such a combined approach is not only to identify the consequences public procurement of school meals have for the animals in industrialized agriculture, but also to analyse whether these are consistent with the intentions of the governing documents, as well as if there are conflicting values and to what extent they are ethically acceptable taking the wider set of values (animal welfare, climate change, ecological sustainability, and global social justice) into account. The strength of such a model is that it can help explain the conditions in food production as well as contribute to the understanding of how well this matches the ideals, norms and values associated with animal welfare and a sustainable society. Further it will contribute to both formulating tools for decision-making, possibly by using an ethical matrix (see Forsberg, 2007 for a critical discussion), and guidelines for persons at municipalities responsible for purchasing meals for official institutions.

Expected outcome

Our belief is that the study will show that an increasing public awareness of the implications food consumption have on animal welfare and ecological sustainability has not yet had any major impact on the food served at preschools and schools, nor on the policies and legislation regarding food at official institutions. A couple of important factors which can contribute to the slow progress in the field can be that: (1) the organization of municipalities and counties, responsible for the handling of publicly provided meals, runs the risk of compartmentalization questions regarding school meals; while (2) the legislation on public procurement can provide an economization of any solution. Altogether this constitutes a reductionist perspective on meals served in public institutions.

1. Municipalities and counties are organized to deal with a wide range of questions where different administrative departments have responsibility over their own specific area. This may be a rational way of addressing well defined tasks within each field but difficulties may occur as all relevant questions doesn't follow these organizational boundaries. This can lead to compartmentalization, where the competence to address multidimensional questions, as the ones regarding food in large institutions, are spread through different departments of the organization. Because of this there is a risk that questions regarding school meals will be labelled in the already existing organizational framework and thus reduced to either issues of public health, social equity, animal welfare, ecological sustainability, or global justice – but scarcely all of these at once.
2. There are also examples of municipalities which have specified requirements, consistent with the animal welfare act, in their public procurement of food intended for pupils of preschools and schools, where the outcome has been appealed throughout the Administrative Court system. Hence, if the procurement procedure doesn't reach the result expected of the strong financial interest involved, they will be appealed before an administrative court. This can be one important reason, for municipalities and counties, to embrace the lowest price alone-clause and avoid requiring certain conditions for purchase, such as demand of animal welfare or ecological sustainability, and thus evade

extensive legal processes. If this is the case, the legislation on public procurement can contribute to a reductionist view on publicly provided meals where the economic aspects are paramount.

We argue that questions regarding publicly provided meals are characterized by a multidimensionality which in many ways defies administrative boundaries as they involve a wide array of assessments. Thus it is hard for any single department to carry special competences in all of these special areas. It might be tempting to resolve questions like these in a reductionist fashion, by making them subject to an economical calculus, where least possible amount of money is supposed to buy nutritious food for all those entitled to it in preschools, schools, and hospitals. We would instead like to embrace the multidimensionality of these questions, by providing practical tools or an ethical matrix, to weigh potentially conflicting values, with increasing concern for e.g. animal welfare and environmental sustainability (Mepham, 2010). Ideally this can contribute to form guidelines for improving procurement skills and contribute to better inclusion of consumer interests for 'ethical food' also at official institutions.

An important feature of this study is to construct a model for how to weigh these different factors, in order to make informed judgements of which kinds of food can be considered morally acceptable in publicly provided meals. A model for well-grounded decisions, regarding publicly provided food at schools, but applicable for other large institutions, is proposed to consider relevant factors e.g: (1) public health; (2) social attitudes towards food; (3) animal welfare; (4) climate change and environmental sustainability; (5) global social justice; (6) juridical issues. They all have to be considered in relation to the legitimate expectations citizens' have on the institutions of a democratic society, their respective views of the problems at hand, as well as possible solutions within an already existing administrative framework. At the same time it is important to emphasize that even small alteration of public policies can have vast implications for the sentient beings and landscapes used in the production of food due to the sheer number of meals served in the concerned institutions. The results of this process will in turn be formulated as tools to evaluate existing guidelines and form new policies for purchasers, school boards, and politicians. The model can be elaborated in order to guarantee certain, ethically relevant, aspects of school meals are met and that they are not subjected to the lowest price only clause. We hope that this will contribute to bridging the gap between citizen expectations, procurement skills and legislation within the field of publicly provided meals.

References

Administrative Court verdicts. Available at: www.upphandling24.idg.se/polopoly_fs/1.383382.1304454787!livsmedel%20dom%20i%20forvaltningsratten.pdf and www.upphandling24.idg.se//polopoly_fs/1.434372.1330077375!kr_sthlm_djurskydd.pdf.

Belasco, W. (2008). Food: the key concepts. Berg publications, Oxford, UK.

Dahl, T. and Jensberg, H. (2011). Kost i skole og barnehage og betydningen for helse og læring en kunnskapsoversikt, Tema Nord 2011:534, Nordisk ministerråd, København, Denmark.

Delormier, T., Frohlich, K.L. and Potvin, L. (2009). Food and eating as social practice – understanding eating patterns as social phenomena and implications for public health. Sociology of Health & Illness 31 215-228.

Educational act, SFS (2010). Food agency: Livsmedelsverkets råd: bra mat i skolan. Available at: www.slv.se/upload/dokument/mat/mat_skola/Bra_mat_i_skolan_2007.pdf.

Forsberg, E.M. (2007). Pluralism, the ethical matrix and coming to conclusions. Journal of Agricultural and Environmental Ethics 20: 455-468.

Gavinelli, A. and Ferrara, M. (2009). Animal welfare: the risk manager's perspective. In Smulders, F.J.M. and Algers, B. (eds.) Welfare of production animals: assessment and management of risks. Volume 5 of the Food Safety assurance and veterinary public health series. Wageningen Academic Publishers, Wageningen, the Netherlands, p. 353-370.

Hume, D. (1985[1739-1740]). A treatise of human nature. Penguin Books, London, UK.

Ilkay, J. and Wallace, S. (2012). Ethical leadership in school lunch program meal offerings. Journal of Business Studies Quarterly 4: 170-176.

Ingenbleek, P.T.M. and Immink, V.M. (2011). Consumer decision-making for animal-friendly products: synthesis and implications. Animal Welfare 20: 11-19.

Keeling, L., Immink, V., Hubbard, C., Garrod, G., Edwards, S.A. and Ingenbleek. P. (2012). Designing animal welfare policies and monitoring process. Animal Welfare 21: 95-105.

Lindström, N. (2012). Förhållandet mellan praxis och teori inom etiken. Lund studies in ethics and theology 16. Lund University, Lund, Sweden.

Mepham, B. (2010). The ethical matrix as a tool in policy interventions: the obesity crisis. In: Gottwald, F.-T., Ingensiep, H.W. and Meinhardt, M. (eds.) Food Ethics. Springer Media New York, NY, USA.

Olofsson, M. (2013). Ekot 050313: Dalarna vill servera skolbarnen egna köttdjur http://sverigesradio.se/sida/artikel.aspx?programid=83&artikel=5483566.

Persson Osowski, C. (2012). The Swedish school meal as a public meal. Collective thinking, actions and meal patterns. Uppsala University, Uppsala, Sweden

Röcklinsberg, H., (2006). Consent and consensus in policies related to food – five core values. Journal of Agricultural & Environmental Ethics 19: 285-299.

Steinfeld, H., Gerber, P., Wasenaar, T., Castel, V., Rosales, M. and De Haan, C. (2006). Livestock's long shadow: environmental issues and options. Food and Agriculture Organisation of the United Nations, Rome, Italy.

Vernon, J. (2005). The ethics of hunger and the assembly of society: the techno-politics of the school meal in modern Britain. The American Historical Review 110: 693-725.

Weaver, G.R. and Trevino, L.K. (1994). Normative and empirical business ethics: separation, marriage of convenience, or marriage of necessity? Business Ethics Quarterly 4: 129-143.

Section 12. Community and culture

Contextualising food policy to the citizen: religions as a paradigm

D.M. Bruce
Edinethics Ltd., 11/6 Dundonald Street, Edinburgh, EH3 6RZ, Scotland, United Kingdom; info@edinethics.co.uk

Abstract

Food is integral to all cultures but is expressed in ways particular to each culture. Many different values are involved in food, its growing, preparation and consumption, beyond those understood by a primarily consumer model, especially in relation to religions. This paper contends that any promotion of more ethical patterns of consumption and lifestyle must be done in the context of cultural and religious factors. This is illustrated by an international study on whether GM foods would be considered 'taboo' in ten major religious traditions, in which the author took part. The responses revealed varied insights about food in different religions. To commend an external set of values, like a universal food citizen codex, would approach the question from the wrong angle. Religions already embody values conducive to addressing the major environmental and social challenges. Experience of environmental action in UK churches suggests that inducing guilt about our misuse of nature disempowers people. It is more effective to encourage people to put more into practice what they believe, focused by global challenges. Remaking the connection between 'what I eat' and 'where food comes from' is best done in the context of people's basic beliefs, religious or otherwise.

Keywords: religions, genetically modified food, food ethics, culture, values

Introduction

The recent focus on food security, as cultures habituated to having food 'on tap' have become aware how vulnerable to disruption are the embedded web of dependencies of our daily food supply, raises wider questions of cultural values in relation to food. Viewing food merely as a commodity in a supply chain profoundly misses the deep cultural embedding and local sensitives of how human beings produce, prepare and consume food. In parallel, concerns for the environment, health, social justice and community have led to the promotion of more ethical patterns of consumption and lifestyle. This paper contends that a centralised ethical concept may be well meaning but of little value, because food is practised in the context of personal, cultural and, in particular, people's religious beliefs. The principles and practices of one's belief system can have a profound effect on how food is viewed – what is done with it and why, on what is deemed acceptable and what is not, and indeed on what is going on when we eat. These have to be understood in relation to any proposed response to such issues as climate change, stresses on land and fresh water, challenges from new agri-biotechnologies. The genetically modified food crisis in the UK and Europe illustrates what may happen if these wider values are ignored by government (Bruce and Bruce, 1998).

Religions and diversity

This is illustrated by an international study on whether GM foods in ten major religious traditions, in which the author took part (Brunk and Coward, 2009). In 2004-05 the University of British Columbia, Victoria, in Canada invited representatives of the major world religions and local indigenous religious communities to explore the question of whether genetically modified food would be considered taboo or unacceptable within their religious tradition. It was explored in two ways. One was to ask experts well versed in the tenets of the religion, to interpret the technology in the light of its values. The other was to explore what particular groups of 'lay' believers thought about the issues via focus groups in the Vancouver area. The responses revealed the varied and profound insights about food in different religions.

'Religion' has many meanings – a belief system articulated by theologians or official bodies; what is commonly believed and practised by local believers; the wider culture which the religion has given rise to; what secularised people *think* that a religion is about. Religions are also hugely diverse, which may be missed when secular policy makers consider 'the religious dimension' to their societies. It often comes as a surprise in Europe that the Roman Catholic hierarchical system of church governance and policy is not typical of world religions, or even Christianity. Some religions have no hierarchy at all. Within most belief systems, several ethical opinions may be commonly held, of which GM food is a good example.

The Buddhist representative in the study, David Loy, writes 'There is no such thing as Buddhism' but rather 'a variety of very different cultural traditions so diverse that it is sometimes difficult to see what they share other than the label 'Buddhist' (Loy, 2009). Other religions like Islam and Christianity have beliefs and practices more uniformly expressed, yet incorporate many significant variations within the basic core belief. Judaism is both a religion and a cultural identity, with a wide range of interpretation of a Jewish way of life between say its Orthodox and Reformed traditions. So rather than simply ask, say, 'What do Buddhists think about GM food?' we also need to consider: 'which Buddhists, and where?'

Christianity, of which I am a believer and therefore the most familiar with, displays an immense variety of cultural expression, including practices about food and attitudes to its genetic modification. It has given more thought to this issue than most of other religions, but a single definitive account of GM food shared by the diverse billions of Christians of the world would be impossible. A whole spectrum from enthusiasm to outright opposition to GMOs can be found, but more often somewhere in between. Indeed, the Catholic church has still not taken an 'official' view on GM food, reflecting the differences on the issue.

The study also found a significant difference in *awareness* and understandings of the issue of GMOs from within the religion. In many, however, rather little formal thinking had been done in the mainstream of the religion, even if some individuals or academics may have reflected on it. In some, the issue had been examined in great depth at a formal level, notably the Church of Scotland's pioneering work and in Jewish rabbinical scholarship.

There may also be a difference between expert and lay understandings of the issues. The Church of Scotland SRT Project did the first in depth expert ethical study of the issue in the English language from 1993-99, and played a significant role in the unfolding national debate in the UK (Bruce and Bruce, 1998). Most Scots were exposed to GM issue via the media, and it was quite widely discussed by lay church people in addition to the expert group, and similarly in Germany (Evangelische Kirche Deutschland, 1991) and the Netherlands (MCKS, 1997). In contrast, and probably more typically, Jewish ethicist Laurie Zoloth observes that while Hebrew scholars have given much thought to GM food, and made pronouncements, generally in favour (Zoloth, 2009), no one in the two Jewish focus groups in Vancouver was aware of these. Their lay perceptions were more critical compared with the expert views. This reflects two common features. One is the practical difficulty of disseminating new insights within a religion. The other is the different ways of framing an issue between the more abstract normative approach of the expert interpreter, well grounded in the logic, rationale and ethics of the religion, and the more situated views of their everyday lives of lay people.

GM food comes to most believers mediated through the filters of secular media, government, public relations material on both sides, and local hearsay. Much depends on how one first heard about GM, who presented the issues and with what bias. Where the debate became politicised and polarised, people want trustworthy information to weigh up the issues for themselves. In Scotland the church was such a source, informed but not aligned politically.

Industrial agriculture arose in a context of a view of the world derived from Christianity, rather than, say, a Hindu, Confucian or Islamic one. The Christianity of the 16th century Protestant Reformation encouraged human enquiry and intervention in the world, which opened a door for early modern science, even envisaging crop modification (Bacon 1620 and 1627). Christianity and Judaism have a history of engagement with emerging technologies. More often religions have come late to these issues, once they are part of wider culture. The Islamic expert on the study, Ebrahim Moosa, noted 'GMOs radically change all the inherited presumptions of a religious tradition like Islam and present us with a dynamic system of nature, one that is a constantly emerging novelty' (Moosa, 2009).

Acceptable genes? – some Buddhist and Christian views

Amongst the five major religious traditions, Christianity and Buddhism had no particular ritual significance for food or prohibitions on different foods. For both there was no intrinsic reason why moving genes into a different species should not be done, but for different reasons. In both cases, nature is considered ethically ambiguous: whether something deemed natural does not of itself constitute an ethical guidance, one way or the other.

For the Buddhist, the issue is more about the *motivation* behind use of food, and the institutional or collective reasons to do modification. It is thus more virtue ethics, than a deontological or consequential approach typical of Christian reflections on the subject. GM might or might not decrease the *dukkha* (suffering) for us and the ecosystem, say, through unintended harm to another species, or helping malnutrition to be addressed? There are no side effects because everything is intrinsically linked. The unwholesome roots of motivation (greed, ill will and delusion) are reflected corporately as well as for the individual. But Loy writes: 'The technological modification of plant and animal species without a much better understanding of their genomes and how all genomes of living creatures affect each other is an especially dangerous example of how our ambitions tend to outrun our wisdom. I am led to conclude that genetic engineering of food as currently practised is probably basically incompatible with basic Buddhist teachings.' The technology is considered neither good nor bad, but cannot be separated from who is applying it, how and why.

Christianity is radical in a different way (Bruce, 2009). No formal ritual practice is required for salvation. Christians believe that Jesus Christ fulfilled a priestly role in dying on a cross for all humanity, meeting all such requirements, cancelling sin, and rising from the dead. He taught that what defiles us is not any food that we eat but the sin already in our hearts. In Mark's gospel the writer concludes: 'In saying this, Jesus declared all foods clean' (Mark 7: 14-23). The taking of a mouthful of bread and wine in the sacrament of Holy Communion or the Mass is symbolic is a memorial or re-enactment of this central event of Christian belief. Voluntary fasting may be done as an aid to devotion and to remind oneself of a right attitude to God and God's creation. A focus group in the Mennonite tradition stressed the significance of eating as a focal point community and family act, far more than just fuel for living.

A conservative study could find no clear biblical basis to prohibit GMOs, neither in the separation of plants and animals into 'kinds' in creation narratives, nor the prohibitions on mixing seeds in the Hebrew scriptures, which are not binding for Christians (Bruce and Horrocks). God encourages human intervention in 'having dominion' over the rest of God's creation (Genesis 1:26-28), but accountable to God, and limited within God's laws by respect for the rest of creation, love for our human neighbour, and care for the disadvantaged. If God then has given humans the skills to make fundamental changes of matter (metallurgy and selective breeding are not problematic in the Bible), GM should not be prohibited as such.

But as with Buddhist concerns, how GM is practised has raised serious concerns amongst Christians. Do finite and morally fallen human beings know enough to alter radically what God has created, without damage and risk? The expression 'playing God' is poor theology, but a useful heuristic. For some it expresses hubris, overstepping our human capacities. The identified risks to the environment and to ourselves indicate this is something we should not do. Similar concerns were indeed expressed in all the religions in this study. But other Christians point out that risk is part of the condition in which God situated humans. Since God invites humans made 'in God's image' to reshape the creation, to do so without risk is an impossible pre-condition for any technology. The question is then how to use GM crops prudently for good reasons, not something to be shunned (Church of Scotland, 1999). The main GM focus remains American and European commercial agricultural production, not human needs or the staple crops of the world's poor. Humanitarian examples like vitamin A rice are still the exception. Even a new drought tolerant crop was developed first for US farmers and only later for Africa. The dominance of a few multinational companies and restrictive uses of genes have also drawn severe criticism on issues of justice, fairness and power. For some, in the light of Jesus' command to love our neighbour as ourselves, GM food is so irredeemably tainted by corporate bad behaviour that it cannot be countenanced. For others, this is true of all technology, and something to be redeemed rather than to reject.

Acceptable genes? – some Islamic, Jewish and Hindu views

Judaism, Islam and Hinduism all presents cases where certain foods are deemed ritually unclean or unacceptable for believers to eat. The use of a pig gene in a tomato, say, would generally be viewed as unacceptable in Islam, but the issue was not so clearcut for Jewish scholars. One reason is that the amount of change is tiny in the organism as a whole. Another opinion stated, 'Judaism does not regard genes as food, so the introduction of a gene or genetic material from non-kosher animals would not render the recipient non-kosher' (Walhrman, 2002). A similar distinction was noted in 1993 in evidence by Islamic and Jewish experts to a UK government committee on GM animals (Bruce and Bruce, 1998). Is a living organism to be identified and evaluated ethically according to its precise genetic blueprint or in more holistic terms? What is it about the change that would render it ritually unclean?

Describing Hindu responses, Vasudha Narayanan, noted a difference between the food a Hindu eats in normal life and food taken on holy days or used as an offering to a deity (Narayanan, 2009). To introduce an animal gene into a vegetable or fruit would ender it impure for the ritual or worship context. It might be permissible to eat GM food on a daily basis if there was no health hazard and other food alternatives were deemed worse. As with Islam, it seems that it is the impure origin of the gene that is the concern rather than the trait.

In Islam the concept of genetically modifying plants or animals does not in itself seem to defy a major precept of the Qur'an. Moosa argued that, in contrast to moral or salvation matters, ethical issues tied to secular pursuits, relying on empirical knowledge should be decided on their scientific or practical merits. But what constitutes good stewardship (*khilafa*) of the responsibilities Allah gives to humans remains open to debate. Some traditional religious authorities and some technocratic Muslims give ethical and legal support to GMOs as manageable risks. Others see it as innovating but at the same time disturbing Allah's creation, and seen by some as a means of power to 'colonise' people. Muslim lay focus groups were ambivalent and more hesitant about supporting GMOs. They also expressed another common factor to most religions, to label GM food to give believers a choice, one way or the other.

Jewish expert thinking is complex and is extensively reviewed by Zoloth, examining texts and interpretations of the Hebrew scriptures. The texts are viewed as binding by many within Jewish traditions. Despite underlying concerns about altering nature and natural kinds, she concludes that genetic modification does not violate prohibitions on mixing seeds or animals. Jewish ethics puts a very high priority to saving life and healing the brokenness of the world. In so far as GM food might in some of its applications achieve such aims it would even constitute a mandate. She notes that more conservative authorities have advocate wider use of GM technologies, but more liberal thinkers advocate more cautions on the use of GM food.

Jewish focus groups were unaware of such scholarly thinking and voiced more concerns about safety, the rules of *kashrut* (kosher) and hubris, and the effects of corporations primarily concerned for profit. Zoloth argues that the rules of kashrut are just given, not rationalised, but can be seen as a heuristic for values such as kindness to animals, respect for the environment, the limitations to unbridled consumption and one's own spiritual discipline. She wrestles with the promise of GM foods and issues of justice and responsibility, and finally quotes the midrashic admonition, and in this summarises what many religious scholars have concluded about GM food, 'When you plow the field, you must take the law with you.' GM food is something to assess case-by-case. As often as you are considering innovating, think how the insights and wisdom of sacred texts, religious tradition and local experience and contexts affect and maybe call for changes to your priorities, your intended means of execution, environmental and social impacts, your motives and the good or ill you could do.

Lessons from religion for ethical consumption

This knowingly limited sample from a large and fascinating study indicates the breadth of viewpoints, circumstances and practices that are brought into consideration when an issue like the genetic modification of food is considered within the world's religions. Yet, for all their diversity certain basic principles and concerns are held in common – concerns about hubris, injustice and wrong use of corporate power, the wrong priorities, the failure to prioritise the poor and hungry of the world in the technologies and their application.

To commend an external set of values, like a universal 'food citizen' codex, would seem to approach the question from the wrong angle. In this GM food example, in most cases the religions already embody values conducive to addressing the major environmental and social challenges. Therefore one would rather ask where in the values which people already hold, are to be found the changed behaviour to which such ideas are rightly aspiring?

An practical example of this was found in a different field, that of encouraging churches in the UK to adopt a scheme for grassroots environmental action, UK Eco-Congregation programme. Historically it has been a neglected subject in the churches, despite our theology, and this is what the programme is intended to rectify. In Scotland alone some 300 churches have signed up to a threefold mixture teaching within the church, practical action in the church and within the local community. But church members were not always naturally inclined to set a high priority to environmental change in their own lives or communally.

We found that to motivate people, starting with global threats or our misuse of nature, is powerful at inducing guilt but it often disempowers people by overwhelming them ideas which they feel power*less* to change. A more effective model was to start where they were, to ask themselves: what do we already do environmentally? why do we do it? and what more could we do? In our case it was a case of encourage people to put more into practice what they believe, contextualised and focused by these wider global challenges.

In conclusion, to enable people to remake the connection between what I eat and 'where food comes from' and its global implications are important. But this is far more likely to be effective if these are located in the wider context of people's basic beliefs, whether religious or otherwise. It is only in this deeper sense of citizenship that one can realistically expect that patterns of consumption can be changed. Codices may be done best bottom-up not top down.

References

Bacon, F. (1620). Novum organum, Book I: Aphorisms 18-26 and 44, transl. Urbach P. and Gibson, J. (1994). Open Court, Chicago, IL, USA and Bacon, F. (1627). New Atlantis. In: Bacon, F.: A critical edition of major works. Vickers, B. (ed). (1996), pp. 480-486, Oxford.

Bruce, D. (2009). Some Christian reflections on GM food. In: Brunk, C.G. and Coward, H.G. (eds.) Acceptable genes? Religious traditions and genetically modified foods. State University of New York Press, Albany, NY, USA, p. 115-134.

Bruce, D. and Bruce, A. (eds.) (1998). Enginering Genesis. Earthscan, London, UK.

Bruce, D. and Horrocks, D. (2002). Modifying creation. Paternoster, Carlisle, UK, p.52-54 and 168-175.

Brunk, C. and Coward, H. (eds.) (2009). Acceptable genes?: religious traditions and genetically modified foods. State University of New York Press, Albany, NY, USA

Church of Scotland (1999). Genetically modified food. Reports to the Church of Scotland General Assembly 1999, pp. 20/4 and 20/93-20/103, Church of Scotland: Edinburgh, UK.

Evangelische Kirche Deutschland (1991 and 1997). Einverständnis mit der Schöpfung, Gerd Mohn, Gütersloh, Germany.

Loy, D. (2009). The karma of genetically modified food. In: Brunk, C.G. and Coward, H.G. (eds.) Acceptable genes? Religious traditions and genetically modified foods. State University of New York Press, Albany, NY, USA, p. 179-195.

MCKS (1997). Biotechnologie: God vergeten? Multi-disciplinary Centre for Church and Society, Utrecht, the Netherlands.

Moosa, E. (2009). Genetically modified foods and muslim ethics. In: Brunk, C.G. and Coward, H.G. (eds.) Acceptable genes? Religious traditions and genetically modified foods. State University of New York Press, Albany, NY, USA, p. 135-158.

Narayanan, V. (2009). A hundred autumns to flourish: Hindu attitudes to genetically modified food. In: In: Brunk, C.G. and Coward, H.G. (eds.) Acceptable genes? Religious traditions and genetically modified foods. State University of New York Press, Albany, NY, USA, p. 159-178.

Wahrman, M. (2002). Brave new Judaism: when science and scripture collide. Brandies University Press, New York, NY, USA, p. 205.

Zoloth, L. (2009). When you plow your field, your Torah is with you: genetic modification and GM food in the Jewish tradition. In: Brunk, C.G. and Coward, H.G. (eds.) Acceptable genes? Religious traditions and genetically modified foods. State University of New York Press, Albany, NY, USA, p. 81-114.

The ethics of consuming: community, agency, and participation in global food systems

D. Beck, M. Ivanovic, S. Noll and I. Werkheiser
Department of Philosophy, Michigan State University, East Lansing, Michigan 48824, USA; beckdan2@msu.edu

Abstract

It is easy enough to think about preparing food, growing food, and eating food as straight-forward affairs, or mere biological necessities. However, these processes are wrapped up in social relations, personal identities, and political power. Broadly critical frameworks can help pull apart the varying dimensions in this complex and dynamic relationship in order to provide guidance for reforming our relationship to the food we consume in a more democratic and participatory direction. To that end, we explore the normative implications of the current organization of the world food system from multiple perspectives in critical theory, the philosophy of science, and political philosophy. We offer ethical and political analyses at the levels of community activism, national food policy, and global consumerist capitalism to show how an approach that sees these multiple levels as interactive can provide concrete reasons for supporting certain effective solutions.

Keywords: food safety, local food movement, critical theory, and strong objectivity

Introduction

The consumption of food is not a merely biological concern; it is inextricably intertwined with cultural and political life. It is imperative to examine from a philosophical perspective how human societies actively shape and are shaped by the ways in which food consumption is socially organized. To that end we explore the normative implications of the current organization of the world food system. We provide ethical and political analyses at the levels of community activism, national food policy, and critical social theory. In the first section, we situate our analyses of consumption within broader paradigms of critical theory that provide a framework for understanding how the formation of subjects happens within the current consumer capitalist system. We lay out our broad theoretical approach to consumption in two paradigms: enslavement/ manipulation and agency/ empowerment. This frame sets up further analyses of how we might make more democratic our relation to food. In the following two sections, we examine the proper role of public participation in U.S. food policy in ethical, political, and epistemological terms. In the final section, we further narrow our gaze to the context of the local food movement in order to show how our macro-level theoretical analysis can help guide local community activism. In whole, our varied contributions show how a critical approach that understands consumptive practices as essentially social and cultural, rather than merely biological, is a powerful framework for developing strategies to strengthen democratic control over such an important realm of meaningful human social existence.

Consumption and its discontents

During the 20th century, we have seen how the concept of consumerism has evolved from within an abstract philosophical discourse to the dominant moral concern of our time. In terms of its importance, consumerism has been a controversial source of legitimation for dominant ideologies of global capitalism, as well as a new source of inspiration for ever increasing transnational activism against inequality. In its basic form these two figures – an irresistible structure of (commodity) power and creative self-expression as political resistance – offer grounds for understanding social relations through an analysis of the ways that different segments of a globalized market (brands, advertisement, profit, etc.) impact the subject's formation of her agential identity. One obvious facet of social reality is an inherently ambivalent relation between subjects and the historical contexts constructed around market relations that exercise social control over human actions and choices. If we want to understand the genealogy of human agency in contemporary consumer societies, we must take a critical perspective on historical events that give rise to the cognitive and material dispositions in which subjects are embedded. Since individuals are not isolated units, but rather products of their sociocultural environment, public representations and advertisements affect citizens' conceptions of social conditions, the natural environments, modes of production, consumption, (in)justice and, finally, the lives of other human beings.

In order to address this concern, this section situates the effects of consumerism and subject formation in contingent, historically changing, and culturally variable practices. Although western philosophy articulates the deep cognitive structures that mold the experiences of individuals, it also obliterate differences of class, race, and gender through universal categories, thereby structuring the experience of the general consumer public as linear and homogeneous. Such an understanding is not capable of disclosing the power relations that guide the structural and symbolic constitution of consumer agency. It provides an inappropriate point of departure for challenging cultural traditions and institutions that are maintained through the rise of global neoliberalism. In contrast to this dominant framework, we take as our starting point of the ways in which specific market interests are rationalized, internalized, and legitimized by the public consumer body. In a society such as ours where there is virtually no area of life that remains uncommodified, and in which commodities play an ever-increasing role in shaping identities and cultural meaning, there is a widespread perception that consumption today remains the domain of 'enslavement' that relies upon a dialectic relationship between public representation (e.g. advertisements) and the branding paradigm. To the extent that every society is defined by values and institutions that presuppose that social actors are capable of influencing the decisions of others, it is important to examine the ways in which power structures set up the contexts in which the cultural production of meaning forms and guides the consumer culture. On the basis of such a premise we conceive subject formation – or the genesis of codes of consumption – and their representations as complementary mechanisms within a frame of techniques of social control and domination. Since the way we conceive reality regards the production of knowledge and the exercise of power simply as effects of the same process, one can easily understand that the goals of an institutional apparatus and the goals of public representation are the same.

There are two distinct paradigms within social theory that support our starting assumptions. Each paradigm provides a historically and discursively constructed perspective on the nature of the relation between markets, consumption, and the individual. Although their main focus is to identify the ways in which commodities and consumption affect human agency, these paradigms also clear the ground for potential ways of resistance in regard to predominant consumer culture. The first paradigm, the 'manipulation and enslavement' discourse, draws from the works of Marx and the first generation Frankfurt School (Izberk-Bilgin, 2010). Their analyses disclose the effects of commodities and consumerism as embodied in an ongoing deformation of our social fabric and human agency, calling attention to a pervasive concern in analyzing market power and its politicization in social discourses and practices. Relying on a denunciation of consumption and a rather cynical approach to market ideologies, adherents of such an understanding see radical social change as a main goal – that is, the abolition of social orders and class structures that are engendered by ever increasing consumerism. A main assumption here is an underlying understanding of the market as a social sphere of domination where people in power, through the media, develop and manage a seductive consumer culture that actively numbs the critical capacities of consumers. In contrast, 'the agency and empowerment' discourse, which can be found in Pierre Bourdieu (1984) and Mary Douglas and Baron Isherwood (1996), conceives consumption as a fundamental form of self-expression necessary for the sustainability of social relations, and even as a way to critically challenge social inequalities (Izger-Bilgin, 2010; Ulver-Sneistrup *et al.*, 2011). This understanding of the relationship between the market, consumption and a subject's agency contends that individuals are indeed capable of interacting with cultural resources such as language, music, and art in general to challenge and modify everyday practices, thus rendering consumption as a domain of social experience wherein unjust aspects of the social order are challenged, negotiated, and transformed. Contra the enslavement discourse, then, the empowerment paradigm offers a more positive view of consumer culture and attempts to reverse the negative effects consumerism has on agents by providing possible ways of resistance. By bestowing social agents with more autonomy in regard to the ideological effects of the 'culture industry' and depicting consumption as an inherent mode of social relations, their focus is on the ways in which individuals use the market and commodities to challenge structural social conditions and overcome structural (and symbolic) inequalities.

With this larger frame in mind, let us now move to specific issues in national food policy where market power, politicization, and consumerism have weakened the power of broad public participation. We will see how specific approaches sensitive to these issues help elucidate the nature of the problems and the effectiveness of possible solutions.

Ensuring food safety through public participation in policy formation

According to the Center for Disease Control and Prevention, 1 in 6 Americans get sick, 128,000 people are hospitalized, and 3000 die each year from food-borne diseases (FDA, 2011). Current food safety policy does not seem to be doing an adequate job of safeguarding the public. One way to help ensure food safety is to directly include the views of the public in the process of food safety policy formation. Citizens are affected on a daily basis by the way that food is grown, processed, and transported. Thus, they have a vested interest in food safety and should have a place at the policy table. This would also guard against special interest groups having too much influence on food policy. Agriculture in America has changed dramatically within the last few decades culminating in a decline in family farms and an increase in large industrial operations (Lyson, 2004). In a report to The National Farmers Union, Heffernan states that a major concern regarding the above concentration in the food system focuses on the control exercised by a handful of firms over decision making (Heffernan, 1999). The apparent result of such control can be seen in the increase of foodborne diseases (FDA, 2011) and the significant recalls of food products due to safety issues (Hassanein, 2011). This highlights two important points regarding food safety policy addressed below.

First, agribusiness may have an inordinate amount of influence over food safety policy. How do we guard against such influence, especially when policies pushed by such groups may not be in the best interest of the consumer? Second, food safety policy deals with risk. Every time a person consumes a food product, there is a slight risk that she will get sick from a foodborne disease. This risk is not due to the types of food a person chooses to eat but is independent of individual choice. Food safety risk is not a situation where risk is 'voluntarily chosen' but one where risk has been 'involuntarily imposed' (Schrader-Frechette, 1991).

These two points highlight important issues regarding food safety. First, because contracting foodborne illness is a risk involuntarily imposed on citizens, there is a *prima facie* moral obligation to increase the role of the public in food safety policy formation. Second, there is the issue of guarding against the inordinate amount of agribusiness control and whether a handful of policy makers should make decisions regarding risk on a society-wide scale. With regard to food safety, if risk increases or decreases depending upon how we grow, process, and transport food products, and a handful of policy makers has an inordinate amount of influence over the shape regulatory policy, then they are, in effect, dictating what levels of risk are or are not acceptable for the public. This leads to the question of whether or not it is acceptable to have involuntarily imposed risk levels dictated by a few powerful people. Regarding this issue, the work of Shrader-Frechette (1991) can be useful.

Shrader-Frechette argues that on the individual level one has the right to make rational decisions pertaining to oneself when faced with risk; however, on the societal level, one does not always have the right to make such decisions regarding others as this might violate their rights. In contrast to individual decision making regarding risk, decision making on a societal level 'requires an ethical rule that takes account of the fairness of the allocational process (for instance, whether potential victims exercise free, informed consent to the risk), not merely the outcomes...' (106). For Shrader-Frechette, the democratic process is more important during risk assessments where probabilities are unknown then when they are known. It seems, then, that there is a moral obligation to increase the role of the general public in food safety policy.

In addition, D. Fiorino (1990) gives the three following reasons why more direct public involvement should be undertaken: First, there is limited accountability for regulatory officials within representative democracy. It is rare for congress members to be voted out of office because they did not chastise a regulatory official for a problematic decision. Second, a direct participatory approach could be useful for policy makers because it could assist them in gaining a greater understanding of local conditions relevant to decision-making. Third, the process could increase the acceptability of the decision by the citizens affected. H. Douglas (2009) also argues for more direct public involvement in risk assessment and agrees with D. Fiorino.

Thus, there appears to be excellent reasons for including the public in the the formation of food safety policy. However, if this is so, then how do we efficiently include the public in the process? This might seem like an excellent idea in theory but will it work in practice? In order to ascertain this we will now turn to the Food Safety Modernization Act (FSMA) as a case study.

In December 2010, the US Congress passed the FSMA (FDA, 2011). This act gives the Food and Drug Administration the power to 'require 'risk-based,' preventative controls across the food supply, rather than to respond only after people have become ill from foodborne pathogens' (Hassanein, 2011). According to Hassanein (2011), outrage over contaminated eggs and other food scares motivated consumer groups to help move FSMA through the Senate. This is especially striking as, for years, Congress failed to pass similar reforms (Delind & Howard, 2008). Both consumer groups and associations representing the food industry agreed that there was need for uniform standards that would employ a 'scientific, risk-based approach' to regulation and to the prevention of food contamination (578). Where the local food movement and agribusiness opinion differed was on the 'Tester Amendment' which presumed that food safety risks are dissimilar at different scales. Thus regulation should adequately reflect these differences. Without this amendment, FSMA could have been ruinous for alternative food networks. The Tester Amendment was included but not without the hard work of local food advocates. What is important for our work is the fact that the Make Our Food Safe Coalition (which represented national consumer groups) and local food advocates directly influenced the final form of FSMA. This direct consumer action serves as a powerful example of one way that direct public action and deliberation can help form food safety policy.

This particular avenue of public participation is not new or contested. However, in addition to the above type of participation, the public should also be involved in the 'scientific, risk-based approach' to regulation outlined in the Act itself (Hassanein, 2011). Under FSMA, food manufacturers are required to develop plans to guard against contamination and to share these plans with the FDA. The FDA is also required to inspect food processing plants frequently, especially ones which handle 'high risk' foods. The FDA is given the power to recall products after contamination is found. In addition, foods produced overseas will have to meet the same standards as US produced foods. All throughout these processes, the FDA will rely on expert advice and on risk assessments. These risk assessments will largely determine how effectively the Act actually protects against food safety. These two avenues would be excellent areas where public participation could be efficiently integrated into food safety policy, both in its creation and in its management. The next question to be addressed, then, is how the public can be better integrated into these processes.

Douglas (2009) describes a useful process for integrating public deliberation directly into the science utilized for policy formation. This approach, called the 'analytic-deliberative' process, was originally developed by the Nuclear Regulatory Commission (NRC) because they recognized that informed decision making requires directing scientific efforts to areas that are most relevant to the issue. In addition, the NRC moved away from thinking of risk characterization as something that occurs after risk assessment is finished. Rather, it should inform the entire process so that the right questions are asked at the beginning. For Douglas (2009), the 'analytic-deliberative' framework incorporates values into the risk assessment process while, at the same time, ensures that values do not play illegitimate roles and makes public participation possible.

The process can be broken down into two parts (Douglas, 2009). First, the 'analytic' part includes using exact, repeatable methods that are evaluated by protocols agreed to by an expert community. These methods should ultimately arrive at answers to actual questions. Second, the 'deliberative' part includes formal and informal processes that facilitate a collective consideration of the topics of concern. These deliberations should be objective and include diverse viewpoints relevant to the assessment. Within the 'analytic-deliberative' process, the two parts should be ongoing and mutually supporting. You do not 'finish' the analytic part and then turn to the 'deliberative' part. Instead, the process is recursive. The two parts influence each other as analysis provides new insights and information and deliberation uncovers new questions and problems to investigate. Within this process, deliberation is needed to ensure that suitable values shape the decisions concerning which analyses are completed. Thus, for Douglas, this is where the public can play a direct participatory role within the process.

However, there is one context of judgment that is overlooked by the NRC report, according to Douglas (2009): when we must assess the sufficiency of evidence when faced with uncertainty. Within this context, such judgments must include assessments of possible consequences of error that include ethical and social values. For Douglas, these deliberative processes would be greatly enhanced by incorporating relevant public values and interests in the process. In addition, when experts are at a loss as to how to proceed or are faced with equally viable options, Douglas argues that they should then consult a broader group within the deliberative process to assist them in deciding how to proceed (163). Thus, for Douglas, the 'analytic-deliberative' approach is one where the public can play a greater role in shaping the risk assessments and where values are not used in an illegitimate way. This is especially relevant because, as argued above, food safety policy deals with uncertainty and there is the possibility of food industry influence.

But, won't the inclusion of public values and public voices threaten the objectivity of food science in general and scientific risk assessments in particular? How can we be sure that adopting something like the analytic-deliberative model does not threaten the neutrality of science? We certainly do not want to base our scientific evidence on what we merely would wish to be true about the world. In order to show how these problems are illusory, we will now turn to an analysis of the specific issue of GMO regulation from the perspective of Sandra Harding's strong objectivity methodology.

Legitimizing scientific authority through strong objectivity

Sandra Harding's (2004) methodological proposal of strong objectivity (SO) represents a particular kind of the critical approach outlined at the beginning, as it is applied to scientific research. SO is billed as a way to actually make scientific practices more objective than was previously the case by challenging the biases and prejudices characteristic of the scientific community at large, not just the biases of individual scientists. After elaborating on the problems at the science-policy nexus and on the resources of SO, we illustrate SO's effectiveness for this problem through the example of GMO regulatory policy. We will show why public participation should not be viewed as a potential enemy for the objectivity of science, but rather as an effective bulwark against systemic biases that can hinder scientific objectivity.

As science and industry have advanced, more complex and less well understood risks facing health have come into existence. Political systems responsible for protecting citizens from undue harm have had to keep up with creating more complex regulatory schemes. These regulations require massive input from scientific advisers for their design and implementation. Thus, the authority and power of scientific advisers has increased greatly. The interaction between three factors makes this increased power politically and ethically problematic: information asymmetry, the need for delegation, and the ideal of the autonomy of science. Information asymmetry exists between scientists on one side and citizens and policy makers on the other side (Guston, 2000). David Guston (2000:17) puts it succinctly, 'Scientists know things about the conduct of research that politicians and administrators do not.' Thus, there is a need for delegation. That is, policy makers need to delegate some authority for the design, implementation, and evaluation of regulatory frameworks to scientific advisers. The democratic legitimacy of this delegated power is further complicated by the prevailing ideal of autonomy for scientific research. This is the ideal that scientific research, in order to be objective, must be completely free from the 'corrupting' influence of social values and private interests. It is also known as the value-free ideal (VFI). Although scientists have been delegated a political power only made legitimate in a democratic society by its effective subservience to the public interests it represents, the scientific ideal of autonomy automatically frowns upon the influence of values and interests in science.

SO will help us break through this impasse between democratic legitimation and value-free science by explaining why the myth of value neutrality works against the objectivity of science. SO is the articulation of the logic of scientific research as influenced by feminist standpoint theory (Harding, 2008: 117). In rationally reconstructing the methods of feminist scientists working in predominantly male scientific communities that were uninterested in research that had value for the real concerns of women, Harding noticed an interesting phenomenon; these feminist researchers did not start off their research from the standard models established in their fields (116). Instead, they started their research off from the real and concrete concerns of women's lives in order to understand why the dominant social and psychological structures, both in their particular scientific disciplines and in the society at large, led to such distortion and falsehood about the objective circumstances of their lives. These researchers were able to reveal truths about the lives of women – truths that male dominated science missed – only because their scientific work was explicitly directed by political values of anti-sexism.

SO's criticism of the VFI is that it fails to achieve the goal it sets out for itself. By holding science to an impartial standard, where values are not supposed to play any role in scientific research, the VFI aims to separate politically motivated bad science from science that gets at the truth irrespective of its consequences for political decisions. Yet, the only values that such a science is able to detect are those values that are not shared by the scientific community at large. Community wide values are made largely invisible by the VFI, and it is among these community values that oppressive biases may be located. While a scientific community might be able to root out bad science that is skewed by the idiosyncratic values of one researcher, the prejudiced values shared by the community – for example, if the community is predominantly male and holds unexamined assumptions about female psychology – will escape detection. A politically aware and critically reflexive science, on the other hand, enables researchers and the interested public to explicitly identify where, what, and when values come into play and, hence, consider whether or not those incursions need adjustment (Harding, 2004: 136-137).

The strong objectivity approach will help illuminate current problems in standards for regulating GMOs. The Coordinated Framework for the Regulation of Biotechnology (CFRB), the overall framework guiding policy on GMOs in the United States, has several problematic features. First, it assumes that there is no 'in kind' difference between risks from GMOs and risks from non-genetically modified organisms. Thus, it suggests no prima facie reason for investigating increased risk from GMOs on the sole basis of being genetically modified. Second, the CFRB follows wider federal standards by requiring policy be based on risk assessments that reference economic cost-benefit analyses and 'neutral' scientific data (Kuzma and Meghani, 2009).

At a first glance, we see how the science that under-girds policy decisions on the comparative risks of GMOs is in fact not as value-neutral as the CFRB seems to assume. Kuzma and Meghani (2009) write, 'All four stages of risk assessment have normative components. Decisions have to be made that involve non-epistemic values and concerns during hazard identification, dose–response assessment, exposure assessment and risk characterization.' This does not mean, however, that the science of risk assessment is invalid or too subjective. This is only a problem if we continue to think that values in science automatically threaten the objectivity of science. Strong objectivity helps reveal why such an assumption is false. If we can become critically aware of the way in which values influence the processes of science – both in the context of discovery and in the context of justification – then we can develop methods for effectively monitoring our practices for the wrong kinds of value influences.

Public participation in risk assessment concerning GMOs is one method that can help strengthen the objectivity of the risk assessment research at the same time as it helps ensure the democratic legitimacy of the policy decisions. Public involvement is not only ethically justified, as our previous section shows, but it is also epistemologically justified. Kuzma and Meghani (2009) provide an example of this: 'An experienced farmer might know better the type and timing of the pests that affect his or her fields than a regulator, even if that regulator is an entomologist.' The specialized knowledge of the practitioner outside of official science provides a missing perspective for research on the ecological effects of GMOs. Without public participation, the scientific community as a whole might have remained blind to the particular value laden perspective that enables the farmer to see with more accuracy the effects of pests on his crops.

One other lacuna in traditional analyses of scientific objectivity is the 'context of discovery.' This is the context in which decisions are made about what to study. It is assumed that, just as long as objective methods are employed, the way in which or reason for why the initial hypothesis was formed does not matter for the overall objectivity of the results. It does not matter, for example, if a study on GMOs was funded by agribusiness invested in genetically modified seeds, just so long as the study adheres to proper methods during research. However, decisions at this stage *do* matter in terms of our overall state of knowledge concerning the world because what we choose to study and what we choose to *not* study determines where the gaps in our knowledge exist. If the knowledge is meant to inform our practical decisions concerning how to regulate GMOs, then the decisions made during the context of discovery are important for the objectivity of our practical judgments. For example, little research has been done on the long term safety risks that GMOs pose to vulnerable populations – children, the elderly, and pregnant women (Kuzma and Meghani, 2009). Our judgments concerning GMO policy will likely not be objective in the sense of free from bias towards a particular group or solution because it will not be reliably sensitive to the facts about GMO's effects on vulnerable populations. If research on GMOs is funded primarily by interested corporate parties, we are likely to get an incomplete picture of the truth that favors the needs and interests of industrial agriculture. It is not as if the research will yield false results. Rather, the particular truths that we get access and the truths we still do not have access to will paint a partially distorted overall picture. Public participation makes it more likely that interested publics can mobilize to get the concerns of vulnerable groups on the agenda and thus ensure a less distorted overall picture.

Yet, how do interested publics mobilize to ensure that national food policies meet their local concerns? Are there specific strategies that are more effective than others? Are some strategies counter-productive? How do local communities fight against unresponsive scientific and political bureaucracies that are not motivated to elicit public participation? This final step in our analysis from the theoretical-political level to the national-political and further on to the local-political level is addressed in the final section through reflections on the local food movement.

The ontology of people, food, and locality within local food

Food can be seen as a basic element in the construction of our own subject-hood and our society. In its most prevalent forms, food, its distribution, creation, and consumption reinforces our integration with the institutions of global capitalism. For many people in the West, the way we get our food reinforces our dependency on distant experts and specialists coordinated by the dominant structures of power. In the dominant paradigm, food is constituted as a consumer good and a tool of power, while we as subjects are seen as consumers, whose only way of making change and asserting our autonomy is via the ultimately pointless choice of which brands we wish to align ourselves with through consumption. This leads to a loss of agency and empowerment, even as we are told that the vast amounts of food being produced (and wasted) are worth it. Similarly, food is seen as being inevitably detrimental to the natural world, as we manipulate an enslaved natural world to produce ever-greater quantities of food.

One increasingly popular way to at least begin to address these kinds of problems that has been widely proposed is the idea of local food. Many see this as a valuable strategy for changing our relationship to food to change our own subjecthood and the means by which we can influence our society. Local food is seen as a ways to help individuals and communities once again be able to understand the food on which they are dependent, and allow them to more directly (in cooperation with their neighbors) influence the nature of this food. Local food might be able to help construct what Dahlberg (1993) calls a 'regenerative food system' which is integrated at multiple levels, where each level has its own goals, but where each level adds to the overall health and resiliency of the system, and which does this while reinforcing 'The three E's-Ecology, Ethics, and Equity.' Local food might be a viable way to move toward this because it can help the health of individuals while also causing them to make more connections with their local environment and neighbors. This undermines our identity as consumers and creates spaces for other forms of identity, such as community member or food provider.

However, there are some problems with this rosey picture. Delind (2011) points out that some ways that local food is advocated for and supported works against the more profound challenges to the industrial capitalist culture in which we live. We argue, however, that this is not an indictment of local food as a whole. Rather, the local food movement can usefully be seen as fairly distinct movements sharing the same moniker of 'local'. These can be differentiated by the way they understand and reinforce identities of people, food, and locality. Some versions of the local food movement sees people as consumers, food as a product for those consumers, and locality as a useful way to improve that product. A typical example of the kinds of actions undertaken by this sub-movement would be consumer movements to get large retailers to carry more local food. This approach to local food does little if anything to challenge the status quo. Rather, local food is seen as another good choice for people with enough privileges to make those choices, and another layer to one's identity as a consumer defined by brands. Perhaps not surprisingly, the version of 'local food' that reinforces rather than undermines dominant power structures, despite being the least effective, is the most discussed and supported in mainstream discourse (Guthman, 2004; Horgen, 2001; Mitchell, 2009; Pollan, 2009, 2010, 2011).

Another kind of movement that would also call itself a part of the 'local food movement' does not think that the pressure we can put on companies by our purchasing choices ought to be its focus. Instead, it focuses on making policy changes at a macro level to encourage the production and consumption of local food. The reasoning for this is that it is often a better means of distribution than centralized and top-down systems, particularly in response to environmental catastrophe. A typical example of actions undertaken within this movement would be organizations like FAO or USAID adopting a program to encourage local production and consumption of at least some food. This movement conceives of people not as individualist consumers but as participating citizens who work to make institutions use local food to be more efficient (or at any rate it sees the white, Western people who work with and for such institutions. The ways in which it conceives of the people who are the object of this aid is much less benign). This version of the local food movement conceives of food as an institutional good, and locality as an efficient way to distribute that good. This discourse on local food has more space for agency, but still only constrained within existing institutions (FAO, 2003; Mount, 2012)

A third way to think of local food conceives of people as members of communities and sees food as being co-constituted with those communities. Locality, then, is a way of building more just communities and inter-community relationships. This version of local food can be seen as the local food emphasis within the food sovereignty movement, which aims to use food as a uniting base to pursue many issues of justice from gender and race issues to environmental justice (Flora, 2011). This version of 'local food' is the least well-defined, but possibly the one with the most radical potential. It is also, not surprisingly, the one least discussed in mainstream discourse (Desmarais, 2008; Lyson, 2004; Schanbacher, 2010; Whittman, 2010).

One thing to notice here is that no version we have identified within 'local food', even the third, argues that all food ought to be local. Rather, to the extent that more food can be localized, the benefits each group advocates are likely to occur. Another thing to notice is that while the above paragraphs may seem to be a critique of consumer activism, and to a certain extent they are, I am not arguing that these different movements are inherently incompatible. There is no reason to think that *strategically*, one might not ally with people advocating other conceptions of local food, nor even pursue goals that come from more problematic concepts, as long as this is done with a clear understanding of what one is doing. The problem comes when, as is the case with many advocates of local food, the various concepts that underly the term are confused with each other, and the extent to which they come from incompatible ontologies and are aimed at incompatible goals is not understood. This leads to confusion, internal conflict, and ultimately ineffectivity. By coming to understand what the different people might mean by 'local food', one can more clearly analyze what the possibilities and challenges in local food.

Conclusion and final reflections

The ethical problems surrounding food consumption, production, and regulation are enormously complex. This complexity is so enormous such that it resists a definitive problem-frame that can be adequately described from a single disciplinary background. It represents one of many 'wicked problems' (Rittel and Webber, 1973). With that in mind, the 'least bad' way of taming such a wicked problem might be through on-going processes of collaboration along diverse disciplinary divides. Our contribution to the ongoing discussion attempts to provide an account of what certain approaches in philosophy might add. Each one of the authors of this contribution bring different philosophical lenses to shed light on different aspects of food policy. We recognize the inherent limitations of our approaches, given that the problem and solution frames we employ do not include equally applicable frames from outside of philosophy. Thus, this contribution is meant to be a provisional foray into the complex world of food policy and should always be considered alongside the contributions from other perspectives. We do not expect our contribution to be definitive or comprehensive. Rather, we expect it to be useful, illuminating, and productive of workable solutions that are mutually justifiable from multiple perspectives and always open to modification in light of on-going experience.

References

Bordieu, P. (1984). Distinctions: a social critique of the judgment of taste. Routledge, London, UK.

Center for Disease Control and Prevention (2011). CDC estimates of foodborne illness in the United States. Available at: www.cdc.gov/foodborneburden/2011-methods.html.

Dahlberg, K.A. (1993). Regenerative food systems: broadening the scope and agenda of sustainability. In: Allen, P. (ed.) Food for the future: conditions and contradictions of sustainability. Wiley, New York, NY, USA.

DeLind, L.B. (2011). Are local food and the local food movement taking us where we want to go? Or are we hitching our wagons to the wrong stars? Agriculture and Human Values 28: 273-283.

Delind, L.B. and Howard, P.H. (2008). Safe at any scale? Food scares, food regulation, and scaled alternatives. Agriculture and Human Values 25: 301-317.

Desmarais, A.A. (2008). The power of peasants: reflections on the meanings of La Vía Campesina. Journal of Rural Studies 24: 138-149.

Douglas, H. (2009). Science, policy, and the value-free ideal. University of Pittsburgh Press, Pittsburgh, PA, USA.

Douglas, M. and Isherwood, B. (1996). The world of goods: towards an anthropology of consumption. Routledge, New York, NY, USA.

Fiorino, D. (1990). Citizen participation and environmental risk: a survey of institutional mechanisms. Science, Technology, and Human Values 15: 226-243.

Flora, C.B. (2011). Review: Schanbacher, W.D.: The politics of food: the global conflict between food security and food sovereignty. Journal of Agriculture and Environment Ethics 24: 545-547.

Food and Agriculture Organization of the United Nations (2003). Trade reforms and food security: conceptualizing the linkages. FAO, Rome, Italy.

Food and Drug Administration (2011). Background on the FDA food safety modernization act. Available at: www.fda.gov/Food/GuidanceRegulation/FSMA/ucm239907.htm.

Guston, D. (2000). Between politics and science: assuring the integrity and productivity of research. Cambridge University Press, Cambridge, UK.

Guthman, J. (2004). Agrarian dreams: the paradox of organic farming in California. University California Press, Berkeley, CA, USA.

Harding, S. (2008). Sciences from below: feminisms, postcolonialities, and modernities. Duke University Press, Durham, NC, USA.

Hassanein, N. (2011). Matters of scale and the politics of the food safety modernization act. Agriculture and Human Values 28(4): 577-581.

Heffernan, W. (1999). Consolidation in the food and agriculture system. Report to the National Farmers Union. National Farmers Union, Denver, CO, USA

Hrding, S. (2004). Rethinking standpoint epistemology: what is 'strong objectivity? In: Harding, S. (ed.) The feminist standpoint theory reader: intellectual and political controversies. Routledge, New York, NY, USA.

Izberk- Bilgin, E. (2010). An interdisciplinary review of resistance to consumption: Consumer movements, activism, and ideology. Consumption Markets and Culture 13(3): 299-325.

Kuzma, J. and Meghani, Z. (2009). The public option. EMBO Reports 10(12): 1288-1293.

Lyson, T.A. (2004). Civic agriculture: reconnecting farm, food, and community. Tufts University Press, Medford, MA, USA.

Mitchell, S. (2009). The corporate co-opt of local. The wire. Available at: www.wirenh.com/Features/Features_general/the_corporate_coopt_of_local_200.

Mount, P. (2012). Growing local food: Scale and local food systems governance. Agriculture and Human Values 29(1): 107-121.
Pollan, M. (2009). A defense of food: an eater's manifesto. Penguin Press, New York, NY, USA.
Pollan, M. (2010). The food movement, rising. The New York Review of Books, June.
Pollan, M. (2011). Food rules: an eater's manual. Penguin Press, New York, NY, USA.
Rittel, H. and Webber, M. (1973). Dilemmas in a general theory of planning. Policy Sciences 4: 155-169.
Roberts, T. (2007). Willingness to pay estimates of the societal costs of US food-borne illness. American Journal of Agricultural Economics 89(5): 1183-1188.
Schanbacher, W. (2010). The politics of food: the global conflict between food security and food sovereignty. Praeger, Santa Barbara, CA, USA.
Schrader-Frechette, K. (1991). Risk and rationality: philosophical foundations for populist reforms. University California Press, Berkeley, CA, USA.
Ulver-Sneistrup, S., Akergaard, S. and Kristensen, D.B. (2011). The new work ethics of consumption and the paradox of mundane brand resistance. Journal of Consumer Culture 11(2): 215-238.
Whittman, H. (2010). Food sovereignty: reconnecting food, nature, and community. Food First Books, Oakland, CA, USA.

Section 13. Teaching veterinarians

Vethics: professional ethics for veterinary officers

H. Grimm and K.L. Weich
Messerli Research Institute, Ethics and Human-Animal-Studies, University of Veterinary Medicine Vienna, Medicine University of Vienna, University of Vienna, Veterinaerplatz 1, 1210 Vienna, Austria; herwig.grimm@vetmeduni.ac.at; kerstin.weich@vetmeduni.ac.at

Abstract

In 2012, the project 'Vethics – Professional Ethics for Vets' has been launched at the Messerli Research Institute. This project is funded by the Austrian Ministry of Health and aims at encouraging and supporting practice-oriented interdisciplinary research in the context of ethics in human-animal interaction. Against the background of changing perceptions of animals in Western societies, the aim of the project is to support official veterinarians with regard to ethical decision-making (1). As veterinarians and official veterinarians play a major role in the field of human-animal interaction, the idea of the project is to explore ethical dimensions of practical day-to-day life of official veterinarians. Here, the link between professional experience and practical problems with ethical theory gains importance. More specifically, some key topics and crucial conflicts emerging in official veterinarians' professional life have been identified that provide the thematic structure of the project (2). Amongst others, 'killing of pets' was one of them and the first one elaborated in two workshops. First results of these workshops and the status quo of the project will be briefly presented in the following (3). Relevant insights into the normative structure of ethical challenges related to 'killing pets' as part of the official veterinarian's professional responsibility will be discussed in brief. The paper concludes with general considerations about the general idea of a veterinary ethics (4).

Keywords: official veterinarians, human-animal interaction, applied ethics, euthanasia

Introduction: human-animal interaction

Significant and obvious change is taking place in our Western societies with regard to human-animal interaction. On the one hand, animals are considered family members (cf. Cohen, 2002; Walsh, 2009). On the other hand, they appear to be mere production units e.g. in the context of farming. The prominent example is the killing of male day-old chicks in gigantic numbers (cf. Aerts *et al.*, 2009: 117). This is due to the fact that they have no (or even a negative) economic value. The most controversially debated Dutch artist Tinkebell puts her finger right on it when she announced to kill 61 day-old chicken in a performance (SZ, 2005). A foreseeable and immense media response was the result. Animal protectionists bought the 61 chicken, therewith rescuing or extending the day-old chicken's lives. Tinkebell identifies a pressing issue: On the one hand, farm animals like pigs, cows and chicken are raised to become tasty food. On the other hand, biologically relatively similar animals are perceived as invaluable family members. This paradox has lead Tinkebell to the statement that our attitude towards animals is pathological. The relationship between man and animals is changing and gathers a lot of media attention. These developments have consequences for professionals dealing with animals. Veterinarians are in the front line whenever and wherever practical decisions have to be made. They are challenged with diverse normative views of biologically similar animals.

Especially veterinary officers often find themselves in a field of tensions between economic limitations, animal protection, political interests, legal requirements and media attention. This is particularly true when it comes to unpopular measures, such as killing healthy animals to prevent epidemic outbreaks, killing healthy animals in overcrowded shelters, killing aggressive or unwanted pets, etc. On the one hand, they are responsible for securing animal interests. On the other hand, they are as well in charge of public interests. In all these instances, one of the main challenges is to mediate between human and animal interests within a social context and within practical constraints.

Since a relevant part of decision-making with regard to human-animal interaction is ethical in nature, the idea of veterinary ethics (cf. Tannenbaum, 1995 [1989]: 2) and the outlined project is to support official veterinarians in ethical decision-making. One of the main ideas of the project is to explore ethical dimension of practical day-to-day life of official veterinarians. Consequently, the link between professional experience and practical problems with ethical theory gains importance. Questions about the normative background of this profession and ways to use this background to solve practical problems arise. As indicated by Carr (1999), professionalism incorporates 'distinct ethical dimensions' as one of the 'principle criteria of professionalism' that can be analyzed: (1) professions provide an important public service; (2) they involve a theoretical as well as practical grounded expertise; (3) they have a distinct ethical dimension which calls for expression in a code of practice; (4) they require organization and regulation for purposes of recruitment and discipline; and (5) professionals require a high degree of individual autonomy – independence of judgment – for effective practice (Carr, 1999: 34).

Since official veterinarians regularly have to deal with decisions that incorporate moral aspects, it is worthwhile to reflect on the ethical background of their professional life. Adequate decision-making requires sophisticated empirical knowledge as well as knowledge about values and principles relevant to the field. In the following, a brief summary of the project 'Vethics', its project design, first results and upcoming topics will be presented.

Vethics: professional ethics for official veterinarians

Official veterinarians are regularly faced with moral questions in their professional life. Against this background, the project 'Vethics – Professional Ethics for Veterinary Officers', which has been launched in November 2012 at the Messerli Research Institute, aimed in a first step at recognizing the wide range of ethical issues in the professional life of official veterinarians.

Project design and planning phase

During the project's planning process, the idea to establish professional ethics for official veterinarians was presented at three conferences in different parts of Austria, where official veterinarians and authorities exchanged and discussed state-of-the art knowledge and developments in their profession. The presentation consisted of a general introduction into the field of ethics in human-animal interaction, a brief outline of veterinary officers' current challenges, and an outline of major characteristics of 'professional ethics'. The idea of 'professional ethics' was presented along the following characteristics: (1) specific to a professional context; (2) problem-oriented and therefore integrating the professionals' knowledge; (3) pragmatic in the sense that it should contribute to better decision-making and can be integrated into future professionals' education; and (4) practice-oriented, meaning easy to pick up and understandable. After the presentation, the audience was invited to discuss and identify pressing issues.

The presentation of the project's aims and ideas and the consequent discussions contributed significantly to the project design. In particular, topics and sub-topics were identified and selected in the planning stage of the project. The official veterinarians were invited to hand in further relevant issues per e-mail. Already in this phase, official veterinarians from all Austrian Federal States were involved and contributed with their ideas and insights into relevant ethical issues of their professional life. Within a timeframe of roughly six months, three main topics have been identified together with Austrian veterinary officers. After that phase, the list with three main topics and sub-topics was closed. As a result, the following list provides the thematic structure of the project for three years:

- 2013: killing animals:
  - euthanizing pets: killing to end suffering/killing to serve public interest;
  - slaughtering: killing for food;
  - culling: killing to deal with epidemic outbreaks
- 2014: use and abuse of animals:
  - farm animals: using animals as living resources;
  - farm animals in touristic regions;
  - new perspectives on animals.
- 2015: psychosocial reasons for animal welfare problems:
  - animal hoarding;
  - anthropomorphizing animals;
  - high tech veterinary medicine.

Workshops and methodology

In spring 2013, two workshops on the first topic 'euthanizing pets' have been carried out. The aim was to have 10 to 12 veterinary officers at maximum per workshop. Besides the veterinary officers, two ethicists and one person for documentation attended the workshops.

As mentioned above, Vethics is designed to develop practice-oriented professional ethics. Therefore, ethical theories and concepts are used to reflect and frame practical ethical problems in specific professional contexts. The guiding understanding of ethics within the project is to reflect practical situations, in which well-established guiding principles or norms lose their plausibility or orienting function, so reflection and normative guidance is needed. In this sense, ethics comes into play when customs, lifestyles and normative institutions lose their guiding function and formerly self-evident normative beliefs fail do direct actions (cf. Höffe, 2008 [1977], 10). In this respect, the project follows the idea that Tannenbaum coined as 'normative veterinary ethics' (cf. Tannenbaum, 1995 [1989]: 41).

Workshop 1: euthanizing pets

The main task of the first workshop was to structure and analyze the ethical problem of euthanizing healthy pets from an official veterinarian's perspective. Therefore, practical experience from 12 participating official veterinarians was reflected, applying ethical methodologies under the supervision of two ethicists. For this purpose, the Ethical Matrix (Mepham *et al.*, 2006) was used to analyze two study cases: (1) aggressive dog; (2) cats in an overcrowded animal shelter. Both cases were 'real life situations' introduced by two official veterinarians, who had to deal with these situations.

The Ethical Matrix was used to compare two strategies of dealing with the study cases in groups of three people: *strategy 1*: euthanizing; *strategy 2*: non-euthanizing. The intention of comparing a euthanizing and a non-euthanizing strategy was to make explicit the 'hot spots' of different strategies in one case. Unsurprisingly, from an official veterinarians' point of view, both strategies lead to different, but nevertheless moral challenges and problems. Whereas the strategies that saved the life of animals lived up to animal protection standards and failed in respect of public interests, euthanizing strategies lived up to public interests and failed in respect of the wellbeing and respect of life.

During a chaired discussion of the results, we elaborated and discussed the 'hot spots' of the various strategies. It became obvious that official veterinarians have to consider not only the norm of the wellbeing (in terms of ending suffering), but also public interests, such as safety or considering monetary costs in human-animal interaction in their professional life. After the discussion we were able to gather relevant aspects in two categories: (1) the individual animal's wellbeing and respect for the animal's life (norm 1); (2) public interest (norm 2). Both categories can be seen as relevant sources of reasons to justify killing animals and have plausibility for veterinary officers.

This first result is particularly relevant in the context of Austrian legislation since the Austrian Animal Protection Act (AAPA) asks for a 'justifying reason' to legally kill animals. Whereas killing an animal in order to gain food is accepted as 'justification', there are not such clear statutory provisions for legally killing dangerous animals or supernumerary animals that have been confiscated. Whether or not further reasons can justify killing of animals, is a highly debated topic in Austria. The AAPA does not identify relevant reasons to kill animals, but intentionally leaves it to debate and within the responsibility of experts what can be considered a justifying reason. Often the 'justifying reason' is found in 'ending suffering' in this debate. From the official veterinarian's perspective, also other reasons are considered relevant. Whereas practical veterinarians are mainly faced with the ethical question of euthanasia in the sense that killing of animals is a measure to end suffering, official veterinarians have to deal with the ethical problem of killing animals for other reasons than ending suffering. Therefore, the official veterinarians' responsibility often differs from the practical veterinarian's responsibility. Different levels of veterinary officer's responsibility with regard to killing animals have been differentiated and listed, using the structure established by Ropohl. He provided a feasible analytic framework to address practical dimensions of responsibility (Ropohl, 1993 [1987]). The focus was first on exploring the limits of individual responsibility of official veterinarians and on ways to deal with actual responsibility.

First results: workshop 1

In the workshop it was clearly indicated by official veterinarians that they find themselves in a situation where not only animal interests and/or their keepers' interests matter. The crucial problems arise because of other interests than the animals' interests that play a major role in their professional life. These are mainly public interests, such as safety in the case of dangerous dogs, spending tax money on shelter animals, public health and so forth. It became clear that official veterinarians *have to* consider public interests as justifying reasons for killing animals. Otherwise, their profession could not be carried out in legal terms. Since official veterinarians are not only held responsible for animal wellbeing and the life of animals, but also for public interests, both norms have to be considered in ethical decision making. This is not an easy task, since the two norms are often in conflict and professional obligations may further conflict with personal attitudes and moral convictions.

Workshop 2: euthanizing pets

The second workshop built on the results of the first one and was organized six weeks later. This time a new group of 10 participants contributed and carried on, where the first group had stopped. Therefore, the main results of the first workshop, which said that official veterinarians had to consider two conflicting norms in the context of euthanizing pets, were transformed into two main tasks for the second workshop:

1. Comparative analyses: Identification of qualifying criteria for decision-making.
2. Ethical position of veterinary officers: are the two norms (animal wellbeing and the life of animals vs. public interests) equally relevant in the professional life of official vets?

Comparative analyses

In a first step, an 'ideal case' of euthanasia was presented in terms of a narrative in order to compare and contrast it with other 'non-ideal' cases: In short, the narrative of 'ideal euthanasia' told the story of Paula, an old, friendly, female Labrador. She had been part of the family for 16 years. One day she behaved strangely, so the father of the family took her to the vet. The practical veterinarian diagnosed cancer.

Luckily, Paula was not suffering. She was taken home from the vet. After a couple of days, the vet visited the family. The time had come. All family members gathered in the living room and said good bye to Paula. Then she peacefully fell asleep with the help of the vet...This story was used to develop criteria about an 'ideal euthanasia'. It turned out that an ideal euthanasia in a normative sense is characterized by the fact that 'nobody can be blamed'. Further, and this is only the same thing in other words, if it 'goes without saying' that 'the time has come' and it is 'obviously clear to everybody' that the dog has to be euthanized, we can talk about an ideal euthanasia. It goes without saying that such criteria are not scientific in nature. However, empirical facts are of importance and provide the basis.

During the discussion, this insight was turned into another direction: The social fact that 'nobody can be blamed' can be read as 'there is nobody who has to take on responsibility'. If everything is entirely clear and no doubts arise that euthanasia is the measure to be taken, we do not speak of a moral decision or a moral problem. Quite the opposite, whenever it is entirely clear what is to be done and acting is according, we do not speak of a moral problem but of 'fulfilling an accepted norm'.

Despite the fact that all participants considered 'Paula' an ideal case, it became transparent that this ideal pervaded and influenced decision-making processes in the field of killing animals as a norm in general. This could be made explicit in the second part of the workshop. In the second part, important characteristics of the study case 'Paula' were contrasted with characteristics of the two study cases of workshop 1: aggressive dog (A); cats in an overcrowded animal shelter (B). The participants were asked to contrast the strategy to euthanize the aggressive dog and cats in an overcrowded shelter with the 'ideal case' Paula along the following questions:

- In which respects do A and B differ from 'Paula'?
- Is there someone to be blamed for, if animals are killed in A and B? If yes, who and why?
- Are A and B forms of euthanasia? If yes, why?

To put it briefly: Similarly to the results of the first workshop, veterinary officers made the point clear that their professional life demands considering not only animal wellbeing and respect for life, but also public interests. In cases like A and B, official veterinarians are asked in their expert role. Since the legal framework does not prescribe how to decide on the case, professional autonomy and ethical decision-making is relevant in this context. The official veterinarians considered it their responsibility to support or even carry out such decision-making processes and contribute to tricky questions with their expert knowledge.

Knowing that both norms are relevant, the participants were asked about the relevance of the two norms in their professional life. Methodically, the official veterinarians discussed three real-life statements from practical veterinarians on 'euthanizing pets'. They were asked to point out whether they agreed or disagreed and whether animal interests or public interests should prevail. In a chaired discussion they stated their opinions on how official veterinarians should position themselves with regard to the two norms.

First results: workshop 2

Regarding the importance of the two conflicting norms, participants came up with very similar statements: A general positioning is not possible. Taking a position is only sensible when taking into account specific characteristics and the facts of the case. The heterogeneity of cases does not allow general decisions. Therefore, criteria like adequate diagnosis, thorough examination of contextual knowledge (animal keeper's condition, costs of alternatives to euthanasia, lifespan of the animal, etc.) and conclusive lines of argumentation are of major relevance.

General considerations

Applied ethics of human-animal interaction deals with describing practical situations and conflicts and their historical and cultural framework – this includes existing moral norms. Just as conflicts in our dealing with animals have to be understood as social heritage and can not be reduced to empirical questions, also animals in their specific apprehension become subject to ethical reflection.

In our opinion, taking into account practical constraints of professional responsibility is vital for the development of professional ethics. During the workshops with official veterinarians, ethical concepts and theories contributed to structure the professional field of veterinary medicine, providing freedom of action and opening room for interpretation. However, when it comes to decision-making, the question of how to consider animal interests as well as public interests in a reasonable way cannot be solved on an abstract level. Hence, the project aims at making these connections explicit in order to present cornerstones and elements of a critical attitude and practice as part of the professional life of official veterinarians.

References

Aerts, S., Bonnen, R., Bruggeman, V., De Tavernier, J. and Decuypere, E. (2009). Culling of day-old chicks: opening the debates of moria? In: Millar, K., West, P.H. and Nerlich, B. (eds.) Ethical futures: bioscience and food horizons. Wageningen Academic Publishers, Wageningen, the Netherlands, p. 117-122.

Carr, D. (1999). Professional education and professional ethics. Journal of Applied Philosophy 16(1): 33-46.

Cohen, S.P. (2002). Can pets function as family members? Western Journal of Nursing Research 24(6): 621-638.

Eurobarometer (2005). Special Eurobarometer 229. Attitudes of consumers towards the welfare of farmed animals. DG Health and Consumer Protection, Brussels, Belgium.

Höffe, O. (2008 [1977]). Lexikon der Ethik. Beck, München, Germany.

Mepham, B., Kaiser, M., Thorstensen, E., Tomkins, S. and Millar, K. (2006). Ethical matrix. manual (www.ethicaltools.info). LEI, The Hague, the Netherlands.

Ropohl, G. (1993 [1987]). Neue Wege, die Technik zu verantworten. In: Lenk, H. and Ropohl, G. (eds.) Technik und Ethik. Reclam, Stuttgart, Germany, p. 149-176.

Süddeutsche Zeitung (2009). Skandalkünstlerin Tinkebell. Available at: www.sueddeutsche.de/leben/skandalkuenstlerin-tinkebell-unser-verhaeltnis-zu-tieren-ist-krank-1.446297.

Tannenbaum, J. (1995 [1989]). Veterinary ethics: animal welfare, client relations, competition and collegiality. Mosby, St Louis, MO, USA.

Walsh, F. (2009). Human-animal bonds II: The role of pets in family systems and family therapy. Family Process 48(4): 481-499.

Assessment of effects of ethics teaching to 1st year veterinary students by using the ethical reasoning tool

T. Dich, T. Hansen and P. Sandøe
Institute of Food and Resource Economics, University of Copenhagen, Rolighedsvej 25, 1958 Frederiksberg C., Denmark; td@ifro.ku.dk

Abstract

Increasingly, courses are set up to teach ethics to veterinary students. However, little is known about the effects of these courses. An earlier paper presented at EurSafe by Dich *et al.* has examined different tools to assess the effects of ethics teaching. The conclusion was that the ethical reasoning tool (ERT) could constitute a valid measure of the changes in student cognition and the effect of the teaching programme. The aim of the present paper is to present a design for using ERT to measure the impact of a first year course in veterinary ethics, held at the University of Copenhagen, and to present a first set of results. The overall aim of the assessed nine-week course in veterinary ethics is to introduce different ethical positions to the students and to give them the opportunity to identify, work with and reflect on ethical questions related to the veterinary profession, as a part of their training to become professionals. Therefore, it is relevant to measure if the learning and teaching environment has brought changes in the students' capacity to undertake ethical reasoning. The results show that there is an increased ability to identify ethical issues and to utilize a framework within which to consider and clarify these issues. Some of the students also improved their decision-making skills and became more aware of their roles as veterinarians. However, a majority of the students didn't show any progress in their ethical reasoning skills.

Keywords: effects of ethics teaching, student learning, attitude change, ethical reflection

Introduction

The ethical reasoning tool (ERT) was originally developed to assess the cognitive reasoning of nurses in response to an ethical practice dilemma. It categorises responses to a selected, relevant case study into three professional response levels (McAlpine *et al.*, 1997):

- Level 1 (traditional): thinking dominated by the use of personal moral values and beliefs and/or conventional moral reasoning.
- Level 2 (traditional/reflective): practical considerations moderated by some use of reflective reasoning, indicating recognition of some of the relevant ethical issues, and the need for consideration of more than one's own personal beliefs.
- Level 3 (reflective/pluralistic perspective): critical thinking about ethical issues, with the use of an ethical framework and recognition of the value of other points of view.

Ethical development is here seen as a cognitive progression from a narrow and self-serving perspective to a potentially reflective and pluralistic perspective. It means that in order to undertake an ethical reflection, a person should be able to perceive and identify ethical issues, apply one or more ethical principles to a given situation, consider alternatives and should come to a personal decision on how she/he would behave, but still be aware that this might just be one of several, equally valid opinions.

This corresponds very well with the theoretical framework applied in the course studied in this paper, and the expected effect on students' learning. The studied course in veterinary ethics is based on the skill-based approach funded on a pluralistic view on ethical values. The purpose is to support the students in developing tools to recognise, analyse and solve moral problems that can be used in the process of ethical decision making and in dealing with ethical disagreements. Another aim of the course is the introduction of the students to different fields of activity in the veterinary profession and to the different ethical dilemmas one could face as a veterinarian (Hansen *et al.*, 2009).

The aim of this study is to present and discuss the results from the use of the ERT to assess the effect of a veterinary ethics course held at the University of Copenhagen in autumn 2010.

Method

Test

In the first class of the course in autumn 2010 a pre-test was given to the 166 present 1st year students. They were informed about the purpose of the exercise and anonymity was guaranteed by assigning a random number to each of their answers – a number which each of them were told to remember and keep by themselves. The students were also asked to inform about their gender and age.

The students were presented to a case, neutralisation of cats (spaying and castration), which included information on behavioural and health consequences both of the surgery and of allowing the cat to remain intact. At first, the students were asked to identify significant ethical issues raised by the case, then to state how they would advise the owner of the cat and to justify their answer, and finally to provide their views on the matter from other perspectives. They were given maximum 45 minutes to write down their answers. At the end of the course, the students were given a post-test which was exactly the same as the pre-test case study with the same directions. After completion of the test they could retrieve their own pre-test, and were asked to re-read both their previous and recent test answers, and to write down their reflections on their own responses to the advice they had given. Only upon completion of the post-test the students were fully informed about the method of assessment, in order not to distort the results by prior knowledge of what was to come.

Analysis

101 students completed both the pre- and post-test, which were then read by the two first authors in order to get an idea of the kind of responses coming from the students. It was found that they were reflecting four different ethical reasoning components: (1) recognition of ethical issues; (2) use of an ethical framework; (3) considerations on own reasoning; and (4) perception of veterinarian/client relationship. On the basis of this finding, combined with the three different levels of moral thinking used by the ERT, the ethical reasoning components were divided into the three levels of ethical reasoning (concept levels), which are also a part of the theoretical framework of the ERT. Exemplar behaviours within each concept level and ERT level were discussed, tested and moderated by the two first authors, and the result is shown in Table 1.

Using this matrix as a tool for further analysis of the responses, a random sample of 50 responses were selected and read by the first author (Figure 1 and 2).

Results

As this way of presenting the study only evaluates the development of the sample as a whole and not the individual student, it is not possible to see neither how many have progressed from level 1 to level 2 or 3 or from level 2 to 3, nor how many have regressed or not moved at all. The data material, however, gives the opportunity to provide such analysis, and the result is presented in Figure 3.

Table 1 Ethical reasoning components and response categories with exemplar behaviours used for the analysis

Ethical reasoning components	Concept level (levels of ethical reasoning)		
	Level 1 (traditional)	Level 2 (traditional/reflective)	Level 3 (reflective/pluralistic)
	• Predominant use of personal beliefs • Right/wrong views • Low/no recognition of ethical issues • Ethical egoism	• Conformity to contextual/social views • May experience cognitive dissonance, but obedience to perceived authority/power • Some recognition of ethical dilemmas	• Critical/reflective thinking about ethical issues • Pluralistic perspective: seeks viewpoint justification • Personal accountability/responsibility for choices • Challenges unethical practices
1. Recognition of ethical issues	• Uses own values • Low/no recognition of ethical issues • Practical aspects dominates	• Some recognition of ethical issues • Focus on welfare aspects	• Recognises ethical issues at a more general level – but relevant to the case • Recognises other's viewpoint
2. Use of an ethical framework	• No use at all • Personal opinion in focus	• Uses some ethical theories in the analysis	• Incorporates the ethical framework in the analysis and advices • Aware of the relevance of an ethical framework
3. Considerations on own reasoning	• No reflection on personal values • Right/wrong answers	• Some awareness of personal values versus ethical reasoning	• Comments on own argumentation and advices
4. Perception of veterinarian-client relationship	• No professional awareness • Personal based advice • Not aware of client perspective • Right/wrong view	• Some awareness of client's perspective and interest • Some recognition of the role of a vet	• Recognises different ways to give advices • Aware of client's perspective and interest • Goes beyond personal values

Discussion

The individually based way of presenting the results (Figure 3) seems to be a better way to illustrate the students' development as this shows how many students have actually moved from one level to another. This is therefore considered to be a better tool for the teacher in evaluating the effect of the teaching than using the ERT framework for the data presentation (Figure 2).

The results show that there is an increased ability to identify ethical questions (26%) and to use the ethical framework introduced at the course (34%). A majority of the respondents gave the same advises but where better in arguing for their viewpoints (36%) and some of them expressed more professional values where the interest of the client was taken into consideration (26%).

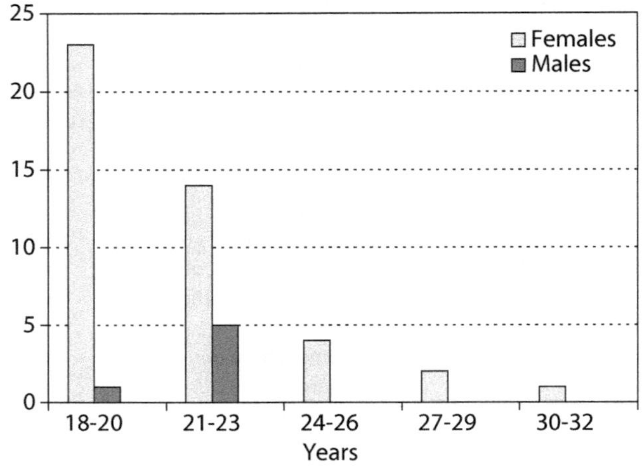

Figure 1. Frequency distribution of demographic variables of the sample (gender, age) (n=50).

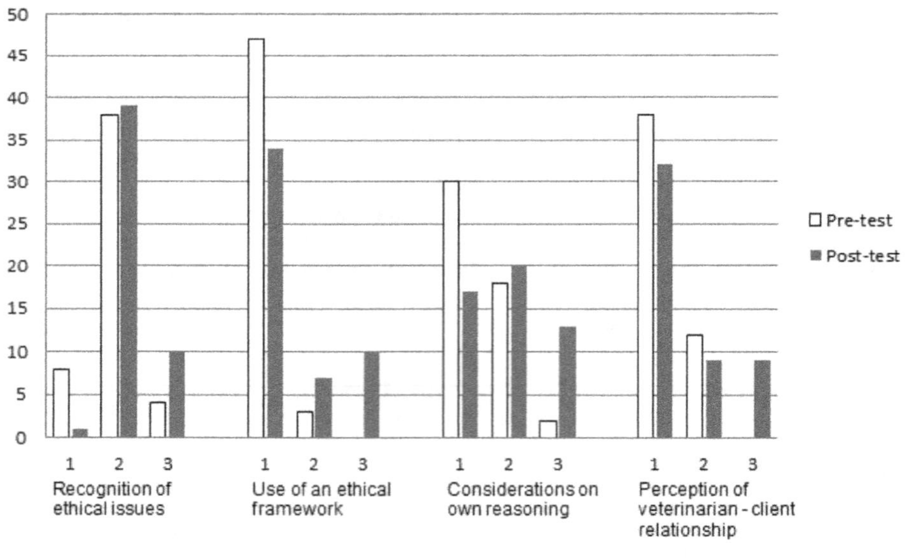

Figure 2. Changes in scores for ethical reasoning components, from pre-test to post-test using the ERT (n=50).

From the results presented in Figure 3 it is also seen that between 62% and 74% of the students did not show any development at all within the four ethical reasoning components. From a teacher's point of view this indicates that there is room for improvement.

It was found that not many students (16%) started on response level 1, the level where thinking is dominated by the use of personal moral values. This might, however, be due to the formulation of question 1 ('Which ethical questions does this case raise?'), where the students are already asked to

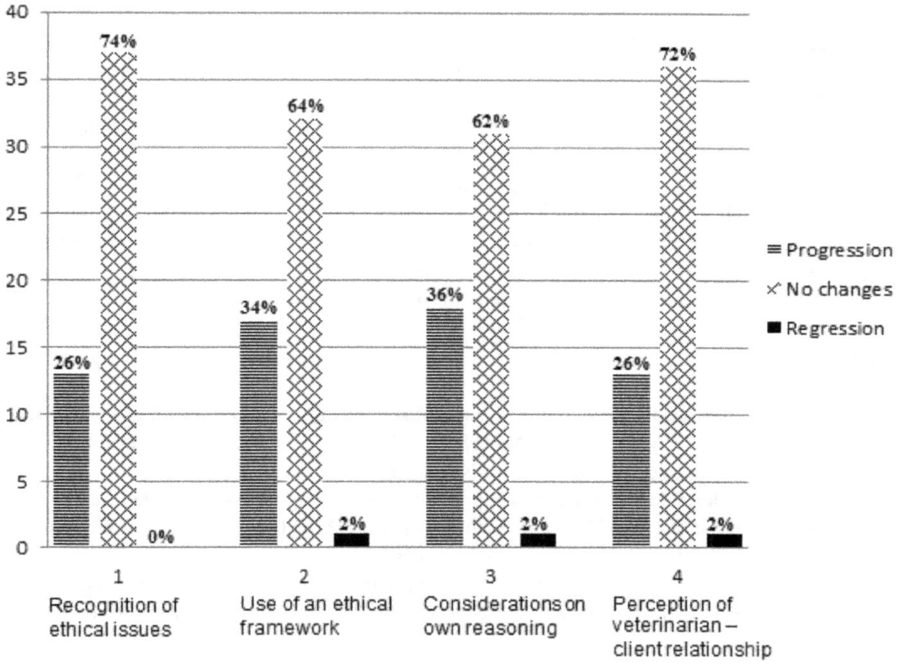

Figure 3. Individual student development from pre-test to post-test (n=50).

identify ethical questions. The question could have been more open without using the term 'ethical', e.g. 'What is at stake here?'. Another factor could also constitute from the fact that many students may have oriented themselves in the course material and course description, which gave them some ideas on ethical issues and ethical reasoning, i.e. they did not start from 'scratch'.

The validity of the outcome of the study is also influenced by the study design. As an explorative, observational study it is not possible to exclude the influence of other factors than the ethics teaching. It is though seen as a minor uncertainty factor as the students only had courses in zoology and chemistry in that period and no hands-on experience with animals or clients, which may have contributed to the students' cognitive development within ethical reasoning.

Another source of uncertainty, especially when it comes to the student's own comments about increased ability to view a case from more perspectives, could derive from the student's desire to demonstrate that she/he has learned something or to please the teacher if she/he considered the test as a kind of assessment or examination, even though everybody were informed that this was not the case.

Interpretation of the responses is one of the main sources of uncertainty. Firstly, the identification of the responses into the exemplar behaviours relies fully on how the reader 'translates' the responses that do not always fall clearly into one particular category. The uncertainty could have been reduced by letting more people read and rate the responses, and to discuss possible discrepancies until consensus was reached. Secondly, the reader who in this case is identical with one of the teachers, might have certain expectations to the outcome which influences the way the responses are read and interpreted. Positive expectation to the students learning as a result of the teaching is something all teachers aim for.

This study doesn't fully represent the gender distribution among the 1st year students. There is a slight over-representativity of males in the study (12% in the study versus 10% of all course attendants), which could influence the results. How this might influence the results could be interesting to analyse, i.e. is there any gender specific difference in the development of ethical reasoning? The same might apply for the age of the students, i.e. are there any correlation between age and ethical reasoning development?

The ERT is a valid framework for measuring changes in student cognition and the effect of ethics teaching programmes. It is, however, challenging to design a test that clearly let the responses fall into the three different levels both when it comes to the formulation of the questions and the interpretation of the responses. One way to deal with this aspect could be to formulate more open questions and to elaborate further on the exemplar behaviours within each concept level.

References

Dich, T., Hansen, T. and Sandøe, P. (2009). How to assess what comes out of ethics teaching in veterinary science programmes. In: Millar, K., West, P.H. and Nerlich, B. (eds.) Ethical futures: bioscience and food horizons. Wageningen Academic Publishers, Wageningen, the Netherlands.

Hansen, T., Dich, T. and Sandøe, P. (2009). A skill-based approach to the teaching of ethics in the life sciences – experiences from Denmark. In: Millar, K., West, P.H. and Nerlich, B. (eds.) Ethical futures: bioscience and food horizons. Wageningen Academic Publishers, Wageningen, the Netherlands.

McAlpine H., Kristjanson, L. and Poroch, D. (1997). Development and testing of the ethical reasoning tool (ERT): an instrument to measure the ethical reasoning of nurses. Journal of Advanced Nursing 25(6): 1151-1161.

Section 14. Poster session

Who is a 'consumer' on food law: some reflections on the notion of consumer and the EU food law

A. Arroyo Aparicio
Universidad Nacional de Educación a Distancia, Commercial Law, C/ Alba de Tormes, 49, Colmenar Viejo, 28770 Madrid, Spain; aliciaarroyoaparicio@gmail.com

Abstract

The notion of consumer as it is known in a legal sense differs from the concept of consumer as used in other matters (e.g. Sociology or Marketing). In legal sense, in a contractual way for instance, a precise definition of the 'consumer' is essential in order to delimit the persons entitled to extended legal protection in relations with traders. The wider the circle covered by the definition of consumer, the more extensive the scope of consumer law provisions is and the less reason there is to speak about consumer law as a special regulation. This is clear so far, but also it's clear that the notion of consumer itself contains several specificities. In this regard, it's possible to consider 'consumer' as 'citizen', what it would be an abstract approach, and also it could be used 'consumer' in several concrete approaches. The article will analyze the notion of consumer and General EU Food Law and the aim is to point out some conclusions about this concept.

Keywords: consumer protection, consumer *acquis*, consumer Policy, food rights, food safety

The notion of 'consumer'

Abstract notion and concrete notions

The primary idea is that there is not only one notion of consumer. Linked to this first basic idea is the view that it can be distinguished an abstract notion and different concrete notions (Bercovitz, 1987: 108).

A very clear statement of an abstract notion can be found in the 93 Special Message of President Kennedy to the Congress on Protecting the Consumer Interest, dated on March 15, 1962. 'Consumers, by definition, include us all' are the very well known opening words of this Message that summarised four fundamental consumer rights: the right to safety, the right to be informed, the right to choose and the right to be heard. It has to be appreciated that the abstract notion considers 'consumer' as a 'citizen'.

The above cited Message was relevant not only in the United States but elsewhere. In the mid 70s, the European Commission adopted its first ambitious consumer policy Program with the Council resolution of 1975 (OJ 1975 C 92/1), clearly inspired in the Message of President Kennedy. In this Program the consumer is no longer seen merely as a purchaser and user of goods and services for personal, family or group purposes 'but also as a person concerned with the various facets of society which might affect him directly or indirectly as a consumer' (n. 3 Council resolution).

The Council set out five basic rights: (1) the right to protection of health and safety; (2) the right to protection of economic interests; (3) the right of redress; (4) the right to information and education; (5) the right of representation (right to be heard). The right to protection of health and safety – relevant to Food Law – can be understood in 1975 as follows: goods and services offered to consumers must be such that, under normal or foreseeable conditions of use, they present no risk to the health or safety of consumers.

As already noted, the abstract notion of consumer is commonly used in policy Programs. The abstract notion is not the appropriate definition within which to address concerns with regard to the individual rights of consumer. Each one of the concrete notions shall ensure that the criteria to grant individual rights are complied. By contrast, because it is abstract, the use of the 'consumer' in policy Programs implies the general recognition of consumer right to education, right to information, etc (Bercovitz, 2012: 635).

Thus, the second basic idea is that there are several concrete notions. These different notions are used to determine the circle of persons entitled to exercise the specific legal protection.

In 1985, the first consumer protection directives, the Directives on Product Liability and Doorstep Selling were enacted. These directives contain two very different definition of consumer.

It can be mentioned that Directive 85/374 concerning liability for defective products applies to damage: caused by death or by personal injuries; caused to an item of property intended for private use or consumption. Article 9 lit (b) ii of this Directive 85/374 states that for the purpose of Article 1, 'damage' means a damage to, or destruction of, any item of property other than the defective product itself, provide that the item of property was used by the injured person 'mainly' for his own private use or consumption. It can be highlighted that this Directive is related to non-contractual obligations and the consumer is the 'injured person'.

A different scope of application can be found in the Directive 85/577 concerning Doorstep Selling. In this case, a 'consumer' is defined as 'a natural person who (...) is acting for purposes which can be regarded as outside his trade or profession'. Even if this notion is commonly used in other consumer contracts directives, some divergences in these legal texts can be pointed out and also it is to be noted that the Member States have chosen different legislative techniques to transpose the definitions.

EC Consumer protection directives

In consumer contracts, a precise definition of 'consumer' is essential in order to delimit the persons entitled to extended legal protection in relations with traders (Kingisepp, Värv, 2011: 44). Thus the notion of consumer has been defined in each directive on consumer contracts.

In a comparative analysis of the notion of consumer in these directives, a consumer is a 'natural person' 'who is acting for purposes which are outside some kind of business, commercial or trade activity'. These two common features can be found in most consumer protection EC Directives and also in European procedural Law (Arts. 15-17 Brussels I Regulation n. 44/2001) and European rules on conflict of laws (Rome I; today Regulation n. 593/2008) (Ebers, 2007). Some exceptions can be found as the Directive 90/314 that contains a very different definition. It defines a consumer as a person who takes or agrees to take the package ('the principal contractor'), or any person on whose behalf the principal contractor agrees to purchase the package ('the other beneficiaries') or any person to whom the principal contractor or any of the other beneficiaries transfers the package ('the transferee'). In this sense, the Directive protects those who conclude package travel contracts even if these contracts are for business related purposes (Arroyo, 2003: 106).

For contracts that serve private and business purpose (mixed purpose transaction, e.g. the acquisition of a domestic coffee machine by an employee for his office), the directives doesn't contain expressed rules. In contrast, in the above cited Directive 85/374 concerning liability for defective products, preponderant purpose prevails (see Art. 9 lit. (b) ii: 'used by the injured person *mainly* for his own private use or consumption').

On this subject, the mixed purpose transactions, the judgment of the ECJ in C-464/01 Gruber –on a purchase of tiles by a farmer for roofing a farm building used partly for private and partly for business purposes- was not clear enough. The ECJ stressed that a person can invoke the special rules of jurisdiction of old Articles 13-15 of the Brussels I Convention (today Art. 15-17 Brussels I Regulation n. 44/2001) in respect of dual use contracts only if the trade or professional purpose is so 'limited as to be negligible' in the overall context of the transaction. It is clear that this decision relates only to European procedural law but it is also clear that the Court adopted a restrictive consumer notion. The Court defined a consumer transaction as a transaction which is concluded by a natural person, who is acting for purposes outside his or her trade or profession, on the one hand, and a party, who is acting for purposes within his or her trade or profession, on the other.

A clear solution for mixed purpose transactions is not offered by the most part of the consumer contracts directives.

In past years, as the EU consumer *acquis* is being revised and the principles of European contract law are being drafted, the authors have mainly focused on the aims and principles of consumer contract directives. The notion of consumer is revised as one of the common structures in these directives. In this context, some authors propose that a harmonized definition for all consumer protecting directives should be found.

A common definition can be already found in the Directive on Consumer Rights. This Directive will replace, as of 13 June 2014, the current Directive 97/7/EC on the protection of consumers with respect to distance contracts and the current Directive 85/577/EEC to protect consumer with respect to contracts negotiated away from business premises (Doorstep Selling). Chapter I of the new Directive contains the common definitions such as 'consumer' and 'trader' and provides for a common set of rules applicable in all Member States, only allowing them to diverge from these rules in a few specific cases. 'Consumer' means any natural person who, in contracts covered by the Directive, is acting for purposes which are outside his trade, business, craft or profession.

Whereas it has been clarified that a common consumer definition will apply in some cases, the existing different concrete notions are to be maintained. In this context Directive 2005/29/EC on Unfair Commercial Practices (the 'UCPD') needs to be stressed. It is a horizontal Directive which applies to all business-to-consumer transactions. The system of protection set up by the Directive consists of a three-tier prohibition. First, the Directive establishes a general clause banning all unfair commercial practices which are contrary to professional diligence and that are likely to distort the economic behaviour of the 'average consumer'. Second, the Directive regulates in detail two key categories of unfair practices, namely misleading (including misleading actions and omissions) and aggressive practices which are likely to affect the 'economic behaviour of consumers'. Third, the Directive provides for a 'black list' of 31 particularly harmful practices which are banned up-front, regardless of the circumstances. Finally, the UCPD provides for special protection to 'consumers who are vulnerable' because of their mental or physical infirmity, age or credulity.

The UCPD is based on full harmonisation, meaning that Member States may not retain or introduce stricter consumer protection rules, except in the areas of financial services and immovable property. All Member States transposed the Directive by the end of 2009.

It should be emphasized that in the UCPD, although the common definition of consumer is being used in the introductory definitions of the text, the specific legal determinations linked to the protection of the consumer are extremely important. It can be said that the notion of consumer is in this sense 'elastic' (Arroyo, 2003: 99). Indeed, the notion of consumer is intimately linked to the aim of the specific legal rule ('functional notion'; Arroyo, 2003: 116). As can be seen, the scope of some legal rules is being determined by the 'economic behavior of the average consumer' or by the 'consumers who are (...) vulnerable'. 'Average consumer' means the consumer 'who is reasonably well informed and reasonably observant and circumspect' according to ECJ case law.

In all these cases and prominently in the UCPD, the notion of consumer is a veritable legal fiction (Bercovitz, 2012: 637). Once the law is applied to the scope define by the notion of consumer, it is not necessary to verify whether a particular consumer, that matches the definition, deserves the protection or not. Specific sociological circumstances -others than the legally determined (e.g. consumers who are vulnerable)- shall not be taken into account to determine whether there is imbalance in the contractual position. It is assumed that the consumer is 'economically weaker and less experienced in legal matters than (their) professional co-contractor' [Case C-96/00, *Rudolf Gabriel,* [2002] ECR I-0000 (11.7.2002)]. Thus, all persons who match the notion of consumer will benefit from the protection: e.g. a lawyer, PhD. specialized on Directive 97/7/CE, remotely buying a present for his daughter, could withdraw the contract just like any other consumer.

Consumer notion on EU general food law

In many countries, the principles concerning food safety and consumer protection are established in national legislation. At EU level, it was not until 28th January 2002 that the European Parliament and the Council adopted Regulation (EC) 178/2002 laying down the General Principles and requirements of Food Law.

In the words of the Health and Consumers DG, the aim of the General Food Law Regulation is to provide a framework to ensure a coherent approach in the development of food legislation. At the same time, it provides the general framework for those areas not covered by specific harmonised rules but where the functioning of the Internal Market is ensured by mutual recognition. It lays down definitions, principles and obligations covering all stages of food/feed production and distribution.

As far as the definitions are concerned, it should be first noted that the aim of the Regulation is to provide the 'basis for the assurance of a high level of protection of human health and consumers' interest in relation to food'. In addition to this, it can be said that 'final consumer' is described as the 'ultimate consumer of a foodstuff who will not use the food as part of any food business operation or activity'.

With regard to the general objectives, food law shall pursue 'one or more of the general objectives of a high level of protection of human life and health and the protection of consumers' interests, including fair practices in food trade, taking account of, where appropriate, the protection of animal health and welfare, plant health and the environment' (Art. 5). The protection of consumers' interest is in Art. 8 developed as follows: food Law shall provide a basis for consumers to make informed choices in relation to the foods they consume. It shall aim at the prevention of: (a) fraudulent or deceptive practices; (b) the adulteration of food; and (c) any other practices which may mislead the consumer.

Of all the aspects mentioned in the Regulation there is one that deserves particular focus: the safety requirements. In the case of food safety requirements, the terms of 'health' or 'human consumption' are used (see Art. 14). In particular, 'food shall be deemed to be unsafe if it is considered to be: (1) injurious to health; (2) unfit for human consumption'. The objective of this Article 14 is clear: to protect 'public health' (Guidelines, 2010). It establishes therefore the factors that need to be taken into consideration when deciding whether food, as defined in Article 2 of the Regulation, is injurious to health or unfit for human consumption. The requirements of Article 14 apply to food that is 'placed on the market'.

The Article does not, however, cover primary production for private domestic use, or the use of food for private domestic consumption, which are exempted by Article 1(3) of the Regulation.

The concept of 'injurious to health' relates to the potential to harm human health. Article 14 (4) (c), for instance, requires that if food is produced for a 'group of consumers' with particular health sensitivities (e.g. intolerant or allergic), then these sensitivities should be taken into account when determining whether a food is injurious to health.

In summary, the objective of the Article 14 to protect public health complies with the responsibility for the risk assessment lies with the food business operators, under the control of national competent authorities once informed, as stated in Article 17.

In addition to the safety requirements the information is extremely relevant, that is why, as it can be recalled, Food Law aims to establish the rights of consumers to safe food (risks analysis) and to accurate and honest information (transparency, labelling, advertising and presentation of food or feed).

Having outlined these basic ideas, in a legal and broader sense, consumer Law is concerned with a right to information in order to balance the asymmetry of the position of the consumer vis-à-vis the professional on the market (Reich: pg. 9). On Food Law, the asymmetry is clear and mandatory information rules are relevant to allow informed choices. It is also clear that the consumers 'are required to make more complex choices' in Commissioner Borg's words. But information is not the only factor, however important it may be. Indeed, as food Law is one area where consumers should be guaranteed certain quality or safety standards, the potential liability of producers and suppliers, and the opportunities for consumers to obtain compensation when injured by defective products is very important. In fact, as far as product liability is concerned, some authors proposed years ago the development of a 'right to the protection of legitimate expectations' as a further step towards consumer protection. It has been stated here that ECJ case law appears to go in the opposite direction (Case C-52/00, Commission France [2002] 25.4.2002).

As seen above, the EU General Food Law covers not only the final consumer or what can be understood as consumer in a restrictive approach of the notion. The EU food Law is to protect 'health' and other consumer interests. In other words, General Food Law covers us all.

Conclusions

In a broader sense, several expressions are used to point out some changes in consumer policy: e.g. in Europe, 'consumer law becomes part of the economic citizenship' (Micklikz); in the US, 'Goodbye consumer, hello citizen' (P. Cottingan, 2009). Specifically on food matters, in the US, 'Eating right here: Moving from consumer to food citizen' (Wilkins, 2004). In any case, it's hard to determine precisely whether there are changes or if so, they have potential legal effects.

But, there are several national initiatives trying to note that there are a number of issues that deserve additional emphasis. For instance, as a reaction to the food scandals of recent years, a French Resolution has been recently promoted to pursue at European level the development of the European consumer Law related to the control and perfect knowledge of the food ('un droit européen du consommateur à la maîtrise et à la parfaite connaissance de son alimentation'; 28th February 2013). In another direction, the eleventh Report of the Environmental Audit Committee (House of Commons, UK, 30th April 2012), related to 'Sustainable Food' concludes that the Government must use the 'Green Food Project to provide a foundation for developing a broader food strategy that takes into account the health, environmental, social and economic consequences of the way that the food we eat is produced, sold and disposed of'.

Two levels of the notion of consumer can be distinguished, an abstract notion and different concrete notions. In consumer contracts, a precise definition of the 'consumer' is essential in order to delimit the persons entitled to extended legal protection in relations with traders. However, Food Law is to protect 'health' and other consumer interests, meaning that it cover us all.

References

Arroyo, A. (2003). Contratos a distancia. Aranzadi, Pamplona, Spain, 414 pp.

Bercovitz, A. (1987). Estudios jurídicos sobre protección de los consumidores. Tecnos, Madrid, Spain, 326 pp.

Bercovitz, A. (2012). Apuntes de derecho mercantil. 13ª, Aranzadi, Spain, 644 pp.

Ebers, M. (2007). EC consumer law compendium. p. 713-732.

Kingisepp, M. and Värv, A. (2011). The notion of consumer in EU consumer acquis and the consumer rights directive – a significant change of paradigm? Juridica International XVIII: 44-53.

Reich, N. (2004). The consumer as citizen – the citizen as consumer – reflections on the present state of the theory of consumer law in the EU. Etudes de droit de la consummation. Liber Amicorum Jean Calais-Auloy, Dalloz, 943 pp.

Food safety and ethical responsibility in a globalized world – the role of private standards

C. Carlsson
AgriFood Economics Centre, Lund University, Box 730, 220 07 Lund, Sweden; cecilia.carlsson@agrifood.lu.se

Abstract

Rules and guidelines on food safety and ethical aspects of production, such as animal welfare and labour conditions, have traditionally been a matter of national legislation. However, in recent years, powerful retailers and food processors have started to impose requirements on downstream producers through private standards. Since private standards are used by private actors and not by states, they are not bound by existing trade rules. This means that private standards can include requirements relating to ethical aspects such as animal welfare and environmental protection on both imports and domestic production, and that they thereby have potential to make production conditions more equal in different countries. Through a case-study approach, the study investigates whether private standards harmonize production conditions across countries. Interviews are performed with three companies active on the Swedish and international food markets to understand which private standards are used and what requirements are made in these. The findings point to the existence of two related, but different, layers of private standards. Standards that are harmonized and used by many actors contain some requirements on ethical responsibility, but focus mainly on food safety. On the other hand, requirements on production methods, relating to animal welfare, environmental protection and labour conditions, are more heterogenic, since maintaining non-harmonized private standards within these areas is an important competitive tool for the individual firm. This, in turn, implies that they have less scope for harmonizing production conditions across countries. Nevertheless, firms do make requirements also within these areas, and the study thereby indicates that the private sector and the companies on the market are important actors in formulating requirements on the inclusion of ethical responsibility in production.

Keywords: private firms, production conditions, harmonization

Background

Private standards within the food sector contain requirements on how production shall be arranged to meet a number of specified criteria. Private standards can be designed either as business-to-consumer (B2C) standards, such as Fairtrade and Rainforest Alliance, to differentiate products and communicate a special feature of the product to consumers through product labelling, or as private business-to-business (B2B) standards, that operate between the actors in the food supply chain and are invisible to consumers. The GlobalG.A.P. standard and the BRC Global Standard for Food Safety are examples of private standards that operate business-to-business.

This study focuses mainly on private business-to-business standards, which emerged as a risk management tool that retailers and food processors can use to specify how production should take place to ensure that the produced foodstuff is safe for consumption. The increasing focus on food safety on behalf of the private sector was to a large extent a consequence of a number of food scares, such as the outbreak of the bovine spongiform encephalopathy (BSE), also called the mad cow disease, which resulted in an increased need to restore consumers' confidence and in the private sector being assigned an increased legal responsibility for ensuring food safety.

Due to the unequal division of market power within the food sector, powerful retailers and food processors can require that suppliers must comply with a private business-to-business standard. This means that if private business-to-business standards are extensively used by purchasers, they can become *de facto mandatory* for suppliers that want to sell their products, although they are not part of legislation (Henson and Humphrey, 2010).

Private actors using private standards are not bound by the existing WTO rules on trade. This means that whereas national legislation on production methods, for example with respect to animal welfare, environmental protection and labour conditions, only applies to domestic production, private standards used by private actors can make such requirements also on imports (Andersson and Gullstrand, 2009; Jobbs, 2010). If they do and if they are extensively used as a requirement for both domestic producers and producers in other countries, private standards can be a driver of harmonization of production conditions relating to ethical aspects such as animal welfare, environmental protection and labour conditions across countries.

To investigate whether this is the case, both the contents of the private standards and their role on the market must be analysed. Do private standards contain requirements related to production methods, such as animal welfare and labour conditions? How stringent are the requirements in comparison to public regulations? And must private standards in general be complied with by suppliers?

Method

This study adopts a case study approach to understand how private standards work in practice and what role they play according to the actors using them. Interviews are undertaken with representatives for the retailers ICA and Bergendahls Food and the food processing company Findus which are active on Swedish as well as on international food markets. That the study is based on three cases only means that the analysis can point to some important aspects, but it is not a comprehensive study of private standards in general. This is clearly a limitation, but there exist few, if any, detailed previous studies building on firm-internal information on retailers' and food processors' work with private standards in Sweden. For this reason, the analysis adds important knowledge to the understanding of the role of private standards in the food sector, despite the limited number of cases.

The choice of including the three companies ICA, Bergendahls Food and Findus Sverige AB can be motivated from a number of aspects. Including ICA with the largest market share provides an indication of the practices employed by the most dominant player on the Swedish food retail market. Bergendahls Food, which is a smaller, family-owned company with approximately 10% of the market in some regions and the owner of the CityGross chain, provides an illustration of the practices of a smaller but nevertheless important actor.[37] Hence, by including both ICA and Bergendahls Food, the perspectives of a large as well as of a relatively small food retail firm are included into the analysis. Finally, the inclusion of Findus Sverige AB, one of the leading companies within the markets for deep-frozen fish, vegetables and semi-finished products, provides insights from a food-processing company operating on the Swedish as well as on the European and international markets. Furthermore, Findus can provide an illustration of the views of a company situated in the middle of the chain between the concentrated retail industry and the producers with less market power.

[37] See Dagligvarukartan (2012) for information on market shares and related aspects.

Results

The results of the study indicate that private standards do contain requirements both on food safety and on animal welfare, environmental protection and labour conditions, but that the requirements are imposed in different ways and for different reasons. This results in two different types of private standards, which are described below.

Harmonized private standards for food safety

Since food scares damage the reputation of the whole industry and not only the confidence in an individual company, firms benefit from co-operating to achieve food safety and as a result, private standards' requirements on food safety are relatively harmonized at the international level through recognition by the international benchmarking harmonization the Global Food Safety Initiative (GFSI). That the imposed requirements are similar further implies that they have the potential to equalizing requirements on producers in different countries. The case studies with two retailers and one food processing company in the Swedish food sector show that third-party certification to a private standard for food safety in general is a requirement on suppliers. Taken together, this implies that private standards contribute to harmonizing requirements on food safety across countries.

The private standards that are approved by the GFSI[38] are commonly used by the three firms included in the case study. These private standards are recognized as equivalent when it comes to ensuring the minimum level of food safety specified by the GFSI, but they can still differ in their requirements on production methods, since these aspects are not subject to benchmarking.

A comparison of the private standards active in GFSI-benchmarking reveals that their coverage of requirements on animal welfare, environmental protection and labour conditions do vary.[39] The compared private standards differ when it comes to their extensiveness in terms of the variety of aspects covered, the stringency of the requirements as well as the accepted time limit for the implementation of the necessary measures. Whereas some cover a relatively small number of aspects, others are more comprehensive. Within the considered aspects for environmental issues, waste disposal is covered by all standards to some extent, whereas the coverage of requirements on soil protection, handling of chemicals and water use, recycling and disposal varies. Similarly, animal welfare-related aspects are covered by all three standards where such requirements are applicable, but the extensiveness differs. Labour conditions are regulated by some standards in terms of training, safety, access to sanitary facilities and similar requirements. However, none of the reviewed standards include requirements on aspects related to labour rights and labour welfare as stated for example in the international labour standards of the International Labour Organization, but two private standards provide additional standards focusing on social responsibility as a voluntary supplement.

The private standards dominating the food sector thus to some extent do extend beyond the core issue of food safety, but the scope and extent to which non-food safety-related issues are covered vary. That the private standards cover different issues to different extents illustrates the complexity of the topic, and further indicates that it might be difficult to provide a clear-cut answer to the question of whether private standards is a driver behind an equalization of requirements on ethical aspects such as animal

[38] See GFSI 1

[39] The comparison is performed with help of the database Standards Map provided by the International Trade Centre (ITC)and includes the private standards BRC Global Food Standard, CanadaGAP, FSSC 22000, GlobalG.A.P.,GRMS, IFS Food version 5, Primus GFS and SQF. The information was retrieved 2012-09-06.

welfare, environmental protection and labour conditions across countries. Production conditions might basically be equalized to different extents depending on which private standard that is used.

A more detailed comparison between animal welfare-requirements for pigs, cattle and poultry in one of the GFSI-benchmarked standards, the GlobalG.A.P., reveals that this private standard imposes requirements that approximately correspond to EU rules, but are below Swedish legislation. An extensive use of this *particular* private standard could thus harmonize requirements on animal welfare on the EU level.[40] It should, however, be kept in mind that differences in how private standards and public regulations are constructed make it difficult to directly compare the requirements. Whereas legislation must be complied with, private standards contain both requirements that must be fulfilled but also requirements that function partly as recommendations. For example, in the GlobalG.A.P. standard, the producer must fulfil 95% of the control criteria classified as a 'minor must', which means that third-party certification can be achieved without compliance with all the control criteria. This makes it difficult to understand what requirements producers face through private standards and how these compare to the requirements within public regulations. The comparison is also aggravated by difficulties in defining animal welfare and perform a comparison of animal welfare-requirements. There is no common view on how to combine different aspects of animal welfare into an overall measure (Hoffmann *et al.*, 2010), and since the same level of animal welfare can be achieved in different ways and under different production systems, one must focus on the outcome, that is the, health, behaviour, condition, etc. of the animal. Making a judgment only by comparing a number of specified requirements in different sets of rules is thus difficult.

Competing private standards for firm profiling

Furthermore, within the areas of animal welfare, environmental protection and labour conditions, the undertaken case studies show that firms use different private standards and that the contents of these private standards are heterogenic, since maintaining non-harmonized private standards within these areas also functions as a profiling tool for the individual firm. For example, ICA bases their requirements on the IP Sigill standard and thereby indirectly on Swedish legislation, and the food processor Findus requires all producers, including producers in non-EU-countries, to follow the EU rules on animal welfare. In this case, a public standard is thus extended and imposed as a private standard for producers in other countries, and thereby contributes to making production conditions more equal across countries. However, in addition to different firms imposing differing requirements, the requirements are not always controlled by an independent third party. This adds to the difficulty in determining what requirements are really made on suppliers and to what degree harmonization of production conditions holds for the food sector in general.

The two types of private standards are illustrated in Figure 1. The harmonized first category of private standards that aim at ensuring food safety and at protecting the reputation of the industry are located at the bottom, whereas the second category of private standards are found further up the pyramid.

However, the dividing line between the two categories of private standards is not static. As time passes and the industry faces new challenges, new areas, here illustrated by the question mark, will be added on the top of the pyramid. The areas that were previously of importance for profiling the company are thereby pushed downwards and become increasingly interesting as objects for harmonization and co-operation.[41] The Global Social Compliance Initiative (GSCP: www.gscpnet.com), which aims at promoting comparability of social, labour and environmental practices, is an example of such beginning

[40] Carlsson and Johansson (2013).

[41] Personal interview with Lena Sparring, Director Product Quality and Safety, ICA, 2012-11-26.

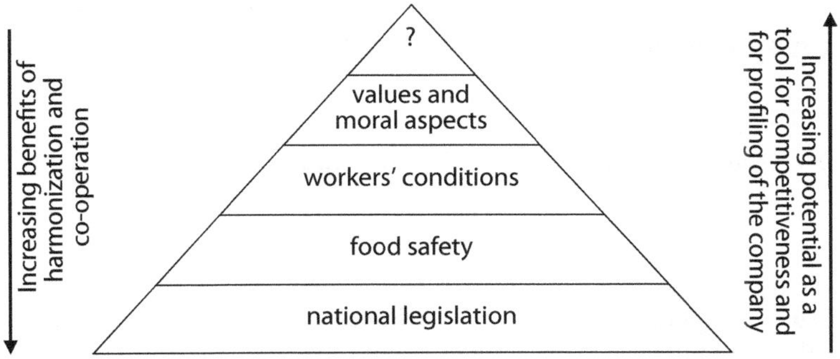

Figure 1. The pyramid of private standards (based on a personal interview with Lena Sparring, Director Product Quality and Safety, ICA, 2012-11-26).

harmonization and benchmarking efforts.[42] This implies that in the longer run, it is likely that private standards will have an increasing scope for equalization also of production conditions such as labour conditions and environmental protection across countries.

More research is needed to be able to judge to which extent private standards in general result in a harmonization of requirements on animal welfare, environmental protection and labour conditions across countries. Nevertheless, despite the small number of firms included in the case study, these show that private firms do impose requirements with an ethical dimension, and private firms are thus important actors in the formulation of requirements on the inclusion of ethical responsibility in production.

References

Andersson, A. and Gullstrand, J. (2009). Certifiering, konkurrens och handel. Konkurrensverket. Uppdragsforskningsrapport 2009:13.

Carlsson, C. and Johansson, H. (2013). Private standards – leveling the playing field for global competition in the food supply chain? Report 2013:2. AgriFood Economics Centre, Lund, Sweden. Available at: www.agrifood.se/Files/AgriFood_Rapport_20132.pdf.

Dagligvarukartan (2012). Delfi Marknadspartner, Stockholm. Available at www.delfi.se/html/publikationer3.html.

GlobalG.A.P. (2012g). Integrated farm assurance. All farm base. Control Points and Compliance Criteria. English Version 4.0. FoodPLUS GmbH, Cologne, Germany.

Henson, S. and Humphrey, J. (2010). Understanding the complexities of private standards in global agri-food chains as they impact developing countries. Journal of Development Studies 46: 1628-1646.

Hoffmann, R., Höjgård, S., Rabinowicz, E. and Andersson, H. (2010). Djurvälfärd och lönsamhet – var står vi idag. Rapport 2010:4. AgriFood Economics Centre, Lund, Sweden.

Jobbs, J.E. (2010). Public and private standards for food safety and quality: international trade implications. The Estey Centre Journal of International Law and Trade Policy 11: 136-152.

SQF (2012). SQF code. A HACCP-based supplier assurance code for the food industry. 7th edition. Safe Quality Food Institute, Food Marketing Institute, Arlington, TX, USA.

[42] See www.mygfsi.com/about-gfsi/gfsi-recognised-schemes.html.

Societal conformity of conventional and diversifying pork production systems in five European countries

K.H. de Greef
Wageningen UR Livestock Research, P.O. Box 65, 8200 AB Lelystad, the Netherlands; karel.degreef@wur.nl

Abstract

In the long term, social acceptability is a *sine qua non* issue in the sustainability of production systems. A practical tool was developed and applied to assess the degree of social acceptance of pork production systems. Societal conformity was defined as the degree to which production systems meet the demands and expectations of society. The tool assesses judgments of informed professionals on nine sustainability themes: Animal health, Animal welfare, Economic sustainability, Environmental Impact, Genetic diversity, Human working conditions, Meat safety, Meat quality, and the overall term Public image. A distinction is made between stakeholders involved in the production chain ('*Insiders*') and those around the chain (NGO-representatives, '*Outsiders*'). The tool was applied in 5 countries (D, ES, F, NL, UK), assessing the stakeholder views of the conventional pork production system and two alternative systems in each country. *Insiders* judged the overall sustainability of the conventional pork system about neutral, with positive values for meat safety and meat quality. Judgment of the *Outsiders* was generally negative. The southern Europe NGO's scored more outspoken negative that their northern colleagues. *Insiders* expected the public to be more negative, whereas *Outsiders* expected them to be more positive than themselves. Both stakeholder groups expected considerable improvements in sustainability of alternative systems, and virtually no undesirable effects of the systems changes in other sustainability themes. *Outsiders* were more outspoken in their views, both for the sustainability of the conventional system and for the degree of improvement of the alternative systems. Results show consistency between the specific sustainability themes and the general parameter Public Image, although the underlying mechanism of *Outsiders* seems to be of another nature than that of the *Insiders*: the data suggest that informed professionals from within the system build up their view from the various sustainability aspects, whereas those from outside are more outspoken and their overall view corresponds with several separate themes, making their view on specific aspects reasonably informative for the other themes.

Keywords: social acceptability, pig production, stakeholder judgement

Introduction

To acquire and maintain a position, production systems need both a market position (for their economic viability) and a societal position (for their licence to produce, ultimately their legal viability). The latter is complicated for animal production systems because of the various potential social concerns that are involved (Kanis *et al.*, 2003). In the latest two decades, European systems have experienced forces from society to adapt towards more sustainable practices. Especially in North West Europe, animal welfare and environmental impact (including local hinder) are key issues. However, less prominent themes may also affect how people also form an overall opinion about the system as a whole.

Within a larger assessment of European pork production systems, in which the themes Animal Health, Meat safety, Meat quality Human working conditions, Animal Welfare, Economic sustainability, Environmental impact, Genetic Diversity (Bonneau *et al.*, 2011) were assessed empirically, a need was felt to assess overall this social acceptability too. In literature, several reports assessing acceptance of production practices are available (eg Eurobarometer, 2006; Te Velde *et al.*, 2002), and generally studies are either qualitatively, based on focus groups or (semi-)quantitative, based on surveys. A practical integrative tool to assess social acceptance empirically which fitted to the practical assessment of the other sustainability themes (Bonneau, 2011) was not found (Edwards, 2007). A simple method to assess the degree to which production systems fit to the expectations and demands of the society was therefore developed and applied. This was labelled as 'Societal conformity'. The term *Societal conformity* labels the concept on the degree to which production systems meet requirements and expectations of society as a whole. The developed tool comprised two distinct parts. In the first, qualitative part, a questionnaire assesses public expressions of societal judgement on animal production systems. Further, observed initiatives of chain actors to explain or improve their practices and products are assessed. The second, semi-quantitative part of the tool approached the stakeholder population as two separate groups: inside and outside the production chain. Present contribution reports on application of the semi-quantitative part of the Societal Conformity tool, in which the conventional pork production system is evaluated in five countries, in each country together with two adapted or alternative systems. Aim is to assess acceptability/acceptance of pork production systems in Europe for stakeholder groups within and outside the production system.

Material and methods

Societal conformity was assessed for the conventional pork production system in five countries and for two alternative systems in each country. Each set of two alternative systems' was composed of an adapted conventional system and one other system (D & UK: Organic; F & ES: traditional; NL: another adapted conventional). Informed professionals from both the society side and the production side were asked to judge on nine sustainability themes: Animal Health [AH], Meat safety [MS], Meat quality [MQ], Human working conditions [HW], Animal Welfare [AW], Economic sustainability [€], Environmental impact [EI], Genetic Diversity [GD] and Public Image [PI]. The professionals were asked to judge the conventional pork production system in their country (*Perfect; Good; Normal; Concerning; Unacceptable; ?*). Next, they were asked to judge the changes on these themes for two alternative systems (*--; -; similar; +; ++*). After scoring this (3×9=27 scores), the respondents were asked to fill in a similar table, but this time from the viewpoint of the public (for example: 'How do you think the *public* judges Animal Health in the conventional system' etc.). The representation group of the society-side ('*Outsiders*')was to be composed of three NGO representatives (animal, environment, consumer) and a journalist of a general newspaper. The representation group of the Production-side ('*Insiders*') was to be composed of a representative that runs or owns one of the systems, one for each of the systems involved, three sector specialists and an agricultural journalist. In several instances, the actual composition of the professionals group deviated from the intended composition due to availability and willingness to cooperate, resulting in 33 *insiders* and 18 *outsiders*. The questionnaires were collected centrally and analysed together. The scores were transformed into numerical values. The resulting data were averaged on group level (Insiders and Outsiders) within country. As the method does not allow full size traditional statistical evaluation of the results, the presentation is primarily numerical and graphical. In the analyses with adequate degrees of freedom (the regressions), standard regression analysis was applied. In some cases, the 9 sustainability themes were converted into 4 major sustainability themes People (MQ, MS, HW, PI), Planet (EI, GD), Profit (€) and Pigs (AW, AH).

Results

Overall judgement of conventional systems over Europe

Figure 1 shows the scores for the conventional systems on all nine sustainability themes separately across countries for the chain involved respondents ('Insiders', blue solid line) and for the societal representatives ('Outsiders', red dashed line). Insiders judge rather neutral (most scores close to 0), except for Food safety which is quite positive (~+1), and Public image, which is clearly negative (~-1).

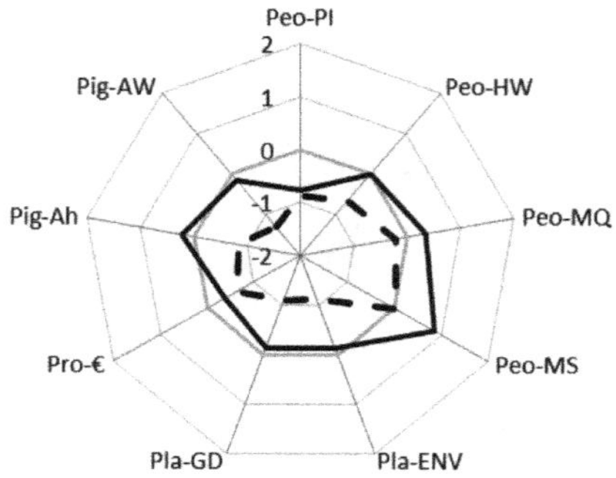

Figure 1. Average judgement across Europe of the conventional pork production systems. Solid line: chain involved representatives ('Insiders'); dashed line: societal representatives ('Outsiders'). Main sustainability categories: 'Pla': Planet; 'Peo': People; 'Pro': Profit; 'Pig': Pigs. Codes: 2: Perfect; 1: Good; 0: Normal; -1: Concerning; -2: Unacceptable.

Outsiders are quite negative on all themes except on the primary product characteristics Meat safety (~neutral) and Meat quality (nearly neutral).

Split out into the four major sustainability categories (People, Planet, Profit and Pigs), there is reasonable similarity between countries. Despite the different levels, variability within countries is similar for the themes, although the standard deviation of the theme People is somewhat lower (μ= -0.18, -0.63, -1.17, 0.59; σ= 0.39, 0.59, 0.57, 0.56, for People, Planet, Profit and Pigs, respectively).

In Figure 2, the scores for the nine sustainability themes were averaged into one overall sustainability number. For the five countries separately, the average judgements are generally around 0 or well below, indicating neutral to unfavourable score for overall sustainability. Across countries, the overall scores (averaged for the four main themes) is -0.39, ranging from -0.09 for the own judgement of the Insiders to -0.81 for the own judgement of the Outsiders. Across countries, the average judgement of the Insiders

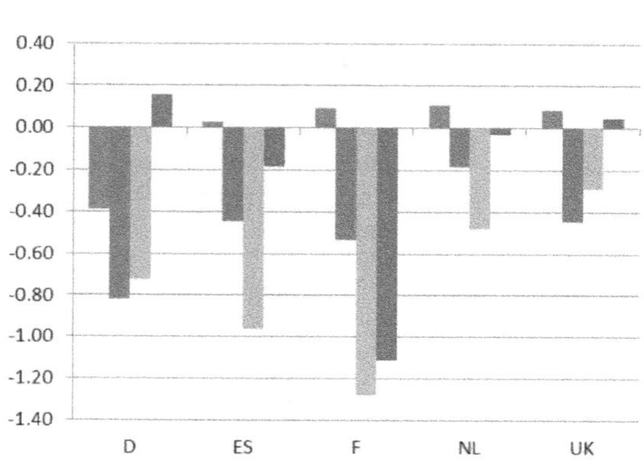

Figure 2. Judgement of degree of overall sustainability of conventional pork production systems in 5 European countries by chain representatives ('Insiders') and others ('Outsiders'). 'Self': own judgement, 'Soc': expected judgment of the public. Per country from left to right: Insiders Self; Insiders Soc; Outsiders Self; Outsiders Soc.

is about neutral, caused by most countries being slightly positive, compensated by a clearly negative judgement in Germany. The most negative overall judgement is the one of the Outsiders, on average they judge sustainability as -0.75, with the extremest values in southern Europe. Insiders think the public is somewhat negative (-0.49), again with the most negative value in Germany. Except in France (-1.1), Outsiders think the public is not very negative about the conventional systems (-0.22). While Insiders think the public judges half a point lower than themselves, NGO's think the public judges half a point better than themselves [-.47 vs +.52]. Comparison of own judgement and expected public judgement also indicates that Insiders are in general more positive than what they expect from the public, whereas Outsiders are more negative than their expectation of the public. Especially in the southern countries (F and ES), the Outsiders (esp. NGO's) are considerably more negative than the other judgements.

Overall judgement of alternative systems: does diversification improve sustainability?

The judgement of the sustainability improvements brought forward by the alternative systems is expressed in Figure 3, in a similar way as the overall judgment of conventional systems earlier in Figure 1. For each country, the scores of the two alternative systems are averaged. The results illustrate that the alternative systems are judged considerably positive for their improvement in overall sustainability. On average, Insiders judge the alternative systems as +.44 better than conventional in overall sustainability, and think the public judges this considerably higher. NGO's (outsiders) judge improvements of +1.6 on average and think the public thinks similar, except for France, where the NGO's think the public will judge the improvement considerably higher than themselves. Overall expectation of the public's judgement is similar for Insiders and outsiders, except again for France, were NGO's (outsiders) have considerably higher expectation from the public than the insiders have, and except the British, where the reverse is true. In the UK, the insiders have a higher expectation of the public view than the outsiders have.

The alternative systems give improvements on all four major sustainability themes in virtually all combinations of theme, country and response groups. The only two substantial exceptions are the British Insiders who think the public expects reduced profitability for the alternative systems and the French Insiders who themselves expect reduced profitability for the alternative systems. In The Netherlands, the improvement is relatively small in all four themes, and highly uniform among the stakeholder groups.

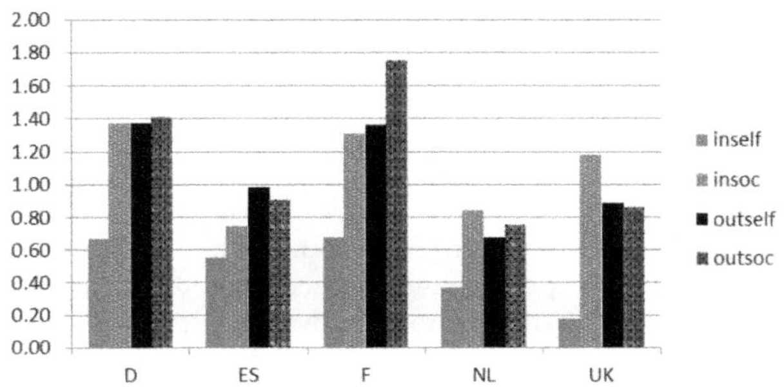

Figure 3. Judgement of improvement of overall sustainability by alternative systems relative to conventional pork production systems in 5 European countries by chain representatives (Insiders) and others (Outsiders). Self: own judgement, Soc: expected judgment of 'the public'. Average of the two alternative systems per country. Per country from left to right: Insiders Self; Insiders Soc; Outsiders Self; Outsiders Soc.

Relation between the score for public image and the overall sustainability score

In the questionnaire, a score for Public Image of each system was collected as one of the nine sustainability themes. A relation can be expected between this score and the overall sustainability score (as built up from the averaged scores of the underlying nine sustainability themes). This was analysed for the assessments of the conventional systems, in order to allow evaluation of consistency between the measure for public image and indirect measures, built up from the other sustainability themes. Only the regressions of the self-scores are significant, implying that the average of sustainability score is a predictor for public image for the professionals themselves, but not for their expected public score. The x-values of the regression indicate that for the insiders (x=0.45), the score for Public image increases slower with increasing overall scores than for the outsiders (x=1.1). Correlations between the score for Public Image and the separate own sustainability scores were calculated. For the Insiders, these correlations are weak: 0.14 on average, no correlation reached 0.50 in this group. The score for public image of the Outsiders correlates strongly with their other sustainability scores (0.61 on average). In the Outsiders data, only the correlations with Economy and Genetic diversity were below 0.50.

Discussion

Degree of sustainability & acceptance of the conventional system

The European pork production system is quite uniform. Similarly, the professionals' judgement of the conventional system is not very different across Europe either. Current assessment indicates that the perceived sustainability of the conventional system is worrisome. The overall score is at best neutral (Insiders' own view) but predominantly negative (the other judgements). Overall, there are moderate worries on the level of sustainability. The contrast between Insiders (esp. chain involved stakeholders) and Outsiders (esp. NGO's) is limited in D, NL and UK. In the south of Europe, there is a clear discrepancy between these two groups. Their Insiders judge similar to the rest of Europe, but their NGO's are far more negative. Meat safety is an exception: the conventional system scores predominantly positive for this.

Improvements by alternative systems

Alternative systems are generally appreciated as considerable improvements compared to the conventional system. It is acknowledged that in a respondents study, this is caused by the method or by the situation itself: systems that are presented as improvements are inclined to be valued more positive. However, the absence of clearly negative judgements is still surprising given the scientific view that there are not only synergies, but also trade offs between sustainability changes (e.g. De Boer *et al.*, 2011). For the present domain, undesired relations (trade offs) between animal welfare, environmental impact and meat safety are the most likely mentioned ones, but were virtually absent in the data. It is therefore interesting that the changed procedures in the contrasted systems do not seem to reduce the trust in meat safety of the professionals. The farmers economic position is to some degree different: in several cases, Insiders had a less positive expectation of the economic advantages of the alternative systems.

Insiders versus Outsiders

Besides the observed differences in the level of scores, the data also demonstrate a difference between Insiders and Outsiders in the relation between the score for Public Image and the other sustainability indicators. Results suggest that informed professionals from within the system have a nuanced view on the sustainability, building their view up from the various sustainability aspects. Outsiders are more outspoken and their overall view corresponds with several separate underlying themes. In other words:

an overall view of the Insiders needs several sustainability parameters, whereas for Outsiders, their view on some aspects is reasonably informative for the rest.

Conclusions

Insiders (Producers) rate the systems neutral to slightly positive, except for Germany. They base their judgement on the various aspects of sustainability, but they fear society has a negative view of pork production. Outsiders (NGOs) rate the systems negatively, based on a general assumption throughout not built up of the various individual sustainability aspects. They think society has a more positive view than they have themselves.

The professionals judge that alternative systems bring improvements on nearly all sustainability themes. According to them, this also the view of the public view. No clear indications for expected drawbacks (unfavourable side-effects of changed practices in the alternative systems) were brought forward.

The tool to assess Societal conformity seems consistent: there are clear correlations between sustainability components and the expected Public Image. But the data suggest that the underlying views or strategies in judgement of the two stakeholder groups (Insiders versus Outsiders) differ.

Acknowledgement and disclaimer

The authors gratefully acknowledge from the European Community financial participation under the Sixth Framework Programme for Research, Technological Development and Demonstration Activities, for the Integrated Project Q-PORKCHAINS FOOD-CT-2007- 036245. The views expressed in this publication are the sole responsibility of the author(s) and do not necessarily reflect the views of the European Commission. Neither the European Commission nor any person acting on behalf of the Commission is responsible for the use, which might be made of the information. The information in this document is provided as is and no guarantee or warranty is given that the information is fit for any particular purpose. The user thereof uses the information at its sole risk and liability.

References

Bonneau, M., Antoine-Ilari, E., Phatsara, C., Brinkmann, D., Hviid, M., Christiansen, M.G., Fàbrega, E., Rodríguez, P., Rydhmer, L., Enting, I., De Greef, K.H., Edge, H., Dourmad, J.Y. and Edwards, S. (2011). Diversity of pig production systems at farm level in Europe. Journal on Chain and Network Science 11: 115-135.

De Boer, I.J.M., Cederberg, C., Eady, S., Gollnow, S., Kristensen, T., Macleod, M., Meul, M., Nemecek, T., Phong, L.T., Thoma, G., Van der Werf, H.M.G., Williams, A.G. and Zonderland-Thomassen, M.A. (2011). Greenhouse gas mitigation in animal production: towards an integrated life cycle sustainability assessment. Current Opinion in Environmental Sustainability 3: 423-431.

Edwards, S.A. (2007). Combined review of the available tools for assessment of sustainability. Report D-II.1.1 of EU-F6-project Q-Porkchains, 122pp.

Eurobarometer, 2006, European Commission, Special Eurobarometer 238. Wave 64.1 – TNS Opinion & Social. European Commission, Brussels, Belgium.

Kanis, E., Groen, A.F. and De Greef, K.H. (2003). Societal concerns about pork and pork production and their relationships to the production system. Journal of Agricultural and Environmental Ethics 16: 137-162.

Velde, H., Aarts, N. and Van Woerkom, C. (2002). Dealing with ambivalence: farmers' and consumers' perceptions of animal welfare in livestock breeding. Journal of Agricultural and Environmental Ethics 15: 203-219.

The legal clash between public health, environmental protection and the free market of GMOs: the Cartagena protocol and the WTO agreements

L. Escajedo San Epifanio
Department of Constitutional Law and History of Political Thought, University of the Basque Country, Bilbao, Spain; leire.escajedo@ehu.es

Abstract

The goal of Safety in the use of Biotechnology (BT) became a priority on the international Agenda since the United Nations Conference on Environment and Development held in Rio de Janeiro in 1992. The challenges we are to face at the time of setting up the regulatory framework for modern BT are many and varied. In addition to the risks, governing norms will have to attend to a great many social interests which are often incompatible among one another. According to official data offered regularly by different UN Agencies and by the Biosafety Clearing House created in the context of the Cartagena Protocol, more than seventy countries have some sort of governing system for Biosafety. In addition, there are about fifty binding and non binding international tools to manage the issue with and the vast majority of them were adopted after 1995/1996, when the commercial expansion of GM began. The present research pays close attention to the Cartagena Protocol, given its worth as a global tool and because of its binding nature. Worth mentioning as well is the elaboration process of the Document because this was the scenario on which the different views on modern Biotechnology were put forth. That should serve to account for the complexity of the issue and for the difficulty of yielding agreements on it. We will also point out the conflicts between the Protocol and the regulations governing international trade of food stuffs and stemming off from WTO agreements, namely.

Keywords: regulatory mechanisms for the safety in the use of biotechnology, Cartagena protocol, SPS agreement, TBT agreement

Global regulatory mechanisms for GMOs: the path walked

If we consider the issues associated to the application of GMOs, the challenges we are to face throughout regulatory frameworks are many and varied. In addition to the risks, governing norms would have to attend to a great many social interests which are often incompatible among one another. We need to conduct the issue of research and applications in such manner that freedom of thought as well as scientific production is warranted. It could be added that juridical security for the progress of scientific and biotechnological innovations is likewise to be provided. But on the other hand, commercialisation of BT products involves significant issues such as the protection of consumers' autonomy, the fulfilment of the human right to food or the free choice in democracies, among others.

The Convention on Biological Diversity (CBD) asked in 1992 for two relevant goals regarding GMOs. The need to secure biosafety was one of them, but the Convention also took the compromise to seek 'fair and equitable participation of the benefits resulting from the use of genetic resources'. What has been the path walked in two decades? The potential risks to health and to the environment associated to that qualitative change have been in fact the core of the international discussion regarding the making of frameworks related to GMOs. According to official data offered regularly by different UN Agencies and by the Biosafety Clearing House created in the context of the Cartagena Protocol, more than seventy countries have some sort of governing system for Biosafety. In addition, there are about fifty binding and non binding international tools to manage the issue with. Less attention was paid to their impact on the models for socio-economic development, as it has been regarding the model of international economic relationships we now enjoy.

In order to get an idea of the tremendous efforts deployed, suffice it to note the dates when the initiatives were launched: a few were implemented at the end of the 80s and early 90s, but the vast majority of them were adopted after 1995/1996, when the commercial expansion of GM crops began. Moreover, three of the most important international tools (the Aarhus Convention, the Cartagena Protocol and the Treaty on Phytogenetic Resources) and over half the national systems today in effect were implemented after 1998 through programmes launched by the OECD, the UN Programme for Development and the Global Environmental Facility. The elaboration process of the Cartagena Protocol was the scenario on which the different views on modern Biotechnology were put forth. And its application still points out the conflicts between the Protocol and other regulations governing international trade of food stuffs.

The Cartagena protocol: nature and legal value

Adopted on 29th January 2000 in Montreal, the Cartagena Protocol was the fruit of very significant efforts and long discussions and negotiations. There are other international agreements which include measures in relation to the safety in the use of biotechnology, but the Cartagena Protocol has some characteristics which make it stand out among all of them.

Its universal vocation should be highlighted, since it inherited from the framework which has served as support for it, that is, the Biodiversity Agreement, and from other measures coming from the United Nations Conference on Environment and Development (UNCED). Nearly all the countries on the Planet participated in its development. However, some finally decided not to adhere to it. The major transgenic crop-producing countries, like the USA, refused signing it. Nevertheless, despite it has not been signed or ratified by all countries, we must bear in mind that more than a hundred have done so already. Note that this also has some non-direct effects to non-signing Countries, since the Protocol was hoped to be applicable to all the movements where at least one of the signing Countries takes part.

In terms of content, it can be criticised that, the Protocol only approached some aspects such as cross-border movements, transit, manipulation and usage of living organisms modified by modern biotechnology (LMOs). Some key aspects are left out, such as the likeliness of harm and damage to human health, animals or ecosystems which might be generated as a consequence by such organisms. More important, each country establishes and implements within its borders the safety levels in the use of biotechnology.

Its main mechanisms are: (1) an information exchange system to provide assistance to the parties in the application of the Protocol provisions and to facilitate the information exchange in relation with the LMOs; (2) previously based agreements regarding LMOs to be freed in the environment; (3) measures applicable to the LMOs for direct consumption or to be processed; (4) control measures for involuntary LMO movements.

The clash between the Cartagena Protocol and the SPS and TBT agreements of the World Trade Organization

Our Global Village, as it is common knowledge, is not subject to a sole government or a democratic or fair one. International institutions that have contributed to initiatives on Biotechnologies are many and varied, and the relationship among these bodies of international governance is not juridical in nature nor politically peaceful. We have institutions like the OCDE and the WTO basically created for the purpose of conducting free international trade. The WTO monitors commercial policies in the different countries and must act as mediator in any dispute arising between member countries (147 in total). Alongside these organisms we find institutions that are committed to the improvement of health, the protection to the environment and to the development of the peoples all over the world. The vast majority of nations on the Planet are represented within a few of them, but the regulatory tools they have to fulfil their ends cannot be compared, neither juridically nor the facto, to those managed by the above-mentioned commercial and agreement-wielding institutions. Such is the case of the FAO, which aims to improve the efficiency in the production, manufacturing, trading and distribution of food and agro-food products. Another institution that is worth mentioning here is the Global Environmental Fund, created within the UN Programme for the Environment to strengthen international cooperation and to finance intervention actions in the event of threats to biodiversity.

There is great tension between the mechanisms deployed to ensure equitable participation in the benefits resulting from genetic resources and the ones that support the global free trade. The same tension is observed regarding GMOs and, for the time being, only free trade is benefiting from it. This is in contrast to the weakness of the *corpus iuris* that was set up to fight the evils of hunger, and in contrast to the tools developed to protect the environment in their joint pursuit to better distribute global resources (such as the CBD).

The regulations applicable to international trade are developed at the headquarters of governance, the WTO, and it is well known that in these forums the individuals involved hardly ever enjoy equal participation conditions. On the other hand, as we have already mentioned before, we count on several institutions committed to health, the environment or the development of the people's of the world, but whose binding strength does not compare with the one of commercial agreements. The vertiginous expansion of GMOs commercial farming has collided interests such as the protection of the environment and the consumers with the international trade regulation rules.

The WTO does not totally discard the application of precautionary approaches to GM farming and GM food trade. However, the agreements applicable to these matters greatly limit the possibility to require additional safety measures based on precautionary criteria with the claim to avoid excessively protectionist practices. The applicable regulation in these cases is collected on the WTO SPS Agreement on Sanitary and Phytosanitary Measures (SPS Agreement, 1994) and TBT Agreement on technical Obstacles to Trade (TBT, Agreement1997).

The SPS Agreement refers to all the sanitary and phytosanitary measures which might directly or indirectly affect the international trade (art. 1). Therefore, although it does not explicitly make reference to the GMOs, the trade operations involving these or the products made with them are understood to be under its scope of application. The Agreement requires that, when a member of the WTO establishes sanitary or measures which may obstruct the international trade, they be based in the existence of risks to human, animal or plant health. The assessment of such risks is referred to the evaluation techniques developed by organizations of international relevance and it is reminded that assessment must lie on the weighing out of numerous factors among which 'available scientific evidence' stands out.

This does not mean that there is no room for precautionary approaches, because the SPS Agreement admits that 'where sound scientific evidence is not enough, a Member State may provisionally adopt sanitary or phytosanitary measures on the basis of the appropriate information available'. The appeal Body in hormone matters understands that we are before the application of a precautionary principle, but the appeal Body for measures affecting farming products developed the conditionings within which the measure, based on this precautionary approach, can be understood as reasonable, limiting considerably the possibilities these criteria have.

In turn, the TBT Agreement related to technical barriers to trade, is an agreement to avoid the unnecessary obstacles to international trade. It states that the restrictions to trade are objectively legitimated and in this sense, it admits that no country should be prevented from establishing the necessary measures to care for its export quality, protecting human, animal or plant health or to prevent unfair practices. This agreement is applicable to all kinds of products and in the case of GMOs it is of special interest as it respects the obligation to label these products. In any case, it cannot be forgotten that, in the cases where human, animal or plant health could be at risk, the application of the SPS agreement is always a priority.

Regarding labelling, the TBT agreement establishes that it will not constitute a technical obstacle to trade whenever it is applied in agreement with the principles of necessity and proportionality, being these two principles of wide recognition on international trade matters.

In opinion of the countries gathered at the Miami Group, the provisions in the Cartagena Protocol admit a broader scope to the precautionary principle than the WTO regulations do. And a clashing conflict results. In this sense, the 10th article in the Protocol seems to be particularly conflictive when stating that:

> Lack of scientific certainty due to insufficient relevant scientific information and knowledge regarding the extent of the potential adverse effects of a living modified organism on the conservation and sustainable use of biological diversity in the Party of import, taking also into account risks to human health, shall not prevent that Party from taking a decision, as appropriate, with regard to the import of the living modified organism {...} in order to avoid or minimize such potential adverse effect.

On the other hand, provisions, like the labelling or traceability requirements, encounter difficulties to be assumed as reasonable or proportional ones.

Which regulation shall prevail? In agreement with International Law, given that the Cartagena Protocol has no independence from international agreements as WTO regulations do, in case of conflict between countries, WTO arbitration shall prevail. However, this does not mean that non-commercial considerations may not affect WTO decisions. In fact, there are more and more external considerations and pressures under which the WTO has to give in because its actions are no longer judged by loyalty to specific theory or ideology, but rather by its effects on the real world.

In this regard, it is not in vain to take into account the Biodiversity Agreement and the Documents that supplement it, like the Cartagena Protocol, as they warn on the mutual independence the Countries have in terms of environmental protection, and as it applies to each one and all of them. The UNCED gave way to the conviction that all countries need one another to achieve an environmentally sustainable social and economic development model and this is a fact assumed as key element which will not, in a very long term, modify the international economic order.

Many discourses, be they political, ethical, juridical or economic in nature, have affirmed in the last decades that modern biotechnology offers an opportunity in terms of a better, fairer living-together. Hopes seemed to have been placed in biotechnology to encompass collective and individual rights to the equitable distribution of the Planet's resources, or the right to share the benefits of technological progress, as well as it urges the commitment towards the preservation of the global environment. But biotechnological progress has not yet found the seeds with which to crop a better world.

Acknowledgements

Research Team Grant of the Directorate on Scientific Policy of the Basque Country 'Multilevel Constitutionalism and Integration of Diversity (Political, Legal-Institutional and Cultural-Religious)';

'Federal Systems between integration of diversity and stability', funded by the Spanish Ministry of Science and Innovation, DER 201020850.

References

Clapp, D. and Fuchs, D. (eds.) (2009). Corporate power in global agrifood governance. The MIT Press, Cambridge, MA, USA.

Escajedo, L. (2007). Towards a healthy and sustainable development of biotechnology, SIBI, Spain.

Escajedo, L. (2010). Vers un cadre régulateur de la sécurité de l'application de la biotechnologie: le chemin parcouru. Revue Générale de Droit International Public 114: 5-34.

Higgins, V. and Lawrence, G. (eds.) (2005). Agricultural governance. Globalization and the new politics of regulation. Routledge, New York, NY, USA.

Mechelm, K. (2004). Food security and the right to food in the discourse of the United Nations. European Law Journal 5: 631-648.

Oosterveer, P. (2007). Global governance of food production and consumption. Issues and challenges. Edward Elgar, Cheltenham-Northampton, UK.

Page, E.A. and Redcliff, A. (2002). Human security and the environment. International comparisons. Edward Elgar, Cheltenham, UK.

Welsh, J. and MacRae, R.J. (1998). Food citizenship and community food security: lessons from Toronto. Canadian Journal of Development Studies 19: 237-255.

Wright, W. and Middendorf, G. (2008). The fight over food: producers, consumers and activists challenge. Pennsylvania State University Press, University Park, PA, USA.

Mediterranean diet and sustainable food habits: the case of Neapolitan children

M.T. Gorgitano and V. Sodano
University of Naples Federico II, Agricultural Department, Via Università 96, 80055 Portici (Naples), Italy; gorgitan@unina.it

Abstract

The Mediterranean diet (MD hereafter) is widely recognized as one of the healthiest dietary patterns, able to benefit biodiversity and food system sustainability. The study was aimed at gathering information on real children's dietary habits in Southern Italy in order: (1) to assess the current adherence to MD, which is considered as a proxy for food consumption sustainability; and (2) to give suggestions on the role of school nutrition education programs in fostering more sustainable food habits. The adherence of Neapolitan children to the MD was assessed using an *ad hoc* Mediterranean Diet Quality Index. Results from the Neapolitan case study demonstrate that eating habits in the area follow unhealthy and unsustainable food models, which are quite far from the traditional Mediterranean diet. The main conclusion of the paper is that a major effort should be made by the Ministry of Education, school managers and educators, in the field of food education intervention.

Keywords: Mediterranean diet, consumption sustainability, nutrition education

Introduction

The Mediterranean diet (MD) is widely recognized as one of the healthiest dietary patterns, able to benefit biodiversity and food system sustainability (Padilla *et al.*, 2012; Romaguera, 2010). Historically, Southern Italy has been characterized by a Mediterranean-style diet, primarily because of the extensive local availability of fruit, vegetables and fish (Keys, 1980). Nonetheless, over the last thirty years taste standardization and the spreading of supermarkets, together with food system globalization, have profoundly changed people's behaviours and attitudes. As a consequence, food habits seem to have shifted away from the Mediterranean model (Piccinelli *et al.*, 2011). Over recent years, in the area, a high prevalence of health problems associated with poor food habits has been found. Particularly worrying is data on childhood obesity due to the fact that this increases the risk of many non-communicable diseases in adult life. While, in Italy, obesity rates are low with respect to most developed countries, they are very high among children, with 1 in 3 children being overweight (OECD, 2011). Southern Italy shows the worst data, with about 40% of children being overweight or obese (Spinelli *et al.,* 2013).

Our research was aimed at gathering information on actual children's dietary habits in Southern Italy in order: (1) to assess the current adherence to MD, which is considered as a proxy for food consumption sustainability; and (2) to give suggestions on the role of school nutrition educational programs in fostering more sustainable food habits.

Section 1 defines food consumption sustainability and presents the Mediterranean-style diet as a model of sustainable diet. Section 2 shows the research results, which allowed an assessment of the degree of adherence to MD in a sample of Neapolitan children. Section 3 draws some conclusions and policy implications, with reference in particular to the role of nutrition education programs.

Sustainable food consumption and the Mediterranean diet

In a broad perspective 'food consumption to be sustainable has to be safe and healthy in amount and quality; and has to be realized through means that are economically, socially, culturally and environmentally sustainable – minimizing waste and pollution and not jeopardizing the needs of others' (Reisch, 2010). According to this definition, policies aimed at food consumption sustainability may include a variety of goals, such as (Power, 2010; Wolff and Schonherr, 2011): protecting the environment and natural resources, with a special focus on water, soil, climate and biodiversity; promoting human health; supporting rural economies; promoting animal health and welfare; preserving socio-cultural food diversity; meeting the nutritional needs of the less well off people. These policies may be carried out also by those private companies who claim to be socially responsible.

A prerequisite for promoting consumer behaviour changes able to achieve such a wide range of goals is a clear understanding of the relationship between sustainability and dietary habits. Since the nineties, economists and industrial ecologists have produced a rich body of theoretical models and field research analysis which clarifies the environmental consequences of diets and offers sound frameworks for evaluating alternative diets from several points of view simultaneously (Duchin, 2005). Early life-cycle studies of food examined the environmental impact of different food consumption baskets in terms of energy and land use. When studying the Swedish case, Carlsson-Kanyama and Faist (2000) found evidence of a higher use of energy in the production of animal compared to plant food, greenhouses in comparison to open-air cultivations, and processed in relation to fresh foods. Comparing European and U.S. diets with respect to land requirements, Gerbans-Leenes and Nonhebel (2002) found a substantially higher use of land for the U.S diet (even standardized to a common energy intake), due to higher meat consumption. Using material flow analysis and Economic Input-Output Life Cycle Assessment (EIO-LCA) methods, successive studies have tried to measure the sustainability of complete food production/consumption systems, taking into account the various activities along the food chain and the effects of dietary changes on the whole economy. Further research efforts have pointed to the worldwide framework, exploring the interconnection between dietary changes in affluent and poor countries (Fuchs and Lorek, 2000). Results from this broad array of research are varying and sometimes contradictory; nevertheless, there is almost complete agreement on at least three points (Duchin, 2005): for food consumption to be sustainable, a shift from a meat-based to a plant-based diet should occur at a global level; reducing food miles and processing helps to reduce energy use and carbon emissions; the most sustainable diet (including environmental, economic and health dimensions) seems to be the Mediterranean-style diet, even when conventional agricultural techniques are assumed (Padilla *et al.*, 2012). At international level, the generally accepted definition of MD pattern (Bach-Faig *et al.*, 2011; Key *et al.*, 1986) refers to a diet which is varied, not very caloric and based on fresh, local and seasonal products, whenever possible. MD is rich in fresh fruit and vegetables and low in meat, added sugar, salt and saturated fatty acids. It is worth noticing that another widely recognized sustainable diets are the Japanese and the Asian diets; nevertheless, we specifically considered the Mediterranean-type diet because it hinges on agricultural products available in many countries and, moreover it is the traditional diet of the region where we carried out our empirical study.

Based on these findings, in the present study a high adherence to MD was considered as a proxy of sustainable food consumption. The degree of adherence to MD was estimated for a sample of children from Southern Italy, in order to gain some insights into the current food consumption sustainability in the area and to understand what could be the role of school nutrition education policies for redirecting food consumption towards more sustainable patterns.

Children and sustainable food consumption: a case study

The study was carried out in the municipality of Naples, which is the largest municipality in Campania and in Southern Italy and the third largest municipality in Italy, with a population of about 1 million. Children aged 9-11 years were the targeted population group. Focusing on a particular population subgroup with homogenous nutritional requirements facilitates the interpretation of results and allows for comparison among individuals within the group. Childhood is an important target for the study of food habits, because correct early food habits perform a key role in shaping dietary choices in later life (Mikkila *et al.,* 2004); moreover, children may be more easily targeted with food education programs aimed at diet improvements.

In the study, performed as a cross-sectional analysis, a random sample of children enrolled in public primary schools were provided with an ad hoc questionnaire aimed at collecting data on eating behaviours and built on the example of previous literature on MD indexes (Bach *et al.,* 2006; Mila-Villarroel *et al.*, 2011; Sofi *et al.*, 2008). On collected data a Mediterranean Diet Quality index (MDQI) was estimated.

Data on children's dietary habits was collected during an extensive survey carried out at the University Federico II of Naples. The survey involved a random sample of public primary schools (34 schools) stratified by territory in such a way as to represent all the ten administrative districts of Naples. 2,127 children aged between 9 and 11, enrolled in the 5th class, were given a questionnaire during a face-to-face interview. The survey was carried out by trained interviewers, in order to overcome response bias due to the limited cognitive ability of many children when estimating the kind and amount of food consumed.

The adherence of Neapolitan children to the MD was assessed using an ad hoc MDQI. In epidemiological studies, indexes evaluating the adherence to MD may be classified in three categories (Bach *et al.*, 2006): (1) those based on a positive or negative scoring of the diet components, with positive scores assigned to dietary components consistent with MD (Lazarou *et al.*, 2009; Sahingoz and Sanlier, 2011; Serra-Majem *et al.*, 2004; Trichopoulou *et al.*, 2003); (2) those that add or subtract standardized components (Alberti-Fidenza and Fidenza, 2004); (3) those that are based on a ratio between components. The MDQI belongs to the first category and is adapted from the KIDMED index previously used by Serra-Majen *et al.* (2004). The MDQI was obtained additively combining two groups of 13 basic indicators (Table 1) equally weighted. The first group (10 indicators of positive quality) includes the indicators, which were assigned a score of +1, associated with the presence of components related positively to the MD. The second group (3 indicators of negative quality) includes indicators signaling harmful dietary habits, with a score of -1. Consequently, the MDQI ranges between -3 and +10. The scores obtained were divided into three sections, each associated with three different assessments of the quality of the diet adhered to by those children interviewed: low diet quality (score between -3 and +3), b) medium diet quality (score between +4 and +6), c) and satisfactory diet quality (between +7 and +10).

Table 2 shows the results. The MDQI absolute frequencies, reported for three ranges of score, clearly show that the Neapolitan children's diet quality is quite far from the ideal MD. 46% of children had a low MDQI result (MDQI score >3), 43,7% had a medium MDQI result (MDQI score between 4 and 6), and only 10,3% showed a high adherence to MD (MDQI score between 7 and 10). The frequency of each of the 13 basic indicators within the three ranges of QMDI scores, helps highlight the main elements of weakness in children's dietary habits. These are the low consumption of fruit and vegetables, the poor breakfast habits, the excessive consumption of cakes and meal destructuration. Particularly worrying is the low consumption of vegetables, with only 49,3% of children eating them once a day on a regular basis and a mean 12,6% easting them more than once a day.

Table 1. Mediterranean diet quality index score: frequency distributions (cumulative values).

	QMDI score														Median	Mode	*P*-value
	-3	-2	-1	0	1	2	3	4	5	6	7	8	9	10			
Absolute values	0	8	53	153	334	612	979	1,328	1,662	1,909	2,024	2,093	2,119	2,127			
Frequency distribution		0.4	2.5	7.2	15.7	28.8	46.0	62.4	78.1	89.8	95.2	98.4	99.6	100			
Gendre[1]																	<0.001
Boys		0.4	3.1	8.5	17.7	31.3	48.4	65.2	81.4	91.7	96.2	98.9	99.6	100	4	3	
Girls		0.4	1.9	5.9	13.8	26.2	43.6	59.6	74.8	87.8	94.1	97.9	99.6	100	4	3	
Mother's educational level[2]																	<0.001
Housewife		0.6	3.1	8.1	18.4	32.7	50.1	65.9	80.6	91.2	96.4	98.8	99.8	100	3	3	
Professional level																	
Low			3.0	8.3	16.1	30.4	46.7	65.2	82.1	93.5	96.4	99.1	99.7	100	4	4	
Medium			0.7	4.5	11.0	21.0	37.5	57.0	71.8	86.3	93.5	98.6	99.3	100	4	4	
High			1.1	4.5	7.9	16.9	35.7	48.9	68.8	82.3	89.8	95.5	99.2	100	5	5	
Parents' educational level and income[2]																	<0.001
Low level																	
Single-income family		0.8	3.4	8.6	19.9	35.1	51.3	68.0	82.2	91.7	96.9	98.9	99.8	100	3	4	
Two-income family			2.2	8.9	16.1	29.5	45.1	63.8	81.7	93.3	96.0	99.1	99.6	100	4	4	
Medium level																	
Single-income family			1.8	5.3	12.1	26.3	46.6	63.3	77.2	90.7	95.7	98.9	100		4	3	
Two-income family			0.8	3.1	12.4	20.9	34.9	58.1	71.3	85.3	93.0	97.7	98.4	100	4	4	
High level																	
Single-income family			0.0	3.7	14.6	24.4	45.1	48.8	72.0	87.8	91.5	97.6	98.8	100	5	5	
Two-income family			2.2	4.3	7.2	15.8	32.4	45.3	66.2	80.6	89.2	95.0	100		5	5	
Low-medium level																	
Two-income family			3.0	6.0	14.9	28.4	46.3	65.7	80.6	91.0	95.5	98.5	100		4	4	
Midium-high level																	
Two-income family		0.3	2.0	7.7	11.8	21.9	41.4	56.6	73.7	86.5	93.3	98.0	99.3	100	4	3	

[1] Mann-Whitney U Test.

[2] One Way distribution-free ANOVA - Kruksal-Wallis test *P*-value 4 3.

Table 2. Basic quality indicators – frequency distributions.

Elementary index	Absolute frequency	QMDI score from -3 to +3	QMDI score from +4 to +6	QMDI score from +7 to +10
	2,127 (100)	979 (46.0)	930 (43.7)	218 (10.2)
Postive score index				
High adherence to Mediterranean diet				
Pulses at least once a week	1,907 (89.7)	812 (82.9)	882 (94.8)	213 (97.7)
Fish at least once a week	1,724 (81.1)	678 (69.3)	832 (89.5)	214 (98.2)
Fruit once a day regularly	1,709 (80.4)	620 (63.3)	872 (93.8)	217 (99.5)
Milk or yoghurt for breakfast	1,534 (72.1)	566 (57.8)	766 (82.4)	202 (92.7)
Pasta or rice at least five times a week	1,348 (63.4)	397 (40.6)	750 (80.6)	201 (92.2)
Vegetables once a day regularly	1,049 (49.3)	282 (28.8)	562 (60.4)	205 (94.0)
Cereals or grains for breakfast	702 (33.0)	157 (16.0)	381 (41.0)	164 (75.2)
Fresh fruit more than once a day	545 (25.6)	61 (6.2)	326 (35.1)	158 (72.5)
Yoghurt more than once a day	334 (15.7)	60 (6.1)	180 (19.4)	94 (43.1)
Vegetables more than once a day	269 (12.6)	24 (2.5)	134 (14.4)	111 (50.9)
Negative score index				
Poor adherence to Mediterranean diet				
Pastries for breakfast	1,317 (61.9)	734 (75.0)	524 (56.3)	59 (27.1)
Cakes more than once a day	1.093 (51.4)	672 (68.6)	393 (42.3)	28 (12.8)
No breakfast always or often	715 (33.6)	474 (48.4)	220 (23.7)	21 (9.6)

Crossing the frequency of QMDI scores with gender and socio-economic characteristics helps detect some factors influencing children's dietary habits. Tables 2 shows cumulative frequency distribution of QMDI scores for the complete sample and for parts of the sample grouped with respect to 3 variables (namely: gender, mother educational level, parents' educational level and income). While gender does not explain many differences in dietary habits, a high educational level of the mother and the belonging to a rich two-income family show a significant positive impact.

Conclusion

In the paper, the need for food education programs has been assessed using data from a case study performed in the Neapolitan area. Results from the Neapolitan case study demonstrate that eating habits in the area follow unhealthy and unsustainable food models, which are quite far from the traditional Mediterranean diet. 46% of children were classified as low adherers to MD and 10% as high adherers. Negative dietary habits included: excessive consumption of snacks and food with low nutritional value (e.g. pastries, snacks and sweets); low consumption of fruit and vegetables; meal destructuring; nutritional imbalance; bad breakfast habits. The poorer dietary habits were found for children with low-income families and mothers with a low educational level. Moreover, a slight gender difference was found, with girls performing better than boys. Our results are consistent with the previous findings for the Spanish case (Serra-Majem *et al.*, 2004) where the substantial abandonment of traditional Mediterranean dietary patterns among children was found to be due to the reduced consumption of fruit, vegetables, pulses and fish and to the habit of between-meal snacking.

The poor diet of Neapolitan children raises serious health and nutrition concerns for the interested area. Poor eating habits during childhood may compromise healthy eating habits of a lifetime with the consequent negative effects on health well-being and food system sustainability. In order to meet such concerns effective corrective policies should be implemented aimed at redirecting children's eating behaviours towards more favourable patterns. The methods commonly used for this purpose are school nutrition education programs (World Health Organisation, 2006). The rationale for school nutrition programs is that providing children with healthy meals when they eat at school and teaching them properties of food and the characteristics of a good diet, may improve children's food habits and tastes. Developing sustainable food school programs is generally an important measure, which may help to promote and implement sustainable food consumption (Llargues *et al.,* 2011).

The present study was the first example of dietary pattern analysis in the municipality of Naples. This is a strong point because it accounts for research originality. Other strong points are: the holistic assessment of the diet based on eating behaviours; the easy use and understanding of results for non experts, teachers and parents; the direct use of results for better targeting nutrition education activities. Nevertheless, it is worth noticing that there are at least three points of weakness. First, being the first study in the area, its results cannot be compared with previous analysis. Second, it was a cross-sectional study and therefore its conclusions cannot be generalized. Third, being focused on the measurement of the quality index diet, the research did not investigate the complete range of variables affecting children's eating behaviour. It can however, still be considered as an insightful explorative study, whose results may be important for future more in depth investigations on nutritional issues in Southern Italy. A particularly interesting result of the study was the correlation found between good eating habits of the children and family's high socioeconomic status. This finding may represent a further motivation for nutrition education programs, with public schools called upon to mitigate social inequalities, providing equal opportunities for children in the field of health and nutrition.

References

Alberti-Fidenza, A. and Fidenza F. (2004). Mediterranean adequacy index of Italian diets. Public Health Nutrition 7: 937-941.

Bach, A., Serra-Majem, L., Carrasco, J.L., Roman, B., Ngo, J., Bertomeu, I. and Obrador, B. (2006). The use of indexes evaluating the adherence to the Mediterranean diet in epidemiological studies: a review. Public Health Nutrition 9: 132-146.

Bach-Faig, A., Berry, E.M., Lairon, D., Reguant, J., Trichopoulou, A., Dernini, S., Medina, F.X., Battino, M., Belahsen, R., Miranda, G. and Serra-Majem, L. (2011). Mediterranean diet pyramid today. Science and cultural updates. Public Health Nutrition 14: 2274-2284.

Carlsson-Kanyama, A. and Faist, M. (2000). Energy use in the food sector: a data survey. AFR Report 291. Swedish Environmental Protection Agency, Stockholm, Sweden.

Duchin, F. (2005). Sustainable consumption of food. Journal of Industrial Ecology 9: 99-114.

Fuchs, D. and Lorek, S. (2005). Sustainable consumption governance. A history of promises and failure. Journal of Consumer Policy 28:3: 261-288.

Gerbens-Leenes, P.W. and Nonhebel, S. (2002). Consumption patterns and their effects on land required for food. Ecological Economics 42: 185-199

Keys, A. (1980). Seven countries a multivariate analysis o death and coronary heart disease. Harvard University Press, Cambridge, MA, USA.

Lazarou, C., Panagiotakos, D.B., Kouta, C. and Matalas, A.L. (2009). Dietary and other lifestyle characteristics of Cypriot school children. BMC Public Health 9: 147-154.

Llargues, E., Rosa, F., Recasens, A., Nadal, A., Vila, M., Pérez, M.J., Manresa, J.M., Recasens, I., Salvador, G., Serra, J., Roure, E. and Castells, C. (2011). Assessment of a school-based intervention in eating habits and physical activity in school children: the AVall study. Journal of Epidem. & Community Health 65(10): 896-901.

Mikkila, V., Rasanen, L., Raitakari, O.T., Pietinen, P. and Viikari, J. (2004). Longitudinal changes in diet from childhood into adulthood with respect to risk of cardiovascular diseases. Eur. J. of Clinical Nutrition 58: 1038-1045.

Milà-Villarroel, R., Bach-Faig, A., Puig, J., Puchal, A., Farran, A., Serra-Majem, L. and Carrasco, J.L. (2011). Comparison and evaluation of the reliability of indexes of adehence to the Mediterranean diet. Public Health Nutrition 14: 2338-2345.

OECD (2011). Obesity and the economics of prevention: fit not fat. OECD, Paris, France.

Padilla, M., Capone, R. and Palma, G. (2012). Sustainability of the food chain from field to plate: the case of the Mediterranean diet. FAO, Rome, Italy.

Piccinelli, R., Arcella, D., Buonocore, P., Capriotti, M., D'Addezio, L., Le Donne, C., Mistura, L., Pettinelli, A., Sette, S., Turrini, A. and Leclercq, C. (2011). L'indagine nazionale sui consumi alimentari in Italia INRAN-SCAI 2005-06. Osservatorio Consumi Alimentari, INRAN. Rome, Italy.

Power, K. (2010). Introducing behavior changes towards sustainable food consumption Corpus 3, December.

Reisch, L. (2010). A definition of 'sustainable food consumption', Corpus 3, December.

Rojas, A., Valley, W., Mansfield, B., Orrego, E., Chapman, G.E. and Harlap, Y. (2011). Toward food system sustainability through school food system change. Sustainability 3: 763-788.

Romaguera, D., Norat, T., Vergnaud, A.C., Mouw, T., May, A.M., Agudo, A., Buckland, G., Slimani, N., Rinaldi, S., Couto, E., Clavel-Chapelon, F., Boutron-Ruault, M.C., Cottet, V., Rohrmann, S., Teucher, B., Bergmann, M., Boeing, H., Tjønneland, A., Halkjaer, J., Jakobsen, M.U., Dahm, C.C. and Travier, N. (2010). Mediterranean dietary patterns and prospective weight change in participants of the EPIC-PANACEA project. American Journal Clinic Nutrition 92: 912-921.

Sahingoz, S.A. and Sanlier, N. (2011). Compliance with Mediterranean diet quality index (KIDMED) and nutrition knowledge levels in adolescents. Appetite 57: 272-277.

Serra-Majem L., Ribas L., Ngo, J., Ortega, R.M., García, A., Pérez-Rodrigo, C. and Aranceta, J. (2004). Food, youth and the Mediterranean diet in Spain. Development of KIDMED Mediterranean diet quality index in children and adolescents. Public Health Nutrition 7: 931-935.

Sofi, F., Cesari, F., Abbate, R., Gensini, G.F. and Casini, A. (2008). Adherence to Mediterranean diet and health status: meta-analysis. BMJ 337: 1344-1353.

Spinelli, A., Lamberti, A., Nardone, P., Bucciarelli, M., Buoncristiano, M., Lauria, L., Andreozzi, S., Pediconi, M., Rubimarca, S., Galeone, D., Spizzichino, L., Menzano, M.T. and Vienna, A. (2013). Okkio alla salute - Sintesi dei risultati 2012. Available at: www.epicentro.iss.it.

Wolff, F. and Schönherr, N. (2011). The impact evaluation of sustainable consumption policy instruments. Journal of Consumer Policy 34: 43-66.

World Health Organisation (2006). A tool for the development of school nutrition programmes in the European Region. WHO, Geneva, Switzerland. Available at: www.schoolsforhealth.eu.

Knowledge synthesis and dissemination in organic research in Sweden: integrating ethics

S. Gunnarsson, P. Fredriksson, R. Hoffmann, B. Johansson, A. Mie, U. Nilsson, E. Röös, E. Salomon, C. Sundberg, K. Ullvén, A. Wallenbeck, C. Winqvist and M. Wivstad
EPOK – Centre for Organic Food and Farming, Swedish University of Agricultural Sciences (SLU), P.O. Box 7043, 75007 Uppsala, Sweden; stefan.gunnarsson@slu.se

Abstract

EPOK (Centre for Organic Food and Farming) at the Swedish University of Agricultural Sciences (SLU) works with knowledge synthesis, communication and research initiation in the field of organic farming. In order to facilitate the networking, researchers at departments of different scientific fields are associated to EPOK on a part-time basis. An important aspect of the knowledge transfer is to integrate the basic ethical concepts of health, ecology, fairness and care of organic farming according to the International Federation of Organic Agriculture Movements into the practical work. In the current EPOK activities the aim is to incorporate the principle of health in the work about animal nutrition and husbandry and in the synthesis work on nutritional aspects of organic food. The ecological principle is implemented in the work concerning cropping systems and interactions with soil fertility, the local flora and fauna and surrounding environment. The fairness principle influence the work on life cycle analysis of organic farming and its effect on climate change, as well as the efforts to give the animals possibility to natural behaviour. The care principle stresses the fact that the current agricultural practises must be developed and refined so that they support ecosystem services that promotes resilient production systems, which is exemplified by the work with biological crop protection methods. Our experience is that this approach to disseminating science to the society is fruitful. Through synthesizing research and experiences by system analysis of the food chain and its environmental impact the ethical principles can be emphasized in a scientific way. Other important activities are to organize workshops and to promote activities that increase the scientific knowledge relevant to organic farming. The approach of EPOK emphasizes the importance of communication through printed leaflets and reports, as well as, electronic newsletters and messages in social media. Furthermore, forums and platforms are created for an enhanced public dialogue.

Keywords: organic principles, communication, stakeholders, system analyses, knowledge synthesis

Introduction

In general, conventional farming is not guided by explicit philosophical or ethical ideas. The basic driving force in conventional agriculture is mainly based on economic considerations related to the market conditions and the general legal framework in the society which to a smaller or larger degree differs between nations and regions. Although organic agriculture also needs to consider such economic considerations, it is in contrast to conventional agriculture guided by general principles that could be sublimated into the principles of health, ecology, care and fairness (IFOAM, 2005). To promote the fulfilling of the organic principles, as well as optimising the productivity, knowledge dissemination for sustainable development is crucial.

Scientific knowledge is stated to be an essential part of the development of the organic agriculture, although IFOAM (2005) also stresses that 'practical experience, accumulated wisdom and traditional and indigenous knowledge offer valid solutions, tested by time.' Conventional agriculture is not, explicitly, guided by any such principle. Scientists are generally trained in scientific methods and rely on results based on these methods. Hence, the emphasis IFOAM puts on practical and traditional knowledge may be challenging to parts of the scientific community. However, EPOK argues that dissemination of scientific knowledge about organic agriculture must be based on understanding the conceptual framework of organic agriculture in order to achieve an impact on the practical farming. That requires constructive ways to transfer scientific results that acknowledge the value of both practical experience and traditional knowledge. This does not imply that knowledge relativism is requested, it simply means that scientists can learn many things from practitioners. Learning from practitioners is especially relevant in the process of identifying questions for future research, since solving problems encountered by practitioners is an important general aim of agricultural research. This process was recently conducted and resulted in a research agenda to be used by Swedish funding bodies in future calls for funding (Centre for Organic Food and Farming (EPOK), 2013).

The IFOAM (2005) principles are in general easy to communicate and understand. However, to be implemented in regulation/certification or used in extension service to the farmers, consumers and other stakeholders, the principles need to be further defined and operationalized (Padel *et al.*, 2009).

The aims of this paper is to describe the background and reflect on the outcome of an approach for dissemination and knowledge transfer in the field of organic farming in Sweden performed by EPOK (Centre for Organic Food and Farming) at the Swedish University of Agricultural Sciences (SLU). Furthermore, examples are given of how the IFOAM ethical principles can be integrated in the process.

The dissemination approach of EPOK

Organic agricultural systems are known to be knowledge intensive as they are based on the principle of working with ecosystems to produce food, and there is a strong need for new knowledge and innovations to increase sustainability in the food chain (EU commission, 2012). Organic agriculture is based on local nature, where climate, soils and ecosystem history cause a large variability in agro-ecosystems. Therefore, a plurality of knowledge development in different production systems is likely to be important as mankind faces an uncertain future with environmental threats, limited resources and great challenges for food supply and food security. Several research goals defined in national and international research strategies for organic agriculture are addressing questions relating to the development of more sustainable food and farming systems in general. They also stress the importance of food security for a growing population. Organic systems can thus be seen as a fore-runner and an innovation system to sustainable food and farming, but needs also to develop a sustainable food supply (EPOK, 2013).

To address the need for dissemination addressing the broad concepts and variability in organic agricultural systems, researchers within different scientific fields are associated to EPOK, working as an interdisciplinary team. EPOK functions as a platform for communicating knowledge concerning organic production and consumption. Important activities are initiating and organising workshops, synthesizing current knowledge and promoting activities that increase the scientific knowledge relevant to organic farming.

Goal conflicts are a natural consequence of ambitious and broad goals/aims, often highlighted when the principles of organic farming are discussed. The work of EPOK includes the challenge of tackling and communicating such goal conflicts through an interdisciplinary approach in EPOKs knowledge syntheses in different research fields (e.g. Röös and Sundberg, 2013; Winqvist, 2012, 2013). The interdisciplinary approach in the synthesis about effects of organic farming on climate change, including putting focus on scientific knowledge as well as identifying knowledge gaps, helps analysis and reflection of the goal conflict between e.g. high production levels in animal production and animal welfare. Previously there have been approaches in natural sciences that conflict of deviating goals are mitigated by isolation of single problem areas that then have been solved scientifically, partly neglecting the interested that have been considered as inferior. However, in many cases there are no clear-cut solutions for the goal conflicts, and the goal conflict may fluctuate over time as knowledge increases and different interests varies over time. The scientific progress gains from keeping focus on the goal conflicts and avoid simplifying of the complex situation.

Integrating the IFOAM ethical principles

The activities performed by EPOK that have been directed towards specific organic research fields can be structured according to the four ethical principles established by IFOAM (2005); health, ecology, fairness and care.

Health

'The health of individuals and communities cannot be separated from the health of ecosystems – healthy soils produce healthy crops that foster the health of animals and people' (IFOAM, 2005). Thus, the health principle reflects a holistic approach to health and the activities at/of EPOK have so far mainly focused on human health related to the consumption of organic products. Organic agriculture does not use pesticides in crop production. If and how this interacts with food choices and human health interests, knowledge both in researchers and in consumers can have a significant impact on policy decisions. Furthermore, an improved animal health in the meat, dairy and egg production is important for fulfilling the health principle concerning animals as well as consumers.

Studies on animal husbandry in research regarding organic production in Sweden place a high emphasis on animal health and welfare and adaptation of production systems to local conditions. One important resource for researchers in Sweden is the unique health data available from the herd monitoring databases created by Swedish farming organisations. The main topics of current research in organic animal husbandry are animal health, welfare and breeding and how these factors interact with different feed sources or feeding strategies that have the potential to mitigate environmental effects. One main conclusion is that healthy animals have less negative environmental impact, due less contamination of zoonotic diseases and medical substances, as well as, microbial genes resistant to antibiotics.

Ecology

In the IFOAM (2005) principle of ecology an emphasis of the use of local resources is made: 'Organic management must be adapted to local conditions, ecology, culture and scale. Inputs should be reduced by reuse, recycling and efficient management of materials and energy in order to maintain and improve environmental quality and conserve resources.'

The principle of ecology is primarily related to crop production systems; for human consumption, as well as, for animal feed (secondary usage). The aim is that 'Organic farming, pastoral and wild harvest systems should fit the cycles and ecological balances in nature' (IFOAM, 2005). Key questions for organic crop production in Sweden are diversification of farming systems cultivars and crops, strengthened ecosystem services and reduced usage of external input such as fossil fuels. The need of knowledge and how it can be put to practice is urgent as the crop production struggles with how to control weeds and how to secure higher yields in a near future with changing climate conditions.

A better understanding of how ecosystem services function and how they can be strengthened is therefore crucial for a sustainable organic crop production. Ongoing Swedish research, concerning these important issues include studies of landscape effects on biological control (Jonsson *et al.*, 2012), pollination services (Andersson *et al.*, 2012; Rundlöf *et al.*, 2008) and how within field habitat manipulation impact pest insects and natural enemies (Nilsson *et al.*, 2012). In a research synthesis published by Winqvist (2013) it was found that e.g. wild plants benefit from organic farming, particularly plants that are insect pollinated, annuals and rare. About 20% greater biodiversity was found on organic farms, when compared to conventional farms.

Moreover, a research synthesis concerning habitat manipulation in vegetables and apple was initiated at EPOK as a consequence of a workshop where farmers, advisors and scientist met to discuss and identify knowledge gaps. This ongoing work will consist of interviews with Swedish organic growers to learn from their experiences and also a review of current international research.

Fairness

The IFOAM principle of fairness 'is characterized by equity, respect, justice and stewardship of the shared world, both among people and in their relations to other living beings'. Furthermore, IFOAM states that 'Organic agriculture should provide everyone involved with a good quality of life, and contribute to food sovereignty and reduction of poverty. It aims to produce a sufficient supply of good quality food and other products' (IFOAM, 2005).

In a comprehensive EPOK synthesis report about organic farming and climate change these concerns were incorporated (Röös and Sundberg, 2013). The report discussed not only the direct emission of greenhouse gases (GHG) from agriculture; the main ones being nitrous oxide from soils, methane emissions from ruminants, carbon dioxide from energy use and emission of nitrous oxide and methane from manure, but it also described the indirect causes of climate change related to the widespread global use of land for agriculture. For example, the most beneficial use of agricultural land for reduced GHG emissions is the cultivation of bioenergy which could be used to replace fossil fuels in the agriculture sector as well as in the transport and energy sector. However, the cultivation of bioenergy is related to several ethical concerns as it competes for land with food production. Although the report did not present an ethical analysis, it provided background information on the potential of bioenergy as an energy source in agricultural systems and listed pros and cons with different types of bioenergy. Correct empirical information is crucial when forming ethical conclusions and by providing this and raising relevant questions it is hoped that the work will stimulate to ethical discussions that can help to develop organic farming in a more fair direction.

The principle of fairness also 'insists that animals should be provided with the conditions and opportunities of life that accord with their physiology, natural behavior and well-being'. In organic production in Sweden, this has been taken even further, and in herds following the standards of the Swedish national incorporated certification organisation (KRAV, 2012), all animals, including ruminants, pigs and poultry, are given the opportunity to graze during the growing season and, if possible, outdoor access all times of the year as far as the health and welfare is not in conflict.

Care

The principle of care, 'Organic Agriculture should be managed in a precautionary and responsible manner to protect the health and well-being of current and future generations and the environment', is also relevant to the work of EPOK, and it explicitly discusses the role of science: 'Science is necessary to ensure that organic agriculture is healthy, safe and ecologically sound' IFOAM, 2005).

In the synthesis described above (Röös and Sundberg, 2013), the principle of care was taken into account. A long-term perspective was chosen, in which a need for large reductions of greenhouse gas emissions is foreseen. This led to the question of the possibilities to achieve large reductions in the climate impact of organic agriculture. As a consequence, it was necessary to discuss not only how, but what, to produce in organic agriculture, because animal foods have a much higher climate impact than plant-based foods. Organic food production in Sweden today is dominated by animal production that uses 90% of the organically managed land. The care of climate indicates that a much lower animal production is needed on a global scale, which might also be true for Sweden. However, this can be in conflict with the principle of ecology as Sweden has 450 000 hectares of semi natural grasslands with high biodiversity, that needs grazing animals to be maintained. Hence the need of interdisciplinary knowledge communicated to policy makers.

Conclusions

Through synthesizing research and experiences by system analysis of the food chain and its environmental impact the ethical principles can be emphasized in a scientific way. Other important activities are to organize workshops and to promote activities that increase the scientific knowledge relevant to organic farming. The approach of EPOK emphasizes the importance of communication through printed leaflets and reports, as well as, electronic newsletters and messages in social media. Furthermore, forums and platforms are created for an enhanced public dialogue. Our experience is that this approach to disseminating science to the society is fruitful.

References

Andersson, G.K.S., Rundlöf, M. and Smith, H.G. (2012). Organic farming improves pollination success in strawberries. PLoS ONE 7:e31599. Available at: http://dx.doi.org/0.1371/journal.pone.0031599.

Centre for Organic Food and Farming (EPOK) (2013). Research agenda for organic agriculture 2013. M. Wivstad (ed.), EPOK, SLU, Uppsala, Sweden, 16 pp.

EU commission (2012). A decade of EU-funded, low-input and organic agriculture research (2000-2012). EU, Brussels, Belgium, 284 pp. Available at: http://dx.doi.org/10.2777/19796.

International Federation of Organic Agriculture Movements (IFOAM) (2005). The principles of organic agriculture. Preamble. IFOAM, Bonn, Germany. Available at: http://ifoam.org/about_ifoam/principles/index.html.

Jonsson, M., Buckley, H.L., Case, B., Hale, R.J., Wratten, S.D. and Didham, R.K. (2012). Agricultural intensification drives landscape-context effects on host-parasitoid interactions in agroecosystems. Journal of Applied Ecology 49: 706-714.

KRAV (2012). Regler för KRAV-certifierad produktion. Utgåva 2012. [Standards for production certified by KRAV. Edition 2012] (In Swedish, English summary), Växjö, Sweden 280 pp.

Nilsson, U., Rännbäck, L.M., Anderson, P. and Rämert, B. (2012). Herbivore response to habitat manipulation with floral resources: A study of the cabbage root fly. Journal of Applied Entomology 136: 481-489.

Padel, S., Röcklinsberg, H. and Schmid, O. (2009). The implementation of organic principles and values in the European Regulation for organic food. Food Policy 34: 245-251.

Röös, E. and Sundberg, C. (2013). Ekologisk produktion och klimatpåverkan. (In Swedish, English summary). EPOK-Centre for Organic Food and Farming, SLU, Uppsala, Sweden. (In press).

Rundlöf, M., Nilsson, H. and Smith, H.G. (2008). Interacting effects of farming practice and landscape context on bumblebees. Biological Conservation 141: 417-426.

Winqvist, C. (2012). More biodiversity on organic farms? Fact sheet, EPOK-Centre for Organic Food and Farming, SLU, Uppsala, Sweden, 4 pp.

Winqvist, C. (2013). Ekologiskt lantbruk, biologisk mångfald och ekosystemtjänster – i ett landskapsperspektiv (In Swedish, English summary). EPOK-Centre for Organic Food and Farming, SLU, Uppsala, Sweden, 51 pp.

Water consumption in rural areas: limits of the ethics of water use – study case of Kurdistan Region, Iraq

R. Harun, F.H. Arion and I.C. Muresan
University of Agricultural Sciences and Veterinary Medicine Cluj-Napoca, Romania; rezhen.harun@usamvcluj.ro; felixarion@usamvcluj.ro; iulia.muresan@usamvcluj.ro

Abstract

All life relies on an essential substance which is water. Where there is water there is life and where water is scarce, life has to struggle because it has no or very limited alternative. Therefore, the question is: how to deal with that no replacement? Water use is a complex issue and involves many actors from local communities to international bodies. Water ethics is being increasingly discussed, especially socio-economic aspects and their integration with policies and practices of water resource management. Although in1992, there was an International Conference which was held in Dublin in principles three and four stated that: 'Women play a central part in the provision, management and safeguarding of water', with 'Water has an economic value in all its competing uses and should be recognized as an economic good'. Water consumption and policy is also an intensively discussed subject. Many researchers try to find an answer regarding the proper use and value of water consumption. This paper presents the particularities of water resources in the rural areas of Kurdistan Region (North of Iraq). During 17th of June and 15th of September, 2012, a research project was conducted in the rural areas of the Kurdistan Region, taking into consideration the precipitation zone so it can be clearly distinguished among the three regions of Kurdistan. The research method used was a survey based on questionnaires. The total number of questionnaires distributed among the farmers was 236 in rural areas of Kurdistan Region. The use of water for irrigation, livestock etc, is estimated as being the cause of more than three quarter of total withdrawal. It is necessary to mention that in the Kurdistan Region there is no regulation regarding the water consumption for rural areas and the water scarcity in the area imposes to adopt a proper legislation regarding the water pricing on one hand, and the ethical principles of water consumption on the other hand. Previous studies revealed the fact that women play a major role in managing the domestic water. Based on this it can be stated that people from rural areas have major problems regarding the water availability and proper measures should be taken.

Keywords: water scarcity, water policy, water value, women

Introduction

This paper presents the difficulties of water use and the problems due to the lack of water in Kurdistan Region in Iraq. Water is one of the most basic and important resources for life. For agriculture, water in rural areas represents an essential production input. The functions of water ethics are right if they enhance the ability of water (Armstrong, 2009) with the ecosystems to maintain life in the study area. Because human activities require water consumption, the problem of water scarcity and the principles of water ethics was wildly debated (Delli, Priscoli, *et al.*, 2004). Climate change is expected to influence the rainfall and water supply, as well as to affect water demand through increased temperatures (Garrick and Jacobs, 2006; Hartmann, 2005). Ethics is based on morality and generally ethics can be defined as the right things to do (Katz, 1991). Water ethics represent a high concern at the international level. Harromoës (2002) presents the main concerns of water ethics: conservation and sanitation and the deprivation of poor communities. The water management is directly influenced by the application of the ethical codes (ECCAP WG14 Report: Water Ethics and Water Resource Management). Ethics assures the normative content of a particular decision by providing reason and justification (ECCAP

WG14 Report: Water Ethics and Water Resource Management). The fundamental principles of water ethics refer to human dignity, participation, solidarity, human equality, common good, stewardship, transparency and universal access to information, inclusiveness, and empowerment (ECCAP WG14 Report: Water Ethics and Water Resource Management).The Principles on Water and Sustainable Development are presented in Dublin Statement (1992): fresh water is a finite and vulnerable resource, essential to sustain life, development and the environment; water development and management should be based on a participatory approach, involving users, planners and policy-makers at all levels; women play a central part in the provision (Wakeman, 1995), management and safeguarding of water; water has an economic value in all its competing uses and should be recognized as an economic good. Rogers *et al.* (1998) analyzed the perception of people regarding the fact that water is an 'economic good'. Women traditionally also play a major role in managing and maintaining communal water supply, they are responsible for the regulation and control of the social use and safe maintenance of water resources, and the same time are rarely involved in the political and legal processes of making strategic decisions regarding water resources management at a level beyond the local one (SIDA, 1994).

Materials and methods

The research was conducted in Kurdistan Region (Figure 1). The Kurdistan Region is composed of three governorates located within the northern part of the Federal Republic of Iraq (Iraqi Constitution, Article 62). The region covers Erbil, Duhok, Sulaimani. The region lies between latitudes 34° 42' N and 37° 22' N and between longitudes 42° 25' E and 46° 15' E. The lowest point in the region is Kifri, which has an elevation of 140 meters above mean sea level which is belongs to Kirkuk/Garmian Governorate, and the highest point is the Peak of Hasarost Mountain in Erbil Governorate, measuring 3,607 meters above mean sea level. The Kurdistan region mainly stretches across the Zagross Mountains up to the Taurus Mountains in Turkey. The region shares its borders with Syria in the west, Turkey in the north and Iran in the east. Absolute population increase figures have not been affected by the economic changes happening throughout the successive stages of development. This situation has been confirmed by the surveys and pertinent population estimates, carried out by the competent technical authorities. Statistical figures indicates that the Region's population increased from about 3,910,329 inhabitants in 2003 to 4,382,167 inhabitants in 2008, at an increase rate of 12.07%. In 2009, population was

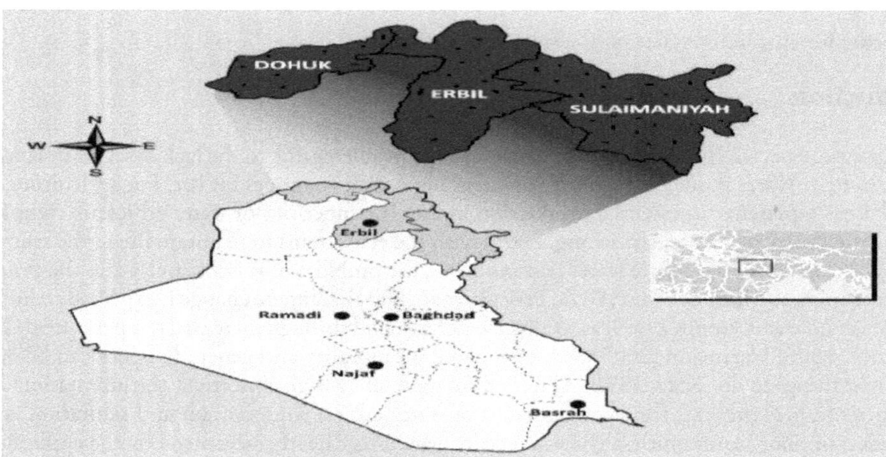

Figure 1. Location of studied area, Kurdistan regional government (KRG-Iraq) (UNDP-Iraq, 2010).

4,698,790, with an increase rate of 7.05%, compared to 2008 and 19.96% in comparison to that of 2003. Population is expected to rise to about 5,601,227 in 2016, if the growth rate remained the same as in the past five years, in terms of birth and mortality rates and other relevant changes. Annual growth population rate by governorate, based on the figures released by KRSO and the results of IHSES, was 3.2% in Erbil, 3.1% in Sulaimani and 2.6% in Duhok, with an overall annual growth rate of 3%. Nearly all inhabitants are ethnically Kurdish and speak the Kurdish language. In the past, the main activities of this population were agricultural cropping and livestock production. Now this economic activity has changed dramatically, as most of the population is engaged in education, government, trade, and industrial businesses. The annual rainfall in the region is not much less than that in most European countries, but its distribution is different: it is exceptional to have rainfall between June and September. Thus, analyses of climatic elements such as rainfall, evapotranspiration and runoff are essential for balancing the ground water. To achieve the purpose of this paper during June and September 2012, 236 questionnaires were applied to farmers from 46 villages.

Results and discussions

In the study area, women's life is particularly concerned with water availability and domestic contexts are very different from those of industrialized ones. Family work usually covers one third to one half of a woman's working day. In rural areas, it includes tiring tasks such as fetching water for domestic use. Women constitute a large portion of the economically active population engaged in agriculture, both as farmers and as farm workers, and play a crucial role in ensuring household food security. Women are the main users of water: for cooking, washing, family hygiene and sanitation. In the same manner as boys enjoy easier access to education than girls. Although women have better understanding of natural variations of water availability because of their role in the family life, as it can be observed in Figure 2a, 37.56% of female are illiterate, 25.99% of them only have basic reading and writing ability. Male literacy and education rates are double those of women. Figure 2b shows the distribution of farmers by gender: 56.22% are women and the rest are men. Also this fact shows the increase in the role of women in water consumption. There is a very clear gender division of roles and resources in all areas of water resources management.

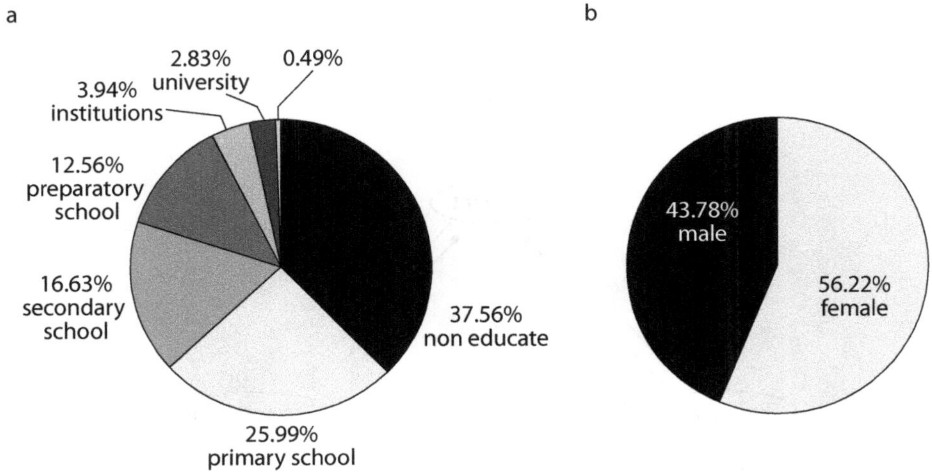

Figure 2. Education level for females (a) and distribution of farmers by gender (b).

The main source of water used for agriculture, as it can be observed in Figure 3, is the precipitation, more than 75% of the cases.

As it can be noticed, the farmers from Kurdistan Region mainly use the precipitation as main source of water. Asked about the source of water for irrigation, 46.18% of them use the canal as source of water for irrigation, and almost 30% of the them use the private tube, 3.4% use the public tube while the rest do not use or did not mention the source. The main method mentioned for irrigation is the furrow (56.35% of the farmers indicated this). The possibility of the maintenance of functions of water requires an ethic code. Allocating water, is effectively allocating life. In particular, taking water from the system for human uses in situations of water scarcity has an immediate negative impact on the life of non-human species. In some areas the water consumption is more than 500 liter/person/day because there is no lack of water, but in some territory water consumption going to 29 liter /person/day because the annual quantity of precipitation is less than 350 mm. The altering flux of water in the landscape affects the functioning of a whole hydro geomorphologic system. That is, its own right a component of the landscape. Consequently, preserving water also preserves elements of the landscape; and altering the distribution of water destroys or disturbs landscape elements. Thus, one needs to consider not only the quantity but also the quality of the water fluxes in the system. To comply with the advent of the concept of (IWRM), more focus was given to the three E's namely, efficient, equitable and environmentally sustainable water resource management (Figure 4). It is constantly being highlighted and advocated that water development and management should target the poorest and marginalized people (especially poor women) and preferentially ensure the coverage of their (fundamental) needs in terms of access and entitlement to secure and adequate water (Global Water Partnership, 2000).

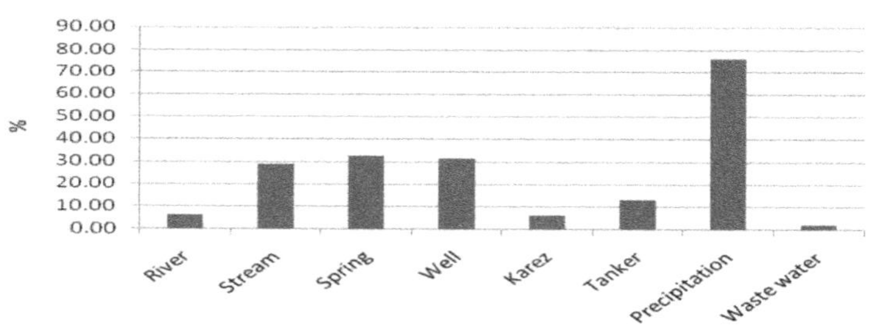

Figure 3. Source of water for agriculture, human activities.

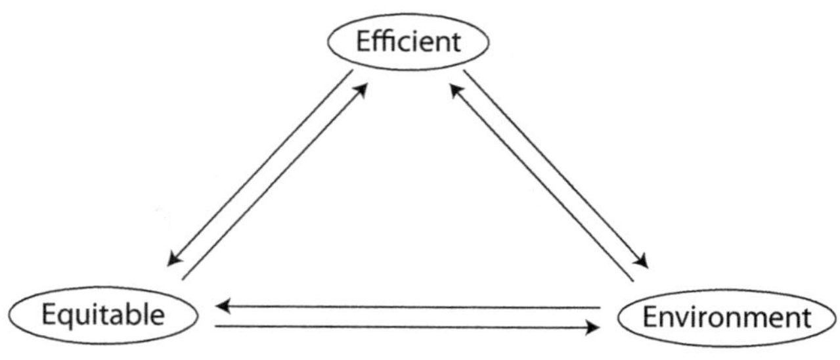

Figure 4. The three E's given to concept of (IWRM).

Conclusions

The study area never has seen the balance between demand and supply of water since the establishment until now because of facts and beliefs that water is source of life, represents (re)birth, cleans the body and by extension purifies it. As a conclusion it can be stated that Kurdistan Region has shortage of water for agriculture, even now for all sector's activities. In the current situation need of water in the subzone of region is quite the same. The marginal willingness to pay for water is different and the price is different. Because of that, the agricultural production costs and the cost of life are different in the regions, but they are competing on the same market, so it is economically and ethically unfair. It was observed that there is lack of strategic plans in general for educating women, that women are traditionally marginalized inside family; they are the main consumers of water. But they are the most uneducated human resources, so it is socially and ethically unfair. Based on these realities, the regional government should change the water policy for assuring the ethical use of water: investments for indiscriminate access to education, prioritized investment on the irrigation system based on the lack of water, subsidizing the agricultural production of the unfavorable regions for assuring a similar competitiveness. Water supply, due to drought, has no surplus water to draw, because of the difference between the regions (mainly, because of the rain). Observing that, the lack of understanding of the economic value of water, quality, level of over and underground water is necessary to take into account every chance to sustain the water's sources, it will become an increasingly valuable tool for negotiations over water among diverse stakeholders.

References

Armstrong, A.C. (2009). Viewpoint - further ideas towards a water ethic. Water Alternatives 2(1): 138-147.

Bogardi J.J. (2007). Water disasters and ethics. In: Ramon Lamas, M., Martinez-Cortina, L. and Mukherji, A. (eds.) Water ethics. Marcelino Botun Water Forum. CRC Press, Baco Raton, CA, USA, p. 315-326.

Delli Priscoli, J., Dooge, J. and Llamas, R., (2004). Water and ethics: overview. The United Nations Educational, Scientific and Cultural Organization (UNESCO), Paris, France.

Garrick, D. and Jacobs, K. (2006). Water management on the Colorado River: from surplus to shortage in five years. Southwest Hydrology 5: 8-9.

Harremoës, P. (2002). Water Ethics – a substitute for over-regulation of a scarce resource'. Water Science & Technology 45(8): 113-124.

Hartmann, H. (2005). Use of climate information in water resources management. In: Anderson, M.G. (ed.) Encyclopedia of hydrological sciences. John Wiley and Sons, Ltd, West Sussex, UK.

Katz, E. (1991). Ethics and philosophy of the environment: a brief review of the major literature. Environmental History Review 15: 79-86.

Rogers, P., Bhatia, R. and Huber, A. (1998). Water as a social and economic good: how to put the principle into practice. Global Water Partnership, Swedish International Development Cooperation Agency, Stockholm, Sweden.

Villholth K.G. (2007). Water and ethics in food production and provision – how to ensure water and food security and equity into the 21st century? In: Ramon Lamas, M., Martinez-Cortina, L. and Mukherji, A. (eds.) Water ethics. Marcelino Botun Water Forum. CRC Press, Baco Raton, CA, USA, p. 81-94.

Wakeman, W. (1995). Gender issues sourcebook for water and sanitation projects. UNDP-World Bank Water and Sanitation Program/PROWWESS. Water and Sanitation Program, World Bank, Washington, DC, USA.

Success and failure of transnational certification regimes

J.-C. Janssen
Cologne Graduate School of Management, Economics and Social Sciences, Richard-Strauss-Str. 2, 50931 Köln, Germany; janssen@wiso.uni-koeln.de

Abstract

The Fairtrade certification mark gains scope and breadth through the certification of new product categories and the growth of sales with varying success across product categories. In this paper, I examine the conditions for successful market penetration of product categories carrying the Fairtrade certification mark in Germany and the United Kingdom. The variation of the degree of ethical consumption between product categories within the same ethical framework is assessed to investigate the role of firms and social movement actors in the process of commensuration. This approach contributes to the literature on commensuration in analyzing successful commensuration without shedding a blind eye on unsuccessful attempts to commensurate within the same ethical framework. From the viewpoint of ethical consumption, it explains why some categories are relevant for ethical affluent consumers while others fail to do so. Firms and social movement actors mobilize consumers and/or apply ethical codes to improve or sustain their relative position in product markets as proposed by Fligstein (2001). Combined, their strategic attempts cumulate in product launches with differing success. Through a narrative of the processes of commensuration of product categories under a common ethical metric, I unravel two different paths for product categories to gain relevance in the eyes of ethical consumers. Highly differentiated product categories get attached with ethical value through the mobilizing efforts of social movements and small firms, which establish a visible niche to encounter incumbent firms in the market. In contrast, the successful launch of undifferentiated products relies heavily on the promoting efforts of large supermarket chains, which offer those products on large scale. In both cases, the product categories have to share characteristics consistent with these strategies. The proposed narrative and the preliminary evidence described will further guide my analysis of successful and unsuccessful product launches.

Keywords: commensuration, political cultural, fair trade

Introduction

Morals or *morality* are not deducible from an externally given logic, but are a socially constructed phenomenon (Beckert, 2011). If this is the case, the evolution of *morality* in markets has to be understood as a social process, in which moral standards or codes are formulated, applied, and possibly adjusted to the conditions of the specific market in question. Certificating their compliance with moral standards by third parties is one possibility for firms to signal credibility to external audiences for their moral claims (Doh *et al.*, 2010). In this process, products and practices are *commensurated* under the same general moral code. Here, *commensuration* is understood as the process of transforming 'different qualities into a common metric' (Espeland and Stevens, 1998: 314). This social process creates new social categories in linking different objects under the same metric (Espeland and Stevens, 1998). While scholars note that most projects of *commensuration* fail, few cases of failure are actually analyzed and fully understood (Huault and Rainelli-Weiss, 2011). *Morality* encoded in certification frameworks offers the possibility to analyze the full range of successful to failed *commensuration* between product categories. Zelizer (2007) rightfully observes that the general idea of Fairtrade as an example for moral codes has affected the public perception of coffee value chains in recent years. In contrast, the same framework taking momentum in the coffee market has problems in the *commensuration* of similar standards to bananas in Germany, for instance. Analyzing *morality* in markets requires the detection of conditions for *morality* in markets with reference to success and failure. While cases in which markets had to transcend moral boundaries are well known (Anteby, 2010; Zelizer, 1978), moral claims themselves have to transcend potential market boundaries. The construction of moral obligations between consumers and producers is a social construction with varying success. When political or ethical consumerism is on the rise (Micheletti, 2003; Shaw *et al.*, 2007), why is the ethical consumer in Germany not interested in bananas but in coffee? Some categories are considered as morally relevant to an audience, while the same audience turns a blind eye on others. Particularly in the case of consumption patterns, scholars argue for a comprehensive *political-cultural* approach to solve those puzzles instead of assuming a growing trend for ethical choice (Holt, 2012) without necessarily addressing firms (Bartley, 2007). The political cultural approach (Fligstein, 1996) suggests an analysis of at least two kinds of actors – incumbents and challengers. Incumbents dominate the markets and exploit their resources to create a stable world and ensure the reproduction of their position. Challengers are more peripheral located actors typically searching for 'niches' to survive and possibilities to encounter the influence of the incumbents. When actors perceive certification as a possibility to encounter incumbents (Bartley, 2007; Weber *et al.*, 2008) and social movements establish visible moral categories (Lounsbury *et al.*, 2003; Weber *et al.*, 2008), they do so with reference to the conditions within specific product markets. Those conditions are addressed in this paper.

Fairtrade offers the possibility to explain the success and failure of the *commensuration* of morality encoded in certification. In contrast to organic products, morality is not related to possible quality characteristics in the case of Fairtrade – the certification considers primary the well-being and the life situations of producers. Furthermore, successful and failed product launches of Fairtrade products are identifiable in approximating success with market share. Two product categories (bananas and coffee) are selected to exemplify the reasoning through a narrative (Langley, 1999: 695f). Many scholars merely apply this method in order to collect data and make sense of it at an early stage, while it also allows for proposing causal linkages across levels and analytical themes from a contextualist perspective (ibid.).

Hutchens (2009: 108ff) argues that national licensers of Fairtrade (and thus national settings) can be located either at the extreme of an 'advocacy' – or 'business model'– approach to Fairtrade. While the 'advocacy approach' is predominant in continental Europe and especially Germany, the licensers in the United Kingdom and the USA are characterized by a strong emphasis on the 'business model' (Hutchens, 2009: 109). To accommodate this finding, one national setting is chosen at each end of the continuum – Germany and the United Kingdom.

Bananas – the relevance of supermarket chains

The worldwide market for bananas is highly oligopolistic with five main multinational companies covering roughly 75 to 78% of the global market share (Morazán, 2012: 20; Smith, 2010a: 19). The globally predominant value chains are either integrated or highly integrated with multinational companies occupying the lead role in packaging, import, export, distribution, and – in part – growing (Morazán, 2012: 10). Even though multinational companies enjoy a major role in the global banana market, national supermarket chains have gained influence over recent years and are able to put pressure even on multinational companies or producer groups (Morazán, 2012: 16). Consequently, national conditions and strategies of major national retailers matter for the understanding of trends in the banana market. In general, bananas show a low potential for product differentiation since quality criteria issue mainly cosmetic features and thus, the products are interchangeable to a certain extant (Hütz-Adams and Ertener, 2012; Shreck, 2002). Therefore, organic and Fairtrade certification offers the opportunity for supermarket chains and companies to differentiate their offerings from the mass market and attract consumers' attention through a widely recognized label. Indeed, compared to other certified product categories, the certification of Fairtrade bananas is regarded as a success story in terms of market share. Two percent of all bananas sold worldwide are Fairtrade certified and maximum national market shares vary from 25% in the United Kingdom (Fairtrade Foundation, 2009) to more than 50% in Switzerland (Max Havelaar Schweiz, 2013). In contrast, the market share of Fairtrade bananas in Germany has shown no considerable growth within the last years and stays at a rather low level of about 2.1% (TransFair e.V., 2012). This strong difference of the factor ten is not mirrored by the sales of other Fairtrade certified product categories between the two national settings and requires further examination.

Banana sales in the United Kingdom and Germany are both affected by the structure of competition between the main supermarket chains screening opportunities to capture market share (Morazán, 2012; Smith, 2010a). These enjoy incumbent positions and thus prefer stability over change as Fligstein proposes. They are the pivotal players in both countries and the market structure is exceptionally dense with five supermarket chains covering about 73% and 71% market share of the grocery sales in Germany and the United Kingdom, respectively (Morazán, 2012: 16; Smith, 2010a: 111). Likewise, price competition characterizes both markets for grocery goods (ibid.). Particularly in the case of bananas, supermarket chains in both national settings favour long term contracts and short value chains to ensure a steady supply of bananas and to control prices (ibid.). In the United Kingdom, all four major supermarket chains offer Fairtrade certified product lines, two among those source all their bananas Fairtrade certified (Smith, 2010a). This is not the case for Germany. Even though competition on price has a significant impact on both markets, the supermarket chains in the United Kingdom do not pursue a strict discount strategy as opposed to Germany. In Germany, the most important discount supermarket chains Lidl and Aldi cover 24% market share of grocery goods (Morazán, 2012: 16). In contrast, discount supermarket chains are less dominant in the United Kingdom. Recently, the four main supermarket chains launched discount product lines to defend their dominant position against Aldi and Lidl's low price policy. Nevertheless, the market share of those two major discount supermarket chains is still rather low and about 7.7% of all grocery sales in the United Kingdom (Wall Street Journal, 2012). The strategies of the large discount supermarket chains are unsuitable for Fairtrade bananas for different reasons. Obviously, the price regulation of FLO limits the discretion for pricing and thus price competition which is the major advantage of the low price policy. Interestingly, low price bananas had been used as one of the main promotional items in the United Kingdom to attract costumers' attention which resulted in fierce price competition (Fairtrade Foundation, 2013; Smith, 2010a: 23). Fairtrade can be regarded as an opportunity to stabilize prices and an alternate way to distinguish the store in the eyes of potential costumers (Fairtrade Foundation, 2013). Furthermore, discount stores offer very limited product ranges. The major supermarket chains in the United Kingdom offer 30 times the product range of discount supermarket chains (BBC News, 2008). Therefore, those supermarket chains are able to certify one single product line to evaluate its success and incrementally extend the amount of supply (Smith, 2010b). Discount supermarket chains lack this opportunity because of their limited product range. In addition, long time storage of fresh fruits is problematic (Barrientos and Smith, 2007) and accordingly failed product launches bear higher risks as it is the case for easy storable goods as coffee, for example. Therefore, fresh fruits need an incremental increase in supply to minimize this risk. Thus, markets in which the strategy of pivotal players includes the provision of a wide range of different product lines seem to be better equipped for the differentiation of formerly rather undifferentiated products through moral labels. The incumbent actors in the United Kingdom preferred certification possibly also to hold their position against the new discount model. In contrast, the German discount supermarket chains already hold an incumbent position and oppose changes away from price competition (at least in cases in which they cannot offer more than one product line).

Coffee – niche strategies

The United Kingdom and Germany share similar levels of market share in coffee, while the higher level of the United Kingdom might be associated to the stronger business emphasis of the national licenser (as argued in Section 1). Thus the second path is paraphrased without stressing the differences between

Germany and the United Kingdom. It is assumed that the successful launch of differentiated product categories relies on already existing niche strategies of small firms facing incumbent companies and social movement activity. Alternative Trading Organizations had long been promoting against the unjust trading relations in the case of coffee, even before formal certification has been granted (Taylor *et al.*, 2005). As the first product acquiring formal certification, coffee became a symbolic product for the whole movement. Consumers were exposed to narratives of the life situations of small scale producers and the consequences of their purchasing decisions (Nicholls and Opal, 2005: 164ff). In addition, the rise of certification took place at a time when small enterprises searched for opportunities to distinguish themselves from the mass market (Raynolds, 2002) in pursuing a challenger strategy. The global mass market for roasting is characterized by an oligopoly of five major multinational companies covering 89% market share (Taylor, 2005: 133). Those can be regarded as incumbents. Fairtrade did not merely offer an appealing label, but its regulative framework fitted well to the already established business model of small enterprises (Raynolds, 2002; Renard, 2005). In part, long term business relations and higher prices had already been established before certification took momentum (Grodnik and Conroy, 2007). This model assures a constant supply of high quality coffee beans (ibid.). Some of the specialty roasters might get trapped by their initial commitment as they grow. For instance, Starbucks sourced only a small proportion of its coffee certified, before criticism evolved around the reproach of window dressing and the proportion has been increased (Taylor, 2005: 150). Today, Starbucks is among the hugest purchasers of Fairtrade certified coffee (Starbucks, 2013). The specialty roasting industry promoted Fairtrade in a manner which is consistent with business interests and thus attracted other enterprises in the process of mainstreaming. In recent years, the Fairtrade certified coffee market is highly competitive with high barriers for new entrants (Farnworth and Goodman, 2008).

Conclusion

The main argument is that the *ex ante* market structure plays a pivotal role in the understanding of differences in the impact of labelling morality on national product markets. For already differentiated product categories, labelling offers an easily understandable cue in the eyes of potential customers (Renard, 2005). In the case of low differentiated products, it may support attempts to differentiate formerly indistinguishable products to avoid competition on the price alone if this is not the main business model. The preliminary evidence for two of the most important Fairtrade products will further guide my attempt to include a wider range of products in a further analysis to foster my argument. From the viewpoint of commensuration, it has been shown that the inclusiveness of the framework alone cannot account for the differences in success as proposed by Weber *et al.* (2008). In addition, the focus on ethical or political consumerism (Micheletti, 2003; Shaw *et al.*, 2007) alone cannot account for those differences. Obviously, differences within the same framework have to be assessed with regard to the actor constellations and strategic opportunities given by the market structure at hand as proposed by the political-cultural approach (Fligstein, 1996). While the preliminary findings are insufficient for inferences to a general process, further, more in-depth analysis will aid in this attempt. A better equipped assessment of social movements' promotional efforts and entrepreneurial activities and strategies of firms will be included in future analysis. Furthermore, the inclusion of firm networks through associations and consumer surveys might enrich and complement such an analysis.

References

Anteby, M. (2010). Markets, morals, and practice of trade: jurisdictional disputes in the U.S. Commerce in cadavers. Administrative Science Quarterly 55: 606-638.

Barrientos, S. and Smith, S. (2007). Mainstreaming fair trade in global production networks. Own brand fruit and chocolate in UK supermarkets. In: Raynolds, L.T., Murray, D.L. and Wilkinson, J. (eds.) Fair trade: the challanges of transforming globalization. Routledge, New York, NY, USA, p. 103-122.

Bartley, T. (2007). Institutional emergence in an era of globalization: the rise of transnational private regulation of labor and environmental conditions. American Journal of Sociology 113(2): 297-351.

BBC News. (2008). Pinch helps discount supermarkets. Available at: http://news.bbc.co.uk/2/hi/business/7739606.stm.

Beckert, J. (2011). Die Sittlichkeit der Wirtschaft. MPIfG Working Paper, 11/8. Max-Planck-Institut für Gesellschaftsforschung, Köln, Germany.

Doh, J.P., Howton, S.D., Howton, S.W. and Siegel, D.S. (2010). Does the market respond to an endorsement of social responsibility? The role of institutions, information, and legitimacy. Journal of Management 36(6): 1461-1485.

Espeland, W.N. and Stevens, M.L. (1998). Commensuration as a social process. Annual Review of Sociology 24: 313-343.

Fairtrade Foundation (2009). Unpeeling the banana trade - a fairtrade foundation briefing paper. Fairtrade Foundation, London, UK.

Fairtrade Foundation (2013). Fairtrade bananas - looking behind the price tag. Fairtrade Foundation, London, UK. Available at: www.fairtrade.org.uk/includes/documents/cm_docs/2008/b/bananas_behind_the_price_tag.pdf.

Farnworth, C. and Goodman, M. (2008). Growing ethical networks: the fair trade market for raw and processed agricultural products: background paper for the World Development Report 2008 (Version 1, November 2006). The World Bank, Washington, DC, USA.

Fligstein, N. (1996). Markets as politics: a political-cultural approach to market institutions. American Sociological Review 61(4): 656-673.

Grodnik, A. and Conroy, M.E. (2007). Fair trade coffee in the United States. Why companies join the movement. In: Raynolds, L.T. Murray, D.L. and Wilkinson, J. (eds.) Fair trade: the challanges of transforming globalization. Routledge, New York, USA, p. 83-102.

Holt, D.B. (2012). Constructing sustainable consumption: from ethical values to the cultural transformation of unsustainable markets. The ANNALS of the American Academy of Political and Social Science 644: 236-255.

Huault, I. and Rainelli-Weiss, H. (2011). A market for weather risk? Conflicting metrics, attempts at compromise, and limits to commensuration. Organization Studies 32(10): 1395-1419.

Hutchens, A. (2009). Changing big business - the globalization of the fair trade movement. Edward Elgar Publishing Limited, Cheltenham, UK.

Hütz-Adams, F. and Ertener, L. (2012). Von der Staude bis zum Konsumenten: Die Wertschöpfungskette der Bananen. SÜDWIND e.V. – Institut für Ökonomie und Ökumene, Siegburg, Germany.

Langley, A. (1999). Strategies for theorizing from process data. The Academy of Management Review 24(4): 691-710.

Lounsbury, M., Ventresca, M.J. and Hirsch, P. (2003). Social movements, field frames, and industry emergence. Socio-Economic Review 1(1): 71-104.

Max Havelaar-Stiftung (Schweiz), Basel, Switzerland. Available at: www.maxhavelaar.ch/de/produkte/informationen/produkt/2/.

Max Havelaar Schweiz (2013). Produktinformationen: Bananen.

Micheletti, M. (2003). Political virtue and shopping. individuals, consumerism and collective action. Palgrave Macmillan, New York, NY, USA.

Morazán, P. (2012). Das krumme Ding mit der Banane: Soziale Auswirkungen des weltweiten Bananenhandels. Die Macht von Supermarktketten in Deutschland. SÜDWIND e.V. – Institut für Ökonomie und Ökumene and Bischöfliches Hilfswerk MISEREOR e.V, Siegburg, Germany.

Nicholls, A. and Opal, C. (2005). Fair trade: market driven ethical consumption. Sage Publications, London, UK.

Raynolds, L.T. (2002). Consumer/producer links in fair trade coffee networks. Sociologia Ruralis 42(4): 404-424.

Renard, M.C. (2005). Quality certification, regulation and power in fair trade. Journal of Rural Studies 21(4): 419-431.

Shaw, D.V., McLeod, D.M., Kim, E., Lee, S.Y., Gotlieb, M.R., Ho, S.S. and Brevik, H. (2007). Political consumerism: how communication and consumption orientations drive 'lifestyle politics'. Annals of the American Academy of Political and Social Science 611: 217-235.

Shreck, A. (2002). Just bananas? Fair trade banana production in the Dominican Republic. International Journal of Sociology of Agriculture and Food 10(2): 13-23.

Smith, S. (2010a). Fairtrade bananas: a global assessment of impact. Institute of Development Studies, University of Sussex, Sussex, UK.

Smith, S. (2010b). For love or money? Fairtrade business models in the UK supermarket sector. Journal of Business Ethics 92: 257-266.

Starbucks (2013). Starbucks und Fairtrade sind Partner seit mehr als 10 Jahren. Available at: www.starbucks.de/responsibility/learn-more/fairtrade.

Taylor, P.L. (2005). In the market but not of it: fair trade coffee and forest stewardship council certification as market-based social change. World Development 33(1): 129-147.

Taylor, P.L., Murray, D.L. and Raynolds, L.T. (2005). Keeping trade fair: governance challenges in the fair trade coffee initiative. Sustainable Development 13(3): 199-208.

TransFair e.V. (2012). Jahresbericht 2011/12. TransFair, Verein zur Förderung des Fairen Handels mit der 'Dritten Welt' e.V., Köln, Germany.

Wall Street Journal (2012). Top U.K. supermarkets in discount challenge. Available at: http://online.wsj.com/article/SB10001424127887324660404578197301166306898.html.

Weber, K., Heinze, K.L. and DeSoucey, M. (2008). Forage for thought: mobilizing codes in the movement for grass-fed meat and dairy products. Administrative Science Quarterly 53(3): 529-567.

Zelizer, V.A. (1978). Human values and the market: the case of life insurance and death in 19th century America. American Journal of Sociology 84(3): 591-610.

Zelizer, V.A. (2007). Ethics in the economy. Journal for Business, Economics and Ethics (zfwu) 1: 8-23.

Sustainability index for beef production in Denmark and Sweden: preliminary results

A. Munk[1], E. Søndergaard[1], T. Kristensen[2], L. Mogensen[2], N.I. Nielsen[1], M. Trinderup[1], R.S. Petersen[1] and H.B. Bligaard[1]
[1]AgroTech, Agro Food Park 15, 8200 Aarhus N, Denmark; [2]Aarhus University, Department of Agroecology, Blichers Alle 20, 8830 Tjele, Denmark; arm@agrotech.dk

Abstract

Production of beef in Denmark and Sweden derives from very varied production systems, from intensive bull production to extensive suckler cow systems based on grazing of semi natural areas. Beef production – in particular extensive suckler cow systems - is regarded as having a high negative impact on the climate compared to meat from pigs and broilers but it may have some advantages in relation to e.g. biodiversity and animal welfare, which are not considered when only the climate impact is in focus. In order to make an overall assessment of the impact of different beef production systems, an indicator-based sustainability index was developed. The aim was to create an index based on data that already exist at the farm or data that was relatively easy to collect. The index should be suitable for benchmarking both between farms and/or production systems but also as documentation or certification to authorities, slaughterhouses or consumers. The index was developed based on scientific literature and practical experience from similar projects, and was tested on two Swedish and three Danish farms. The index includes seven sub-indices including animal welfare, environmental impact, climate impact, use of resources, social responsibility, economy and biodiversity. An on-line platform was developed for handling data collection on farm, calculation and presentation of the index. The test showed that the index was able to assess the sustainability of various beef production systems. However the index needs to be evaluated before it is ready for general use on farms in Denmark and Sweden.

Keywords: indicators, cattle, environmental impact, biodiversity, animal welfare

Introduction

Production of beef in Denmark and Sweden derives from very varied production systems, from intensive bull production to extensive suckler cow systems based on grazing of semi natural areas. Beef production – in particular extensive suckler cow systems – is regarded as having a high negative impact on the climate compared to meat from pigs and broilers but it may have some advantages in relation to e.g. biodiversity and animal welfare, which are not considered when only the climate impact is in focus (Cederberg and Stadig, 2003).

Tools for evaluating the sustainability of beef production in a broader perspective than just climatic impact are an important constituent in the process of evaluating and enhance the overall sustainability of beef production.

The overall aim of the Danish, Swedish and Norwegian Interreg IV A project REKS – Regional nöt- och lamnköttsproduktion – en tillväxtmotor (Regional beef and lamb production – a growth promoter) is to sustain the economic growth in the region of Oresund, Kattegat and Skagerrak by developing and profiling a sustainable production of beef and lamb.

The purpose of this working package was to develop an indicator based model of a sustainability index, which may be used to evaluate the sustainability of beef production across various production systems and across various elements like nutrient balance, animal welfare, biodiversity and climatic impact.

The aim was that the sustainability index should be based on existing data on farms or data easy to collect on farms. Data should be collected and analysed to express the sustainability of beef production at farm level and in a given period, typically a year.

Methods and the process

The first step was to identify and define the attributes (sub indexes) that should be included in a sustainability index for beef production. This was done in a literature review and at workshops. In addition, there were two transnational seminars with participation of invited experts and stakeholders from Norway, Sweden and Denmark.

System borders for the sustainability index for beef production systems

In this study, it was decided that the sustainability assessment of beef production included the specialized bull production based on male calves from dairy and the production of beef in suckler cow systems. Thus, it was a requirement that the same tool should be able to handle both types of beef production.

Production of beef from dairy cows and heifers was not included in this study. As the sustainability assessment of this beef production is already included in that for milk production (see e.g. Bélanger *et al.*, 2012).

The sustainability assessment of beef production was limited to farm level and, if possible, only the part of the farm which is directly linked to the production of beef was included. A sustainability assessment of slaughterhouses, cutting plants, transport of meat, retailers as well as the companies that produce feedstuffs for primary production, were not included in this study.

The seven sub-indexes included in the sustainability index for beef production

The index included seven sub-indexes: animal welfare, climate impact, use of resources, environmental impact, biodiversity, social responsibility and economy. The seven sub-indexes were subdivided into 19 components, each of which was assessed by the selected indicators. Table 1 shows a summary of sub-indexes, components and selected indicators in the sustainability index of beef production.

The sub-index for animal welfare was inspired by the Welfare Quality® (Blokhuis *et al.*, 2009) and described the parameters that affect the welfare of the animals from birth until the animals leave the farm. The assessment of animal welfare was divided in four areas: Housing, Feeding, Health and Behaviour.

The climate impact of the beef production, was calculated based on data on the total feed consumption and the feed materials used to the herd, and methane production in the rumen, while other minor climate contributions were recognized as a constant (Kristensen *et al.*, 2011). Contribution from soil carbon change or from deforestation caused by imported feed was not included.

The sub-index for use of resources recognized land use both on-farm and off the farm. Areas with low alternative use (natural areas and permanent pasture) were not included in land use.

Table 1. Summary of sub-indexes, components and selected indicators in the sustainability index of beef production.

Sub-indexes/components	Indicators
Animal welfare	
Behaviour	Access to pasture, the animals can get outdoors, cow/calf relationships, separating animals for treatment, fearful animals, etc.
Feeding	Body condition score, access to and purity of drinking water and feed
Housing	Area/animal, bedding, lying down behaviour, At pasture: access to shelter, cattle routes
Health	Mortality rates, disease frequency, lameness, hair condition, cleanliness, damage to the hocks
Climate impact	
Global warming potential from beef production	Total feed consumption for the production of beef distributed on feed materials. Calculation of carbon footprint of the feed and methane emissions from rumen fermentation per herd per year. Kg CO_2-equivalents per kg 'meat' here defined as slaughter weight.
Use of resources	
Land use related to beef production	Land use, m^2 per kilogram of 'meat' (=slaughter weight). Calculated as land used to production of both imported and locally grown cattle feed, areas with permanent grass and semi natural areas were not included
The total feed consumption per herd per year	Feed consumption, kilogram dry matter per kg of 'meat' (=slaughter weight)
Environmental impact	
Ammonia emissions	Ammonia emissions in the housing and storage of manure was calculated based on data on animal housing practice, storage of manure, use of grazing, per cent NH_3-emission
Nitrogen (N) and phosphorus (P) balance of utilized agricultural area	Added N and P on the field minus N and P removed with crops; kilogram per hectare
Farm use of pesticides for food crops	Achieved treatment index in relation to defined targets
Biodiversity	
Production density	Applied nitrogen, kilogram per hectare
Pesticide use on forage area	Proportion of pesticide treated forage area (included purchased feed) compared to total forage area; per cent
Grassland utilization	Effect on biodiversity depends on utilization. Rank: Grazing (best), hay and silage
Permanent pasture and semi natural areas	Proportion of these areas compared to the total area, per cent
Small biotopes	Proportion of small biotopes compared to the total area, per cent
Social responsibility	
Responsible production of beef	Consideration for neighbours, environment, labour and social issues, recycling, waste, etc. (based on 17 questions)
Economy	
Gross margin for meat	DKK per kilogram of produced 'meat' (=slaughter weight)
Variable costs	Costs for feed, bedding, veterinarian DKK per kilogram of produced meat slaughter weight
Fixed capital (interest rate 5 %)	Cost of capitals of the herd, land, buildings and other facilities for production DKK per kilogram of produced meat slaughter weight

Regarding environmental impact, ammonia emissions increase the risk for acidification of sensitive areas such as Natura 2000 areas. When calculating the indicator 'ammonia emissions in housing and storage of manure', the housing system including floor type, technology used such as slurry acidification and covering of storage facilities were taking into account (Louwagie *et al.*, 2012; Meal *et al.*, 2008a).

Biodiversity was defined as 'the diversity of life and habitats', so a high index expressed that production contributed positively to biodiversity in the farm environment and promoted biodiversity in the cropped area. This implies that the objective of the biological status depends on the land use, so that the use of land for farming in itself is not negative (Noe and Reddersen, 2005).

Biotopes such as plantings that break the cultivated areas, fences, small plantations and lakes in relation to the area increases both the total diversity and the diversity of the cultivated area. Areas with aside, insect ridges and wild streaks relative to the total area were included as an indicator. Forest, scrub and similar maximum of 20 meters from the boundary were also included (Langer and Frederiksen, 2008; Louwagie *et al.*, 2012).

A high index of responsible production expressed that the farmer carried out his production in a responsible manner. Impact of production was considered in relation to how neighbours and others nearby were affected by the production. Indicators were manure management, location of field piles, public access to farm land and production facility appearance and tidiness on the farm. These parameters are also discussed in Good Farming Practice 2005 (Anonymous, 2000), and the tidiness of the holding as an indicator of the production plant appearance (Louwagie *et al.*, 2012).

Conditions concerning farmers' social responsibility included employment and work safety (Courville *et al.*, 2012). Exercising production in an ethical manner is an essential part of responsible production. Indicators included the use of renewable energy and disposal of waste for recycling (Courville *et al.*, 2012).

A high index of economic sustainability expressed that the production was able to generate a profit. The assessment of economic sustainability was divided into three areas: Gross margin from beef production, variable costs and capital tied up in the production of beef. Similar parameters have been used in the literature, see e.g. Gaspar *et al.*, 2009; Meal *et al.* 2008a; Van Calker *et al.*, 2003.

Results and discussion

In the Internet-based tool 'AnalysePlatformen' (www.analyseplatformen.dk), a module for on-line data recording on the farm was developed as well as a calculation module which presented a profile of the sustainability based on the individual sub-indices.

The sustainability index was tested in two herds with suckler cows and three herds with intensive bull production. The results were presented to farmers using radar diagrams. Figure 1 shows the profile of sustainability assessment for two of the tested herds.

The given values of the index were scored from 0 (worst) to 10 (best). This was similar to the principle used by Louwagie *et al.* (2011). The scores were given based on defined correlations between the indicator value and the score for each indicator. In the current version this relationship was based on information about the variation of indicator values found in the literature, and did not necessarily reflect the variation within the current beef systems in Sweden and Denmark.

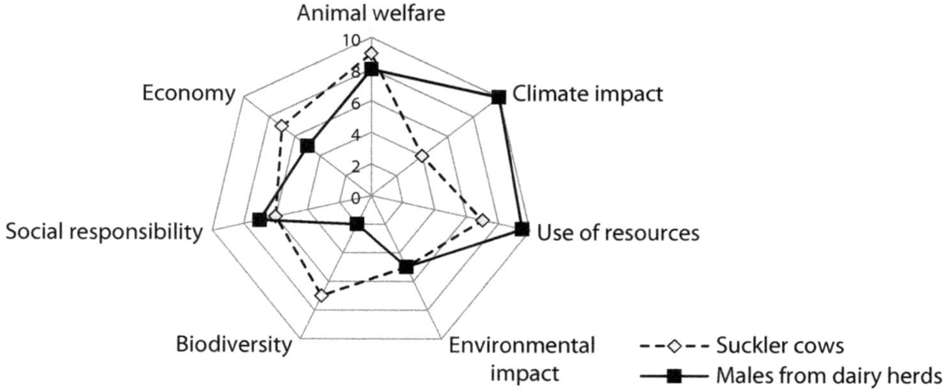

Figure 1. Sustainability index (test results on farms) from two different production systems: a suckler cow herd and a herd with males from dairy herds slaughtered at 10 months. The given values of the index were scored from 0 (worst) to 10 (best).

It is important to emphasize that the scale from 0 - 10 in this study applies to beef production. For example this means that a top score of 10 climate impact is the best achievable for beef production, and that this scale cannot be used to compare the climate impact from beef with meat from pigs and broilers.

Figure 1 shows, that the intensive bull calves production had the best sustainability in terms of climate impact and energy consumption. The reason for a top score of 10 is that the bull calves were slaughtered less than 10 months old and had a high daily weight gain. Thus, the carbon footprint was less than 10 kg CO_2- equivalents per kg 'meat' here defined as slaughter weight.

In contrast, the carbon footprint was 20 kg CO_2 per kg meat for the herd with suckler cows. The reason is that this production includes meat from cows, heifers and bulls and is more extensive than production of bull calves.

Both herds scored high values for Animal Welfare. The main reason for, that the suckler cow herd got the highest score was that these animals were on pasture in the summer. Both herds were in loose housing in the winter.

There was no difference in nitrogen and phosphorus balance of utilized agricultural areas and in ammonia emissions in the houses and storage of manure between the two herds.

The herd with suckler cows got the highest score for Biodiversity. This is mainly due to that the production was based on grazing of semi natural areas.

The main reason for that the herd with bull calves got the lowest score for Economy was high costs of capital for buildings (new stables and other facilities). In Denmark and Sweden there is often a better economy of intensive bull calves production compared with suckler cow systems.

Conclusion

The above results presented in the radar diagram in Figure 1 are preliminary. By using a score between 0 and 10 the aim is to score five to indicate a medium level of production. But with only five test results we do not know the medium level and the variation in the beef production between farms in Denmark, Norway and Sweden. And so far there has been no weighting of the individual indicators, areas and sub-indices in the sustainability index. Therefore more research is required. This will include data collection and interpretation of results from a larger number of farms, and the inclusion of stakeholders from relevant parts of the value chain.

Acknowledgement

This study was funded through the ÖSK Interreg IVA Programme project 'REKS – Regional nöt och lammköttsproduktion – en tillväxtmotor' funded by the European Union.

References

Anonymous (2000). Good farming practice 2005. The Danish Agricultural Societies & Danish Family, 15 pp.

Bélanger, V., Vanasse, A., Parent, D., Allard, G. And Pellerin, D. (2012). Development of agri-environmental indicators to assess dairy farm sustainability in Quebec, Eastern Canada. Ecological Indicators 23: 421-430.

Blokhuis, H.J., Winckler, C., Algers, B., Reenen, K.V. and Leruste, H. (2009). Welfare Quality® -Assesment protocol for cattle. Welfare Quality consortium, Lelystads, the Netherlands, 182 pp.

Cederberg, C and Stadig, M. (2003). System expansion and allocation in life cycle assessment of milk and beef production. The International Journal of Life Cycle Assessment 8: 350-356.

Courville, S., Grenz, J., Guttenstein, E., Jawtusch, J., Kaufmann, Loh, J., Nemes, N., Schader, C., Stalder, S. and Thalmann, C. (2012). SAFA – Sustainability assesment of food and agriculture systems (guidelines). Natural Resources Management and Environment department, FAO, Rome, Italy, 108 pp.

Gaspar, P., Mesias, F.J., Escribano, M., and Pulido, F. (2009). Sustainability in Spanish extensive farms (Dehesas): an economic and management indicator-based evaluation. Rangeland Ecology & Management 62: 153-162.

Halberg, N. (1999). Indicators of resource use and environmental impact for use in a decision aid for Danish livestock farmers Agriculture, Ecosystems & Environment 76: 17-30.

Kristensen, T, Mogensen, L. Knudsen, M.T. and Hermansen, J.E. (2011). Effect of production system and farming strategy on greenhouse gas emissions from commercial dairy farms in a life cycle approach. Livestock Science 140: 136-148.

Langer, V. and Frederiksen, P. (2008). Natur i landbrugslandskabet - samfundets forventninger til det økologiske jordbrug: kapitel 15 'ICROFS-rapport, no. 1: 449-461.

Louwagie, G., Northey, G., Finn, J.A., and Purvis, G. (2012). Development of indicators for assessment of the environmental impact of livestock farming in Ireland using the agri-environmental Footprint Index. Ecological indicators 18: 149-162.

Meul, M., Van Passel, S., Nevens, F., Dessein J., Rogeg E., Mulier, A. and Van Hauwermeiren, A. 2008. MOTIFS: a monitoring tool for integrated farm sustainability. Agronomy for Sustainable Development 28: 321-332.

Noe, E, Halberg, N. and Reddersen, J. (2005). Indicators of biodiversity and conservational wildlife quality on Danish organic farms for use in farm management. Journal of Agricultural and Environmental Ethics 18: 383-414

Van Calker, K.J., Berentsen, P.B.M., Giesen, G.W.J. and Huirne R.B.M. (2005). Identifying and ranking attributes that determine sustainability in Dutch dairy farming. Agriculture and Human Values 22: 53-63.

Stress-free slaughter of outdoor cattle

K.J. Schiffer[1], S.K. Retz[1], M. von Wenzlawowicz[2] and O. Hensel[1]
[1]University of Kassel, Faculty of Organic Agricultural Sciences, Department of Agricultural Engineering, Nordbahnhofstr. 1a, 37213 Witzenhausen, Germany; [2]bsi Schwarzenbek, P.O. Box 1469, 21487 Schwarzenbek, Germany; katrin.schiffer@uni-kassel.de

Abstract

During the last decades, many improvements regarding transport conditions, preslaughter handling and stunning of cattle at commercial abattoirs have been developed but still a lot of difficulties do occur. Especially animals not accustomed to human-animal contact, such as frequently the case in free range beef cattle, can be even more challenging to handle at the day of slaughter. In 2011, free-bullet rifle shooting has become an allowable on-farm slaughter method for outdoor cattle in Germany. To enhance free-bullet rifle shooting to become a more professional method is the superior aim of our studies and certain practice recommendations could already be concluded. Our recommendations are mainly based on four studies of a pilot project, which are partly under evaluation and partly ongoing. The first study is aimed at gaining information about the effectiveness of different types of weapons and ammunition. Isolated heads of post-mortem cattle (German Angus, Galloway and crossbreeds, total n=44) were shot in different ways and sections of the skulls were done, focusing mainly on the shot location, shooting angle, penetration depth and transformation of the bullet as well as on the penetrated brain tissue. In the second study, live cattle (German Angus, Galloway and Highland Cattle, total n=31) were shot according to the results of the first study while behavioural observations and video monitoring have been done. To control stun quality, the same pathological recording as in the first study, but also including brain haemorrhages, was performed. The third and fourth studies (ongoing) focus on key parameters of meat quality and on behavioural signs of pre-mortal stress, respectively. One group (Galloway, n=50) gets slaughtered by using the free-bullet rifle shooting method compared to another group of the same herd (n=50), slaughtered ordinarily at the abattoir (captive bolt stunning). However, free-bullet rifle shooting of outdoor cattle might be a solution for completely stress-free slaughter on certain farms willing to fulfil all the requirements. Most attention has to be paid to the shooter's capabilities.

Keywords: free-bullet, stun quality, animal welfare, work safety

Introduction

During the last three decades, many improvements regarding transport conditions, preslaughter handling and stunning of cattle at commercial abattoirs have been developed (Grandin, 1998, 2012). Still, technical design of preslaughter facilities as well as stunning devices in commercial cattle abattoirs vary considerably (Atkinson and Algers, 2009) and the amount of cattle not properly stunned via a captive bolt pistol is about 9% as a field study on German, Austrian and Swiss abattoirs revealed (Von Wenzlawowicz *et al.,* 2012). Further, the transportation of cattle causes stress and compromises animal welfare (Wernicki *et al.,* 2006). Whereas certain patterns of habituation could be observed in cattle during long-time road transportation (Gebresenbet *et al.,* 2012), various acute stress situations can occur at the abattoir itself (Terlouw *et al.,* 2012). Moreover, the negative impact on meat quality due to pre-mortal stressors is well-known, e.g. Dark-Firm-Dry meat (DFD) in cattle (Ferguson and Warner, 2008). Every year, significant losses to the meat industry are caused by DFD-meat (Shen *et al.,* 2009). Also bruising, lacerations or other superficial blemishes can lead to downgrading of carcasses, due to unsuitable transport conditions (Jarvis *et al.,* 1996). Especially livestock not accustomed to humans, such as frequently the case in free range beef cattle, may get heavily strained by the entire handling and transport processes at the day of slaughter. However, keeping cattle outdoors has become more important in the context of maintaining farming activities in disadvantaged regions (Le Neindre *et al.*, 1996), in the context of landscape management in certain nature conservation areas (Dierking, 2011) and in the context of species appropriate livestock husbandry such as in organic farming systems. A solution which principally avoids pre-mortal stress for slaughter cattle is to transfer the slaughter process from the abattoir to the familiar environment of the cattle like their pasture. In 2011, free-bullet rifle shooting has become an allowable slaughter method for outdoor cattle in Germany but no further constitutional instructions such as for shooting of farmed game (BMJ TierSchlV, 1997) do exist up to now. Thus, the superior aim of our pilot project is to enhance free-bullet rifle shooting to become a more professional and easier controllable method. Some practice recommendations could already be concluded from our long-term investigations, mainly implemented on a farm with 600 Galloways in Northern Germany.

Legitimate background in the EU and in Germany

In the regulation (EC) No 1099/2009 on the protection of animals at the time of killing, free-bullet rifle shooting is listed as an allowable stunning method (Annex I, Chapter I, Table I). On the contrary, regulation (EC) No 853/2004 on the hygiene of foodstuffs (Annex III, Section I, Chapter IV, 2.b) states that only live animals may enter a slaughter plant.

Until recently, also the analogical national legislation in Germany has been ambiguous regarding free-bullet rifle shooting in a similar manner (TierSchlV, 1997 resp. Tier-LMHV, 2007). In November 2011, the German government passed an amendment in the *Tierische Lebensmittel-Hygieneverordnung* (BMJ Tier-LMHV, 2007, §12, Section 3) that clarifies free-bullet rifle shooting as a legal stunning and killing method for outdoor cattle, not longer only in emergency cases but also for commercial sale. However, free-bullet rifle shooting of outdoor cattle is still only allowed after clearance with the responsible authorities such as regulatory agencies and district veterinary offices.

Materials and methods

The first study of the current project aimed at gaining information about the effectiveness of different types of weapons and ammunition in order to reduce the wide range of shooting possibilities. Isolated heads of post-mortem cattle (German Angus, Galloway, crossbreeds, total n=44) were shot in different ways and sections of the skulls were done, focusing mainly on the shot location, shooting angle, penetration depth, damaged brain tissue and transformation of the bullet. The heads originated from an electrical stunning system, which means they did not have any shooting holes yet. In the second study, live cattle (German Angus, Galloway, Highland Cattle, total n=29) were shot according to the results of the first study while behavioural observations and video monitoring have been done. To control stun quality, the same pathological recording as in the first study, but also including an investigation of the brain haemorrhages, was performed. The third and fourth studies (ongoing) focus on key parameters of meat quality and on behavioural signs of pre-mortal stress. One group (Galloway, n=50) gets slaughtered by using the free-bullet rifle shooting method compared to another group of the same herd (n=50), slaughtered ordinarily at the abattoir (captive bolt stunning). All data has not yet been published.

Recommendations for a proper setting and preliminary first results

The following recommendations are mainly based on practice experience which evolved from our long-term investigations (partly under evaluation and partly ongoing).

Preconditions regarding cattle

According to prior consultation with the responsible authority, on-farm ante-mortem inspections have to be done in time. It turned out to be advantageous to inspect a whole group of animals even if only one animal is meant to be shot. Thus, the shooter has a free choice and the individual standing in the most appropriate position can be shot. In order to increase calmness, cattle earmarked for free-bullet rifle shooting have to be familiar with the surroundings where the shooting will occur. During our studies, this has been achieved by re-grouping the animals in time and permitting free access to the shooting area. Also separation of single animals has to be avoided. Because the animals tended to get suspicious while shooting four individuals per day, it is recommendable to shoot not more than two animals of the same herd per day and week.

Preconditions regarding place and work-safety

Free-bullet rifle shooting of cattle can only take place if all safety measures are strictly maintained. No shooting should be performed on an open pasture land where it might become impossible to reach an animal after a failed shot, but within an enclosed area like a paddock instead. The fence should be wooden, sufficiently strong and with an earth wall as a bullet catcher behind. Within the paddock, no facilities must be used that might become dangerous in terms of failed shots or diverted bullets. The shooter should be located on a raised hide within or next to the paddock. Less unexpected head movements and thus, less failed shots, could be achieved when the shooter and other operators were out of sight of the animals and no other sources of interference were around.

Recommendations regarding shooting capabilities and right targeting

Being a certified and skilled hunter does not automatically qualify for free-bullet rifle shooting of cattle neither are the same laws as for hunting valid. In contrast to ordinary hunting distances, shooting of cattle is preferably done within a distance of only a few meters and the region at the forehead that has to be hit is comparably small. Thus, a specified proof of competence of the shooter`s capabilities is necessary and has to be certified by an authorised shooting teacher. Documented marksmanship should be renewed yearly as Von Wenzlawowicz (2012) suggests and, additionally, a test shot before each kill should become mandatory. The short distance between a shooter and cattle (the authors strongly recommend <30 m) requires technical adjustment of the weapon as well as special shooter`s training and must not be neglected. In our studies, the best aim point for an effective shot has been assumed to be the same as known for captive bolt stunning, namely on the forehead of cattle, 2 cm above the intersection of two imaginary lines drawn from each eye to the opposite horn. However, targeting at the ear instead of the forehead is not recommendable. Our first study (earshot isolated heads n=13) revealed that 54% of the shots did not even damage the brain tissue due to bad accuracy.

Recommendations regarding suitable weapons and ammunition

As compiled by Von Wenzlawowicz (2012), suitable weapons for stunning and killing of cattle are hunting rifles, especially repeating rifles or semi-automatic rifles that allow a quick additional shot if necessary. Also back-up stunning equipment, such as a captive bolt pistol, must be ready at hand. The penetration of the free bullet into the cranial vault has to evoke a fatal and irreversible damage of the brain. Thus, the calibre has to be sufficient but, bullets with oversized calibres can leave the body again, still featuring a deadly amount of energy and extreme caution has to be exercised. Expanding bullets (in contrast to full-metal jacked bullets) showed better preliminary results in our studies, due to the increasing diameter of the bullet and a maximum energy transfer within a short distance. The implementation of a silencer turned out to be unnecessary regarding the conspecifics' reactions on the shot and disadvantageous regarding stun efficiency, due to a reduced velocity of the bullet.

How to salvage and check stun quality

Salvage should be implemented as fast but also as calm as possible without frightening the conspecifics. This worked best in our studies by opening the shooting paddock on its back side automatically or via a tackle, which facilitates backing off of the animals. Stun quality has to be controlled immediately and no vital reactions such as righting reflex, corneal reflex, movements of the eyeball, spontaneous blinking, rhythmic breathing or vocalisation must appear. If any doubts regarding the effectiveness of the stun occur, the animal has to be shot directly a second time. For salvage from the paddock, the stunned or killed animal can be raised with a front loader by means of strong belts fixed at the back legs.

How to bleed

According to the German law, the maximum time span for the stun-to-stick interval for cattle stunned via a captive bolt pistol is 60 seconds (BMJ TSchlV, 1997, Annex 2, regarding §13, Section 3). Due to a lack of information concerning the probability of an immediate death after a free-bullet rifle shot, the interval of 60 seconds should serve as a rule (Von Wenzlawowicz, 2012) in order to maintain insensitiveness until a death occurs during bleeding. Bleeding may only be implemented by a skilled person and should be done via a chest stick. The whole process has to be performed in a hygienic manner. This involves exchanging the knife between opening the skin and cutting the main arteries. According to law in force, all blood has to be collected and disposed in an appropriate way (Von Wenzlawowicz, 2012). The bleeding success turned out to be better when lifting the animal only by the back legs instead of all four legs with elevated back legs. The collected blood accounted for an average of 5.9% of the cold body weight of the carcasses (n=16).

Transportation

If hygienically sound, a car trailer can be used to transport the dead animal to a slaughterhouse where gutting and further dressing take place. The sticking wound may not be contaminated and the trailer has to be closable and watertight as no liquids must leak out (Von Wenzlawowicz, 2012). According to the amendment in the German law (BMJ Tier-LMHV 2007), the transport must not take longer than one hour. Before any shooting is done, the handover at the slaughterhouse has to be formally and technically adjusted. In our studies, special trolleys have become necessary in order to bring the animals into the abattoir.

Discussion and conclusion

Recently, aspects of animal welfare at slaughter have increasingly attracted the public interest and media. Accordingly, German cattle farmers who are already implementing free-bullet rifle shooting, report about customers in their direct marketing who show a high willingness to pay more for meat of stress-free slaughtered cattle. However, meat from cattle slaughtered via free-bullet rifle shooting will probably remain a niche product and the authors agree with Gilliam *et al.* (2012), describing free-bullet rifle shooting as effective when properly applied, but also reminding of safety concerns which urgently need to be focused on.

Regarding work safety, special attention has not only to be paid to the shooting itself but also to the salvage. Even if conspecifics are neither frightened by the sound of the shot nor the collapsing animal, they might get stressed if a tractor and more people appear in the shooting paddock. To avoid accidents when the cattle needs to back-off, training courses in animal handling techniques are highly recommended. Moreover, dangerous situations can evolve during fixation of the belts at the hind legs of the shot animal and the person in charge has to know about clonic/tonic muscle contractions and spasms that may occur. However, the danger of being hurt by convulsions of a stunned animal can occur after captive bolt stunning in a similar way at abbatoirs.

Insufficient captive bolt stuns at abattoirs are often caused by service related problems with the stunning weapons (unclean devices, worn out parts, use of damp ammunition), even though they have to be serviced well and regularly in order to ensure a proper stunning (Atkinson and Algers, 2009). Not only high time pressure that the slaughterhouse personnel hast to cope with but also a lack of shooting accuracy due to unrestrained animals or disability or fatigue of the shooter often lead to failed shots or poor stun quality (Grandin, 1998; Gregory and Shaw, 2000). Failed shots that occurred during the authors' studies highlighted the importance of necessary maximum marksmanship as well as a mandatory test shot before the kill in order to detect possible weapon-related technical defaults in time. Moreover, the need of being mentally and organizationally able to discontinue a slaughter procedure becomes obvious, if the animals, for any reason, are not calm enough, even though shooting might have been planned for that particular day.

As a conclusion, the authors suggest that free-bullet rifle shooting of outdoor cattle is only justifiable in case of high professionalism of all actors involved whereupon most attention has to be paid to the shooter's capabilities. A resolution could be that not necessarily only farmers themselves but rather specialized units, e.g. from slaughterhouses, offer free-bullet rifle shooting as an additional service, with personnel trained in bleeding and also shooting. In addition to the animal welfare aspect of stress-free slaughter and its marketing benefit, an economic surplus could be achieved if there was also a measurable improvement in meat quality. However, free-bullet rifle shooting as a slaughter method suitable for shy cattle, should not be misunderstood as a permission to generally neglect outdoor cattle in terms of routine handling or veterinary care.

Acknowledgements

The studies mentioned are funded by the German Federal Office for Agriculture and Food (BLE). We are also grateful for scientific support to Prof. Dr. Bo Algers, SLU, Sweden, for revision of the text to Katerina Boudnikova, Praha, Czech Republic, and especially to Gerd Kaemmer and staff, Bunde Wischen e.V., Germany, for their patience and their dedication to our project.

References

Atkinson, S. and Algers, B. (2009). Cattle welfare, stun quality and efficiency in 3 abattoirs using different designs of stun box loading, stun box restraint and weapons. Project report. Faculty of Veterinary Medicine, Department of Animal Environment and Health, Swedish University of Agricultural Sciences (SLU). Skara, Sweden, 35 pp.

Bundesministerium der Justiz (BMJ), German Federal Ministry of Justice (1997). Verordnung zum Schutz von Tieren im Zusammenhang mit der Schlachtung oder Tötung (TierSchlV). Revised 2006. Available at: www.gesetze-im-internet.de/tierschlv/index.html.

Bundesministerium der Justiz (BMJ), German Federal Ministry of Justice (2007). Verordnung über Anforderungen an die Hygiene beim Herstellen, Behandeln und Inverkehrbringen von bestimmten Lebensmitteln tierischen Ursprungs (Tier-LMHV). Revised 2011. Available at: www.gesetze-im-internet.de/tier-lmhv/index.html.

Dierking, U. (2011). Zum Einstieg. In: Deutscher Verband für Landschaftspflege (DVL) and Molfsee Stiftung Naturschutz Schleswig-Holstein (eds.) Wilde Weiden zwischen Nord- und Ostsee. Husum Druck- und Verlagsgesellschaft, Husum, Germany, p. 9-15.

European Commission (EC) (2004). Regulation (EC) No 853/2004 of the European Parliament and of the Council of 29 April 2004 laying down specific hygiene rules for food of animal origin. Available at: http://eurlex.europa.eu/LexUriServ/LexUriServ.do?uri=OJ:L:2004:226:0022:0082:EN:PDF.

European Commission (EC) (2009). Council Regulation (EC) No 1099/2009 of 24 September 2009 on the protection of animals at the time of killing. Available at: http://eurlex.europa.eu/LexUriServ/LexUriServ.do?uri=OJ:L:2009:303:0001:01:EN:HTML.

Ferguson, D.M. and Warner, R.D. (2008). Have we underestimated the impact of pre-slaughter stress on meat quality in ruminants? Meat Science 80: 12-19.

Gebresenbet, G., Wikner, I., Bobobee, E.Y.H., Maria, G. and Villarroel, M. (2012). Effect of transport time and handling on physiological responses of cattle. Journal of Agricultural Science and Technology A2: 800-814.

Gilliam, J.N., Shearer, J.K., Woods, J., Hill, J., Reynolds, J., Taylor, J.D., Bahr, R.J., Crochik, S. and Snider, T.A. (2012). Captive-bolt euthanasia of cattle: determination of optimal-shot placement and evaluation of the Cash Special Euthanizer Kit® for euthanasia of cattle. Animal Welfare 21(S2): 99-102.

Grandin, T. (1998). Objective scoring of animal handling and stunning practices at slaughter plants. Journal American Veterinary Medical Association 212: 36-39.

Grandin, T. (2012). Developing measures to audit welfare of cattle and pigs at slaughter. Animal Welfare 21(3): 351-356.

Gregory, N. and Shaw, F. (2000). Penetrating captive bolt stunning and exsanguination of cattle in abattoirs. Journal of Applied Animal Welfare Science 3(3): 215-230.

Jarvis, A.M., Messer, C.D.A. and Cockram, M.S. (1996). Handling, bruising and dehydration of cattle at the time of slaughter. Animal Welfare 5: 259-270.

Le Neindre, P., Boivin, X. and Boissy, A. (1996). Handling of extensively kept animals. Applied Animal Behaviour Science 49: 73-81.

Shen, Q.W., Du, M. and Means, W.J. (2009). Regulation of postmortem glycolysis and meat quality. In: Du, M. and McCormic, R.J. (eds.) Applied muscle biology and meat science. CRC Press, Boca Raton, FL, USA, p. 175-194.

Terlouw, E.M.C., Bourguet, C. and Deiss, V. (2012). Stress at slaughter in cattle: role of reactivity profile and environmental factors. Animal Welfare 21(S2): 43-49.

Von Wenzlawowicz, M. (2012). Kugelschuss auf der Weide als Betäubungs-/ Tötungsverfahren zur Schlachtung von Rindern. Tierärztliche Vereinigung für Tierschutz e.V. (TVT) Arbeitskreis 3, Betäubung und Schlachtung (ed.) Merkblatt Nr. 136. Bramsche, Germany. Available at: www.tierschutz-tvt.de/fileadmin/tvtdownloads/merkblatt136_2012.pdf.

Von Wenzlawowicz, M., Von Holleben, K. and Eser, E. (2012). Identifying reasons for stun failures in slaughterhouses for cattle and pigs: a field study. Animal Welfare 21(S2): 51-60.

Wernicki, A., Urban-Chmiel, R., Kankofer, P., Mikucki, P., Puchalski, A. and Tokarzewski, S. (2006). Evaluation of plasma cortisol and TBARS levels in calves after short - term transportation. Revue de Medecine Veterinaire 157(1): 30-34.

Meating agriculture

E. Schmid
Department Veterinary Public Health Federal Administration County Vorarlberg, Landhaus, 6901 Bregenz, Austria; erik.schmid@vorarlberg.at

Abstract

Changing a system has three preconditions: dissatisfaction, vision, first step. The system of agriculture in western countries seems to get less satisfying the more intensive production is growing. Modern agriculture fails to ensure safe nutrition of global population but contributing to massive climate change. In 'Food Crash' Felix zu Löwenstein (2011) resumes that we will feed the world organic or not. To take first steps at the point of sale consumers have to be well informed about the whole process of food production. In industrial countries knowledge about farming is disappearing not only because of euphemistic pictures in advertising. The perception of highly emotional problems like animal welfare and environmental protection is differing significantly between farmers and consumers threatening communication and mutual understanding. Farmers tend to define the quality of a product in technical terms. Consumers want fulfilment of their expectations also within processing. Data of Eurobarometer are proving that the purchasing and nutrition behaviour of consumers significantly correlates with their knowledge and background of experience with local animal husbandry systems. Pupils as the consumers of the future need to be guided along the process from farm to fork. Within the program 'animal welfare goes to school' a class of an urban secondary school was invited to formulate their expectations on fair and regional meat products. In three excursions they experienced a wide range of animal husbandry systems including poultry, pigs and cattle. The final visit took them to a local producer of cheese and a retailing company. A concept of labelling of meat worked out by the pupils themselves was handed over to the marketing division. The whole process was also documented by a professional film-team producing a 'making of' DVD named 'Meating agriculture'. The management of the retailing company was deeply impressed by the process and involvement of the young people. The company recently rented one of the biggest farms belonging to the abbey Mehrerau at the lake of Konstanz to build a 'lighthouse' for organic farming analogous to Gut Rheinau in Switzerland. This concept perfectly matches the vision of the government of Vorarlberg to reach autonomy of energy until 2050. A first step taken by an important player on the market.

Keywords: nutrition, education, farm visits, labelling

Introduction

Austrian Animal Protection Act has a probably unique obligation in comparison to other EU- countries concerning promotion of animal welfare (§2): 'Federal government, provinces and communities are bound to awake and deepen the understanding of the public and especially of the youth for animal welfare...'.

The County of Vorarlberg took action by promoting an educational project for three years. It was situated in Dornbirn at the 'inatura', a small local natural history museum. The current results from Special Eurobarometer prove a considerable interest in more knowledge about animal husbandry and a demand for more and clearer information, especially about the welfare conditions behind the products. The 'European Union Strategy for the Protection and Welfare of Animals 2012-2015' argues the need for a strategy for animal welfare by stating 'consumer`s lack of appropriate information on animal welfare aspects' and announcing strategic action in 'providing consumers and the public with appropriate information'. A simplified EU legislative framework for animal welfare would consider the use of outcome-based animal welfare indicators, criteria developed by the Welfare Quality® project (www.welfarequality.net). These indicators are also recognised at international level by organisations such as the World Organisation for Animal Health (OIE).

Scientific based and detailed assessment protocols are not suitable for information of consumers. The results of the evaluation process should be translated into a simple labelling system for products like in eggs (3 = cage, 2 = barn, 1 = free range, 0 = organic). The general debate about the cage system of laying hens, the veal or farrowing crates lasted for decades resulting in public awareness of the animal welfare problems. The discussion about husbandry conditions of poultry, turkey crowded on deep litter, fattening pigs and cattle on slatted floor has started recently. Knowledge about the systems is poor, information and labelling almost missing. Periodical Eurobarometer, consumer polls and monitoring give clear evidence that experience with living animals and adventure on farm are the most effective tools to change consumer attitudes and finally habits.

The educational concept

'Tierschutz macht Schule' (www.tierschutzmachtschule.at) is a platform for persons, institutions and organisations that represent a balanced and respectful communication of animal welfare relying on scientific findings. The process of transfer of animal welfare knowledge to young people is started as early as possible in school (Schmid, 2012). Respecting the highly individualised motivation within the profession of teachers to deal with additional work and ethical problems a double strategy is offered: Teachers with average engagement can order the education materials for free and use it in individual manner. Highly motivated teachers are invited to join a post graduate course to qualify as specialist teacher in animal welfare (8 ECTS). These teachers can give assistance to colleagues with poor interest in animal welfare or felling not fit enough for the challenging item. Practical experience and feedback shows that distribution and use of educational material is increasingly accepted. The number of the specialised animal welfare teachers is growing slowly (actual number of 19). Not surprising the mass distribution of educational material is nice to have but less effective in comparison to lectures of specialised teachers which needs big effort on the other hand.

In summer 2011 the 6a class of the Bundesgymnasium in Dornbirn (high school) started a unique project in a voluntary class in biology. The pupils aged 16-17 simply wanted to get to know more about how food especially meat is produced, how animals are kept for. Living in an urban area they had almost no idea and only vague assumptions about it. But they were extremely interested. First they got basic information about local and international agriculture, legal regulations and function of the market. Not to forget the ethical aspects of using animals in very different ways resulting in 'split ethics' between the treatment of farm animals and pets. The final product of the project should be a proposal for labelling of poultry, pork and beef using the existing labelling of eggs as an example. To get a realistic impression about keeping farm animals three days of visits on farms were necessary.

The farm visits

Corresponding to the three products the visits were to poultry, pig and cattle farms. The pupils were taken to conventional holdings (mainly deep litter in poultry, slatted floor in pig and cattle) but also to free range farms to get a possibility for direct comparing. Such a full program is very exciting and interesting but also very impressing and partly shocking. To get in touch and direct contact with curious sniffing pigs, calm licking bulls and girlfriend cows was an unexpected and highly impressive experience.

Even the farmers themselves were very surprised about the interest of the young people from town, even more about their lacking fear and spontaneous reactions. The farmers' explanations and their authentic communication were far more impressive than any booklet or video. But also climbing into a milking parlour and breathing the air of a deep litter barn cramped with 25,000 hens was very memorable. The periodical data of Eurobarometer (2009) showing that individual gained experience is longer lasting than any story telling could not be better shown in a pedagogical context. But such a field trial needs high input: in time and money. The overall costs were minimum one week of time and € 2,000 for travelling costs. Therefore it is impossible to implement such a project in everyday school routine. In a reduced form and adaptation to local possibilities it should be possible everywhere at least to a certain extend. All in all it is a good investment in better understanding of local and regional agriculture and food production including retailing.

The teacher of the class summarised the project very honest stating that he himself has learned a lot and that the excursions imparted much more knowledge to his pupils than he ever could tell in the classroom. He also admitted that the experience of the farm visits has changed his life-style of consumption.

The making of 'Meating agriculture'

According to the high input and costs of this educational project the organisers concluded to document the whole history of origins by producing a 'making of'. This idea resulted in another significant raise in the total costs as the media-company chosen to make the DVD-documentation is working at highest technical and professional level. The pupils were invited to make their own documentary film about their own project with the support of professionals. One of the initial considerations was to let them do also the filming themselves but calculating the nearly equal costs to the benefit in technical quality of the film the decision to prefer professional support was clear. Another advantage was that the pupils could concentrate on their editorial comments which were honest and authentic in a very impressing way. The excitement at the beginning soon changed to experienced comments. The biggest challenge was to find a smart title accurate to the exceptional project and attractive to young people.

Some weeks ago in Dornbirn the premiere of a documentary film from Manu Coeman criticising intensive farming took place. Cover: 'loveMEATender' (www.at-production.com/lovemeatender/video.php). So why not: 'MEATing Agriculture'? The affirmative voting of the pupils was spontaneous and unanimous.

Results, messages

The interviews showed not only their progress in knowledge but also their engagement with the problems of intensive farming and their growing empathic status with the adverse fate of many animals. Not only the teacher experienced a significant change in mind. The starting point of passive consumers altered to surprising consternation leading to active involvement.

The farm visits and the interviews showed a big difference in the perception of the problem of animal welfare between the farmers and the pupils, corresponding with the investigations of Vanhonacker (2008). The DLG (German Agricultural Society) published the results of a questioning concerning welfare labelling in her report Nr 8 in 2010 asking the farmers the most important criteria in assessing 'animal friendly' husbandry systems. Not surprisingly the farmers are thinking in technical terms due to their vocational training putting the housing and management at top with big gap to animal health and almost neglecting animal behaviour, where they should be experts in dealing with on the basis of daily work. This could be a result of the technical orientation of their vocational education or simply a sign for getting indifferent to arising problems. Asking the pupils after the farm visits their ranking it was the complete opposite picture. Animal behaviour and animal health are their most important criteria assessing the welfare quality of an animal husbandry system (Figure 1).

This different perception of animal welfare by farmers in a more resource based way and by the public in a more animal based approach is probably the main problem in finding a simple and clear concept for welfare labelling. Including animal based indicators for assessing animal welfare is one of the main conclusions of the study 'Consequences and perspectives of a European animal welfare label from a German point of view' published by Georg-August-University Göttingen (2010).

The labelling concept

The students wanted to create their own labelling-system. In workshops they gathered information and discussed the multitude of existing labels which they felt to be more confusing than helpful. Being very familiar with the well-known system of labelling of eggs they decided to use a three step model. The ground level was fixed by the legal minimum standards, the premium level should be the best husbandry system, free range and organic. The interviews with many of the farmers showed the difference between the minimum legal standard and free range to be very large, not only economical in terms of housing and management, but also psychological in mind. To facilitate entering the quality-area and not getting frustrated by not reaching the top level from the beginning an intermediate level for beginners should be helpful. The experience of the farm visits finally led to the conclusion that it is not possible to improve a system by technical means when the system itself is not suitable in principle. The decisive example was the slatted floor in fattening pigs. The arguments of the pupils were as simple as convincing, even for the farmers. Perhaps the pupils had the advantage not to think as professionals or experts, but just have a look at the animals and their behaviour. For them it was just common sense that you must not keep an intelligent and inquiring animal like a pig on a slatted concrete floor. No toys or 'enrichment material'

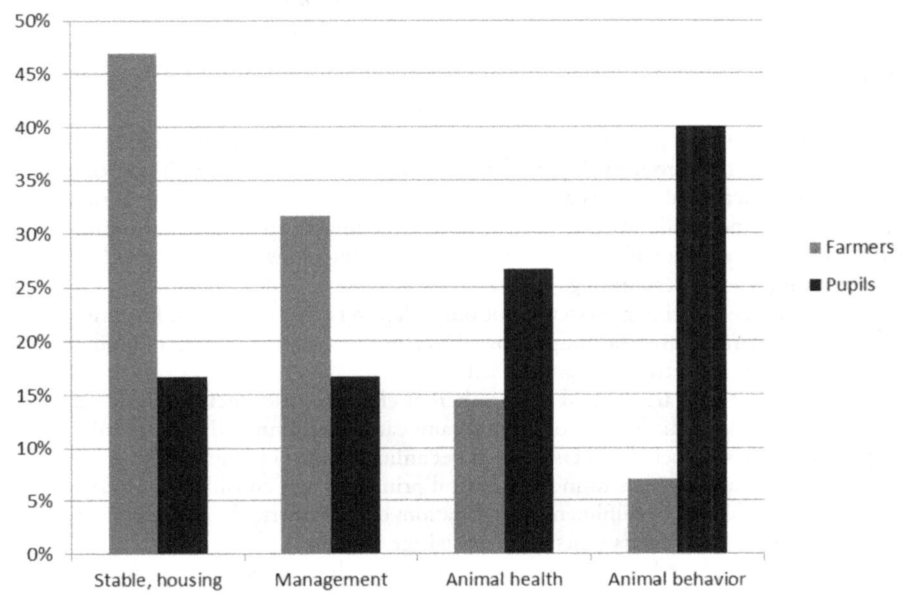

Figure 1. Criteria assessing animal welfare; importance for farmers and pupils.

at all can compensate the boring living conditions causing deprivation, harm and damage (biting tails). In this empathic challenging situation they got a 'feeling' for behavioural needs. Finally the tail of the pigs was the tipping point: you must not adapt the animals to the existing system, but you have to adapt the system to the existing needs of the animals. This aspect primarily formulated by behavioural science and approved by ethical considerations seems to result in practical experience. The routine mutilation of docking the tail is symbolic for the unsuitable system of slatted floor. The threshold for the intermediate level was clear: no slatted floor, but ring-tailed pigs!

At the end of the project the group visited the head office of a retailing company specialised in marketing of regional products. The labelling concept was given to the marketing director who was very impressed by the work of the pupils stating that he never had a visiting group more interested.

Actual development

One year after the farm visit the largest pig-farm in Vorarlberg changed the housing system from slatted floor to straw bedding system. The farmer created an own marketing-label using the ring-tail as leading argument (www.die-mit-dem-ringelschwanz.at).

The retailing company visited by the class rented one of the largest dairy-farms for 66 years announcing to change to organic: feeding no concentrates, no dehorning, no antibiotics when drying off; following the concept of Gut Rheinau in Switzerland (www.fintan.ch).

The Animal Welfare Association of Germany (www.tierschutzbund.de) and FOUR PAWS-foundation (www.vier-pfoten.de/service/tierschutzlabel) created animal welfare-labels very similar to the concept of the pupils: an intermediate beginner category and a premium sector. The retailing company mentioned above decided to implement the animal welfare label of FOUR PAWS-foundation.

It would not be realistic presuming all this happening because of the project of the 6a-class of BG-Dornbirn, but the pupils did their part and they can be proud of it.

Conclusions

- Farm visits give a lot of experience to pupils and are able to change attitudes and habits.
- The effect is dependent from the input of time and possible numbers of visits. Within the current very differing educational systems every possibility to get pupils in touch with agriculture should be used. Farmers and public are thinking different concerning the perception of animal welfare. The 'technical' and the 'emotional' approach have to be taken into consideration by both sides to enable communication and understanding.
- There is urgent need to change systems, not only adapting existing systems. Increasing health and welfare problems in intensive farming followed by almost countless scandals in food industry seem to be caused by the industrialised system itself.
- The needs of an animal are defined by its inherent characteristic: their intrinsic value has to be respected. Otherwise a fair balance of mutual gains cannot be found. The use of animals must not result in an ethical split between personalised pet animals and anonymised farm animals.
- Information and labelling has to include ethical principles and considerations. The quality of a product is mainly defined by fulfilment of expectations of consumers. These are dominated by welfare and health aspects, both weak points of industrial agriculture.
- Each system has simple indicators: slatted floor = dock-tail; straw-bedding = ring-tail.
- Ethics of consumption is watching animals trough glasses of citizens, market and law.

References

DLG (2010). Machen die Landwirte mit? Mitteilungsblatt 8: 75-77.

European Commission (2012). European Union strategy for the protection and welfare of animals 2012-2015: COM(2012) 6 final/2. EC, Brussels, Belgium.

Georg-August-Universität-Göttingen (2010). BLE Projektkennzahl 08HS010: consequences and perspectives of a European animal welfare label from a German point of view, Abstract, p. 12-13.

Löwensten, F. (2011). Food Crash; Wir werden uns ökologisch ernähren oder gar nicht mehr. Pattloch Verlag, München, Germany.

Schmid, E. (2012). Food ethics: new religion or common sense? In: Potthast. T. and Meisch, S. (eds.) Climate change and sustainable development. Wagenigen Academic Publishers, Wageningen, the Netherlands, p. 373-378.

Special Eurobarometer 270 Wave 66 (2009). Attitudes of EU citizens towards animal welfare. Available at: www.ec.europe.eu/public_opinion.

Vanhonacker, F. and Verbeke, W. (2008). Do citizens and farmers interpret the concept of farm animal welfare differently?, Livestock Science 116(1-3): 126-136.

Statistics on crimes committed to animals and changes in the animal welfare control in Sweden

B. Staaf Larsson[1]*, L.J. Keeling*[2] *and E. Hjerpe*[3]
[1]*Swedish Centre of Animal Welfare, Swedish University of Agricultural Sciences, P.O. Box 7068, 750 07 Uppsala, Sweden;* [2]*Department of Animal Environment and Health, Swedish University of Agricultural Sciences, P.O. Box 7068, 750 07 Uppsala, Sweden;* [3]*Department of Animal Breeding and Genetics, Swedish University of Agricultural Sciences, P.O. Box 7023, 750 07 Uppsala, Sweden; birgitta.staaf.larsson@slu.se*

Abstract

This article presents statistics on violations of the Animal Welfare Act and cruelty to animals according to the Penal Code, as well as changes in the frequencies and outcomes from official animal welfare control following its transition from the Swedish municipal to the county administration boards. The Swedish Council for Crime Prevention (Brå) presents statistics on these animal welfare violations and the cruelty to animal cases. Each year the Swedish Board of Agriculture presents statistics on the number of cases where animals are taken into custody and how many people are prohibited from keeping or handling animals. There is no specific code for crimes committed to animals and to get statistics on prosecutions for this study a special search from the database at Brå had to be requested. Statistics showed that during the previous twelve year period on average 63 individuals per year were prosecuted for violations of the Animal Welfare Act, of which fewer than four people were sentenced to prison, and 68 people were prosecuted for cruelty to animals in the Penal Code, of which 43 were sentenced to prison. When the county administration boards took over the control responsibility from local authorities in the beginning of 2009, the number of inspections carried out was reduced to almost half (54%) of the number during the previous five years (2004-2008). But regarding cases where animals were taken into custody, it doubled the year the county administration boards took over the responsibility and in 2010 the amount more than tripled. The last two years these cases also increased, but not that much. Since the county administration boards seem to focus on the worst cases, this change of responsibility has had a great impact on animal welfare. It can therefore be concluded that the intention of Art. 3 in the Control Regulation (EC) No 882/2004, is fulfilled.

Keywords: animals taken into custody, cruelty to animals, animal welfare act, animal welfare inspections, county administration board

Introduction

The second paragraph of the Swedish Animal Welfare Act (1988: 534) states that animals must be treated well and protected from unnecessary suffering and disease. The fourth paragraph states that animals must be kept and maintained in a satisfactory environment and in a manner that promotes their health and allows them to behave naturally. It is said that Sweden has the strictest animal protection law in the world, but how respected is it in practice? Is the number of crimes against animals decreasing? How many people have received a prison sentence? And what has the move of operational control from the smaller municipalities to the larger county administrative boards meant?

Crime statistics

The Swedish Government and the EU requires the central authorities to submit reports on animal welfare control under Art. 44 in Regulation (EC) No 882/2004, the so-called 'Control Regulation'. The official statistics, presented at the website of the Swedish Board of Agriculture (2013), include the number of animal welfare control visits by the county administrative boards and their decisions following the visit, including prohibition to take care of animals.

Brå has the official responsibility for crime statistics in Sweden and publishes crime statistics for the whole country. The statistics are based on information from police, prosecutors and courts. Prosecution statistics are based on convictions in the District Court and information from prosecutors about punishment and prosecution. Conviction in the District Court is included in the statistics on convictions even if a higher court frees the case. On the other hand, a person who has been acquitted in the District Court, but convicted in the Court of Appeals, is not included in the statistics. In the official statistics, the number of prosecutions against the cruelty to animals part (16th Chapter, 13th paragraph) of the Penal Code are included, but the number of prosecutions for violations of the Animal Welfare Act are not included in the statistics. This makes it difficult to get an overview of the situation regarding animal welfare crimes.

Previous studies

Carl-Gustaf Alm (1965) presented a Swedish study on cruelty to animals in the period 1858-1962. The first animal protection association in Sweden was founded in 1869 and during the latter half of the 1900s century it was followed by a host of new animal welfare associations. To coordinate these local associations, the Swedish Welfare Associations Federation was founded in 1897 (Alm, 1965). Two years later, in 1899, 1,243 people were prosecuted for cruelty to animals.

The increasingly comprehensive animal welfare control and the strengthening of sanctions against cruelty to animals, in combination with the decreasing use of draught animals probably all contributed to the slow onset of the decline in the number of people prosecuted after 1910-1912. To get an idea of how the number of people prosecuted has varied over time, the comparable statistics on cruelty to animals for the period 1965-2010 reports have been added to Alms data for the period 1858-1962 (Figure 1).

After introduction of imprisonment in 1901, until the year 1962, 232 people out of a total of 24,737 prosecuted for cruelty to animals were sentenced to prison, i.e. less than 1% (Alm, 1965). On average, 3.74 people were sentenced to imprisonment per year. For the period 1965-1998, it was 63 of 649 people convicted for cruelty to animals that received prison sentences, i.e. approximately 10%, and the average number of people sentenced to prison over this period was 1.85 per year.

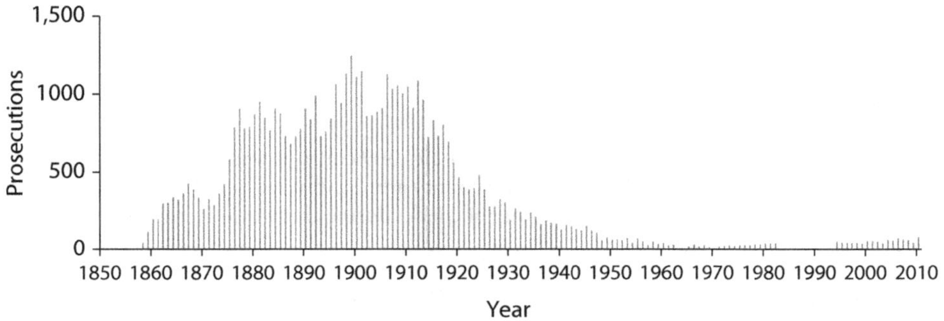

Figure 1. Number of persons found to be guilty for cruelty to animals during the years 1858-2010. Except year 1963-64, 1970, 1983-93. (Alm, 1965; Statistics Sweden, 1970, 1974, 1993; L. Klemeti, personal communication).

Materials and methods

At the website of the Swedish Board of Agriculture (2013) the number of animal welfare inspections are presented, the number of decisions where a person is prohibited from having animals, the number of cases where animals are taken into custody and the number of suspected violations presented to the police/prosecutors by the county administration boards or, before the year 2009, by the municipalities.

Brå publishes Sweden's official crime statistics. The statistics include the number of reported crimes and persons deemed to be guilty of crimes through a so-called prosecution decision of the court (a conviction in the District Court) or by a prosecutor. When a notice of a suspected violation comes in, the police register the case and categorize it by crime codes in accordance with rules made by Brå in consultation with the police, prosecution, economic crimes bureau and the customs service. Unfortunately the number of crimes related to the Animal Welfare Act and cruelty to animals are not reported separately from other crimes in the crime statistics. All reports of suspected violations of the Animal Welfare Act ends up in the statistics of 'other violations of special criminal provisions for which imprisonment is included in the range', crime code No 4013. Reports of suspected cruelty to animal offenses are reported together with riots, mutiny and sedition in crime code 1606 in statistics called 'other violations of Penal Code Chapter 16' (Swedish Council for Crime prevention, 2012a). The total number of notifications of suspected violations of the Animal Welfare Act/cruelty to animals is not reported. In the published statistics only prosecution statistics reported in the case of cruelty to animals offenses are given, no statistics for violating the Animal Welfare Act. To obtain prosecution statistics for violating the Animal Welfare Act, and more detailed information about cruelty to animal offenses a special request had to be made to Brå. Penalties imposed (such as prison sentences length) and sexes of those convicted were some of the additional information requested.

A person prosecuted several times during a year is reported as one person for every prosecution time (Swedish Council for Crime prevention, 2012b). Connected with the prosecution decision will be information about the criminal code to Brå. The Criminal Code is a public document, but as soon as the information is received by Brå it is under the Secrecy Act (1980:100) and the Act (2001:99) on official statistics. Information about matters discussed in the Court of Appeal can only be obtained if the name or social security number of the 'offender', the target number in the Court of Appeal, or the day the decision was taken are known. It is not possible to search for the statute. No follow-up of cases from notification to final treatment is therefore feasible based on the received statistics. Thus the statistics shown here are from decisions in the District Courts, and do not take into account whether the judgment was modified in the higher courts.

Results

Control statistics

Statistics show that in the period 2000-2012 a total of 254,309 controls were performed. Once animal welfare control shifted from municipalities to the county administration boards in 2009, the number of checks decreased to 54% of the average number of checks during the previous five years (2004-2008) (Figure 2).

Of all the controls during the period 2000-2012, 13,548 (5.3%) resulted in an injunction of improvements or a decision to ban the person from having animals in a particular building, 2,271 (0.9%) in notifications to the indictment and 5,486 (2.2%) in the animals being taken into custody (Figure 3). Regarding cases where animals were taken into custody, it doubled the year the county administration boards took over the responsibility and in 2010 the amount more than tripled. The last two years these cases also increased,

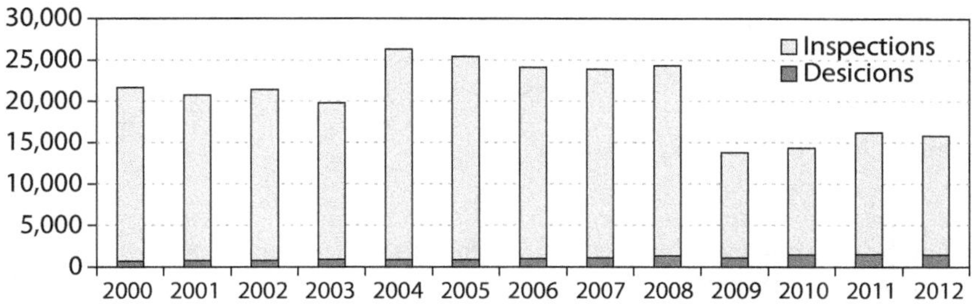

Figure 2. Number of animal welfare inspections and decisions during years 2000-2012.

but not that much. Art. 3 in the Control Regulation (EC) No 882/2004[43] states that controls should be risked based and the county administration boards seem to focus on the worst cases. The number of cases where animals are taken into custody are in total 1,790 for the years 2000 to 2012 (Figure 3).

Animal welfare act

A total of 751 prosecutions for violations of the Animal Welfare Act were recorded for the period 1999-2010. This corresponds to an average of 63 prosecutions a year. These prosecutions include convictions for both major (651) and secondary (100) crimes. The most common penalty for violating the Animal Welfare Act was a fine (58%). For violations of the Animal Welfare Act the maximum penalty was increased from one to two years in prison on 1 January 2003. However, less than four persons have been sentenced to prison and for durations of no more than six months. The exact number and the penalty could not be obtained because of confidentiality.

Cruelty to animals

A total of 812 prosecutions for cruelty to animals during the years 1999-2010 were recorded, of which 98 are penal, 10 prosecution and 704 convictions. This corresponds to an average of 68 prosecutions a year. Of these, 812,654 prosecutions were as a penalty for the principal offense.

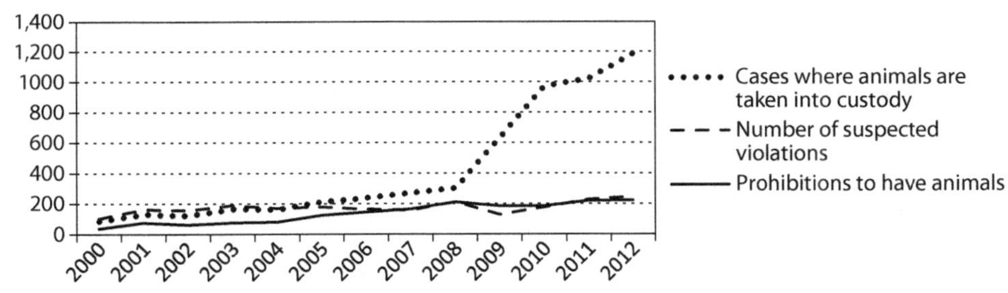

Figure 3. Cases where animals are taken into custody, number of suspected violations and prohibitions to have animals during years 2000-2012.

[43] The corrected version published in EUT L 191, 28.5.2004, p 1, Celex 32004R0882R(01).

The most common penalty for cruelty to animals was a fine (44%). Prison sentences were given in 7.3% of the cases. Even though the maximum sentence for cruelty to animals is two years, for those persons sentenced to jail imprisonment varied between less than one and twelve months.

Discussion

The number of cases where animals were taken into custody more than doubled the year the county administration boards took over the responsibility for supervision and the following year there were more than three times as many as before. The last two years these cases also increased, but not that much. Since the county administration boards seem to focus on the worst cases, this change of responsibility has had a great impact on animal welfare. It can therefore be concluded that the intention of Art. 3 in the Control Regulation (EC) No 882/2004 was fulfilled, since the worst cases have been remedied. But the number of prosecutions notifications have only increased slightly over this period, despite the sharp increase in the number of cases where animals were taken into custody. This is remarkable, but may have to do with the fact that the county administration boards had not previously been responsible for the primary control of animal welfare. They have the tradition of distributing EU subsidies to farmers and working with rural development. Another major contributory factor is lack of time, when urgent matters take priority over writing a notification of a suspected violation to prosecutors or police.

It is also noteworthy that when the animal welfare control was transferred to the county administration boards in 2009, the number of checks decreased to about half (54%) of what it had been during the previous five years. Admittedly, the number of employees decreased slightly, but not nearly to the extent that the number of inspections decreased. One explanation could be that the county administration boards introduced at the same time a new case management system, which was time consuming for administrators. Another is that a number of time-consuming pending animal welfare cases were also transferred (Swedish Agency for Public Management, 2011). Distances to the animal keepers became longer after the transition, because almost all county administration boards include more than one municipality. At the same time, the Control Regulation changed to require unannounced inspections with the corresponding risk that there are no keepers at home. Another likely explanation may be that it is very time consuming to take animals into custody compared to carrying out a normal animal welfare control. Given that the number of these cases increased so sharply, county administrative boards may not have had time to carry out so many controls. But, since the checks that have been carried out have resulted in so many cases where animals have been taken into custody, this must indicate that the risk-based control has worked and that control efforts have been focused on the keepers who had the worst animal welfare.

Suspected violations of the Animal Welfare Act and cruelty to animals can be reported to police or a prosecutor at the county administration board, municipality, or any private person. The number of suspected violations reported by the Board of Agriculture in Figure 3, only includes the notifications made by municipal or county administration boards to police/prosecutors. It therefore corresponds to neither the total number of notifications of suspected violations of the Animal Welfare Act/cruelty to animals nor the total number of notifications that go to trial. The number of cases in the District Courts of course depends on how many suspected offenses are reported to the police/prosecutor that led to prosecution. Many animal welfare issues are dropped already by the county administration boards and police because lack of resources. As mentioned above, the number of prosecutions reports from authorities has remained at about the same over the past 10 years.

Remarkably, only two people have been sentenced to prison for more than six months for cruelty to animal offenses in Sweden, both in 2002 and none for violating the Animal Welfare Act. The penalty scale two years is therefore not fully used. When the penalty for violating the Animal Welfare Act increased from one to two years, the idea was that for serious violations of the Animal Welfare Act only imprisonment would be considered. The legislature's intent is thus not observed and an educational effort directed at police and prosecutors would be desirable to increase the understanding and knowledge of these crimes. If knowledge of crimes against animals increases in the courts, perhaps even criminal penalties would be used more in the future. A proposal in the investigation of a new Animal Welfare Act is to introduce a crime of 'serious violation of the Animal Welfare Act' that can provide up to four years in prison (SOU, 2011: 75). When the penalty amount is increased for violations of the Animal Welfare Act, this will mean that these crimes have a higher priority by prosecutors and police.

References

Alm, C.G. (1965). Cruelty to animals. Applied Studies Criminal Justice. Malmö, Sweden.

SOU (2011). New animal welfare act. Fritzes Publications, Stockholm, Sweden.

Statistics Sweden. From 1965 to 1969. (1970). Bureau of Statistics Yearbook, Part 2, court and prosecution statistics. SCB. Stockholm, Sweden.

Statistics Sweden. From 1971 to 1973. (1974). Bureau of Statistics Yearbook, Part 2, court and prosecution statistics. SCB. Stockholm, Sweden.

Statistics Sweden. From 1975 to 1992. (1993). Entitled Statistical Yearbook 1975-1992. SCB. Stockholm, Sweden.

Swedish Agency for Public Management (2011). Animal welfare control development 2011:23. Stockholm, Sweden.

Swedish Board of Agriculture (2013). Animal welfare control 2000-2010. Available at: www.jordbruksverket.se/amnesomraden/djur/djurskydd/rapporterochstatistik/statistikoverdjurskyddskontroller.4.2e937121386c0f24 3f8000127.html.

Swedish Council for Crime prevention (2012a). Coding of crime. Version 11.0. National Crime Prevention Council. Information and publishers. Stockholm, Sweden.

Swedish Council for Crime prevention (2012b). Persons found guilty of offenses. Final statistics for 2011.

Swedish Government (2002). 2001/02:93. Amendments to the Animal Welfare Act, etc.

Author index

www.ingramcontent.com/pod-product-compliance
Lightning Source LLC
Chambersburg PA
CBHW061150240926
51846CB00006B/974
* 9 7 8 9 0 8 6 8 6 2 3 1 3 *